PREFACE TO THE SECOND EDITION

It gives us great pleasure and satisfaction to present this revised edition of Molecular Biology and Biotechnology for undergraduate students. During last four years the book was liked by the students. On the basis of suggestions and syllabus of different universities, several additions and alteration have been made and new chapters on **Biosensors** and **Nanotechnology** have been incorporated. New material or definitions have been incorporated in the sections like colorimeter, centrifuge, industrial production of insulin, BCG vaccine, interferon using recombinant technology and large number of short answer questions for theory paper or viva voce.

We are sure the students that they will find the book more useful. All suggestions are welcomed for further improvement of the book.

Udaipur

Prof. K.G. Ramawat
(email ID: kg_ramawat@yahoo.com)
Dr. Shaily Goyal

PREFACE TO THE FIRST EDITION

We are pleased to present a new book on 'Molecular Biology and Biotechnology' designed to meet the requirement of students. A full course in 'Molecular Biology and Biotechnology' has been introduced at Bachelor degree level to familiarise the modern science to the students. This book contains 31 Chapters covering all the topics prescribed in the syllabus of undergraduate students. This requires comprehensive subject coverage of topics in simple and understandable language, as most of the topics are new to students. Our full effort was to make it very simple and well illustrated to develop interest in students about science in general and biotechnology in particular.

Working past 34 years in the field of biotechnology, we have been instrumental in developing these courses. Now, we are presenting the students a comprehensive book to meet the requirement of course. As a researcher and teacher of biotechnology for the last 34 years, it is our humble endeavour to provide student latest information on this frontier area of biology and students will find latest information on topics such as Types of RNAs and DNA sequencing by Nyren's method. All the chapters are arranged as a continuous and progressive theme for clear understanding of the modern science.

We hope students and teachers find this new book on 'Molecular Biology and Biotechnology' useful. Suggestions to improve the book are always welcomed and entertained wholeheartedly. We are thankful to S. Chand & Co., New Delhi to publish this book.

Udaipur

Prof. K. G. Ramawat
Dr. Shaily Goyal

Contents

MOLECULAR BIOLOGY AND BIOTECHNOLOGY

(For Undergraduate Courses)

Prof. K.G. RAMAWAT
Former Head
Department of Botany
M.L. Shukhadia University
Udaipur, Rajasthan

Dr. SHAILY GOYAL
M.Sc. Ph.D.

S Chand And Company Limited
(ISO 9001 Certified Company)

S Chand And Company Limited

(ISO 9001 Certified Company)

Head Office: Block B-1, House No. D-1, Ground Floor, Mohan Co-operative Industrial Estate, New Delhi – 110 044 | Phone: 011-66672000

Registered Office: A-27, 2nd Floor, Mohan Co-operative Industrial Estate, New Delhi – 110 044 Phone: 011-49731800

www.**schandpublishing.com**; e-mail: **info@schandpublishing.com**

Branches

Chennai	:	Ph: 23632120; chennai@schandpublishing.com
Guwahati	:	Ph: 2738811, 2735640; guwahati@schandpublishing.com
Hyderabad	:	Ph: 40186018; hyderabad@schandpublishing.com
Jalandhar	:	Ph: 4645630; jalandhar@schandpublishing.com
Kolkata	:	Ph: 23357458, 23353914; kolkata@schandpublishing.com
Lucknow	:	Ph: 4003633; lucknow@schandpublishing.com
Mumbai	:	Ph: 25000297; mumbai@schandpublishing.com
Patna	:	Ph: 2260011; patna@schandpublishing.com

First Edition 2010
Second Edition 2014; Reprint 2016, 2019
Reprint 2021

ISBN : 978-81-219-3512-8 **Product Code:** H6MBB68BTCH10ENAB14O

PRINTED IN INDIA

By Vikas Publishing House Private Limited, Plot 20/4, Site-IV, Industrial Area Sahibabad, Ghaziabad – 201 010 and Published by S Chand And Company Limited, A-27, 2nd Floor, Mohan Co-operative Industrial Estate, New Delhi – 110 044.

CHAPTER 1

Organization of Genomes

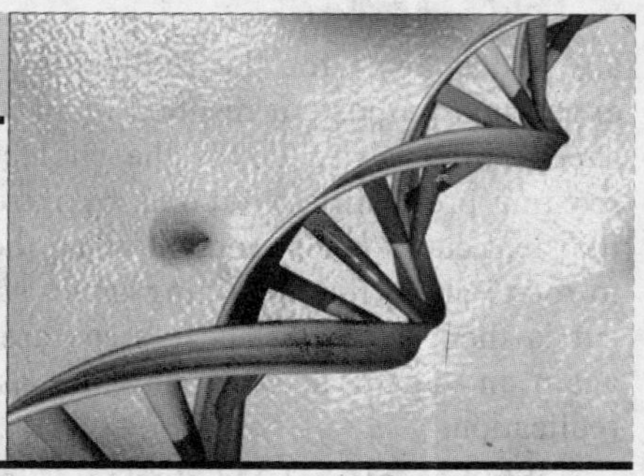

1. BACTERIAL GENOME

The total of cell's DNA is referred to as genome. The prokaryotic genome is an excellent example of how DNA can be packaged into a highly organized and condensed structure. Most prokaryotes possess only one double stranded DNA molecule.

The prokaryotic bacteria cell is much smaller than an eukaryotic cell, ranging from 1 to 5 μm in diameter. Most eukaryotic cells range in size from 10 to 100 μm. In fact the size of a bacterial cell is even smaller than the nucleus of most of eukaryotic cells. Traditionally it is known that typical prokaryotic genome is a single circular DNA molecule, localized within the nucleoid, the highly staining region of the prokaryotic cell. A nucleoid is composed of about 80% DNA, 10% protein and 10% RNA and occupies about 25% of the volume of the cell. Most of the studies have been done on *E. coli.* A key feature of packaging large, circular prokaryotic into small volumes is the twisting of the DNA molecule (Fig. 1.1). This twisting is referred to as supercoiling and involves rotating a free end of one of the two stands around the other and reattaching the ends. Supercoiling occurs when additional turns are introduced into the DNA double helix (positive supercoiling or right handed twisting) or if turns are removed (negative supercoiling or left handed twisting).

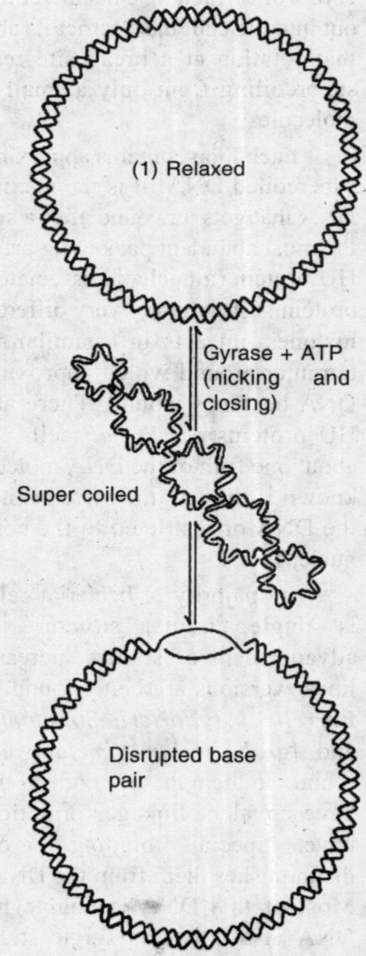

Fig. 1.1. DNA coiled and relaxed.

The direction has important impact on the structure of DNA because the stress of negative supercoiling tends to make it easier to open the DNA into single stranded regions, while the positive supercoiling makes the unwinding more difficult. The supercoiling in *E. coli* nucleoids is negative.

1.1. Process of Supercoiling

Supercoiling of DNA is controlled by some enzymes called topoisomerases. Two types of topoisomerases are found in prokaryotes and eukaryotes (*a*) Topoisomerase I—which introduces temporary single stranded cuts in DNA (*b*) Topoisomerase II (also called as DNA gyrase) which introduces temporary double stranded cuts in DNA (Fig. 1.2). Topoisomerases are required in all reactions that deal with double stranded DNA. When strands of double helical DNA are pulled apart for the replication and gene expression purpose, it resulted in an increase in supercoiling away down the molecule. This twisting problem is overcome by introducing a transient cut in one strand, allowing the free end to rotate about the remaining intact strand. The cut is then resealed. These steps are repeated again and again as DNA unwinds as in DNA replication.

Studies of isolated nucleoids of *E. coli* suggest that its DNA molecule does not have unlimited freedom to rotate once a break is introduced. The most likely explanation for this is that the bacterial DNA is attached to protein core from which 40-50 supercoiled loops radiate out into the cell, that restrict its ability to relax, so that rotation at a break site results in loss of supercoiling from only a small segment of the molecule.

Each loop contains approximately 100 kb of supercoiled DNA. It is the maximum amount of DNA that gets unwound after a single cut. One of the most abundant packaging protein in *E. coli* is HU protein (a nucleoid associated protein). This protein is structurally very different to eukaryotic histones but acts in a similar way, forming a tetramer around which approximately 60 bp of DNA becomes wound. There are some 60,000 HU proteins per *E. coli* cell, enough to cover about one fifth of the DNA molecule, but it is not known if the tetramers are evenly spaced along the DNA or restricted to the core region of the nucleus.

In majority of bacterial cells studied DNA is single circular structure. But with the advancement of studies increasing number of linear versions are being found. For example in *Borrelia burgdorferi, Streptomyces coelicolor* and *Agrobacterium tumefaciens* linear DNA is found. In them the free ends of the linear DNA have covalent linkage formation or the ends have special binding proteins which distinguishes them from the DNA break points. Most of this DNA is double helix, however DNA also exists as single stranded in some viruses and under experimental conditions. This ssDNA tend to become a hairpin like double helix due to complementary base pairing.

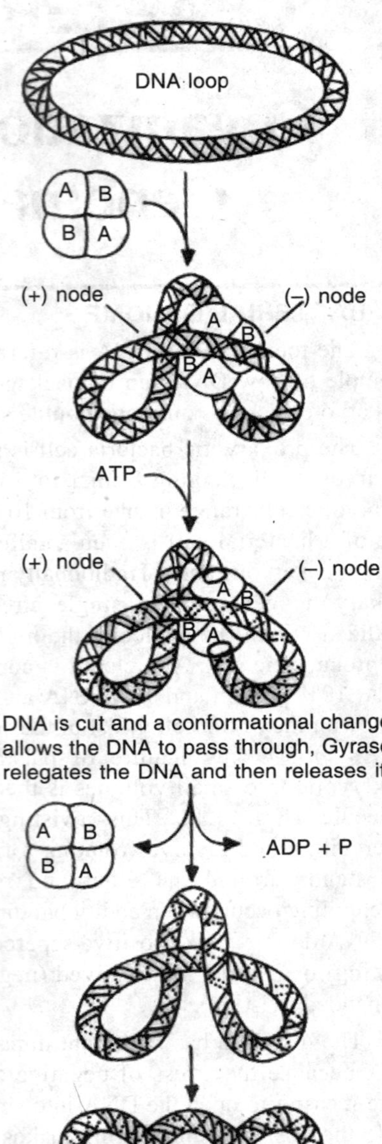

Fig. 1.2. A simple model for the action of bacterial DNA gyrase (topoisomerase II). The A-subunits cut the DNA duplex and then hold onto the cut ends. Conformational changes occur in the enzyme that allow a continuous region of the DNA duplex to pass between the cut ends and into an internal cavity of the protein. The cut ends are then re-ligated, and the intact DNA duplex is released from the enzyme. The released intact circular DNA now contains two negative supercoils as a consequence of DNA gyrase action.

1.2. Multipartiate Genome

A multipartiate genome means that the genome is divided into two or more DNA molecules. With such genome sometimes it is hard to distinguish between the original and the plasmid. A plasmid is a small piece of DNA which is often but not always circular, that co-exists with the main genome of the bacteria. Some types of plasmids are able to integrate into the main genome, but others are thought to be permanently independent. Plasmids carry genes that are not usually present in the main chromosome like antibiotic resistance etc. Plasmids are independent entities and should not be included in bacterial main genome. Many plasmids are able to transfer from one cell to another and sometimes found in the cells of the bacteria of other species.

2. EUKARYOTIC GENOME: DNA

DNA was discovered in 1869 by Johann Friedrich Miescher, a Swiss biochemist working in Germany. He called it "nuclein". Somewhat later, he isolated a pure sample of the material now known as DNA from the sperms of salmonfish, and in 1889 his pupil, Richard Altmann, named it "nucleic acid".

Deoxyribonucleic acid (DNA) is a nucleic acid that contains the genetic instructions used in the development and functioning of all known living organisms and some viruses. The main role of DNA molecules is the permanent storage of information. It contains the instructions needed to construct other components of cells, such as proteins and RNA molecules. The DNA segments that carry these genetic information are called genes.

Chemically, nucleic acids are macromolecules constructed as a long chain (strands) of monomers called as nucleotides. A nucleotide consists of three parts (*i*) 2'Deoxyribose- This is a pentose sugar composed of five carbon atoms.

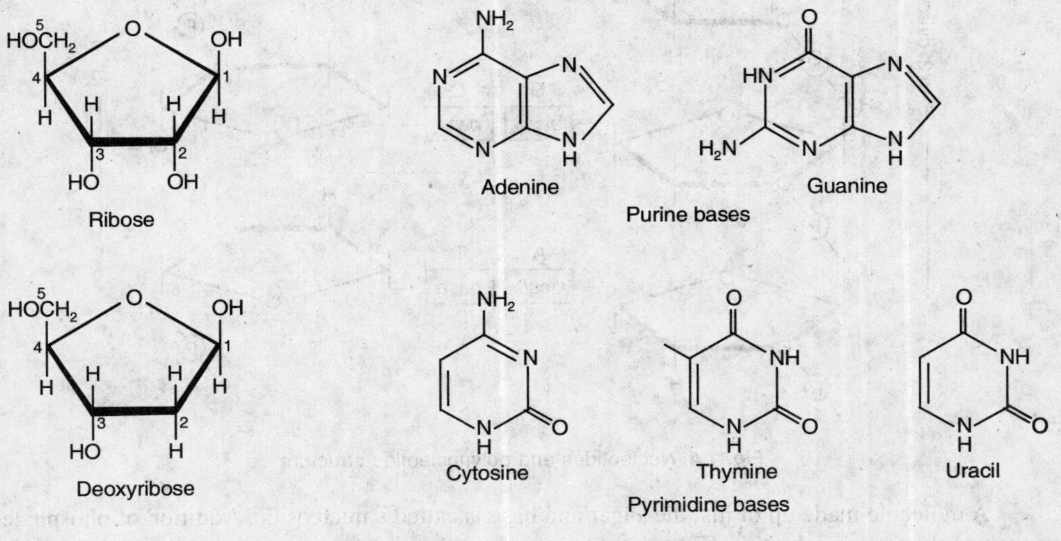

Ribose

Adenine

Guanine

Purine bases

Deoxyribose

Cytosine

Thymine

Uracil

Pyrimidine bases

Fig. 1.3. Ribose, deoxyribose sugar, prurines and pyrimidines.

(*ii*) A nitrogenous base- There are two types of nitrogenous bases occur in nucleic acid, purines and pyrimidines. Purines are larger consisting of two rings and pyrimidines are smaller molecules, consisting of a single ring. DNA has two different purines, adenine and guanine and two different pyrimidines, thymine and cytosine. (In RNA uracil is present instead of thymine). (*iii*) A phosphate group- It comprises of one, two or three linked phosphate units attached to the 5′-carbon of the sugar (Fig. 1.3).

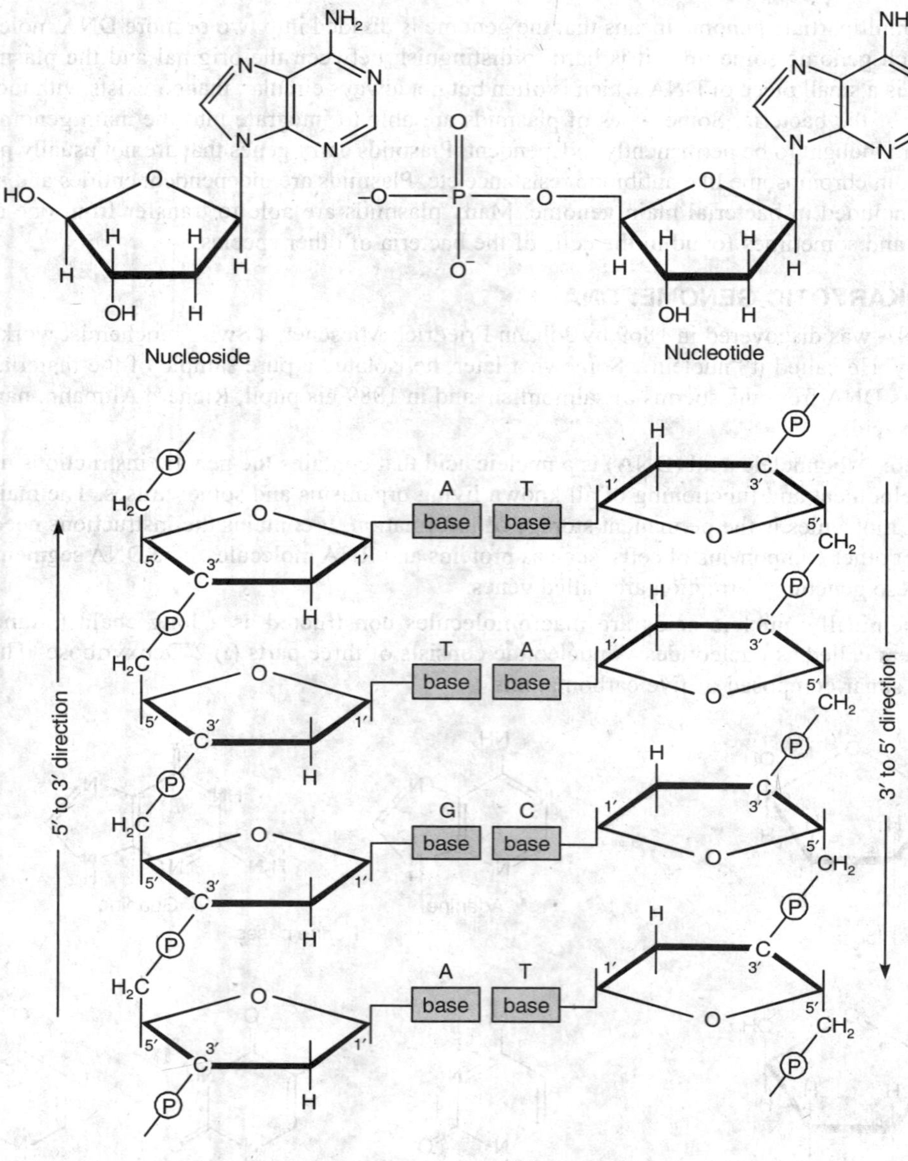

Fig. 1.4. Nucleotides and polynucleotide structure.

A molecule made up of just the sugar and base is called a nucleoside. Addition of phosphates converts this to a nucleotide (Fig. 1.4). The nucleoside triphosphate act as substrate for DNA synthesis. The full chemical name of the four nucleotides that polymerize to make DNA are:

2' Deoxyadenosine 5'-triphosphate (dATP)

2' Deoxycytidine 5'-triphosphate (dCTP)

2' Deoxyguanosine 5'-triphosphate (dGTP)

2' Deoxythymidine 5'-triphosphate (dTTP)

In a polynucleotide, individual nucleotide are linked together by phosphodiester bonds between 5′ and 3′ carbons of two nucleotides. Base pairing between the two strands involve the formation of hydrogen bonds between an adenine on one strand and thymine on the other strand, and between a cytosine and a guanine.

3. DNA DOUBLE HELIX

Utilizing X-ray diffraction data of Wilkins and Rosalind Franklin, obtained from crystals of DNA, James Watson and Francis Crick proposed a model for the structure of DNA in 1953. This model predicted that DNA would exist as a helix of two complementary antiparallel strands, wound around each other in a rightward direction and stabilized by H-bonding between bases in adjacent strands. James Watson and Francis Crick got Noble prize for the discovery of DNA structure. It included the following components:

1. The molecule is made of two chains (strands) of nucleotides.
2. The two chains spiral around each other to form a pair of right handed helices.
3. The two chains make one double helix run in opposite directions, that is they are antiparallel. Thus, if one chain is lined up in 5′→3′ direction, its complementary chain must be lined up in 3′→5′ direction.
4. The sugar-phosphate-sugar-phosphate-backbone of each strand is located on the outside of the molecule and the bases attached to them on the inner side of the two strands. The phosphate group present on the DNA makes it negatively charged. 3′ and 5′ (called as third prime and fifth prime) end of a helix is determined by 3rd carbon and 5th carbon of ribose sugar.
5. The bases are attached on planes (sugar-phosphate backbone) that are approximately perpendicular to the long axis of the molecule. Hydrophobic interactions and van der Waal forces between the stacked planar bases provide the stability for the entire DNA molecule.
6. The two strands are held together by the hydrogen bonds between the bases of the two strands (Fig.1.5). There are two hydrogen bonds between A and T, while three hydrogen bonds between G and C.
7. A pyrimidine in one chain is always paired with purine in the other chain.
8. The spaces between the two adjacent turns of the helix form two grooves of different width- a wider major groove and a more narrow minor groove- that spiral around the outer surface of the double helix.
9. The double helix makes one complete turn every 10 base pairs (3.4 nm).
10. As the G on one strand always binds with C of other strand and A of one to T of another, the nucleotide sequence of the two strands are always fixed relative to one another. There is no limitation of any type of nucleotide bases. This relationship between the two strands of double helix is called as complementarity. A is complementary to T and G to C. Therefore, purines are equal to pyrimidine nucleotides (A = T and G = C or A+G = C+T), this is known as Chargaff's rule.

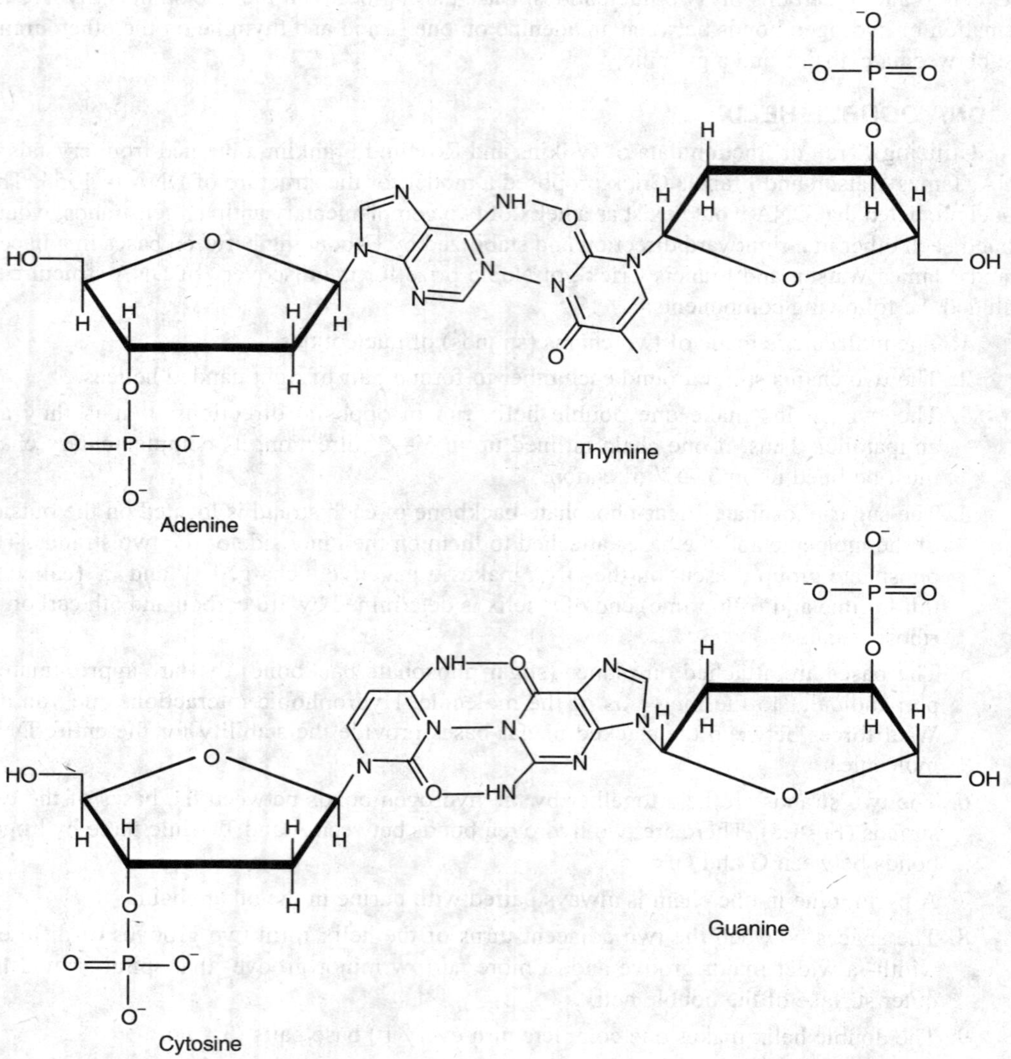

Fig. 1.5. Bonding pattern between purines and pyrimidines.

Different types of DNA structures

The double helix described by Watson and Crick was called the B-form of DNA (Fig. 1.6). This form of DNA is predominantly present in living cells. The characteristic features of B-DNA are (*i*) it has helical diameter of 2.0 nm. (1nm ==10Å) (*ii*) Distance taken up by a complete turn of the helix (pitch) is 3.4 nm. (*iii*) Each turn of the helix has 10 base pairs in it, and each base pair is 0.34 nm distance apart.

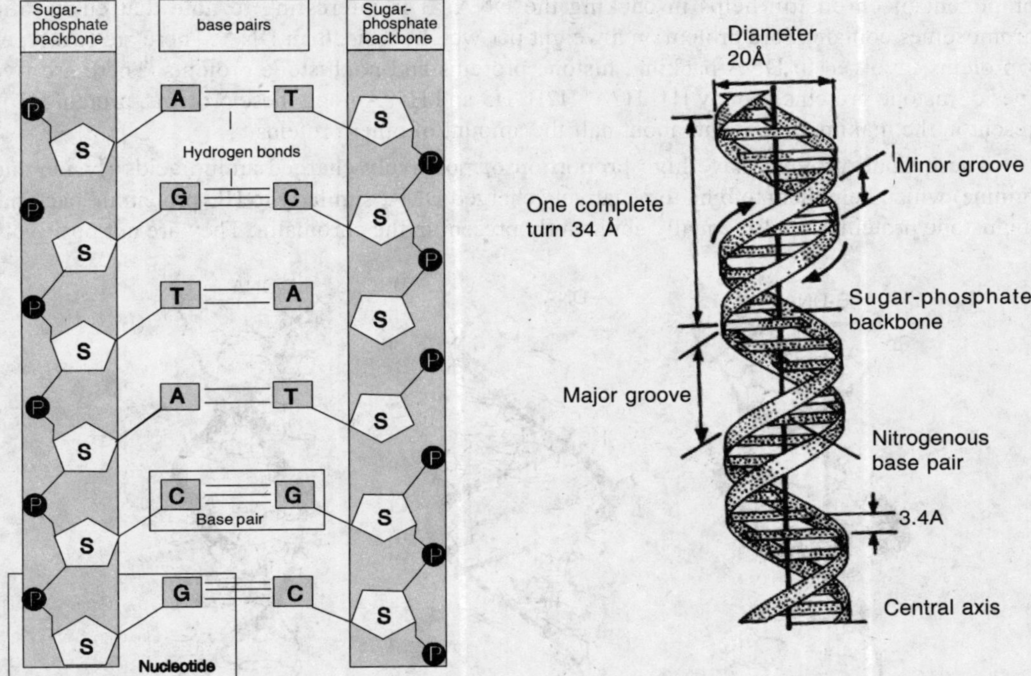

Fig. 1.6. DNA double helix, chemical and physical structure.

There is another feature found in the DNA that is rotation within individual nucleotides. This leads to major changes in the overall structure of the helix. It has been observed that these changes occur when fibres containing DNA molecules are exposed to different relative humidity. The modified version of the double helix called the A-form has the following characteristics (*i*) It has helical diameter of 2.55 nm. (*ii*) Distance taken up by a complete turn of the helix (pitch) is 3.2 nm. (*iii*) Each turn of the helix has 11 base pairs in it, and each base pair is 0.29 nm distance apart. Other variations include B′, C, C′, C″, D, E, and T-DNAs. All these are right handed helices like the B-form. A more extreme type of DNA organization is also possible, leading to the left handed Z- DNA with a diameter of only 1.84 nm, which make it thin in comparison to other forms of DNA (Fig. 1.7; Table1.1).

Table1.1. Difference between different types of DNA.

Feature	B-DNA	A-DNA	Z-DNA
Types of helix	Right handed	Right handed	Left handed
Helical diameter (nm)	2.0	2.55	1.84
Rise per base pair	0.34	0.29	0.37
Distance per complete turn (pitch) (nm)	3.4	3.2	4.5
Number of base pairs per complete turn	10	11	12
Topology of major groove	Wide, deep	Narrow, deep	Flat
Topology of minor groove	Narrow, shallow	Broad, shallow	Narrow, deep

4. DNA PACKAGING

The increased level of coiling is necessary to make compacted structure (chromosomes) and to pack huge length of DNA into the small volume of the visible chromosome. The protein component of chromatin helps in packing the DNA. It is interesting to note that eukaryotic chromosomes contain more protein on a weight per weight basis, than DNA. There are two types of proteins involved in DNA packing, histone proteins and nonhistone proteins. There are five types of histone proteins namely H1, H2A, H2B, H3 and H4. Among these proteins, protein H1 is present in the maximum amount, about half the amount of other proteins.

These small proteins have high proportion of positively charged amino acids (lysine and arginine) which help them to bind to negatively charged DNA, similarly to HU protein in bacteria. Nonhistone proteins are also equally abundantly present in the chromatin. They are not positively

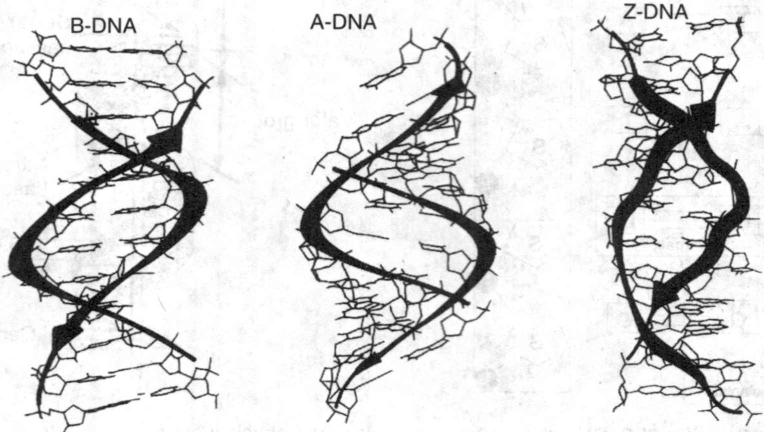

Fig.1.7. Coiling of different types of DNA.

charged like histones. Moreover, these proteins are quite heterogeneous. Many nonhistone proteins are involved in maintaining the coiled structure of chromosomes, and many are associated with regulation of gene expression and replication of DNA.

Different order of DNA coiling is as follows:

1. The structure of chromatin is complex and highly regular. It consists of repetitive units of protein complexed with DNA. First order of coiling process involves that about 200 bp long DNA coils round the small protein body containing two each of histones H2A, H2B, H3 and H4. This DNA-protein unit is called nucleosome core particle. This is about

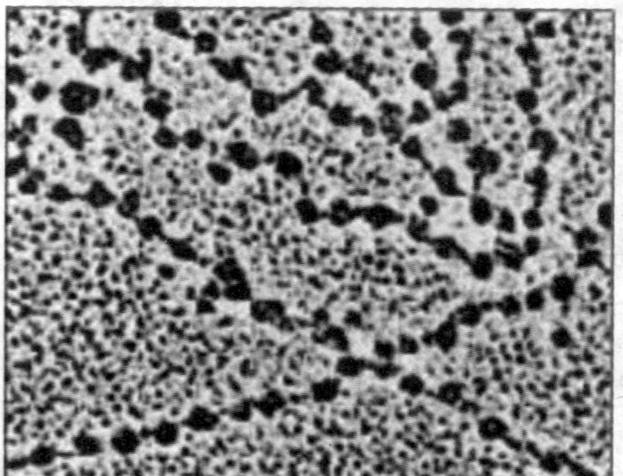

Fig. 1.8. Electron micrograph of *Drosophila melanogaster* chromatin after swelling reveals the presence of nucleosomes as 'beads on a string'.

11nm in diameter. A single molecule of H1 protein is linked to the DNA between these core particles. This H1 linked protein and the nucleosome core particle together are called nucleosome. When chromosomes are seen under electron microscope at this stage they look like a thin string of beads (Fig. 1.8). It is observed that DNA between nucleosomes is 50-60 bp long and that the DNA surrounding the nucleosome is 146bp in length.

2. After the first level of coiling DNA forms second order coiling of nucleosome which makes filament 34 nm in diameter, known as a solenoid. Each turn of solenoid contains 6 to 8 nucleosomes. Histone H1 is an elongated molecule and play an important role in coiling 11nm filament into the 34 nm solenoid (Fig. 1.9).
3. Besides this, DNA is separated into loops that are attached to a protein linker every 20,000 to 100,000bp of double stranded DNA. The protein linker and the base to which they are attached are called scaffold.
4. In metaphase chromosomes 300 nm fibres are coiled within the chromosomes and represent the final level of coiling. The total diameter of a chromatid is about 700 nm.

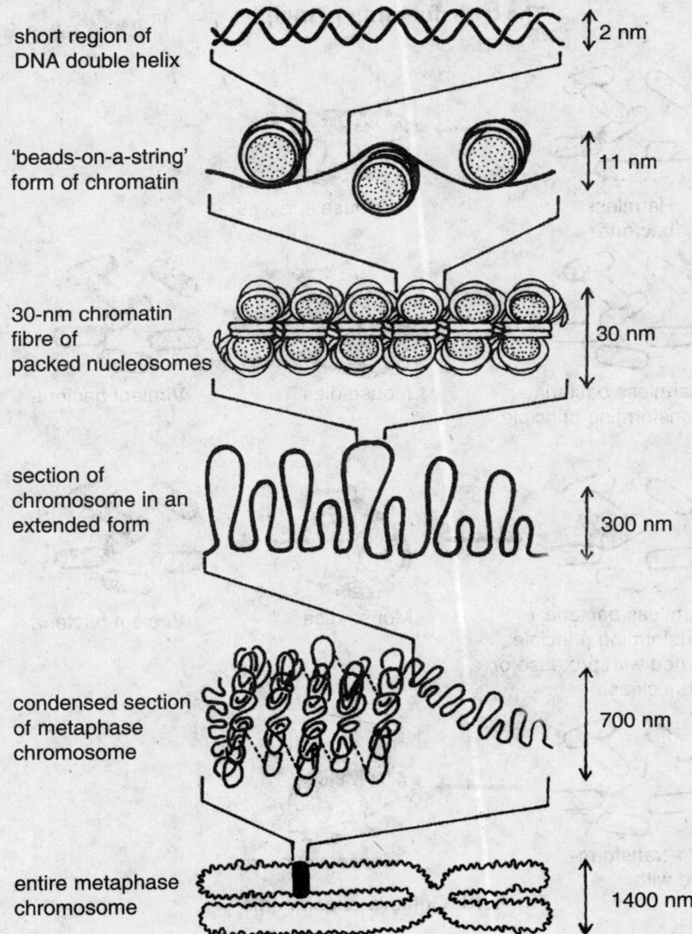

Fig. 1.9. Packaging of DNA in to a chromosome like structure involving several levels of folding.

5. DNA, THE GENETIC MATERIAL

DNA is the genetic material. This was established by the experiments conducted by many of the scientists. First set of experiments was carried out by Griffith (Fig. 1.10). It was observed that a bacterium *Streptococcus pneumoniae* when grown in laboratory on a nutrient agar medium in a Petri dish, the wild type bacteria form a colony that is large and smooth. If these smooth colony bacteria are injected into a mouse, they are capable of killing the mouse. However, a mutant of *S. pneumoniae* produces a small, rough colony (as a result of the loss of an extracellular polysaccharide coat) and this mutant is not lethal in mice. Griffith laid the foundation of DNA as a genetic material by discussing that rough non-virulent strain when mixed with heat killed virulent strain they become permanently virulent (smooth). And this change was passed from generation to generation, thus was heritable. This process came to be known as transformation. These experiments raised the possibility that heat killed virulent bacteria liberated some genetic material onto the medium which passed through the cell wall of non-virulent bacteria, and then this material was incorporated into the genetic material of non-virulent bacteria making it permanently virulent.

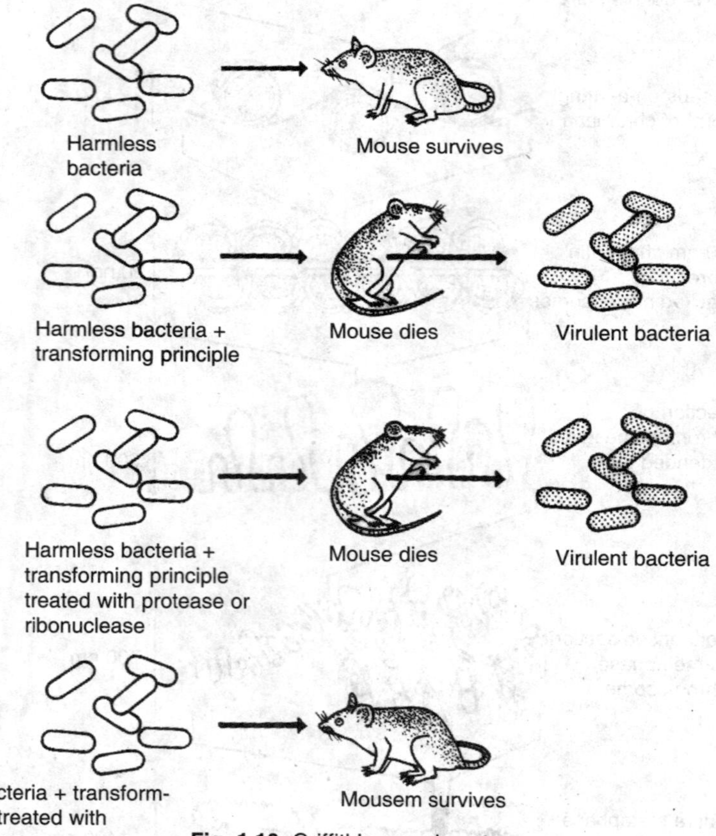

Fig. 1.10. Griffith's experiments.

In 1944 after 10 years of work, the American microbiologist Oswald T. Avery, Colin M. Macleod and Maclyn McCarty (working at the Rockfeller Institute in New York) further proved that DNA from the dead *S. pneumoniae* cells was responsible for the change from a non-virulent to a virulent state. They came to this conclusion by the following experiments they conducted.

Firstly, they purified DNA from the *S. pneumoniae* virulent stain that transformed the non-virulent strain to a virulent strain. Secondly, they destroyed the protein present in the extract by protein digesting enzymes (proteases). After this treatment also the extract was virulent. This suggested that proteins were not the component responsible for making non-virulent strains virulent. Furthermore, they digested the virulent extract by ribonuclease, an enzyme that destroys RNA. Again, they found that the extract was still virulent. Then they observed that dexoyribonuclease (DNase), an enzyme that digests DNA, destroyed the transforming material, and the extract was no more virulent. This experiment leads them to conclude that DNA was the transforming material that transformed the non-virulent strains into the virulent strains.

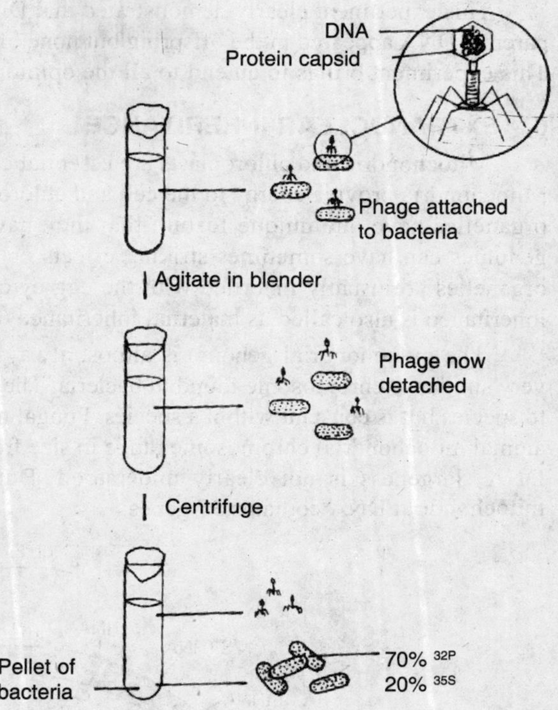

Fig. 1.11. Hershey-Chase's experiment.

The controversy that whether DNA is the genetic material or not came to an end after Alfred D. Hershey and Martha Chase experiment in 1952 (Fig. 1.11). They did experiments on bacteriophage/phage, the virus which infects bacteria. This bacterial virus is composed of only protein coat called as head surrounding the DNA molecule inside it. They labelled the protein coat of the infecting phage with ^{32}S, a radioactive isotope of sulphur. Proteins contain sulphur in the amino acids methionine and cystine but DNA does not contain sulphur. The DNA of the phage was labelled with the radioactive phosphorous isotope ^{32}P, because phosphorous is abundant in DNA, and not found in the protein coat of the phage. These labelled phages were used to study the destination of protein and DNA during infection and multiplication.

1. The protein coat labelled phages were used to infect the bacteria.
2. As per the procedure of infection, the DNA of the phage was transferred into the bacteria.
3. The bacteria were then sheared in blender to remove the attached phages. After centrifugation most of the radioactivity was seen in the supernatant solution which contain the phage protein.
4. Then it was observed that little or no radioactivity was observed in the daughter phages, which were released by the bacterial cell lysis.
5. This demonstrated that very little or no phage protein even entered the bacterial cell.

In the second set of experiment

1. The DNA labelled phages were used to infect the bacteria.
2. Phage DNA transferred into the bacteria during infection.
3. The little radioactivity was measured in the supernatant solution. Then it was observed that all the daughter phages exhibited ^{32}P labelled DNA in them.

This experiment clearly demonstrated that DNA is the genetic material. Most of the labelled parental DNA appeared in the offspring but none of the labelled protein was found in the offspring. This experiment brings to an end to all the opinions that protein can be the genetic material.

6. EXTRANUCLEAR INHERITANCE

Mitochondria and chloroplasts are essential component of the cytoplasm of eukaryotic cells. Mitochondria provide energy to the cell and chloroplasts are the site of photosynthesis. These two organelles have one unique feature that they have their own genomes and mutations in these genomes can have sometimes striking effects on the organism′s phenotype. These cytoplasmic organelles are usually inherited with the egg cytoplasm from the maternal parent. This type of inheritance is also called as maternal inheritance or cytoplasmic inheritance.

The mitochondrial genome is almost always organized into single, circular chromosomes, very similar to chromosome found in bacteria. The size of this genetic material varies form species to species but is constant within a species. Fungal mitochondrial chromosomes are about 75kb, and animal mitochondrial chromosome range in size from 13-18 kb. The reason of plant mitochondrial DNA largeness is not clearly understood. But it has small number of genes. The human mitochondrial DNA contains 37 genes.

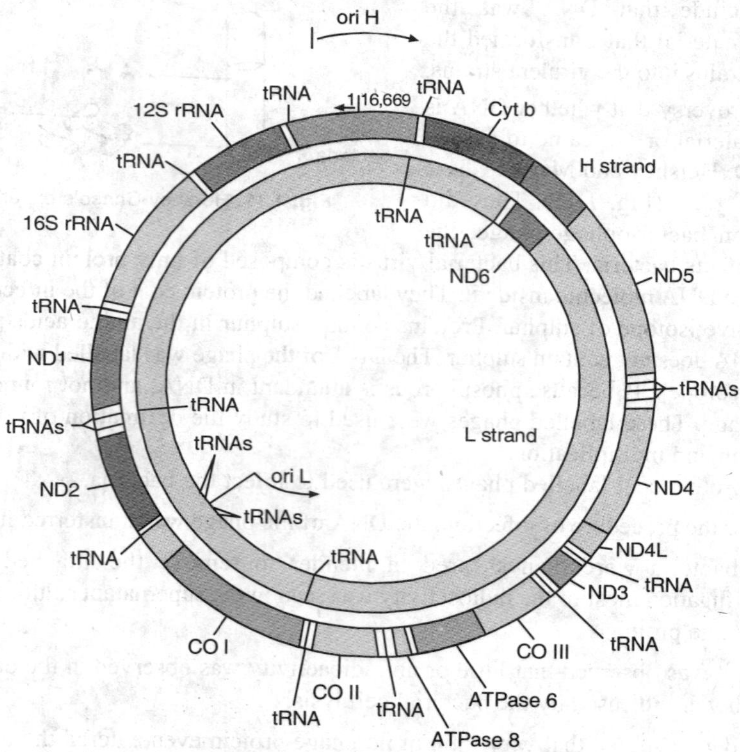

Fig. 1.12. Human mtDNA.

Each mitochondrion contains between 5 and 20 copies of the mitochondrial chromosomes, and most cells contain many mitochondria (Fig. 1.12). The number of mitochondria may vary with the cell type, for example skin cells contain about 100 mitochondria, human liver cells contain about 1000 mitochondria and a mature egg may contain 10 million mitochondria.

Although individual mitochondrial DNA is small, but together they make a significant fraction of total DNA.

The chloroplast genome is also organized into a single, circular chromosome. Chloroplast genome varies in size with species to species. This genome is about 8-10 times larger than animal's mitochondrial genome. Chloroplast genome size varies from 130-150 kb. Chloroplasts in general have more genes in their genome than the mitochondria's genome. A typical chloroplast genome contains about 110 genes. Out of which 50 produces products involved in photosynthesis. A characteristic feature of this genome is that it contains repeated regions and duplicate copies of gene.

Chloroplast also contains multiple copies of the chloroplast chromosome, and each plant cell contains multiple chloroplasts. The number of chloroplasts varies per cell varies from species to species and in different tissues within a plant. In a typical maize leaf cell chloroplast DNA is about 15% of the total DNA. The inheritance pattern of genes on mitochondria and chloroplast genome are not Mendelian. Their inheritance is based on the inheritance pattern of the organelle and the number of mitochondrial/ chloroplast present in the cell. In most species, individuals inherit their cytoplasmic organelles from only one parent (the mother). In these species, the mitochondria and/ or chloroplasts are inherited along with the egg cytoplasm. This type of inheritance is termed maternal inheritance or uniparental inheritance. The distinguishing features of such type of inheritance are that the genotype of the offspring is determined entirely by the parent that contributes the cytoplasmic organelle. For example, if a female with mutant organelle genome is crossed to a male with normal organelle the progeny will also have mutant organelle genome. Recent investigations show that male mitochondria are sometimes passed to the progeny but are excluded or degraded.

The genes in the organelle also interact with the genes in the nucleus. One well known example is cytoplasmic male sterility (CMS) in maize. CMS was discovered in 1933 by Marcus Rhoades. CMS plants fail to produce viable pollen in its tassel and is thus male sterile. However, these plants are fertile as females.

CMS results from a mutation of a mitochondrial gene whose function is vital for proper pollen maturation. CMS has considerable economic importance. Producers of hybrid seed corn, make extensive use of CMS in their breeding program as in a controlled cross, the breeder wants to prevent unwanted pollination of certain plants. So they use these male sterile plants. But later it was found that CMS plants are highly susceptible to infection of Sothern corn leaf blight (a fungal infection).

Other utilities of these genomes are that mitochondrial DNA (mtDNA) analysis is useful in forensic cases in which nuclear DNA is insufficient for short tandem repeat (STR) typing. Body remains, head hairs with no cellular material (hair follicle) attached to the root bulb and aged skeletal remains are the samples most commonly analyzed for mtDNA because nuclear DNA is not recoverable from these tissues. Usually a cell has hundreds or thousands of mitochondria which can occupy up to 25% of the cell's cytoplasm, and each mitochondrion contains 1-10 mtDNA molecules. The high copy number of mtDNA molecules found in each cell is one reason why mtDNA is recoverable from hairs and old skeletal remains.

Origination of mitochondria and chloroplasts

1. The endosymbiont theory on the origin of mitochondria and chloroplast is that they were originally free living bacteria that developed symbiotic relationship with primitive eukaryotic cells.
2. There are many evidences in favour of this theory for example these two organelles are not assembled by the cell rather arise by the division of existing mitochondria and chloroplasts. The division is by simple fission process as in the case of bacteria.
3. Organelle DNA synthesis takes place in all stages of interphase (G1, S1 and G2). This continuous division indicates that organelle division is not regulated by the same genes that regulate the nuclear and cell division cycles.

4. Mitochondrial membranes structure resembles bacterial membrane structure. Mitochondria and Chloroplast both have single, circular chromosomes, like bacteria. The molecular structure of the organelle enzymes that control DNA synthesis, RNA synthesis and gene function resemble bacterial enzyme more closely than they do with eukaryotic enzymes.

QUESTIONS

1. Describe the structure of DNA as given by Watson and Crick.
2. How DNA is packaged into a chromosome like structure ?
3. Write short notes on:
   (a) Hershey and Chase Experiment.
   (b) Different types of DNA.
   (c) How 3′ and 5′ ends are determined ?
   (d) What is Chargaff's rule ?
   (e) Nucleosomes-Solenoid model.
   (f) Mitochondrial DNA
   (g) Chloroplast DNA

CHAPTER 2

Concept and Nature of Gene

1. NATURE OF GENE

The characteristic of every organism is governed by factors of inheritance termed as genes. As centuries have passed the scientists have learned more and more about the nature of inheritance, and our concept of gene has undergone a remarkable evolution. Over the following century, these hereditary factors were shown to reside on chromosomes and consist of deoxyribonucleic acid; DNA, a macromolecule with extraordinary properties. Out of the total DNA in the genome of an organism about 25% of DNA define genes, but only 1.5% codes directly for proteins, the rest is made up of RNA genes and non-coding sequences, which often either serve no function or its function is still unknown. Possibly the largest part of genome (over 50% with larger eukaryotes) is not transcribed and is partially functionless.

In this brief introduction of gene, our aim is to make it clear, what is a gene in physical sense, a functional sense, how it makes proteins or governs a character. These questions are important to understand how gene functions, and to understand that different components like regulator, promoter, structural gene including start and stop signals make a gene functional (Fig. 2.1) and differentiate it from a simple DNA fragment or polynucleotide chain. Then only, we can obtain a gene, make copies (cloning) and transfer to another organism (transgenic) with a promoter which can regulate expression.

Details of historical developments are given to understand the evolution of science of genetics and our understanding about functioning of genes. The gene has been defined as the unit of genetic information that controls a specific aspect of the phenotype. Such a description, though accurate, does not provide a precise, unambiguous definition that can be used to identify a gene at the molecular level.

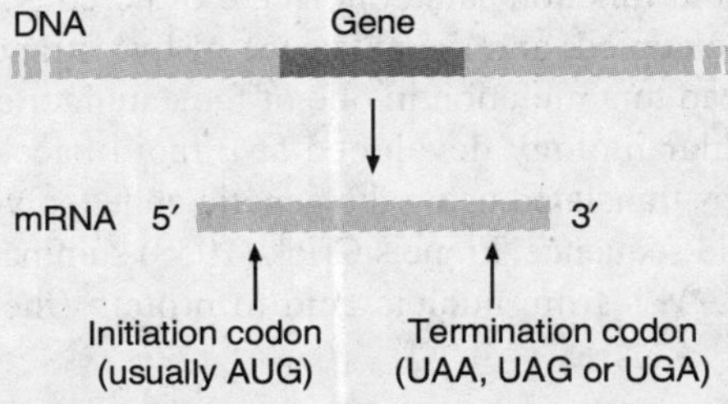

Fig. 2.1. Location of a gene on DNA and its role in transcription.

Definition 1860s–1900s: Gene as a discrete unit of heredity

The concept of the "gene" has evolved and become more complex since it was first proposed. There are various definitions of the term, although common initial descriptions include the ability to determine a particular characteristic of an organism and the heritability of this characteristic. In particular, the word *gene* was first used by Wilhelm Johannsen in 1909, based on the concept developed by Gregor Mendel in 1866.

Definition 1910s: Gene as a distinct locus

In the next major development, the American geneticist Thomas Hunt Morgan and his students were studying the segregation of mutations in *Drosophila melanogaster*. They were able to explain their data with a model that genes are arranged linearly, and their ability to crossover is proportional to the distance that separated them. The first genetic map was created in 1913 and Morgan and his students published *The Mechanism of Mendelian Inheritance* in 1915. To the early geneticists, a gene was an abstract entity whose existence was reflected in the way phenotypes were transmitted between generations. The methodology used by early geneticists involved mutations and recombination, so the gene was essentially a locus whose size was determined by mutations that inactivated (or activated) a trait of interest and by the size of the recombining regions. The fact that genetic linkage corresponded to physical locations on chromosomes was shown later, in 1929, by Barbara McClintock, in her cytogenetic studies on maize.

Definition 1940s: Gene as a blueprint for a protein

Beadle and Tatum (1941), who studied *Neurospora* metabolism, discovered that mutations in genes could cause defects in steps in metabolic pathways. This was stated as the "one gene, one enzyme" view, which later became "one gene, one polypeptide."

Definition 1950s: Gene as a physical molecule

The fact that heredity has a physical, molecular basis was demonstrated by the observation that X rays could cause mutations. Griffith's (1928) demonstration that something in virulent but dead *Pneumococcus* strains could be taken up by live nonvirulent *Pneumococcus* and transform them into virulent bacteria was further evidence in this direction. It was later shown that this substance could be destroyed by the enzyme Dnase (Avery et al. 1944). In 1955, Hershey and Chase established that the substance actually transmitted by bacteriophage to their progeny is DNA and not protein. A practical view of the gene was that of the cistron, a region of DNA that in *trans* could not genetically complement each other.

Definition 1960s: Gene as transcribed code

It was the solution of the three-dimensional structure of DNA by Watson and Crick in 1953 that explained how DNA could function as the molecule of heredity. Base pairing explained how genetic information could be copied, and the existence of two strands explained how occasional errors in replication could lead to a mutation in one of the daughter copies of the DNA molecule. From the 1960s on, molecular biology developed at a rapid pace. The RNA transcript of the protein-coding sequences was translated using the genetic code (solved in 1965 by Nirenberg and co-workers into an amino acid sequence. Francis Crick (1958) summarized the flow of information in gene expression (Fig. 2.2) as from nucleic acid to protein (the beginnings of the "Central Dogma").

Definition 1970s–1980s: Gene as open reading frame (ORF) sequence pattern

The development of cloning and sequencing techniques in the 1970s, combined with knowledge of the genetic code, revolutionized the field of molecular biology by providing a wealth of information on how genes are organized and expressed. The first gene to be sequenced was from the bacteriophage MS2, which was also the first organism to be fully sequenced. The parallel development of computational tools led to algorithms for the identification of genes based on their sequence characteristics. In many cases, a DNA sequence could be used to infer structure and function for the gene and its products. This situation created a new concept of the "nominal gene," which is defined by its predicted sequence rather than as a genetic locus responsible for a phenotype. The identification of most genes in sequenced genomes is based either on their similarity to other known genes, or the statistically significant signature of a protein-coding sequence. In many cases, the gene effectively became identified as an annotated ORF in the genome.

Definition 1990s–2000s: Annotated genomic entity, enumerated in the databanks

The current definition of a gene used by scientific organizations that describe genomes still relies on the sequence view. Thus, a gene was defined by the Human Genome Nomenclature Organization as **"a DNA segment that contributes to phenotype/function. In the absence of demonstrated function a gene may be characterized by sequence, transcription or homology".** Recently, the Sequence Ontology Consortium reportedly called the gene a "locatable region of genomic sequence, corresponding to a unit of inheritance, which is associated with regulatory regions, transcribed regions and/or other functional sequence regions".

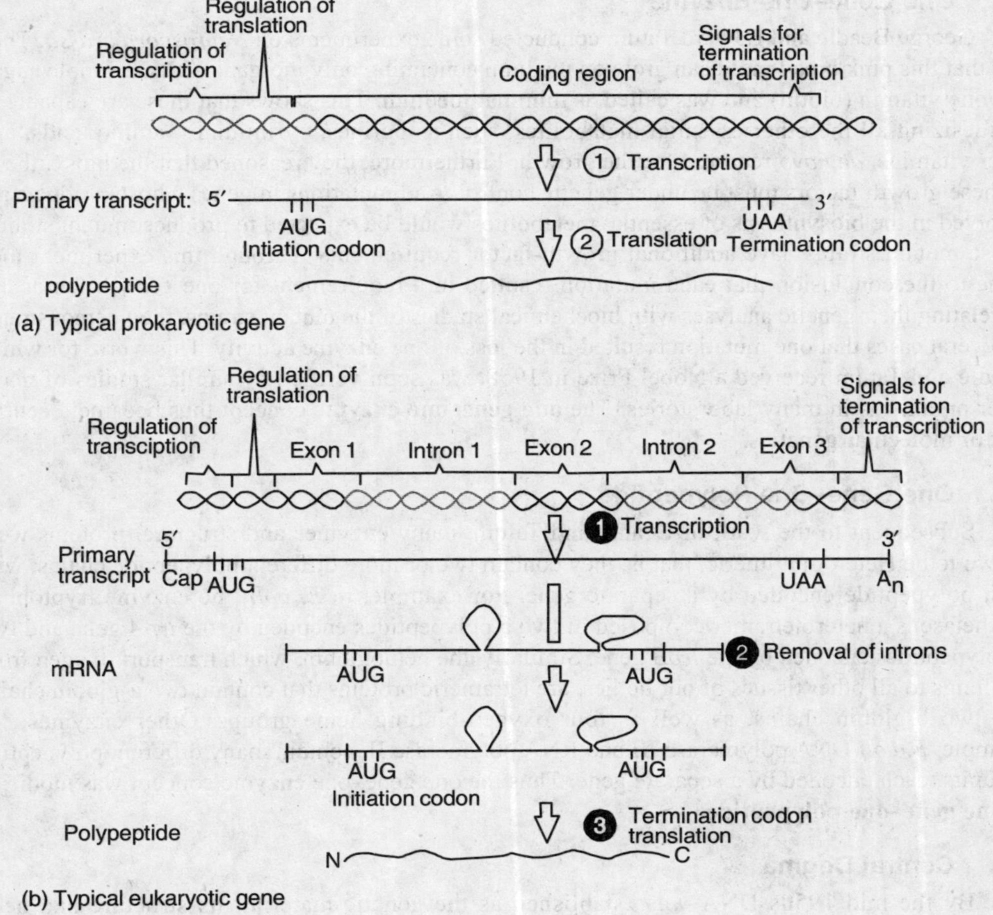

Fig. 2.2. Process of expression of a gene in prokaryotes and eukaryotes.

A current computational image

Genes as "subroutines" in the genomic operating system

Given that counting genes in the genome is such a large-scale computational attempt and that genes fundamentally deal with information processing, the dictionary of computer science naturally has been increasingly applied to describing them. In particular, people in the computational biology community have used the description of a formal language to describe the structure of genes in very much the same way that grammars are used to describe computer programs—with a precise syntax of upstream regulation, exons, and introns. Moreover, one image that is increasingly popular for describing genes is to think of them in terms of subroutines in a huge operating system (OS). That is, insofar as the nucleotides of the genome are put together into a code that is executed through the process of transcription and translation, the genome can be thought of as an operating system for a living being. Genes are then individual subroutines in this overall system that are repetitively called in the process of transcription.

2. CONCEPTS OF GENE

2.1. One Gene–One Enzyme

George Beadle and Edward Tatum conducted some experiments on *Neurospora crassa*. They saw that this pink bread mold can grow on medium containing only inorganic salts, a simple sugar, and one vitamin (biotin) and was called as minimal medium. This shows that these are capable of synthesizing all the other essential metabolites, such as purines, pyrimidines, amino acids, and other vitamins, *de novo* required for the growth. Furthermore, they reasoned that the biosynthesis of these growth factors must be under genetic control. And mutations in genes whose products are involved in the biosynthesis of essential metabolites would be expected to produce mutant strains. These mutant strains have additional growth-factor requirements. Through this experiment they came to the conclusion that each mutation resulted in a requirement for one growth factor. By correlating their genetic analyses with biochemical studies of the mutant strains, they demonstrated in several cases that one mutation resulted in the loss of one enzyme activity. This work, for which Beadle and Tatum received a Nobel Prize in 1958, was soon verified by similar studies of many other organisms in many laboratories. The **one gene–one enzyme** concept thus became a central part of molecular genetics.

2.2. One Gene–One Polypeptide

Subsequent to the work of Beadle and Tatum, many enzymes and structural proteins were shown to be Heteromultimeric, that is, they contain two or more different polypeptide chains, with each polypeptide encoded by a separate gene. For example, in *E. coli*, the enzyme tryptophan synthetase is a heterotetramer composed of two a polypeptides encoded by the *trpA* gene and two b polypeptides encoded by the *trpB* gene. Similarly, the hemoglobin, which transport oxygen from our lungs to all other tissues of our bodies, are tetrameric proteins that contain two a-globin chains and two b-globin chains, as well as four oxygen-binding heme groups. Other enzymes, for example, *E. coli* DNA polymerase III and RNA polymerase II, contain many different polypeptide subunits, each encoded by a separate gene. Thus the one gene–one enzyme concept was modified to **one gene–one polypeptide**.

2.3. Central Dogma

By the mid 1950s DNA was established as the genetic material, its structure had been analyzed by Francis Crick and James Watson (1953). Crick had stated the **'Central Dogma'** of molecular biology that the linear sequence of nucleotides in a segment of a DNA molecule determines the linear sequence of nucleotides in an RNA molecule (termed as transcription), while

RNA molecule in turn determines the sequence of amino acids in a protein by 'informational specificity' (termed as translation; Fig. 2.3). Whole process is known as information flow from DNA to protein or central dogma. This is one of the functions of DNA molecule (the other is replication). This principle of flow of information from DNA to protein is well established. However, after the discovery of reverse transcriptase by Temin and Baltimore and the proteinaceous infectious particles called as Prions, scientists are trying to find evidence for complete reversed Central Dogma. We know that there are several RNA viruses (e.g., TMV) that can produce DNA with the help of enzyme reverse transcriptase. How prions are multiplied is not known. If this is known then both steps of central Dogma will be known to function in both directions. Temin, Baltimore and Dulbecco got Noble prize of medicine in 1975.

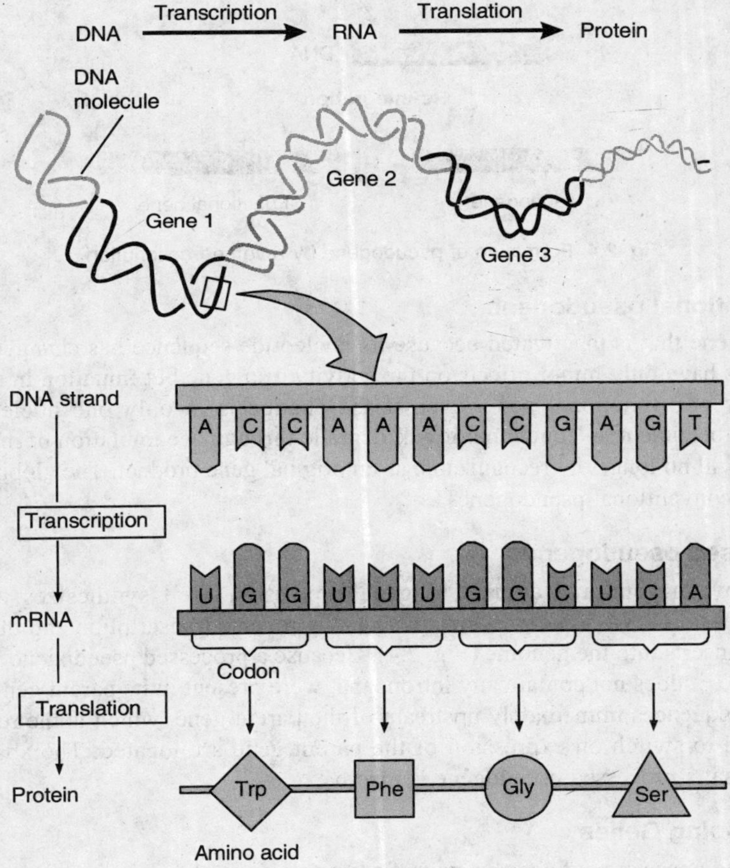

Fig. 2.3. Central dogma flow of information from DNA to protein.

3. DIFFERENT TYPES OF GENES

3.1. Pseudogenes

Pseudogenes are copies of nonfunctional genes. These are non-functional relatives of known genes that have lost their protein-coding ability or are otherwise no longer expressed in the cell. Although they may have some gene-like features (such as promoters, CpG islands, and splice sites), they are nonetheless considered nonfunctional, due to their lack of protein-coding ability resulting from various genetic changes (stop codons, frameshifts, or a lack of transcription) or their inability to encode RNA (such as with rRNA pseudogenes). For example human globin gene clusters contain five genes that are no longer active. Pseudogenes are the indication that genomes are continually undergoing change. There are mainly two types of pseudogenes:

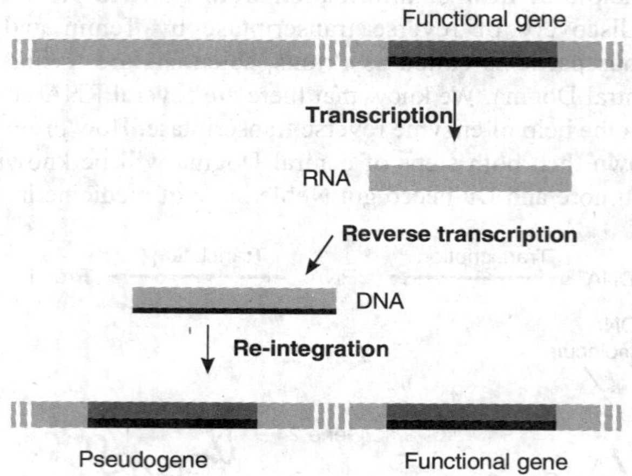

Fig. 2.4. Formation of pseudogene by reverse transcription.

3.1.1. Conventional pseudogene

This is a gene that is inactivated because its nucleotide sequence has changed by mutation. Many mutations have only minor effects on the activity of a gene but mutation in some gene can result in totally non-functional gene, even if the change is in only one nucleotide. Once a pseudogene has become non- functional it will degrade through accumulation of more mutations, and eventually will no longer be recognizable as an original gene product. The globin pseudogenes are example of conventional pseudogenes.

3.1.2. Processed pseudogenes

It arises by an abnormal adjunct to gene expression. Cell synthesizes mRNA during transcription. These mRNA sometimes make cDNA by reverse transcription and this cDNA copy subsequently reinserts into the genome (Fig. 2.4). Because a processed pseudogene is a copy of an mRNA molecule, it does not contain any introns that were present in its parent gene. It also lacks the nucleotide sequences immediately upstream of the parent gene, which is the region in which the signals used to switch on expression of the parent gene are located. The absence of these signals means that a processed pseudogene is inactive.

3.2. Overlapping Genes

Overlapping genes are defined as a pair of adjacent genes whose coding regions are partially overlapping. In other words, a single stretch of DNA codes for portions of two separate proteins. For two genes to overlap, the signal to begin transcription for one must reside inside the second gene, whose transcriptional start site is further "upstream." In addition, the "stop" signal for the second gene must not be read by the ribosome during **translation**, using the RNA copy of the gene. This is possible because RNA is read in triplets, meaning that it can contain three separate sequences that can be "read" by the cell's protein-making machinery. Such sequences of nucleotide triplets are called reading frames, and they are different in the RNA transcripts of the overlapping genes.

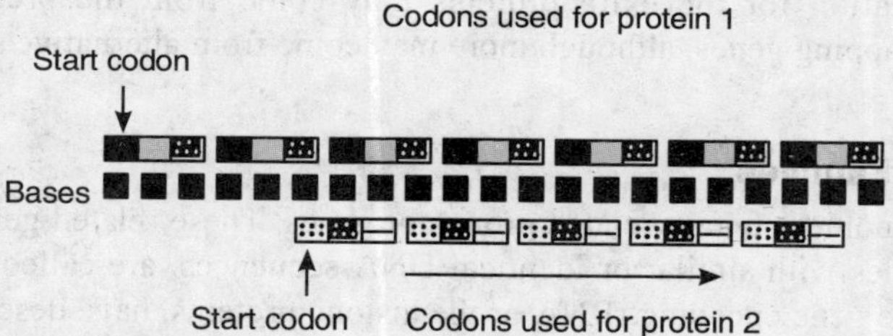

Fig. 2.5. Two genes may share the same sequence by reading the DNA in different frames.

Two genes overlap in a finer manner when the same sequence of DNA is shared between two non-homologous proteins (Fig. 2.5). This situation arises when the same sequence of DNA is translated in more than one reading frame. In cellular genes, a DNA sequence usually is read in only one of the three potential reading frames. In some viral and mitochondrial genes, however, there is an overlap between two adjacent genes that are read in different reading frames. The distance of overlap is usually relatively short so that most of the sequence representing the proteins retains a unique coding function. In some cases, a single gene may generate a variety of mRNA products that differ in their content of exons. The difference may be that certain exons are optional i.e. they may be included or spliced out. There can also be such that one or the other is included, but not both. The alternative form produces proteins, in which one part is common and the other part is different.

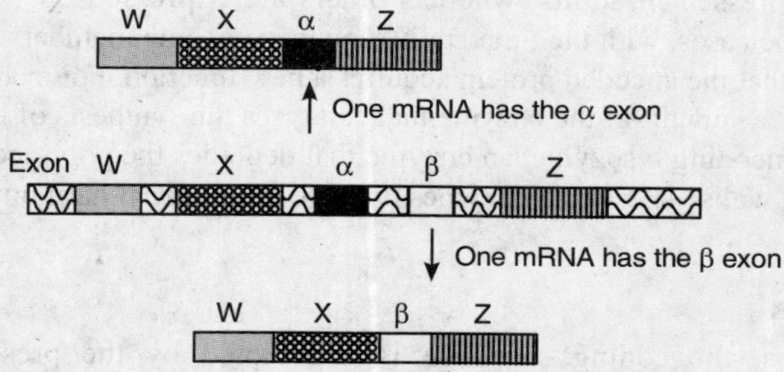

Fig. 2.6. Alternate splicing generates the α and β variants of troponin T.

In some cases one exon is substituted for other. In the Fig. 2.6, the proteins produced by the two mRNAs contain sequence that overlap extensively but are different within the alternatively spliced region. The 3' half of the troponin T gene of rat muscle contain five exons, but only four are used to construct an individual mRNA. Three exons WXZ, are the same in both expression patterns. However, in one pattern the α exon is spliced between X and Z and in the other pattern, the β exon is used. The α and β forms of troponin T, therefore differ in the sequence of the amino acids present between sequence W and Z, depending on which of the alternative exons can be used to form an individual mRNA, but both cannot be used in the same mRNA.

Some of the benefits of overlapping genes are that they enable the production of more proteins from a given region of DNA than is possible if the genes were arranged sequentially. Indeed, for the bacteriophage PhiX174, overlapping of genes is necessary. The amount of DNA present in the circular, single-stranded DNA genome of this virus is not sufficient to encode the eleven bacteriophage proteins if transcription occurrs in a linear fashion, one gene after another.

The genome economy afforded by overlapping genes extends to the human genome. The recently completed sequencing of the human genome has revealed between 30,000 and 70,000 genes. Yet evidence suggests that the human genome encodes 100,000 to 200,000 proteins. At least part of the information for the extra proteins may come from the presence of up to now undiscovered overlapping genes, although more may come from alternative splicing of exons in a single gene.

3.3. Multigene Families

Eukaryotic genomes contain related groups of genes. These related gene groups, consisting of two or more genes with similar or identical DNA sequences, are called gene families. Gene families, such as the gene encoding rRNA or the histone proteins, have descended by duplication and divergence from common ancestral genes. The DNA sequence similarity within a family can range from identical, or nearly identical, to quite different. In fact, within a family, some sequences may have only 50% DNA sequence identical yet still be similar enough to have clearly evolved from a common ancestral gene. Most members of a gene family are clustered in close chromosomal proximity to one another, however, some are located on different chromosomes. These dispersed gene family members are presumed to have been translocated to their different locations subsequent to, or possibly during, the process of gene duplication.

Generally, members of gene family have the same or related functions. For example, all the members of the mammalian hemoglobin gene family encode proteins whose work is to carry oxygen. However, even when members of gene family have the same basic function, they are not always expressed at the same time during development. Different members may be expressed at different life stages and/or in different tissues, reflecting the fact that evolutionary divergence has occurred at the level of gene regulation. For example, some of the members of the mammalian gene family are expressed in adults, whereas others are expressed only at the fetal stage of development. In some cases, with the time DNA sequence of some members of a gene family may diverge to the point that the encoded protein acquires a new function. For example, the lactalbumin gene, which encodes a subunit of the enzyme that catalyzes the synthesis of lactose, is in the same family as the gene-encoding lysozyme, an enzyme that degrades the polysaccharide component of certain bacterial cell walls. These two enzymes do have a functional harmony, however, they both act on carbohydrates.

3.4. Split Genes

In some genes, the coding sequence is interrupted by the presence of non-coding (untranslated) sequences known as introns. Such genes are known as split genes and the parts of these genes that are translated are known as exons. Split genes are rare in prokaryotes, although they are more common in archaebacteria than in eubacteria. These genes are commonly present in eukaryotes, but the number of such genes, and the number and size of introns per gene, increase with genome complexity. For example, chicken collagen gene has over 50 exons, the human dystrophin gene has 78 introns, and the Dscam gene in *Drosophila* has over 100 introns. Furthermore, in these organisms the introns are much larger than the exons. The dystrophin gene is the most extreme known example of this: the gene has a size of 2.5Mb but the coding sequence is only 14 kb in length. The longest human intron is 480 kb and this is similar in size to the smallest bacterial genomes.

In genes that are related by evolution the exons are of similar size although the genes themselves may differ greatly in length. This means that the introns must be in the same position but can be of different sizes. Furthermore, if a split gene has been cloned it is possible to sub-clone either the exon or intron sequences. If these clones are used as probes in genomic southern blots, one can determine if these same sequences are present elsewhere in genome. Often the exon sequence are present elsewhere in the genome. Often the exon sequence of one gene are found to be related to sequences in one or more other genes. Multiple copies of an exon also may be found in several apparently unrelated genes. Exons that are shared by several unrelated genes are likely to encode polypeptide regions (domains) that provide the unrelated proteins with related properties, e.g., ATP or DNA binding. Some genes appear to be mosaics that were constructed by patching together copies of individual exons recruited from different genes, a phenomenon known as exon shuffling.

There is a second degree of complexity originating from split genes. After split genes are transcribed the introns are excised and the exons spliced together by a complex of snRNA and some 145 different proteins known as the splicesome. As the number of introns within a gene increases it is possible for the pre-mRNA (unspliced messenger RNA) to be spliced in different ways. This is known as alternate splicing and provides a mechanism for producing a wide variety of proteins from a small number of genes. For example, Drosophila's Dscam gene contains 108 exons and alternative splicing theoretically could generate 38,016 different proteins.

3.5. Cryptic Gene

Cryptic genes are the phenotypically silent DNA sequences that are not normally expressed during the life cycle of an organism. However, in a few individuals of population they may be reactivated by various genetic mechanisms like mutation, recombination and transposition. Since cryptic genes are not expressed, and thus do not contribute to fitness, it is expected that they would be permanently inactivated by accumulated mutations, and would thus be rare in populations. Some scientists showed that, in yeast, gene function can be accurately predicted by examining the levels of various known metabolites. Using sophisticated statistical methods and NMR for complete analysis of metabolite concentrations can be determined. When the metabolic profile of yeast containing silent genes is compared with a normal yeast, we can detect the presence of cryptic genes. The observation that cryptic genes are commonly found in microorganisms indicates that there is selection for their preservation in microbial populations. Cryptic genes differ from other silent DNA sequences, collectively referred to as 'selfish DNA' which spread in the genome by making additional copies of themselves, without conferring any phenotypic advantage. They also differ from pseudogenes, which are homologous copies (often derived from mRNA due to reverse transcription) of active genes, but may carry many mutations, so that their reversion to active form is not possible. An example of cryptic gene is that the wild-type *E. coli* K12 does not utilize (to grow) any β-glucoside sugars as sole carbon and energy sources. Mutations in several loci can activate the cryptic bgl operon and allow growth on the aryl β -glucosides, arbutin and salicin.

4. BEADLE AND *NEUROSPORA* GENETICS

George Wells Beadle (1903–1989) grew up on a 40-acre farm near the small town of Wahoo, Nebraska (USA). Beadle (Fig. 2.7) might have become a farmer himself had it not been for the influence of his high school science teacher, Bess MacDonald, who persuaded him to enroll at the University of Nebraska College of Agriculture. Beadle entered Cornell in 1927 and joined Rollins Adams Emerson's laboratory to work on the cytogenetics of maize. Over the next 5 years he published 14 papers dealing with his investigations on maize, all initiated while he was a graduate student at Cornell. With the completion of his graduate work in 1931, Beadle headed off to the California Institute of Technology to work with future Nobel laureate Thomas Hunt Morgan. There he became interested in *Drosophila* and began doing research on genetic recombination. In 1934, Boris Ephrussi, a Rockefeller Foundation Fellow from Paris, came to Morgan's laboratory at Caltech to study *Drosophila* genetics. Beadle and Ephrussi teamed up and began examining eye pigment development in *Drosophila* after devising a method for larval embryonic bud transplantation. These studies were performed in Ephrussi's laboratory in Paris. From these experiments, they proposed that eye color changes in mutant strains of *Drosophila* could be caused by inactivation of specific proteins, acting in a single biosynthetic pathway. This suggested that development could be broken down into a series of gene-controlled biochemical reactions and laid the foundation for the one gene-one enzyme theory that Beadle would eventually propose and make famous. The idea that specific proteins were produced by specific genes was first mentioned in 1909 by Sir Archibald Garrod, an English physician. Garrod proposed that alkaptonuria, an inherited condition in humans in which the urine is black due to the presence of homogentisic acid, was associated with a recessive gene.

Fig. 2.7. George Beadle.

In 1937 Beadle accepted an appointment as Professor of Biological Sciences at Stanford University and invited Tatum to join him as a research associate. For his experimental organism, Beadle chose the red bread mold *Neurospora crassa*, whose life cycle had been characterized, making it an ideal organism for genetic study. He and Tatum knew from the studies of others that *Neurospora* could grow on a minimal medium composed of a sugar, salts, and the one vitamin, biotin. Then they used X-rays to attempt to produce *Neurospora* mutants that had lost the ability to grow on their minimal medium. The 299th mutagenized culture they tested proved to be the lucky one. It did not grow in their minimal medium, but it did survive and grow when vitamin B6 was added. To prove that a single gene had been mutated, Beadle and Tatum performed a genetic cross between the mutant strain and a wild type strain and tested cultures derived from the eight single spores that were the progeny of a single meiosis. Their tests showed that cultures from four progeny spores required vitamin B6 whereas the other four did not, confirming that a single gene had been mutated. Before this finding, mutants requiring amino acids, purines, and pyrimidines were also found, and the science of biochemical genetics was born. However, when Beadle first presented this theory few scientists accepted the concept that one gene specifies the sequence of one enzyme. The one gene-one enzyme theory was eventually verified and accepted, when subsequent investigations by others established that genetic material was DNA and that DNA had a double helical structure, and determined how genetic material is replicated and how it functions in protein synthesis. Tatum later applied their methods to produce bacterial mutants. Using these mutants, Tatum's graduate student, Joshua Lederberg, demonstrated genetic recombination in *Escherichia coli*, thereby founding the field of bacterial genetics. In 1958, Beadle, Tatum, and Lederberg shared the Nobel Prize in Physiology or Medicine for their pioneering studies with *Neurospora* and *E. coli.*

Investigation of the *Neurospora* genome began in the mid-1920s, when C. L. Shear and B. O. Dodge named it and described its life cycle (Fig. 2.8). They also showed that the two mating types of *Neurospora crassa* segregate so as to give 1:1 ratios in randomly isolated ascospores and 4:4 ratios in individual unordered asci. Dodge found that morphological variants also showed Mendelian segregation. In the 1930s, Carl Lindegren used morphological mutants, centromeres, and mating type to construct the first genetic maps, which consisted of six loci in linkage group I and four in linkage group II. Markers suitable for mapping became abundant in the 1940s, when mutagens were used effectively to obtain auxotrophs and an array of other mutant types and techniques were devised for mutant enrichment. By 1949, six linkage groups had been established and the seventh was soon added. From eight loci mapped in 1937, the number had increased to about 50 in 1949, 75 in 1954, and 500 in 1982. Now, at the millennium, the number of mapped loci exceeds 1,000.

Physical knowledge of the genome increased in parallel with the growth of genetic knowledge. Barbara McClintock showed in 1945 that the seven tiny *Neurospora* chromosomes can be identified individually using light microscopy. She and J. R. Singleton went on to describe chromosome morphology and behavior during meiosis and mitosis in the ascus. Genetic evidence of chromosome rearrangements was confirmed cytologically. Genetically mapped rearrangements soon enabled linkage groups to be assigned to each of the seven cytologically distinguishable chromosomes. The discovery of insertional and quasiterminal rearrangements made it possible to map genes by duplication coverage. The complete meiotic karyotype was reconstructed in three

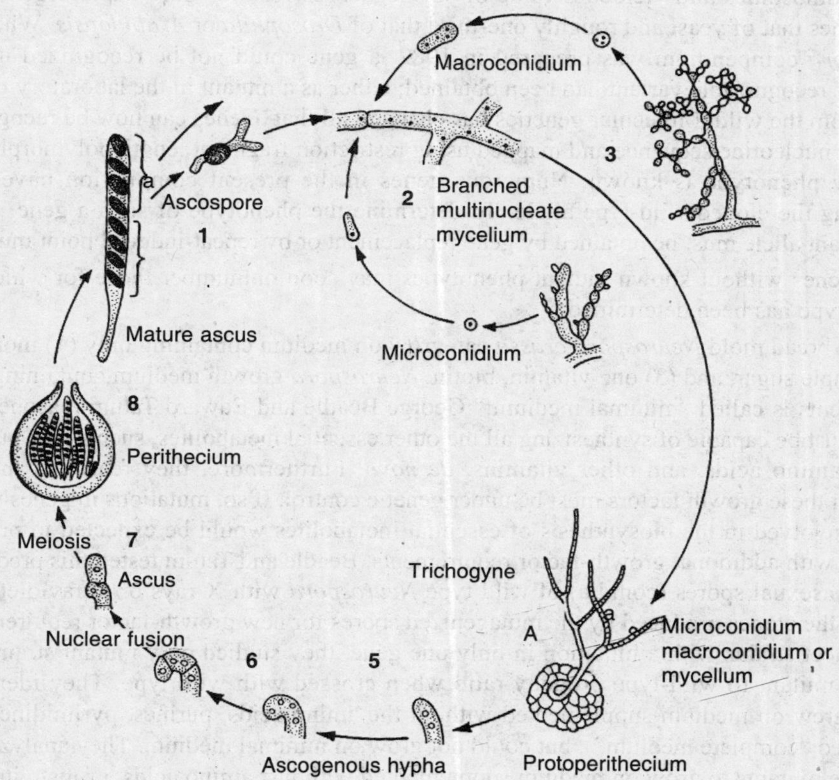

Fig. 2.8. The *Neurospora crassa* life cycle. Stages 1-4, from ascospore germination to fertilization, constitute the vegetative phase, and stages 4-8, from fertilization through ascospore maturation, constitute the sexual phase. Nuclear fusion and meiotic prophase occur 4-5 days after fertilization at 25ºC. Ascospores are shot from the perithecium from the ninth day. (1) The black ascospores, approximately 17 x 26 mm, can be isolated manually from an agar surface without the use of micromanipulation apparatus. Ascospore dormancy is broken by heat or chemicals. A germinating ascospore is shown, with hyphae growing from both ends. (2) The mycelium consists of branched, threadlike hyphae made up of multinucleate cells separated by perforate cross-walls through which nuclei and cytoplasm can pass readily. Linear growth of hyphae can exceed 5 mm/h at 37ºC. (3) Vegetative spores are of two developmentally distinct types: powdery orange conidia 6-8 mm in diameter, mostly multinucleate, and smaller microconidia, which are uninucleate. (4) Protoperithecia are formed by coiling of hyphae around ascogonial cells. Specialized hyphae, the trichogynes, are attracted to cells of the opposite mating type, from which they pick up nuclei and transport them to the ascogonium within the protoperithecium. Fertilizing nuclei may originate from macroconidia, microconidia, or mycelium. Usually only one fertilizing nuclear type contributes to the contents of a perithecium, but exceptions can occur and are revealed when the fertilizing parent is heterokaryotic. (5) Upon fertilization, the haploid A and a nuclei do not fuse but proliferate in heterokaryotic ascogenous hyphae. (6) The final conjugate division before ascus formation occurs in a binucleate hook-shaped structure, the crozier. Nuclei of opposite mating type fuse, and the diploid zygote nucleus immediately enters meiosis. The fusion nucleus is the only diploid nucleus in the life cycle. Until ascospore walls are laid down following a postmeiotic mitosis, the ascus remains a single, undivided cell within which the two parental genomes are present in a 1:1 ratio in a common cytoplasm. (7) Asci within a perithecium do not develop synchronously. Numerous asci are initiated successively from the same ascogenous hypha. Mature eight-spored asci measure about 20 x 200 mm. (8) The perithecial wall is maternal in origin. Each mature perithecium, 400-600 mm in diameter, develops a beak that terminates in an osteole through which ascospores are forcibly shot in groups of eight, each comprising the contents of an ascus.

dimensions using the synaptonemal complex with its associated recombination nodules. Electrophoretic separation of whole-chromosome DNAs provided estimates of the DNA content of individual chromosomes and yielded a value of 43 megabases for the entire haploid genome – about three times that of yeast and roughly one-third that of *Drosophila* or *Arabidopsis*. When the first *Neurospora* compendium was prepared in 1982, a gene could not be recognized until a phenotypically recognizable variant had been obtained, either as a mutant in the laboratory or as a novel allele from the wild. Molecular genetics has changed all that. Genes can now be recognized on the basis of nucleotide sequence and mapped using restriction fragment length polymorphisms, all before any phenotype is known. Numerous genes in the present compilation have been identified using the cloned wild-type allele. To determine the phenotype of such a gene, a null allele or a mutant allele must be obtained by gene replacement or by repeat-induced point mutation (RIP). Genes without known mutant phenotypes may soon outnumber those for which the mutant phenotype has been determined.

The pink bread mold *Neurospora crassa* can grow on medium containing only (1) inorganic salts, (2) a simple sugar, and (3) one vitamin, biotin. *Neurospora* growth medium containing only these components is called "minimal medium." George Beadle and Edward Tatum reasoned that *Neurospora* must be capable of synthesizing all the other essential metabolites, such as the purines, pyrimidines, amino acids, and other vitamins, *de novo*. Furthermore, they reasoned that the biosynthesis of these growth factors must be under genetic control. If so, mutations in genes whose products are involved in the biosynthesis of essential metabolites would be expected to produce mutant strains with additional growth-factor requirements. Beadle and Tatum tested this prediction by irradiating asexual spores (conidia) of wild-type *Neurospora* with X-rays or ultraviolet light, and screening the clones produced by the mutagenized spores for new growth-factor requirements. In order to select strains with a mutation in only one gene, they studied only mutant strains that yielded a 1:1 mutant to wild-type progeny ratio when crossed with wild type. They identified mutants that grew on medium supplemented with all the amino acids, purines, pyrimidines, and vitamins (called "complete medium"), but could not grow on minimal medium. They analyzed the ability of these mutants to grow on medium supplemented with just amino acids, or just vitamins, and so on. For example, Beadle and Tatum identified mutant strains that grew in the presence of vitamins but could not grow in medium supplemented with amino acids or other growth factors. They next investigated the ability of these vitamin- requiring strains to grow on media supplemented with each of the vitamins separately. In this way, Beadle and Tatum demonstrated that each mutation resulted in a requirement for one growth factor. By correlating their genetic analyses with biochemical studies of the mutant strains, they demonstrated in several cases that one mutation resulted in the loss of one enzyme activity. The **one gene–one enzyme** concept thus became a central part of molecular genetics.

It is evident from the above account that *Neurospora* is a very suitable fungi producing 8 ascospores after meiosis in a very short time. There is direct relationship between gene and its product and phenotypic product can easily be separated as ascospores. It is easy to create mutations by X-rays. Therefore, this material is very suitable to study genetic changes by mutations and changes in resulting proteins. At the same time new molecular techniques are helpful to determine altered sequences and to correlate them with characters.

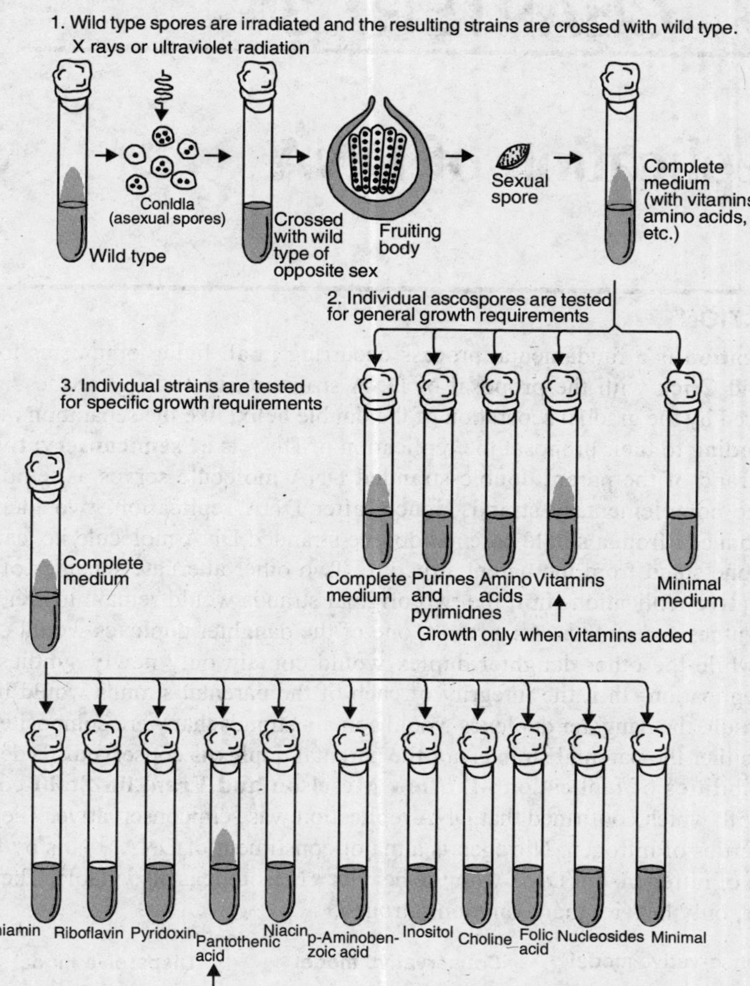

Fig. 2.9. Mutation and nutritional requirement in *Neurospora.*

QUESTIONS

1. What is a gene? Give its structural and functional definition.
2. Define following terms: gene, central dogma, reverse transcriptase, overlapping, pseudo and cryptic genes.
3. Write short notes on :
   (a) One gene one polypeptide
   (b) Central Dogma
   (c) Split genes
   (d) Expression of phenotypic characters
   (e) Contribution of Beadle and Tatum
   (f) Mutation in *Neurospora*
   (g) One gene one enzyme
   (h) Work of Beadle
4. Describe the use of *Neurospora* in understanding genetical principles.

CHAPTER 3

Replication of DNA

1. INTRODUCTION

DNA replication is a fundamental process occurring in all living organisms to copy their DNA. Watson and Crick with the proposal of DNA structure, in 1953 gave the view that the replication occurred by the gradual separation of the double helix, like the separation of two halves of a zipper. According to their proposal the replication of DNA is a "**semiconservative**" process, in which each strand of the parent double-stranded DNA molecule serves as template for the production of the complementary strand. Hence, after DNA replication, two identical DNA molecules are produced from a single parental double-stranded DNA molecule i.e. each daughter duplex contains one stand from the parent structure. Two other alternate schemes of replication were (*i*) Conservative replication- In it the two original strands would remain together, as would the two newly synthesized strands. As a result, one of the daughter duplexes would contain only parental DNA, while the other daughter duplex would contain only newly synthesized DNA. (*ii*) Dispersive replication- In it the integrity of each of the parental strands would be disrupted (Fig. 3.1). As a result, the daughter duplexes would contain strands that were composites of old and new DNA; i.e. neither the parental strands nor the parental duplex is conserved. To decide among these three possibilities of replication **Matthew Meselson and Franklin Stahl** conducted an experiment in 1958, which confirmed that DNA replication was semiconservative. The experiment utilized the properties of nitrogen. Nitrogen is a major constituent of DNA. ^{14}N is by far the most abundant isotope of nitrogen, but DNA with the heavier ^{15}N isotope is also viable. The ^{15}N isotope is not radioactive, only heavier than common nitrogen.

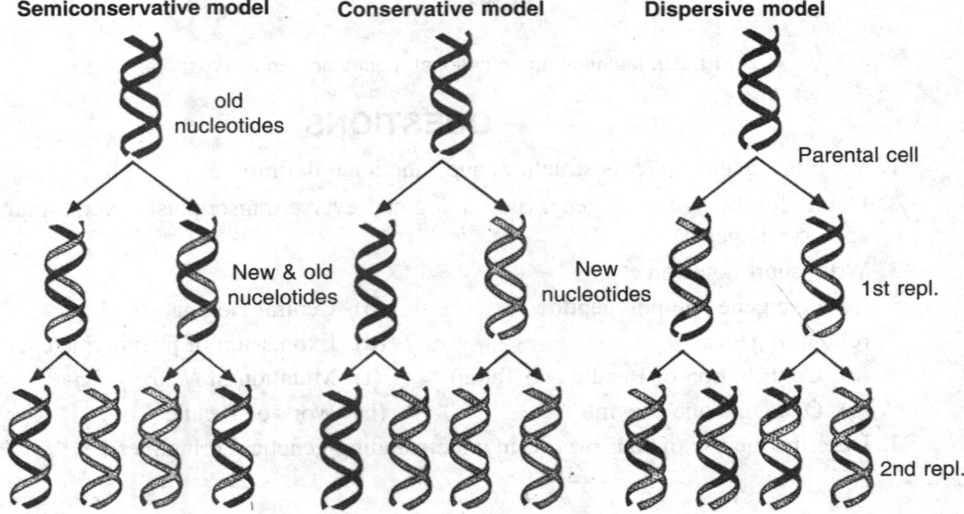

Fig. 3.1. Different schemes of DNA replication.

During the experiment *E. coli* were grown for several generations in a medium containing ^{15}N as the sole nitrogen source. As a result, the nitrogen containing bases of the DNA of these cells contained only the heavy nitrogen isotope. After that, *E. coli* cells with only ^{15}N in their DNA were put back into a ^{14}N containing medium and were allowed to divide in some samples for only once and in some for several generations (Fig. 3.2).

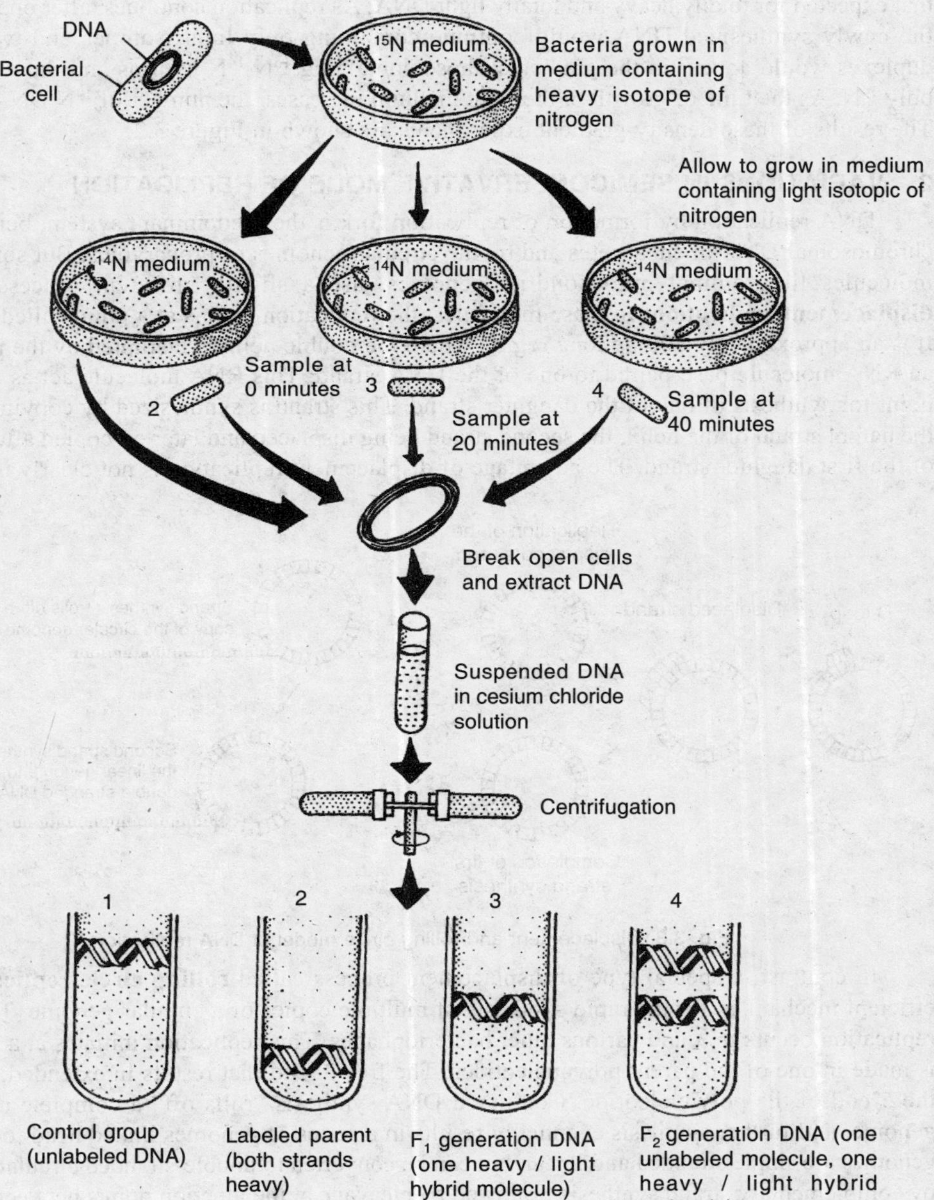

Fig. 3.2. Meselson and stahls' experiment.

When DNA is extracted from ^{15}N grown cells and centrifuged on a salt density gradient, the DNA separates out at the point at which its density equals that of the salt solution. The DNA of the resulting cells had a higher density (was heavier). DNA was then extracted from one generation grown ^{14}N cell and its density was compared to DNA from ^{14}N DNA and ^{15}N DNA. It was found to have close to the intermediate density. This result was in concordance with semiconservative replication. If replication is semiconservative one would expect that the density of DNA molecules would gradually decrease during culture in the ^{14}N containing medium. The density would decrease as light strand were synthesized in association with heavy strands. After one generation, all the DNA molecules would be ^{15}N-^{14}N hybrids and their buoyant density would be halfway between that expected for totally heavy and totally light DNA. As replication continues, after one generation the newly synthesized DNA would continue containing only light isotopes and two types of duplexes would appear in the gradients; those containing ^{15}N-^{14}N hybrids and those containing only ^{14}N. As the time of growth in the light medium increases, the number of ^{14}N DNA increases. The results of these density-gradient experiments are shown in Figure 3.2.

2. VARIATIONS IN SEMICONSERVATIVE MODE OF REPLICATION

DNA replication by formation of replication fork is the predominant system, being used by chromosomal DNA in eukaryotes and by the circular genomes of prokaryotes. But some circular molecules like, animal mitochondrial genomes, use a slightly different process called as **displacement replication**. In these molecules, the replication begins at a point called as D-loop. It is an approximately 500 bp long region where the double helix is disrupted by the presence of an RNA molecule base-paired to one of the DNA strand. This RNA molecule act as the starting point for synthesis of one of the daughter strand. This strand is synthesized by copying of one of the parent strand of the helix, the second strand being displaced and later on copied after synthesis of the first daughter strand. The advantage of displacement replication is not clearly understood.

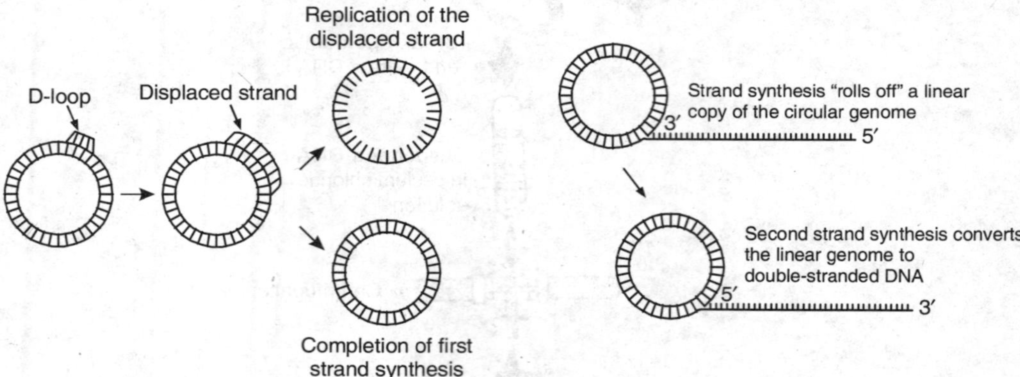

Fig. 3.3. Displacement and rolling circle model of DNA replication.

In contrast, a special type of displacement process called **rolling circle replication** is an efficient mechanism for the rapid synthesis of multiple copies of a circular genome. This type of replication occurs in λ and various other bacteriophages. This replication initiates at a nick which is made in one of the parent polynucleotides. The free 3' end that results in extended, displacing the 5'end of the polynucleotide. Continued DNA synthesis "rolls off" a complete copy of the genome, and further synthesis eventually results in a series of genomes linked head to tail. These genomes are single stranded and linear, but can be converted to double-stranded circular molecules by complementary strand synthesis, followed by cleavage at the junction points between genomes, and circularization of the resulting segments (Fig. 3.3).

3. PROTEINS OF DNA REPLICATION

DNA exists in the nucleus as a condensed, compact structure. To prepare DNA for replication, a number of proteins help in the unwinding and separation of the double-stranded DNA molecule. These proteins are required because DNA must be single-stranded before replication can proceed.

1. **DNA Helicases** - These proteins bind to the double stranded DNA and stimulate the separation of the two strands. This helps in the formation of replication fork by hydrolysis of ATP. The energy liberated is used in the breakage of hydrogen bonds between the bases.
2. **DNA single-stranded binding proteins** - These proteins bind to the DNA as a tetramer and stabilize the single-stranded structure that is generated by the action of the helicases. Replication is 100 times faster when these proteins are attached to the single-stranded DNA.
3. **DNA Gyrase** - This enzyme catalyzes the formation of negative supercoils that is thought to aid with the unwinding process. It creates a nick in one strand, turns the helix and make it straight ladder (as compared to twisted ladder). The nick is again sealed.
4. **DNA Polymerase** - DNA Polymerase I (Pol I) was the first enzyme discovered with polymerase activity, and it is the best characterized enzyme. Although this was the first enzyme to be discovered that had the required polymerase activities, it is not the primary enzyme involved with bacterial DNA replication. The enzyme involved in DNA synthesis in a major way is DNA Polymerase III (Pol III). Three activities are associated with DNA polymerase I;
   - 5′ to 3′ elongation (polymerase activity)
   - 3′ to 5′ exonuclease (proof-reading activity)
   - 5′ to 3′ exonuclease (repair activity)

   The second two activities of DNA Pol I are important for replication, but DNA Polymerase III (Pol III) is the enzyme that performs the 5′-3′ polymerase function.
5. **Primase** - The requirement for a free 3′ hydroxyl group is fulfilled by the RNA primers that are synthesized at the initiation sites by these enzymes.
6. **DNA Ligase** - Nicks occur in the developing molecule because the RNA primer is removed and synthesis proceeds in a discontinuous manner on the lagging strand. The final replication product does not have any nicks because DNA ligase forms a covalent phosphodiester linkage between 3′-hydroxyl and 5′-phosphate groups.

4. THE REPLICATION PROCESS

Replication in prokaryotes and eukaryotes occurs by very similar mechanisms, and thus most of the information presented here for bacterial replication applies to eukaryotic cells as well. It is composed of three phases;

(*a*) **Initiation**- It involves recognition of the positions on a DNA molecule where replication will begin.

(*b*) **Elongation**- It includes the events occurring at the replication fork, where the parent polynucleotides are copied.

(*c*) **Termination**- It is less understood. It occurs when the parent molecule has been completely replicated.

4.1. Initiation

In a cell, DNA replication begins at specific locations in the genome, called "origins". In case of *E. coli* the origin of replication is a sequence of approximately 245 base pairs (bp) called *oriC*. Origins contain DNA sequences recognized by replication initiator proteins (e.g. DnaA in *E. coli* and the Origin Recognition Complex in yeast), these proteins bind to start the process of replication. The initiator proteins recruit other proteins to separate the two strands and initiate replication forks. Unwinding of DNA at the origin, and synthesis of new strands, forms a **replication fork**. The replication fork is a structure which forms when DNA is being replicated. It is created through the action of helicase, which breaks the hydrogen bonds holding the two DNA strands together. The resulting structure has two branching "prongs", each one made up of a single strand of DNA. In bacteria, which have a single origin of replication on their circular chromosome, this process eventually creates a "theta structure" (resembling the Greek letter theta: θ). In contrast, eukaryotes have longer linear chromosomes and initiate replication at multiple origins and whose replication forks progress for shorter distances. For example yeast has about 322 origins, which corresponds to 1 origin per 36 kb of DNA, and humans have some 20,000 origins, or 1 origin for every 150 kb of DNA. Once initiated, two replication forks can emerge from the origin and progress in opposite direction along the DNA. Replication is therefore **bidirectional** with most genomes (Fig. 3.4).

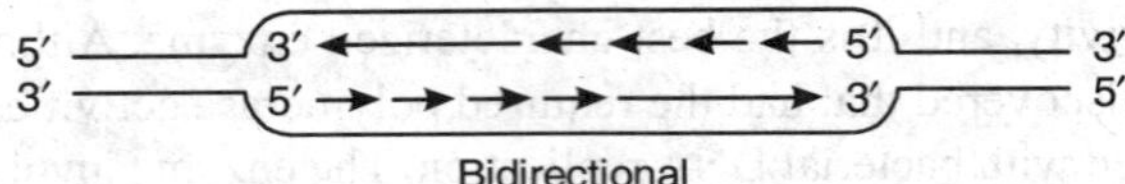

Fig. 3.4. Bidirectional DNA replication.

4.1.1. Initiation of DNA replication in microorganisms (*E. coli*)

We know substantially more about DNA synthesis in prokaryotes than in eukaryotes. As we have discussed that *oriC* of *E.coli* spans 245 bp of DNA. Sequence analysis of this segment shows that it contains two short repeat motifs, one of nine nucleotides and the other of 13 nucleotides. The five copies of nine nucleotide repeat motif are present dispersedly throughout *oriC*. These regions are the binding site for a protein called DnaA. As there are five copies of the binding sequences, it might be imagined that five copies of DnaA attach to the origin, but in fact bound DnaA proteins cooperate with unbound molecules until some 30 copies are associated with the origin. Attachment occurs only when the DNA is negatively supercoiled, as is the normal situation for the *E. coli* chromosome.

The result of DnaA binding is that the double helix opens up (melts) within the tandem array of three AT-rich, 13 nucleotide repeats located at one end of the *oriC* sequence. The exact mechanism is unknown but DnaA does not appear to possess the enzymatic activity needed to break base pairs, and it is therefore assumed that the helix is melted by torsional stresses introduced by attachment of the DnaA proteins. An attractive model imagines that the DnaA proteins form a barrel-like structure around which the helix is wound. Melting the helix is promoted by HU, the most abundant of the DNA packaging proteins of *E. coli*.

Melting of the helix initiates a series of events that construct a new replication fork at either end of the open region. The first step is the attachment of a prepriming complex at each of these two positions. Each prepriming complex initially comprises 12 proteins, six copies of DnaB, and six copies of DnaC, but DnaC has a transitory rōle and is released from the complex soon after it is formed, its function probably being simply to aid the attachment of DnaB. The latter is a helicase, an enzyme which can break base pairs. DnaB begins to increase the single-stranded region within the origin, enabling the enzymes involved in the elongation phase of replication in *E. coli* as the replication forks now start to progress away from the origin and DNA copying begins.

4.1.2. Initiation of DNA replication in yeast

Origins identified in yeast are called autonomously replicating sequences, or ARSs. A typical yeast origin is shorter than *E. coli oriC*, being usually less than 200 bp in length. In it four subdomains are recognized. Two of these – subdomains A and B1- make up the origin recognition sequence, a stretch of some 40 bp in total that is the binding site for the Origin recognition complex (ORC), a set of six proteins that attach to the origin. ORCs have been described as yeast versions of the *E. coli* DnaA proteins, but this interpretation is probably not strictly correct because ORCs appear to remain attached to yeast origins throughout the cell cycle. Rather these are genuine initiator proteins. It is more likely that ORCs are involved in the regulation of genome replication, acting as mediators between replication origins and the regulatory signals that coordinate the initiation of DNA replication with the cell cycle.

There are similar sequences in yeast to that of *oriC* of *E. coli*. This leads us to the two other conserved sequences in the typical yeast origin, subdomains B2 and B3. Our current understanding suggests that these two subdomains function in a manner similar to the *E. coli* origin. Subdomain B2 appears to correspond to the 13-nucleotide repeat array of the *E. coli* origin, being the position at which the two strands of the helix are first separated. This melting is induced by torsional stress introduced by attachment of a DNA-binding protein, ARS binding factor 1 (ABF1), which attaches to subdomain B3. As in *E. coli*, melting of the helix within a yeast replication origin is followed by attachment of the helicase and other replication enzymes to the DNA, completing the initiation process and enabling the replication forks to begin their progress along the DNA. Replications origins in higher eukaryotes have not been much understood.

4.2. Elongation

Once replication has been initiated; the replication forks progress along the DNA and participate in the synthesis of new strand. At the chemical level, the template dependent synthesis of DNA is very similar to the template-dependent synthesis of RNA that occurs during transcription, but the two processes are quite different.

1. Discontinuous strand synthesis and the priming problem- During DNA replication both strands of the double helix must be copied. However, DNA polymerase enzymes are only able to synthesize DNA in the $5' \rightarrow 3'$ direction. This means that one strand of the parent double helix, called the leading strand, can be copied out in a **continuous** manner, but replication of the lagging strand has to be carried out in a **discontinuous** fashion, resulting in a series of short segments that must be ligated together to produce the intact daughter strand. These short segments of polynucleotides are called as **Okazaki fragments**. These fragments were first isolated from *E. coli* bacteria in 1969. Okazaki fragments are 1000-2000 nucleotides in length, but in eukaryotes the equivalent fragments appear to be much shorter, perhaps less than 200 nucleotides in length.

2. The another feature of DNA replication is that DNA polymerase cannot initiate DNA synthesis on a molecule that is entirely single stranded: there must be short single stranded region to provide a 3' end onto which the enzyme can add new nucleotides. Nucleotides are added at a rate of 50,000 bases per minute. The choice of nucleotide is determined by complementary nature. At this rate chances of error are one in one thousand base pair replicated. However, actual rate is quite low (one in one billion). This is equal to about one error per genome per one thousand bacterial replication cycles. This error is further corrected by proof reading (Removal of mismatch nucleotide by DNA polymerase III). This means that primers are needed, one to initiate complementary strand synthesis on the leading polynucleotide, and one for every segment of discontinuous DNA synthesized on the lagging strand. As DNA polymerase cannot deal with an entirely single stranded template, RNA polymerases have no difficulty in this respect, so the primers for DNA replication are made of RNA. In bacteria, primers are synthesized by primase, a

special RNA polymerase with each primer being 4-15 nucleotides in length and most starting with the sequence 5′-AG-3′. Once the primer has been completed, strand synthesis is continued by DNA polymerase III. In eukaryotes the situation is more complex because the primase is tightly bound to DNA polymerase α, and cooperates with this enzyme in synthesis of the first few nucleotides of a new polynucleotide. This primase synthesizes an RNA primer of 8-12 nucleotides, and then hands over to DNA polymerase α, which extends the RNA primer by adding about 20 nucleotides of DNA. After completion of DNA-RNA primer, DNA synthesis is continued by the main replicative enzyme, DNA polymerase δ. Priming needs to occur just once on the leading strand, within the replication origin, because once primed, the leading-strand copy is synthesized continuously until replication is completed. On the lagging strand, priming is a repeated process that must occur every time a new Okazaki fragment is initiated.

3. Replication fork elongation-As with the attachment of DnaB helicase, followed by extension of the melted region of the replication origin, the initiation phase ends. After the helicase has bound to the origin to form prepriming complex, the primase is involved, resulting in the primosome, which initiates replication of the leading strand. It does this by synthesizing the RNA primer that DNA polymerase III needs in order to begin copying the template. More than one helicase is known and this enzyme is involved in various processes, such as transcription, recombination besides replications.

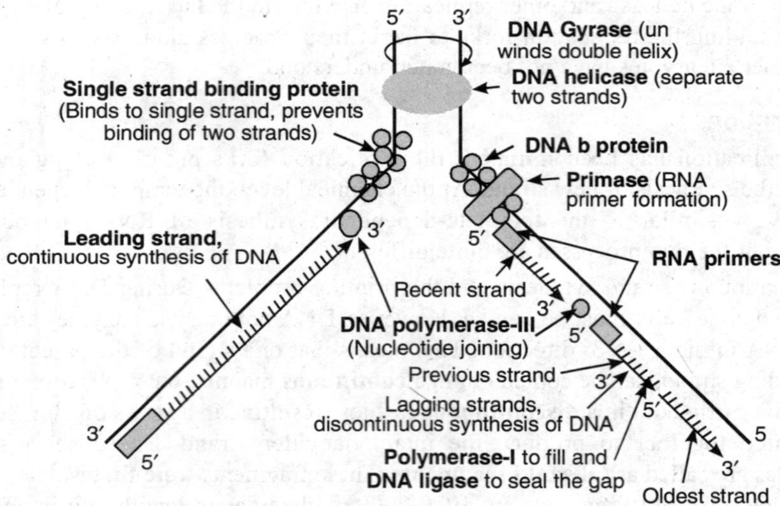

Fig. 3.5. Process of DNA replication in detail showing role of different enzymes.

4. Complementary strands of a DNA tend to become duplex. During the process of replication, these sticky single stranded DNA are prevented to become duplex by special proteins called as single strand binding proteins (SSBs). Once 1000-2000 nucleotides are added in the leading strand, synthesis of lagging strand or Okazaki fragments began. This also requires an RNA primer and DNA polymerase III similar to leading strand.

5. DNA polymerase I is involved in removing the RNA primer from Okazaki fragments, having 5′ → 3′ exonuclease activity. Gap created by primer is filled by adding nucleotides at 3′ end. The nick between two Okazaki fragments is sealed by DNA ligase by the formation of phosphodiester bonds (Fig. 3.5).

4.3. Termination

Because bacteria have circular chromosomes, termination of replication occurs when the two replication forks meet each other on the opposite end of the parental chromosome. *E coli* regulate this process through the use of termination sequences which, when bound by the **Tus** protein, enable only one direction of replication fork to pass through. As a result, the replication forks are constrained to always meet within the termination region of the chromosome.

Eukaryotes initiate DNA replication at multiple points in the chromosome, so replication forks meet and terminate at many points in the chromosome; these are not known to be regulated in any particular manner.

QUESTIONS

1. Describe the process of DNA replication in prokaryotes.
2. Briefly describe the role of different enzymes involved in DNA synthesis.
3. Compare the followings:
   (a) Semiconservative and conservative replication
   (b) Continuous and discontinuous synthesis
   (c) Leading and lagging strands
   (d) Circular and linear DNA replication
4. Write short notes on:
   (a) Meselson and Stahl Experiment
   (b) SSB proteins
   (c) DnaA proteins

CHAPTER 4

Genetic Recombination in Bacteria

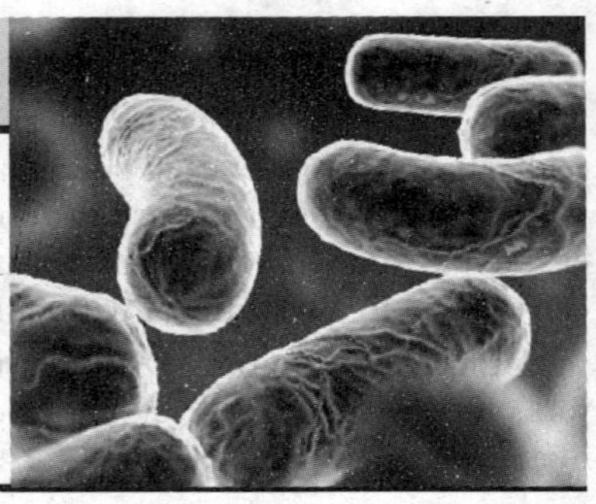

1. RECOMBINATION OF GENETIC MATERIALS

Like meiotic crossing over in eukaryotes, genetic recombination in bacteria provided the basis for the development of methodology for chromosome mapping. The term genetic recombination, as applied to bacteria and bacteriophages, leads to the replacement of one or more genes present in one strain with those of genetically distinct strain. This is somewhat different from our use of genetic recombination in eukaryotes. In eukaryotes, the term describes crossing over that results in reciprocal exchange events. The overall effect is the same in both the system; genetic information is transferred from one chromosome to another resulting in an altered genotype. In bacteria there are mainly three phenomenons responsible for transfer of genetic information-transformation, transduction and conjugation.

2. TRANSFORMATION

Transformation was first discovered in *Streptococcus pneumoniae* and led to the discovery that DNA is the substance of the genes. The path leading to this history making discovery began in 1928 with the work of an English bacteriologist, Fred Griffith. Now transformation is also seen in other bacterial genera like *Haemophilus, Neisseria, Bacillus*, and *Staphylococcus*.

In transformation, pieces of DNA released from donor bacteria are taken up directly from the extracellular environment by recipient bacteria. This leads to a stable genetic change in the recipient cell. DNA entry is thought to occur at the limited number of sites on the surface of the bacterial cell. Passage across the cell wall and membrane is an active process requiring energy and specific transport molecule. During the process of entry, one of the two strands of invading DNA molecule is digested by nucleases, leaving only a single strand to participate in transformation. The active single strand aligns with its complementary region of the host bacterial DNA. This process involves several enzymes and the inserted DNA segment replaces its counterpart from the host genome, which is excised and degraded.

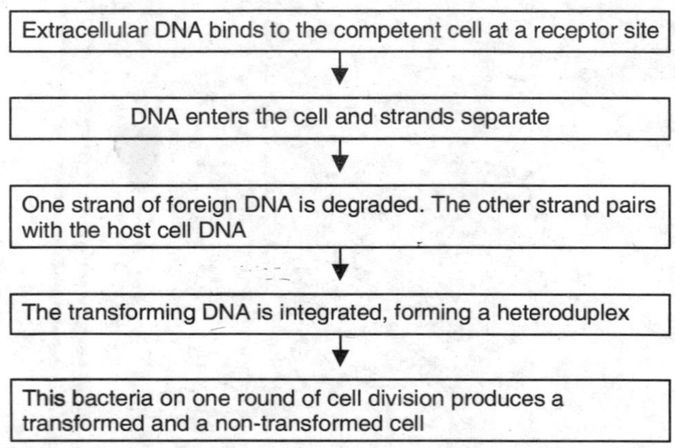

The ability of bacteria to take up extracellular DNA and to become transformed called transformation. This competence varies with the physiologic state of the bacteria. Many bacteria that are not usually competent can be made to take up DNA by laboratory manipulations, such as calcium shock or exposure to a high-voltage electrical pulse (electroporation). In some bacteria (including *Haemophilus* and *Neisseria*) DNA uptake depends on the presence of specific oligonucleotide sequences in the transforming DNA, but in others (including *Streptococcus pneumoniae*) DNA uptake is not sequence specific. Competent bacteria may also take up intact bacteriophage DNA (transfection) or plasmid DNA, which can then replicate as extrachromosomal genetic elements in the recipient bacteria. In contrast, a piece of chromosomal DNA from a donor bacterium usually cannot replicate in the recipient bacterium unless it becomes part of a replicon by recombination. Historically, characterization of "transforming principle" from *S. pneumoniae* provided the first direct evidence of DNA as a genetic material. For detecting the recombination, the transforming DNA must be derived from the different strains of bacteria, bearing some genetic variation. Once it is integrated into the chromosome, the recombinant region contains one host strand and one mutant strand. As these strands are from different sources, this helical region is referred to as a heteroduplex. Following one round of replication, one chromosome is restored to its original configuration, identical to that of the recipient cell and the other contains the mutant gene. Following cell division, one non-mutant (untransformed) cell and one mutant (transformed) cell are produced.

For DNA to be effective in transformation, it must include between 10,000 to 20,000 nucleotides pairs, about 1/200 of the *E. coli* chromosome. This size is sufficient to encode several genes. Genes that are adjacent or very close to one another on the bacterial chromosomes can be carried on a single segment of DNA of this size. Because of this fact, a single transformation event can result in the co-transformation of several genes simultaneously. Genes that are close enough to each other to be co-transformed are said to be linked. Here the linkage refers to the proximity of the genes. Linked bacterial genes were first demonstrated in 1954 during studies of *Pneumococcus* by Rolin Hotchkiss and Julius Marmur. They were examining transformation at the streptomycin and mannitol loci. Recipient cells were Streptomycin sensitive (str^s) and could not ferment mannitol (mtl^-). Cells that were streptomycin resistant (str^r) and could ferment mannitol (mtl^+) were used to derive the transforming DNA. If the *str* and *mtl* genes are not linked, simultaneous transformation for both genes will occur with such a minimal probability as to be nearly undetectable. However, a low but detectable rate of co-transformation did occur. This suggests that the genes responsible for these characters are linked.

In addition to establishing the linkage relationships, relative mapping distances between linked genes can also be determined from the recombination data provided by transformation experiments.

3. TRANSDUCTION

In transduction, bacteriophages function as vectors to introduce DNA from donor bacteria into recipient bacteria by infection. For some phages, called generalized transducing phages, a small fraction of the virions produced during lytic growth are aberrant and contain a random fragment of the bacterial genome instead of phage DNA. Each individual transducing phage carries a different set of closely linked genes, representing a small segment of the bacterial genome. Transduction mediated by populations of such phages is called generalized transduction, because each part of the bacterial genome has approximately the same probability of being transferred from donor to recipient bacteria. When a generalized transducing phage infects a recipient cell, expression of the transferred donor genes occurs. Abortive transduction refers to the transient expression of one or more donor genes without formation of recombinant progeny, whereas complete transduction is characterized by production of stable recombinants that inherit donor genes and retain the ability to express them. In abortive transduction the donor DNA fragment does not replicate, and among the progeny of the original transductant only one bacterium contains the donor DNA fragment. In all other progeny the donor gene products become progressively diluted after each generation of bacterial growth until the donor phenotype can no longer be expressed. On selective medium upon which only bacteria with the donor phenotype can grow, abortive transductants produce minute colonies that can be distinguished easily from colonies of stable transductants. The frequency of abortive transduction is typically one to two orders of magnitude greater than the frequency of generalized transduction, indicating that most cells infected by generalized transducing phages do not produce recombinant progeny.

Specialized transduction differs from generalized transduction in several ways. It is mediated only by specific temperate phages, and only a few specific donor genes can be transferred to recipient bacteria. Specialized transducing phages are formed only when lysogenic donor bacteria enter the lytic cycle and release phage progeny. The specialized transducing phages are rare recombinants which lack part of the normal phage genome and contain part of the bacterial chromosome located adjacent to the prophage attachment site. Many specialized transducing phages are defective and cannot complete the lytic cycle of phage growth in infected cells unless helper phages are present to provide missing phage functions. Specialized transduction results from lysogenization of the recipient bacterium by the specialized transducing phage and expression of the donor genes. Phage conversion and specialized transduction have many similarities, but the origin of the converting genes in temperate converting phages is unknown.

3.1. Zinder and Lederberg Experiment

When certain proviruses turn into virulent viruses, they causes lysis of bacterial cell, and during this they may carry with them segments of bacterial's DNA.

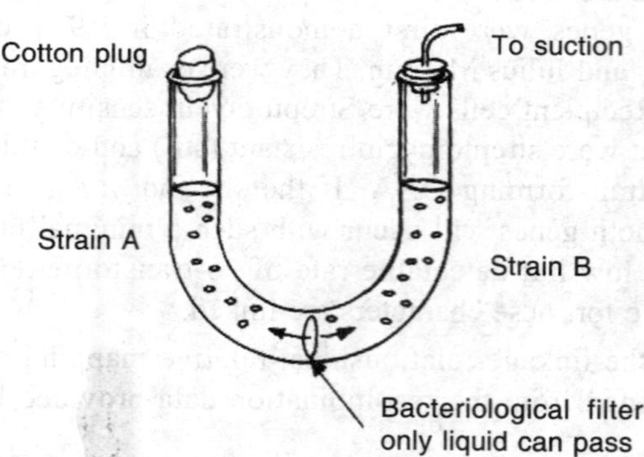

Fig. 4.1. Lederberg's U tube experiment.

If such virus particles carrying a segment of DNA from one bacteria infects another bacteria then the second bacteria can acquire some of the properties of first one due to transfer of genetic material through the virus. This process is called as transduction and was first discovered in 1952 by Norton Zinder and Joshua Lederberg. Zinder and Lederberg discovered transduction through an experiment popularly called U tube experiment. In this experiment a U tube was partitioned into two arms by a filter, so that only virus particles and not the bacterial cells could pass from one arm to another. One arm contained lysogenic bacteria carrying latent virus and the other arm contained bacteria susceptible to this virus. Rarely lysogenic bacteria could undergo lysis and release virus particles which crossed the filter and infected the other susceptible strain, which got destroyed releasing virus particles. These released particles carried genetic material from susceptible strain and through the filter again reached the lysogenic strain and transferred the genetic material to this lysogenic strain. Since two bacterial strains could not come in contact, no conjugation was possible. An enzyme was also used which destroyed free DNA, so that the possibility of the transformation was also eliminated. Regions as large as one percent of the bacterial chromosome may become enclosed in the viral head. In either case, the ability to infect is unrelated to the type of DNA in the phage head, making transduction possible (Fig. 4.1).

3.2. Bacteriophage

Bacteriophages (bacterial viruses, phages) are infectious agents that replicate as obligate intracellular parasites in bacteria. Bacteriophage T4 is one of a group of related bacterial virus. Extracellular phage particles are metabolically inert and consist principally of proteins and nucleic acid (DNA or RNA, but not both). The proteins of the phage particle form a protective shell (capsid) surrounding the tightly packaged nucleic acid genome. Phage genomes vary in size from approximately 2 to 200 kilobases per strand of nucleic acid and consist of double-stranded DNA, single-stranded DNA, or RNA. The DNA is sufficient in quantity to encode more than 150 average sized genes. Phage genomes, like plasmids, encode functions required for replication in bacteria, but beside that, they also encode capsid proteins and nonstructural proteins required for phage assembly. Several morphologically distinct types of phage have been described, including polyhedral, filamentous, and complex. Complex phages have polyhedral heads connected to a tail that contains a collar and a contractile sheath surrounding a central core (Fig. 4.2). Tail fibers, which protrude from the tail, contain binding sites in their tips that specifically recognize unique areas of the outer surface of the cell wall of the bacterial host, *E. coli.*

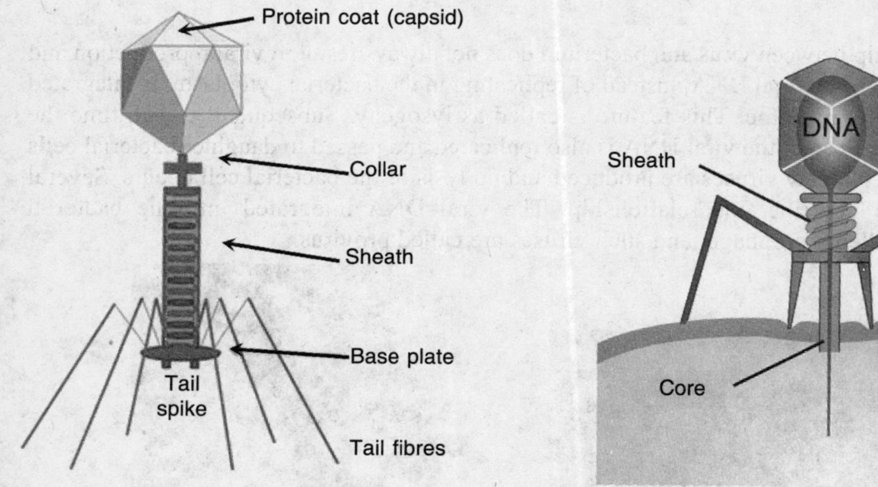

Fig. 4.2. Structure of bacteriophage and bacteriophage injecting DNA into bacteria.

Infection is initiated by adsorption of phage to specific receptors on the surface of susceptible host bacteria (Fig. 4.3, 4.4). Then contraction of the tail sheath causes the central core to penetrate the cell wall. The DNA in the head is extruded, and then it moves across the cell membrane into the bacterial cytoplasm. The capsid remains at the bacterial cell surface. Within minutes, all bacterial DNA, RNA and protein synthesis is inhibited and synthesis of viral molecule begins. At the same time, degradation of the host DNA is initiated. After that viral DNA/RNA components of the head, tail and tail fibres are synthesized. For most phages, assembly of progeny occurs in the cytoplasm, and release of the progeny occurs by bacterial cell lysis by the action of lysozyme (a phage gene product). The latent period is the interval from infection until extracellular progeny appear, and the rise period is the interval from the end of the latent period until all phage is extracellular. The average number of phage particles produced by each infected cell, called the burst size, is characteristic for each virus and often ranges between 50 and several hundreds.

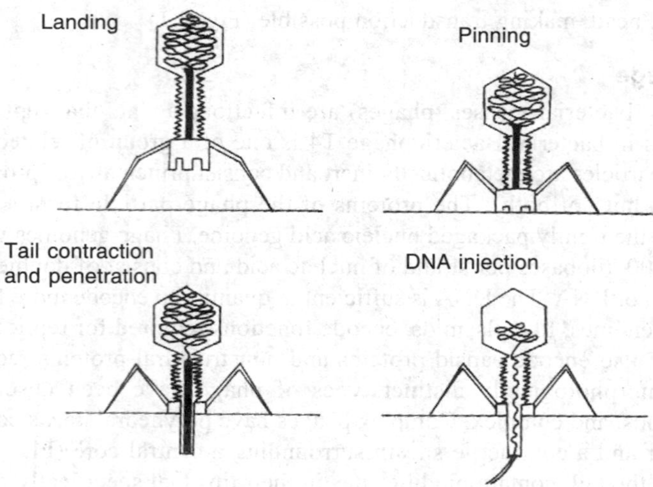

Fig. 4.3. Process of bacteriophage infection.

3.3. Prophage

The relationship between virus and bacterium does not always result in viral reproduction and lysis. In some viruses, the viral DNA, instead of replicating in the bacterial cytoplasm, is integrated into the bacterial chromosome. This feature is called as lysogeny. Subsequently, each time the bacterial DNA is replicated; the viral DNA is also replicated and passed to daughter bacterial cells following division. No new viruses are produced and no lysis of the bacterial cell occurs. Several terms are used to describe this relationship. The viral DNA integrated into the bacterial chromosomes is called a prophage and such viruses are called provirus.

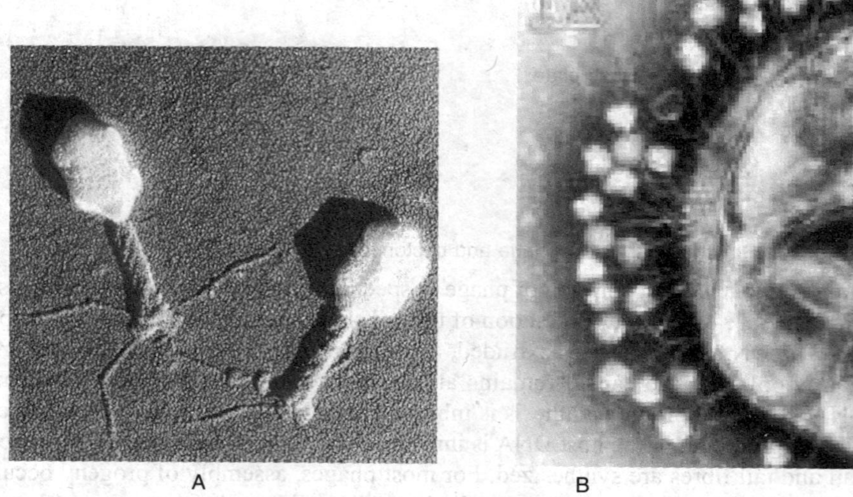

Fig. 4.4. The structure of T4 bacteriophage: (A) An electron micrograph of bacteriophages, and (B) bacteriophages attached to a bacterial cell.

However, in response to certain stimuli, such as chemical, ultraviolet light treatment, or depletion of nutrients, the viral DNA may lose its integrated status and initiate replication, phage production, and lysis of the bacterium. Viruses that can either lyse the cell or behave as a prophage are called temperate. Those that can only lyse the cell are referred to as virulent. A bacterium harboring a prophage that can be lysogenized, is said to be lysogenic; that is, it is capable of being lysed as a result of induced viral reproduction. The viral DNA, which can replicate either in the bacterial cytoplasm or as part of the bacterial chromosome, is sometimes classified as episome (Fig. 4.5).

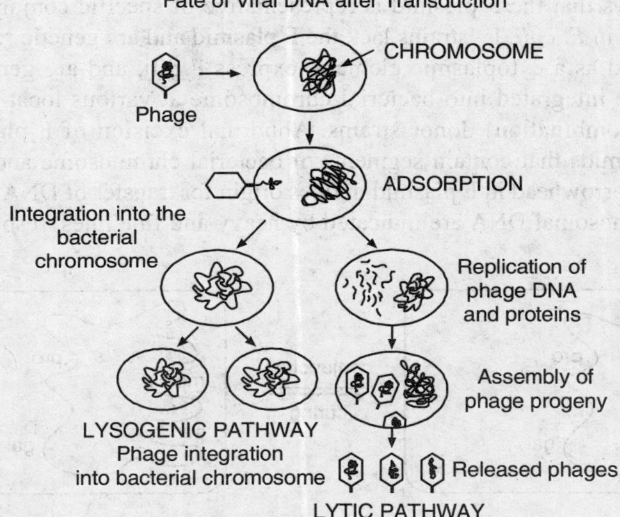

Fig. 4.5. Lytic and lysogenic pathway.

4. CONJUGATION

In 1946, Joshua Lederberg and Edward Tatum showed that the bacteria undergo conjugation, a parasexual process in which the genetic information from one bacterium is transferred to, and recombined with that of another bacterium. In conjugation, direct contact between the donor and recipient bacteria leads to establishment of a cytoplasmic bridge between them and transfer of part or all of the donor genome to the recipient. Donor ability is determined by specific conjugative plasmids called fertility plasmids or sex plasmids.

The F plasmid (also called F factor) of *E. coli* is the prototype for fertility plasmids in Gram-negative bacteria. Strains of *E. coli* with an extrachromosomal F plasmid are called F+ and function as donor, whereas strains that lack the F plasmid are F– and behave as recipients. The conjugative functions of the F plasmid are specified by a cluster of at least 25 transfer (*tra*) genes which determine expression of F pili (conjugation tube), synthesis and transfer of DNA during mating. Each F+ bacterium has 1 to 3 F pili that bind to a specific outer membrane protein (the ompA gene product) on recipient bacteria to initiate mating. An intercellular cytoplasmic bridge is formed, and one strand of the F plasmid DNA is transferred from donor to recipient, beginning at a unique origin and progressing in the 5′ to 3′ direction. The transferred strand is converted to circular double-stranded F plasmid DNA in the recipient bacterium, and a new strand is synthesized in the donor to replace the transferred strand. Both of the exconjugant bacteria are F+, and the F plasmid can therefore spread by infection among genetically compatible populations of bacteria. In addition to the role of the F pili in conjugation, they also function as receptors for donor-specific (male-specific) phages.

The F plasmid in *E. coli.* can exist as an extrachromosomal genetic element or be integrated into the bacterial chromosome. Because the F plasmid and the bacterial chromosome are both circular DNA molecules, reciprocal recombination between them produces a larger DNA circle consisting of F plasmid DNA inserted linearly into the chromosome. *E. coli* contains multiple copies of several different genetic elements called insertion sequences (see section on transposons for more detail), at various locations in its chromosome and in the F plasmid. Homologous recombination between insertion sequences in the chromosome and the F plasmid leads to preferential integration of the F plasmid at chromosomal sites where insertion sequences are located. The chromosomal sites where insertion sequences are found vary among strains of *E. coli.* The Figure 4.6 shows that the F plasmid is representative of specific conjugative plasmids that control donor ability in *E. coli.* F– strains lack the F plasmid and are genetic recipients. F+ strains harbor the F plasmid as a cytoplasmic element, express F pili, and are genetic donors. The F plasmid can become integrated into bacterial chromosome at various locations to produce Hfr (high-frequency recombination) donor strains. Abnormal excision of F plasmid can result in formation of F′ plasmids that contain segments of bacterial chromosome and the corresponding bacterial genes. The arrowhead in F plasmid defines origin for transfer of DNA during conjugation. F plasmid and chromosomal DNA are indicated by heavy and fine lines, respectively.

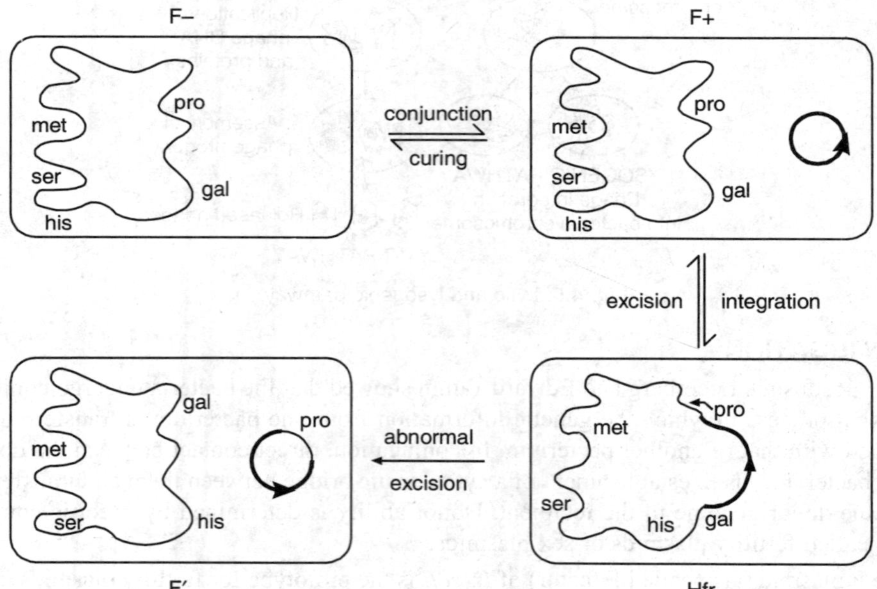

Fig. 4.6. Role of F plasmid in determining donor and recipient states of *E. coli.*

An *E. coli* strain with an integrated F plasmid retains its ability to function as a donor in conjugal matings. Because donor strains with integrated F factors can transfer chromosomal genes to recipients with high efficiency, they are called Hfr (high-frequency recombination) strains. Transfer of single-stranded DNA from an Hfr donor to a recipient begins from the origin within the F plasmid and proceeds as described above, except that the transferred DNA is the hybrid replicon consisting of F plasmid integrated into the bacterial chromosome. Transfer of this entire replicon, including the bacterial chromosome, requires approximately 100 minutes. The identity of the first chromosomal gene to be transferred and the polarity of chromosomal transfer are determined by the site of integration of the F plasmid and its orientation with respect to the bacterial chromosome. Because the mating bacteria usually separate spontaneously before the entire chromosome is transferred, conjugation typically transfers only a fragment of the donor chromosome into the recipient. The probability that a donor gene will enter the recipient bacterium during conjugation decreases, as its distance from the F origin (and therefore the time of its transfer) increases. Mating cells can also be broken apart experimentally by subjecting them to strong shearing forces in a mechanical blender; this is called interrupted mating. Formation of recombinant progeny requires recombination between the transferred donor DNA and the genome of the recipient bacterium. Analysis of progeny from matings that are interrupted after different intervals demonstrates which chromosomal genes are transferred first by particular donor strains, the sequential times of entry for genes that are transferred subsequently, and the progressively lower probability that genes transferred later will appear in recombinant progeny. The circularity of the genetic map of *E. coli* was originally deduced from the overlapping, circularly permuted groups of linked genes that were transferred early by individual donor strains in which the F factor was integrated at different chromosomal locations.

In matings between F+ and F– bacteria, only the F plasmid is transferred with high efficiency to recipients. Chromosomal genes are transferred with very low efficiency, and it is the spontaneous Hfr mutants in F+ populations that mediate transfer of donor chromosomal genes. In matings between Hfr and F– strains, the segment of the F plasmid containing the *tra* region is transferred last, after the entire bacterial chromosome has been transferred. Most recombinants from matings between Hfr and F– cells fail to inherit the entire set of F plasmid genes and are

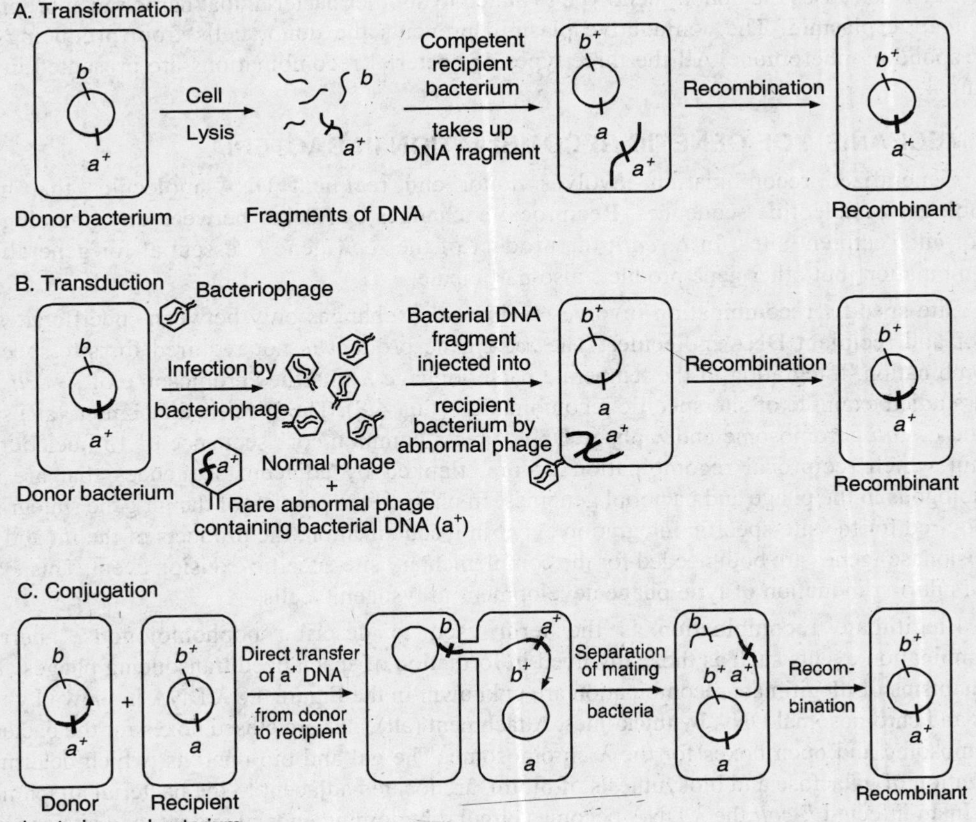

Fig. 4.7. Different types of bacterial recombination.

phenotypically F–. In matings between F+ and F– strains, the F plasmid spreads rapidly throughout the bacterial population, and most recombinants are F+.

Integrated F plasmids in Hfr strains can sometimes be excised from the bacterial chromosome. If excision precisely reverses the integration process, F+ cells are produced. On rare occasions, however, excision occurs by recombinations involving insertion sequences or other genes on the bacterial chromosome that are located at some distance from the original integration site. In such cases segments of the bacterial chromosome can become incorporated into hybrid F plasmids that are called F′ plasmids. By similar processes, segments of the bacterial chromosome can sometimes become incorporated into R plasmids to produce hybrid R′ plasmids. Conjugative R′ plasmids can function as fertility plasmids because they can integrate into the bacterial chromosome by homologous recombination and mediate transfer of chromosomal genes during matings with recipient bacteria. F′ plasmids, R′ plasmids, specialized transducing phages, and recombinant plasmids or phages constructed by gene cloning are hybrid replicons that can include segments of the bacterial chromosome. Therefore, any of these genetic elements can be used to construct the partially diploid bacterial strains that are required for complementation tests and other purposes.

Conjugation also occurs in Gram-positive bacteria. Gram-positive donor bacteria produce adhesins that cause them to aggregate with recipient cells, but sex pili are not involved. In some *Streptococcus* species, recipient bacteria produce extracellular sex pheromones that attracts another bacteria. This causes the donor phenotype to attach to another bacteria, that harbor an appropriate conjugative plasmid. The conjugative plasmid prevents the donor cells from producing the corresponding pheromone. All the three types of bacterial recombinations are presented in the Figure 4.7.

5. MECHANISM OF GENETIC RECOMBINATION IN BACTERIA

Generalized recombination involves donor and recipient DNA molecules that have homologous nucleotide sequences. Reciprocal exchanges can occur between any homologous donor and recipient sites. In *E. coli*, the product of the *recA* gene is essential for generalized recombination, but other gene products also participate.

Site-specific recombination involves reciprocal exchanges only between specific sites in donor and recipient DNA molecules. The *recA* gene product is not required for site-specific recombination. Integration of the temperate bacteriophage λ into the chromosome of *E. coli* is a well-studied example of site-specific recombination (Fig. 4.8). The specific attachment (att) sites on the *E. coli* chromosome and λ phage DNA have a common core sequence of 15 nucleotides, within which reciprocal recombination occurs, flanked by adjacent sequences that are not homologous in the phage and bacterial genomes. In phage λ the product of the int gene (integrase) is required for the site-specific integration event in lysogenization; the products of the int and xis (excisionase) genes are both needed for the complementary site-specific excision event. This event occurs during induction of lytic phage development in lysogenic cells.

Illegitimate recombination is the term used to describe nonhomologous, aberrant recombination events such as those involved in formation of specialized transducing phages. The mechanisms of illegitimate recombination are unknown. In the Figure 4.8 λ DNA is shown by thin lines and chromosomal DNA by thick lines. Attachment (att) sites are closed boxes for the bacterial chromosome and open boxes for the λ chromosome. The gal and bio operons, which determine utilization of galactose and biosynthesis of biotin, are located adjacent to the bacterial attachment site. In an infected *E coli* the λ DNA becomes circular by joining ends m and m′, and site-specific recombination between phage and bacterial att sites results in insertion of the λ genome into the bacterial chromosome. The arrangement of the prophage DNA (m and m′ located internally) is, therefore, a circular permutation of λ virion DNA (m and m′ located terminally).

Genetic mapping of bacterial chromosome is described in the chapter 21 along with genetic mapping.

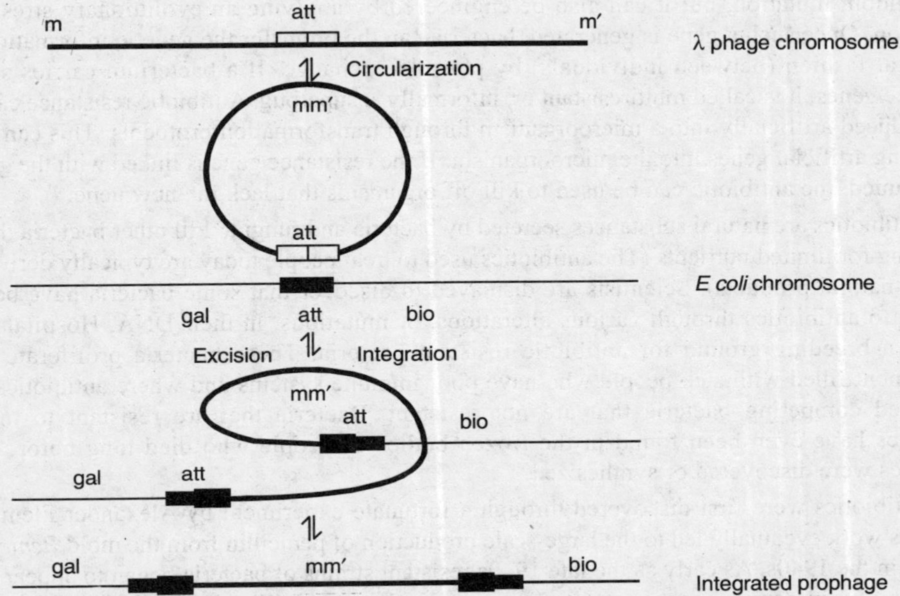

Fig. 4.8. Excision and integration of bacteriophage λ DNA with bacterial DNA.

6. ANTIBIOTIC RESISTANCE

Bacteria are single-celled microorganisms, and most bacterial species are either spherical (called cocci) or rod-shaped (called bacilli) (Fig. 4.9). The extraordinary ability of certain bacteria to develop resistance to antibiotics has been a hot topic on the minds of doctors, hospital staff, reporters, and the general public for several years. It is also used as textbook example of evolution in action. These bacteria are being studied by evolutionary scientists with the hope that they will reveal secrets as to how molecules-to-man evolution could have happened.

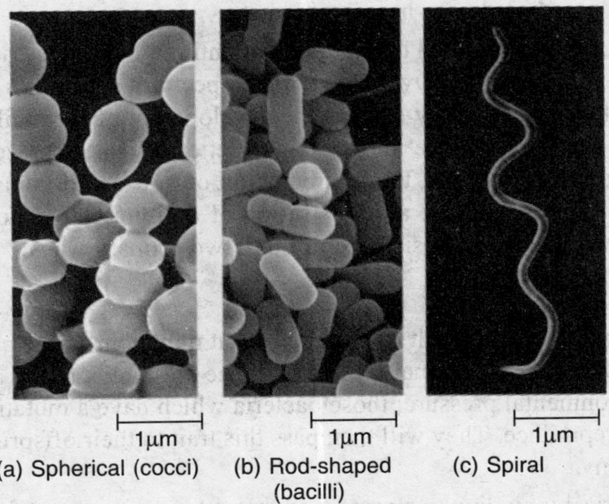

(a) Spherical (cocci) (b) Rod-shaped (bacilli) (c) Spiral

Fig. 4.9. The 3D pictures showing different types of bacteria.

Antibiotic resistance is the ability of a microorganism to withstand the effects of antibiotics. It is a specific type of drug resistance. Antibiotic resistance evolves via natural selection acting upon random mutation, but it can also be engineered by applying an evolutionary stress on a population. Once such a gene is generated, bacteria can then transfer the genetic information in a horizontal fashion (between individuals) by plasmid exchange. If a bacterium carries several resistance genes, it is called multiresistant or, informally, a superbug. Antibiotic resistance can also be introduced artificially into a microorganism through transformation protocols. This can aid in implanting artificial genes into the microorganism. If the resistance gene is linked with the gene to be implanted, the antibiotic can be used to kill off organisms that lack the new gene.

Antibiotics are natural substances secreted by bacteria and fungi to kill other bacteria that are competing for limited nutrients (The antibiotics used to treat people today are typically derivatives of these natural products). Scientists are dismayed to discover that some bacteria have become resistant to antibiotics through various alterations, or mutations, in their DNA. Hospitals have become a breeding ground for antibiotic resistant bacteria. These bacteria proliferate in an environment filled with sick people who have poor immune systems and where antibiotics have eliminated competing bacteria that are not resistant. Bacteria that are resistant to modern antibiotics have even been found in the frozen bodies of people who died long before those antibiotics were discovered or synthesized.

Antibiotics were first discovered through a fortunate experiment by Alexander Fleming in 1928. His work eventually led to the large-scale production of penicillin from the mold *Penicillium notatum* in the 1940s. As early as the late 1940s resistant strains of bacteria began to appear. After four years, when drug companies started to mass produce penicillin, in 1943, the first signs of penicillin-resistant bacteria started to show up. The first bacteria that fought penicillin was called *Staphylococcus aureus*. This bug is usually harmless but can cause an illness such as pneumonia. In 1967, another penicillin-resistant bacteria found. It was called pneumococcus and it broke out in a small village in Papua New Guinea. Other penicillin resistant bacteria found are *Enterococcus faecium* and a new strain of gonorrhea. Currently, it is estimated that more than 70% of the bacteria that cause hospital acquired infections are resistant to at least one of the antibiotics used to treat them. Antibiotic resistance continues to expand for a multitude of reasons, including over-prescription of antibiotics by physicians, non-completion of prescribed antibiotic treatments by patients, use of antibiotics in animals as growth enhancers (primarily by the food industry), increased international travel, and poor hospital hygiene.

Several studies have demonstrated that patterns of antibiotic usage greatly affect the number of resistant organisms which develop. Overuse of broad-spectrum antibiotics, such as second- and third-generation cephalosporins, greatly hastens the development of methicillin resistance. Other factors contributing towards resistance include incorrect diagnosis, unnecessary prescriptions, improper use of antibiotics by patients, the impregnation of household items and children's toys with low levels of antibiotics, and the administration of antibiotics by mouth in livestock for growth promotion. Bacteria can gain resistance through two primary ways:

6.1. Mutation

Antibiotic resistance can be a result of unlinked point mutations in the pathogen genome and a rate of about 1 in 108 per chromosomal replication. The antibiotic action against the pathogen can be seen as an environmental pressure; those bacteria which have a mutation allowing them to survive will live on to reproduce. They will then pass this trait to their offspring, which will result in a fully resistant colony.

The antibiotic works by binding to a protein so that the protein cannot function properly. The normal protein is usually involved in copying the DNA, making proteins, or making the bacterial cell wall. Proteins are involved in all important functions for the bacteria to grow and reproduce. If the bacteria have a mutation in the DNA which codes for one of those proteins, the antibiotic cannot bind to the altered protein; and the mutant bacteria survive. In the presence of antibiotics, the process of natural selection will occur, favoring the survival and reproduction of the mutant bacteria.

During the anthrax scare shortly after the September 11, 2001, attacks in the U.S., Ciprofloxacin (Cipro) was given to potential victims. Cipro belongs to a family of antibiotics known as quinolones, which bind to a bacterial protein called gyrase, decreasing the ability of the bacteria to reproduce. This allows the body's natural immune defenses to overtake the infectious bacteria as they are reproducing at a slower rate. Quinolone-resistant bacteria have mutations in the genes encoding the gyrase protein. The mutant bacteria survive because the Cipro cannot bind to the altered gyrase. This comes at a cost as quinolone-resistant bacteria reproduce more slowly. Resistance to this family of antibiotics is becoming a major problem with one type of bacteria which causes food poisoning. This bacteria increased its resistance to quinolones 10-fold in just five years.

Recent research has shown that the SOS pathway may be essential in the acquisition of bacterial mutations which lead to resistance to some antibiotic drugs. The increased rate of mutation during the SOS response is caused by three low-fidelity DNA polymerases: Pol II, Pol IV and Pol V. Researchers are now targeting these proteins with the aim of creating drugs that prevent SOS repair. By doing so, the time needed for pathogenic bacterial to evolve antibiotic resistance could be extended, and thus improve the long term viability of some antibiotic drugs.

6.2. By horizontal Gene Transfer Share Resistance Genes

Bacteria can also become antibiotic resistant by gaining mutated DNA from other bacteria. Bacteria can exchange DNA. No new DNA is generated it is just exchanged. This mechanism of exchanging DNA is necessary for bacteria to survive in extreme or rapidly changing environments like a hospital (or like those found shortly after the Flood). The mechanisms of mutation and natural selection aid bacteria populations in becoming resistant to antibiotics. The four main mechanisms by which microorganisms exhibit resistance to antimicrobials are:

1. Drug inactivation or modification: e.g. enzymatic deactivation of Penicillin G in some penicillin-resistant bacteria through the production of β-lactamases.
2. Alteration of target site: e.g. alteration of PBP—the binding target site of penicillins—in MRSA (Methicillin resistant *Staphylococcus aureus*) and other penicillin-resistant bacteria.
3. Alteration of metabolic pathway: e.g. some sulfonamide-resistant bacteria do not require para-aminobenzoic acid (PABA), an important precursor for the synthesis of folic acid and nucleic acids in bacteria is inhibited by sulfonamides. Instead, like mammalian cells, they turn to utilizing preformed folic acid.
4. Reduced drug accumulation: by decreasing drug permeability and/or increasing active efflux (pumping out) of the drugs across the cell surface.

Prevention of antibiotic resistance and development of new drugs

Since antibiotics became so prosperous, all other strategies to fight bacterial diseases were put aside. Now since the effects of antibiotics are decreasing and antibiotic resistance is increasing, new research on how to battle bacteria has started. Improving infection control, discovering new antibiotics, and taking drugs more appropriately are ways to prevent resistant bacteria from spreading. Rational use of antibiotics may also reduce the chances of development of opportunistic infection by antibiotic-resistant bacteria due to dysbacteriosis. In developing nations, approaches are being made to control infections such as hand washing by health care people, and identifying drug resistant infections quickly to keep them away from others. The World Health Organization has begun a global computer program that reports any outbreaks of drug-resistant bacterial infections.

Phage therapy, an approach that has been extensively researched and utilized as a therapeutic agent for over 60 years, especially in the Soviet Union, is an alternative that might help with the problem of resistance. Phage Therapy was widely used in the United States until the discovery of antibiotics, in the early 1940s. Bacteriophages or "phages" are viruses that invade bacterial cells and, in the case of lytic phages, disrupt bacterial metabolism and cause the bacterium to lyse. Phage Therapy is the therapeutic use of lytic bacteriophages to treat pathogenic bacterial infections.

Bacteriophage therapy is an important alternative to antibiotics in the current era of multidrug resistant pathogens. British studies also demonstrated significant efficacy of phages against *Escherichia coli*, *Acinetobacter* spp., *Pseudomonas* spp and *Staphylococcus aureus*. Phage therapy may prove as an important alternative to antibiotics for treating multidrug resistant pathogens.

The ability of microorganisms to exchange their genetic material has provided them an armament to fight new drugs and pose serious problem to human health. By understanding the mechanism of genetic recombination we can use modern tools of biology to control bacterial diseases and use them for production of useful drugs.

7. PHAGE GENOME AND RECOMBINATION

The genome of phage A is a double-stranded DNA molecule about 47,000 base pairs in length. In the phage particle, A DNA has singlestranded, complementary ends 12 bases in length, termed mature or cohesive ends m and m′. Within an infected cell, DNA forms a circle through pairing of the single-stranded DNA, and is replicated and transcribed as a circular molecule during the replication-oriented early phase of a development. After this early stage, a development may proceed along the productive (or lytic) pathway or along the alternative lysogenic pathway. The encapsulation-oriented late stage of the productive pathway involves a transcription switch to synthesis of head, tail, and lysis proteins and a replication switch to a rolling-circle mode that generates multimeric genomes (concatemers) that are the obligatory precursors for the cleaved, linear molecules packaged into a phage head. The lysogenic pathway involves a repression of transcription and a site-specific recombination event that inserts A DNA into the host *E. coli* genome. This integrative recombination between the phage and host attachment sites (att) generates a genetic structure that is permuted from the linear order found in the phage particle because the phage attachment site (a a″ or P P) is approximately in the center of the mature DNA molecule. The prophage has structurally distinct attachment sites: a left attL site (b a′ or B P′) and a right attR site (a b′ or P B′); these in turn can recombine to detach the prophage DNA when the virus is induced to lytic development, regenerating the phage att P site (a a′ or P P′) and the original host att B site (b b′ or B B′).

Viruses can undergo genetic recombination. It was first described by A.D. Hershey and R. Rotman in 1949. It was discovered during mixed infection experiments, in which two distinct mutant strains were allowed to simultaneously infect the same bacterial culture. In it two loci are involved therefore recombination is referred to as intergenic. For example, if two strains of a virus having genotypes a^+b^- and a^-b^+ are allowed to infect the same host cell, the resulting progeny of virus will also have recombinants a^+b^+ and a^-b^- , in addition to parental combinations a^+b^- and a^-b^+ . This phenomenon of genetic recombination has been used in preparing linkage maps of viruses. Linkage maps in T2 and T4 phages have been found to be circular, as in the case in *E. coli* chromosome.

8. GENETIC MAPS OF VIRUSES

The viruses, being very small particles, are often considered not suitable for inheritance studies. However plaque (clear area produced on opaque lawn of bacteria on the surface of a petri plate of solid medium) morphology like large *vs.* small, fuzzy *vs.* sharp, host range and virulence are characteristics features of bacteriophages, which have been used for inheritance and recombination in viruses can be illustrated by a cross in T2 phage, attempted by Hershey. This cross ($h^- r^+ \times h^+ r^-$) involved host range and plaque morphology. Here h^+ denotes that these viruses can infect only strain 1 and h^- can infect both strains 1 and 2. r^+ lyses slowly producing small plaques and r^- lyses rapidly producing large plaques. For producing genetic map bacterial strain 1 was infected by both phage genotypes (mixed infection or double infection) and the lysate was analysed by spreading it onto a bacterial lawn composed of a mixture of strain 1 and 2 (h^- will produce clear and h^+ will produce cloudy plaques). Four plaque types were distinguishable (Fig. 4.10);

1. clear and small : h^-r^+
2. cloudy and large : h^+r^-
3. cloudy and small : h^+r^+
4. clear and large : h^-r^-

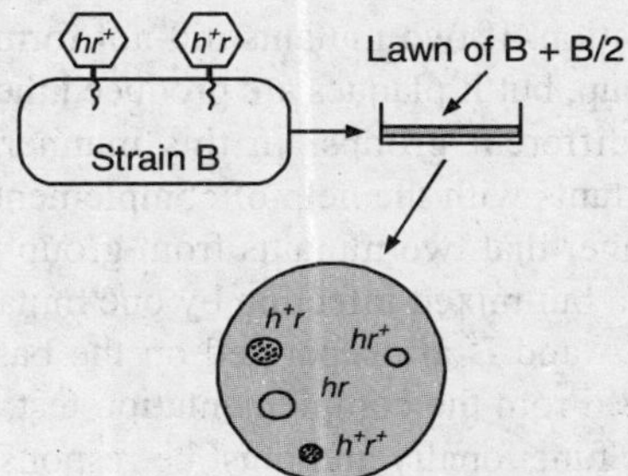

Fig. 4.10. Different types of plaque formation by viral recombination.

The first two of these four are parental phenotypes and last two are recombinants, so that the recombination frequency (RF) can be calculated as follows: RF= $(h^+r^+) + (h^-r^-)$ / total plaques. In order to prepare the linkage map, several rapid lysis genotypes (r1, r2, r3, in order of discovery) were available and the coresoponding strains were called r_a, r_b and r_c. Utilizing these three strains with h^+ and h^- genotypes, following crosses were made: $r_a^- h^+ \times r_a^+ h^-$, $r_b^- h^+ \times r_b^+ h^-$, $r_c^- h^+ \times r_c^+ h^-$, $r_c^- r_b^+ \times r_c^+ r_b^-$. The recombination frequencies were calculated and following two probable linear orders were available: r_a^- r_c^- h^- r_b and r_c^- h^- r_b^- r_a. This can be explained only if the circular map shown in circular genetic map for a T-even phage (Fig. 4.11). S. Benzer resolving the fine structure of *rII* locus has also studied recombination in T4.

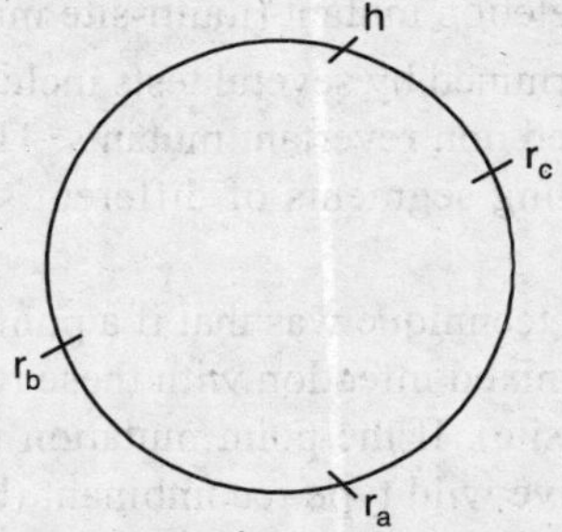

Fig. 4.11. Circular genetic map of bacteriophage T2.

9. FINE STRUCTURE OF *rII* LOCUS IN T4 PHAGE

T4 phage *rII* locus for the rough plaque phenotype of the phage colonies was most extensively studies for the fine structure of gene. The most refined analysis of a single gene ever conducted is the one undertaken during 1950's by Seymour Benzer for a locus in T4 bacteriophage infecting *E. coli*. This locus is known as *rII* locus and a mutant at this locus is responsible for the formation of rough plaques or colonies. This locus had largest number of rapid lysing (r) mutants, and is called *rII* locus. It can be distinguished from other r loci, by the inability of *rII* mutants to produce plaques on lysogenic 'K' strain of *E. coli*, which carries lamda prophage. The *rII* mutant make large plaques with fuzzy edges on *E. coli*, strain B. The wild type phage T4 rII^+ will make small plaques with sharp edges, both on B and / k strains. Further, when 'K' was infected simultaneously by rII^+ and *rII*. Large plaques were formed, since rII^+ helps in lysis. So that *rII* may express. These distinguishing features enabled Benzer to distinguish mutants and wild type phages with high efficiency.

9.1. Complementation Test

To determine complementation relations between *rII* mutant alleles, Benzer used two different *rII* mutants, arbitrarily designated as rII_x and rII_y. He allowed mixed infection of K strain by these two mutants. In most cases, this did not result into lysis and plaque formation; in some cases it did lead to plaque formation. If two mutants did not formed plaques on mixed infection, they were placed in the same group, but if plaques are produced the two mutants involved in mixed infection were placed in two different groups. In this manner, two groups A and B can be established in *rII* region. All mutants with the help of complementation test could be classified in these two groups, in such a manner that two mutants from group A or two mutants from group B could not cause plaque formation but mixed infection by one mutant from each group could cause plaque formation. Since group A and B are separated on the basis of *cis-trans* test, these were called as cistron A and cistron B. From the complementation test, it is clear that in *rII* region two cistron A and B are independent functionally and must be responsible for sequential synthesis for two separate products (different proteins). Therefore, all mutants belonging to one cistron share a common deficiency, which is different from deficiency in the second cistron. When two mutants of same group are mixed, deficiency persists in them. When two mutants of different groups are mixed, they complement each other and may express wild phenotype i.e. lysis and plaque formation.

9.2. Recombination Test

Once *rII* mutants were classified into cistron A and cistron B, Benzer analysed the mutants belonging to same cistron. Mutants were subjected to recombination test to find out whether they are located on same site or different sites separable by a distance. This distance can be resolved by recombinations. Benzer classified all the mutants into two categories (1) revertant type-point mutations (2) non revertant type-deletion mutant (multi-site mutations).

Deletion mutations were determined by several tests including their non-revertant nature. For recombination analysis, Benzer used non revertant mutants. These deletions were arranged in set of overlapping deletions representing segments of different sizes in *rII* region as shown in the Figure 4.12.

The principle involved in this technique was that if a point mutation lies within the region of a deletion in another mutant, then mixed infection with these two will not be able to give rise to a wild type (both defective at same site). If the point mutation falls outside the deletion region of second mutant, it will be able to give wild type recombinant (both defective at different sites).

Using non revertant mutants having successive overlapping deletions of samller lengths, one could locate a mutant to a fairly small region. All point mutations located in this particular small

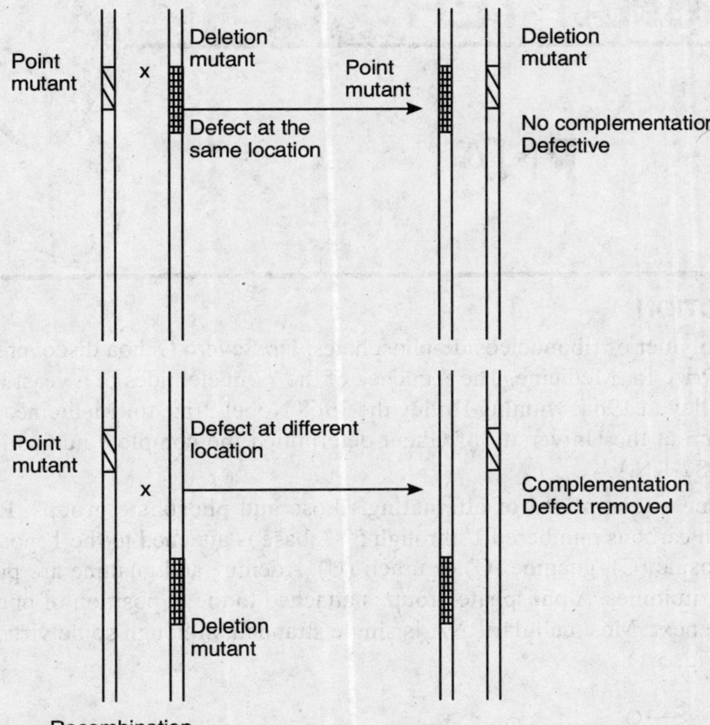

Fig. 4.12. Recombination test of Benzer showing recombination in *rII* locus.

segment of *rII* region, could then be subject to recombination test. For this purpose, two mutants at a time were used for mixed infection on *E. coli* B strain. The lysate produced on plaques was used for infection on K strain to find out the frequency of wild type phage particles produced, because this will represent the recombination frequency. Benzer estimated recombination frequencies as low as 0.0001%. Thus, Benzer was able to divide *rII* region into cistrons A and B and same cistron into hundreds of mutational sites separable due to recombination. By this method ~2400 mutations were mapped in *rII* region and were found to occupy 300 different sites called mutational sites. There were regions having high mutational frequencies called as hot spots whereas other sites having no mutations were called zero sites.

QUESTIONS

1. Describe the fate of lytic and lysogenic pathway. How genetic exchange takes place in bacteria?
2. Describe the method of deletion mapping used by Benzer for constructing a map of *rII* locus of T4 phage. How *rII* locus could be divided into two cistrons.
3. Write short notes on:
   (a) Transformation
   (b) Conjugation
   (c) Transduction
   (d) Antibiotic resistance
   (e) Genetic recombination in viruses
   (f) Bacteriophage
   (g) Mechanism of recombination during transduction
   (h) Structure of rII locus
   (i) Prophage
   (j) Lederberg's experiment

CHAPTER 5

RNA

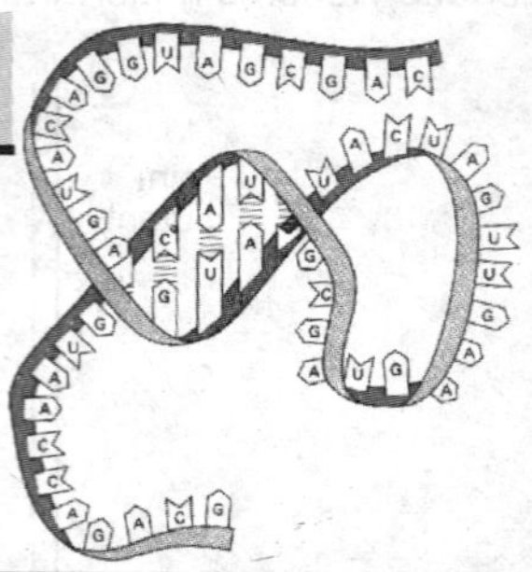

1. INTRODUCTION

RNA is a polymer of ribonucleoside-phosphates. Dr. Severo Ochoa discovered the RNA and got 1959 Nobel Prize for Medicine. The sequence of the 77 nucleotides of a yeast RNA was found by Robert W. Holley in 1964, winning Holley the 1968 Nobel Prize for Medicine. In 1976, Walter Fiers and his team at the University of Ghent determined the complete nucleotide sequence of bacteriophage MS2-RNA.

It's backbone is comprised of alternating ribose and phosphate groups. Ribose is a five carbon sugar with carbons numbered 1′ through 5′. A base is attached to the 1′ position, generally adenine (A), cytosine (C), guanine (G) or uracil (U). Adenine and guanine are purines, cytosine and uracil are pyrimidines. A phosphate group is attached to the 3′ position of one ribose and the 5′ position of the next. Most cellular RNA is single stranded, although some viruses have double stranded RNA (Fig. 5.1).

Fig. 5.1. Chemical structure of RNA.

Several other bases are occasionally found in RNAs including: thymine, pseudouridine and methylated cytosine and guanine. However, there are also numerous modified bases and sugars found in RNA that serve many different roles. Pseudouridine (ψ), in which the linkage between uracil and ribose is changed from a C-N bond to a C-C bond, and ribothymidine (T), are found in various places (most notably in the TψC loop of tRNA). Thus, it is not technically correct to say that uracil is found in RNA in place of thymine. Another notable modified base is hypoxanthine (a deaminated Guanine base whose nucleoside is called Inosine). Inosine plays a key role in the Wobble Hypothesis of the Genetic Code. There are nearly 100 other naturally occurring modified nucleosides, of which pseudouridine and nucleosides with 2′-O-methylribose are by far the most common. The specific roles of many of these modifications in RNA are not fully understood. However, it is notable that in ribosomal RNA, many of the post-translational modifications occur in highly functional regions, such as the peptidyl transferase center and the subunit interface, implying that they are important for normal function. The single RNA strand is folded upon itself, either entirely or in certain regions. In the folded region a majority of the bases are complementary and are joined by hydrogen bonds. This helps in the stability of the molecule. In the unfolded region the bases have no complements. Because of this RNA does not have the purine pyrimidine equality that is found in DNA.

Inside of cells, there are three major types of RNA: messenger RNA (mRNA), transfer RNA (tRNA) and ribosomal RNA (rRNA). There are a number of other types of RNA present in smaller quantities as well, including small nuclear RNA (snRNA), small nucleolar RNA (snoRNA) and the 4.5S signal recognition particle (SRP) RNA. Novel species of RNA continue to be identified. RNA serves a multitude of roles in living cells. These include: serving as a temporary copy of genes that is used as a template for protein synthesis (mRNA), functioning as adaptor molecules that decode the genetic code (tRNA) and catalyzing the synthesis of proteins (rRNA). There is much evidence implicating RNA structure in biological regulation and catalysis. Interestingly, RNA is the only biological polymer that serves as both a catalyst (like proteins) and as information storage (like DNA). For this reason, it has be postulated RNA, or an RNA-like molecule, was the basis of life early in evolution. RNA is a nucleic acid polymer consisting of nucleotide monomers that plays several important roles in the processes that translate genetic information from deoxyribonucleic acid (DNA) into protein products; RNA acts as a messenger between DNA and the protein synthesis complexes known as ribosomes, forms vital portions of ribosomes, and acts as an essential carrier molecule for amino acids to be used in protein synthesis.

2. DIFFERENCE BETWEEN DNA AND RNA

RNA is very similar to DNA, but differs in a few important structural details: RNA nucleotides contain ribose sugars while DNA contains deoxyribose (a type of ribose that lacks one oxygen atom) and RNA uses predominantly uracil instead of thymine present in DNA.

RNA is usually single-stranded, while DNA is usually double-stranded. RNA is transcribed from DNA by enzymes called RNA polymerases and further processed by other enzymes. But there are some exceptions where RNA is double stranded and DNA is single stranded. Example of double stranded RNA are reovirus, retrovirus, hepatitis-B virus etc and the single stranded DNA is present in phage FX174 and parvovirus. Unlike DNA, RNA is not the genetic material of most of the organisms. RNA is the genetic material of only some of the plant, animal and bacterial viruses. Such RNA is called as genetic RNA. However, RNA is present in all the organisms but perform different functions during protein synthesis. In them the genetic message are given by DNA and such RNAs are called as non genetic RNA. RNA serves as the template for translation of genes into proteins, transferring amino acids to the ribosome to form proteins, and also translating the transcript into proteins (Table 5.1, Fig. 5.2).

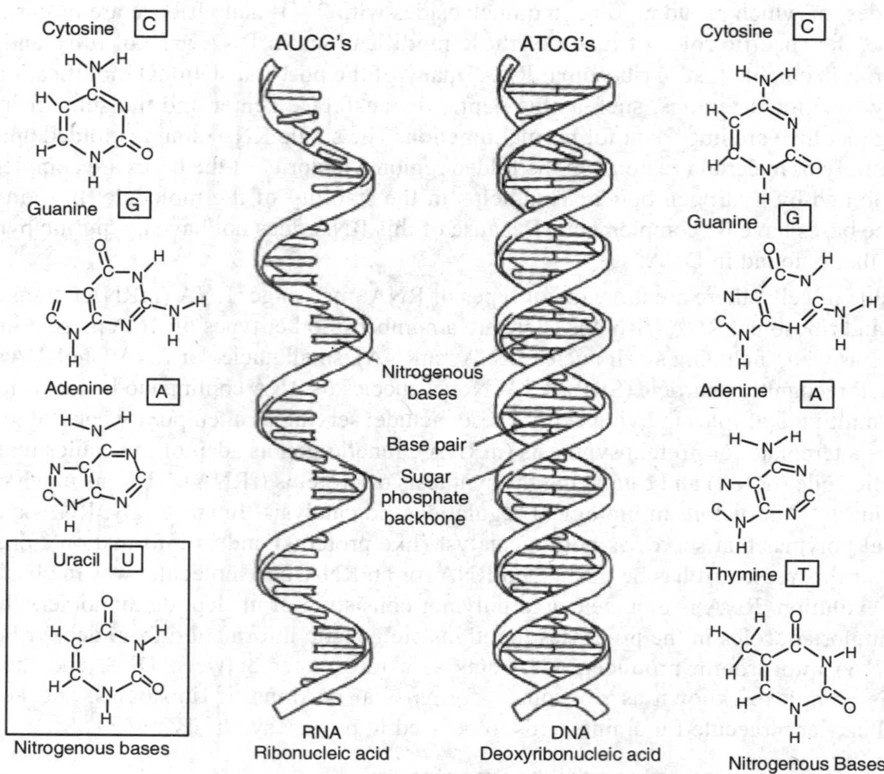

Fig. 5.2. Difference between RNA and DNA.

Table 5.1. Difference between RNA and DNA

RNA	DNA
RNA is single stranded except in some viruses	DNA is double stranded except in few viruses
RNA have ribose sugar	DNA have deoxyribose sugar
Bases present are adenine, guanine, cytosine and uracil.	Bases present are adenine, guanine, cytosine and thymine.
Adenine pairs with uracil	Adenine pairs with thymine
Purine is not equal to pyrimidine	Purine is equal to pyrimidine (Chargaff's rule)
Regions having complementary nucleotides, pairs, and form hair pin loop like structure and helical.	Complementary nucleotides are present throughout the length of the DNA.
RNA is genetic material in some viruses.	DNA is the genetic material in all living organisms.
Length of RNA is short consisting of only few thousands nucleotides.	Length of DNA is quite large consisting of millions of nucleotides.
Three types of RNA are present in an organism: mRNA, rRNA, tRNA.	DNA occurs only in one form in an organism.
mRNA occurs in nucleolus, rRNA and tRNA occur in cytoplasm.	DNA occurs in nucleus, nucleolus, and extrachromosomal DNA in mitochondria and chloroplast.

3. SYNTHESIS OF RNA

There are three types of RNA; messenger RNA (mRNA) or template RNA, ribosomal RNA (rRNA) and soluble RNA (sRNA) or transfer RNA (tRNA). Ribosomal and transfer RNA comprise about 98% of all RNA. All three forms of RNA are made on a DNA template. Transfer RNA and messenger RNA are synthesized on DNA templates of the chromosomes, while ribosomal RNA is derived from nucleolar DNA. The three types of RNA are synthesized during different stages in early development. Most of the RNA synthesized during cleavage is mRNA. Synthesis of tRNA occurs at the end of cleavage, and rRNA synthesis begins during gastrulation.

Synthesis of RNA is usually catalyzed by an enzyme RNA polymerase using DNA as a template, a process known as transcription. Initiation of transcription begins with the binding of the enzyme to a promoter sequence in the DNA (usually found "upstream" of a gene). The DNA double helix is unwound by the helicase activity of the enzyme. The enzyme then progresses along the template strand in the 3′ to 5′ direction, synthesizing a complementary RNA molecule with elongation occurring in the 5′ to 3′ direction. The DNA sequence also dictates where termination of RNA synthesis will occur. RNAs are often modified by enzymes after transcription. For example, a poly (A) tail and a 5′ cap are added to eukaryotic pre-mRNA and introns are removed by the spliceosome. There are also a number of RNA-dependent RNA polymerases that use RNA as their template for synthesis of a new strand of RNA. For instance, a number of RNA viruses (such as poliovirus) use this type of enzyme to replicate their genetic material.

4. TYPES OF RNA

4.1. Messenger RNA (mRNA)

Messenger RNA (mRNA) is only 5-10% of total RNA present in the cell. In 1961, Francis Jacob and Jacques Monod for the first time proposed the name mRNA. This RNA carries genetic information from DNA to the ribosome. Since mRNA is transcribed on DNA (genes), its base sequence is complementary to that of the segment of DNA on which it is transcribed. Usually each gene transcribes its own mRNA. Therefore, there are approximately as many types of mRNA molecules as there are genes. There may be 1,000 to 10,000 different species of mRNA in a cell. These mRNA types differ only in the sequence of their bases and in length. Messenger RNA is first synthesized by genes as nuclear heterogeneous RNA (hnRNA), being so called because hnRNAs varies enormously in their molecular weight as well as in their nucleotide sequences and lengths, which reflects the different proteins they are destined to code for translation. The coding sequence of the mRNA determines the amino acid sequence in the protein. In eukaryotic cells, once precursor mRNA (pre-mRNA) has been transcribed from DNA, it is processed to mature mRNA.

This removes its introns i.e., non-coding sections of the pre-mRNA. The mRNA is then exported from the nucleus to the cytoplasm, where it is bound to ribosomes and translated into its corresponding protein form with the help of tRNA. In prokaryotic cells, which do not have nucleus and cytoplasm compartments, mRNA can bind to ribosomes while it is being transcribed from DNA. After a certain period of time the RNA degrades into its component nucleotides with the assistance of ribonucleases. So, mRNA is short lived. When one gene (cistron) codes for a single mRNA strand the mRNA is said to be monocistronic. In many cases, however, several adjacent cistrons may transcribe an mRNA molecule, which is then said to be polycistronic or polygenic (e.g., in prokaryotes). This polycistronic mRNA encode multiple proteins that are separately translated from the same mRNA molecule (Table 5.2). The eukaryotic mRNA is typically monocistronic i.e., only one species of polypeptide chain is translated per mRNA molecule. Most mRNAs contain a significant noncoding segment, that is, a portion that does not direct the assembly of amino acids. For example, approx. 25% of each globin mRNA consists of noncoding, nontranslated regions. Noncoding portions are found on both the 5' and 3' ends of a messenger RNA and contain sequences that have important regulatory roles.

Messenger RNA is always single stranded. It contains mostly the bases adenine, guanine, cytosine and uracil. There are few unusual substituted bases. Although there is a certain amount of random coiling in extracted mRNA, there is no base pairing. In fact base pairing in the mRNA strand destroys its biological activity. It is heterogeneous in size and sequence. The cap (5′-G) is added to the mRNA after transcription. The addition of 5′ G is catalyzed by a nuclear enzyme, guanylyl transferase. The cap is linked to the 5′ terminus of the mRNA through an unusual 5′,5′-triphospahe linkage. The 5′ cap is formed by condensation of a molecule of GTP with the triphosphate at the 5′ end of the transcript. The guanine is subsequently methylated at N-7 to form 7-methylguanosine. Additional methyl groups are added to the 2′ hydroxyls (—OH) of the first and second nucleotides adjacent to the cap. The methyl groups are derived from S-adenosylmethionine. This cap serves to identify this RNA molecule as an mRNA to the translational machinery. In addition, most mRNA molecules contain a poly-Adenosine tail (poly 'A' tail) at the 3′ end. Both the 5′ cap and the 3′ tail are added after the RNA is transcribed and contribute to the stability of the mRNA in the cell. Therefore, the molecular weight of mRNAs varies from some hundreds to

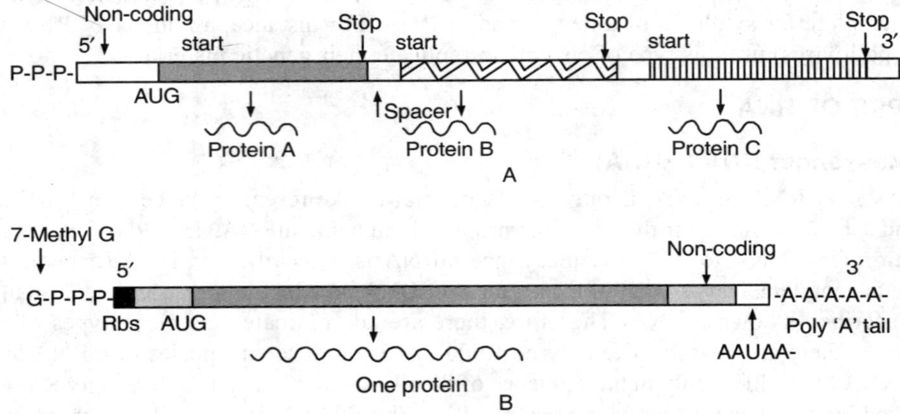

Fig. 5.3. Structure of prokaryotic (A) and eukaryotic (B) mRNA showing gene product. This requires processing of mRNA in eukaryotes before translation.

thousands of Daltons. The molecular weight of an average sized mRNA molecule is about 500,000, and its sedimentation coefficient is 8S. mRNA varies greatly in length and molecular weight. Since most proteins contain at least a hundred amino acid residues, mRNA must have at least 300 (100X3) nucleotides on the basis of the triplet code. In *E. coli* the average mRNA strand has 900 to 1,500 nucleotide units which would code polypeptide chains of 300-500 amino acids. Molecules containing 12,000 nucleotide units are also known. The structure of prokaryotic and eukaryotic mRNA is as follows:

1. **5′ Cap:** At the 5′ end of the mRNA molecule in most eukaryote cells and animal virus molecules is found a 'cap'. This cap is formed by the methylation of any of the four nucleotides. The cap helps mRNA to bind to the ribosomes. Without the cap mRNA molecules bind very poorly to the ribosomes.

   The bacterial mRNA does not have 5′cap. But they have specific ribosome binding site about six nucleotide long, which occurs at several places in the mRNA molecules. These are located at four nucleotides upstream from AUC.
2. **Non-coding regions:** As the name indicate these regions do not code for protein. There are two non-coding regions. First non-coding region (NC1) is followed by a 5′cap and is

10 to 100 nucleotides in length. This region is rich in A and U residues. The second non-coding region (NC2) is followed by termination codon and is 50-150 nucleotide long and contains an AAUAAA residues.

3. **Initiation codon:** Initiation codon is AUG in both prokaryotes and eukaryotes. Bacterial ribosomes bind directly to the AUG region of the mRNA to start the protein synthesis, whereas this is not there in the case of eukaryotes.
4. **The coding region:** It consists of about 1,500 nucleotides. This region is responsible for coding protein with several ribosomes. The combination of mRNA strand with several ribosomes is called polyribosomes.
5. **The termination codon:** The termination codon is required to give the signal to stop protein synthesis.
6. **The poly (A) sequence:** The non-coding region II is followed by poly (A) sequence in the eukaryotic mRNA. The prokaryotic mRNAs lack poly (A). The poly (A) sequences of 200-250 nucleotides are present at 3′OH end of mRNA. Poly (A) sequences are added when mRNA is present inside the nucleus. The function of poly (A) sequence in translation is unknown (Fig. 5.3).

Table 5.2. Difference between prokaryotic and eukaryotic mRNA

Prokaryotic mRNA	Eukaryotic mRNA
Translation begins when the mRNA is still being transcribed on DNA.	Translation begins when the transcription is completed.
Prokaryote mRNAs are very short lived. It constantly under goes breakdown to its constituent ribonucleotides by ribonucleases.	Eukaryotic mRNAs are long lived. Thus are metabolically stable.
In prokaryotes mRNAs are polycistronic.	In eukaryotes mRNAs are monocistronic.
The mRNAs undergo very little processing after being transcribed.	The mRNA undergoes several processing after being transcribed such as polyadenylation, capping and methylation.
Prokaryotic mRNA do not have poly (A) tail.	Eukaryotic mRNA have poly (A) tail.

4.2. Ribosomal RNA (rRNA)

Ribosomal RNA is extremely abundant and makes up to 80% of RNA found in a typical eukaryotic cytoplasm. Ribosomal RNA consists of a single strand twisted upon itself in some regions. It has helical regions connected by intervening single strand regions. The helical regions may show presence or absence of positive interaction. In the helical region most of the base pairs are complementary, and are joined by hydrogen bonds. In the unfolded single strand regions the bases have no complements.

In the cytoplasm, ribosomal RNA and protein combine to form a nucleoprotein called a ribosome. Ribosomal RNA (rRNA) is a component of the ribosomes. The base sequence of rRNA is complementary to that of the region of DNA where it is synthesized. Ribosomal RNA is formed from only a small section of the DNA molecule, and hence there is no definite base relationship between rRNA and DNA as a whole. Primarily there are two types of ribosomes, one is 70s (prokaryotes) and the other is 80s (eukaryotes). The 70S ribosome of prokaryotes consists of a 30S subunit and a 50S subunit. The 30S subunit contains 16S rRNA, while the 50S subunit contains 23S and 5S rRNA. The 80S eukaryote ribosome consists of a 40S and a 60S subunit. In vertebrates the 40S subunit contains 18S rRNA, while the 60S subunit contains 28-29S, 5.8S and 5S rRNA. In plants and invertebrates, the 40S subunit contains 16-18S RNA, while the 60S subunit contains 25S and 58 and 5.8S rRNA.

4.3. Transfer or tRNA

The tRNA molecules are key to the translation process of the mRNA sequence into the amino acid sequence of proteins (at least one type of tRNA for every amino acid). To be precise, the amino-acyl-tRNA-synthase proteins are the 'true' translators of the genetic code into an amino acid sequence. These synthetases *acetylate tRNA* molecules with the proper amino acid that corresponds to the anti-codon in the structure of the tRNA molecule. The anti-codon later recognizes the codon, the triple base sequence which 'codes' for the amino acid along the mRNA strand. A failure of properly acetylating the tRNA with the right amino acid results in an amino acid mutation even though the DNA sequence has not been changed. tRNA molecules are small nucleic acids of 60-95 nucleotides, mostly 76, with a molecular weight 18-20kD, with the secondary structure resembling a clover leaf. Here are a few common features shared by all tRNA molecules found in various organisms.

(1) 5′ terminus is always phosphorylated.

(2) 7 bp stem, may have non-Watson & Crick pairing (like GU) acceptor or amino acid stem at 3′ terminus in which last three nucleotides are CCA-3′-OH. These are added after transcription and amino acylation occurs at 3'-OH group of last base 'A'.

(3) 3-4 bp stem and loop contains the base dihydrouridine (D) [D- arm.]

(4) 5 bp stem and loop containing anti-codon triplet [anti-codon arm].

(5) 5 bp stem and loop contains sequence T C, standing for [T- arm] pseudouridine.

(6) variable arm (between anti-codon and T-arm) length, 3 -21 nucleotides.

(7) contains numerous modified bases (up to 25%) which are all post-transcriptionally modified.

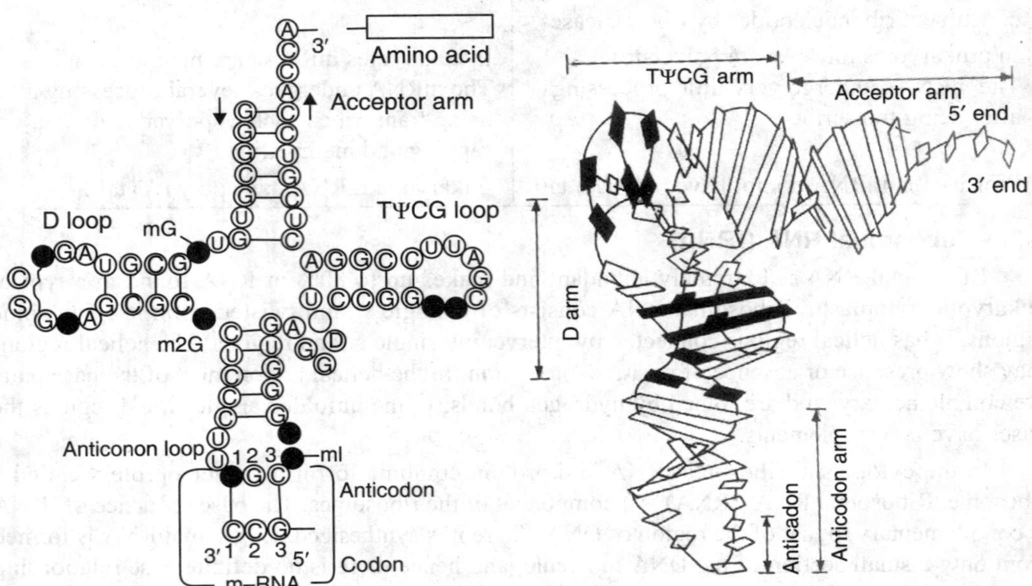

Fig. 5.4. A 3D structure of tRNA: Clover leaf model

The three dimensional structure of tRNA resembles an L-shaped molecule with the D-arm and anti-codon loop building one stretch and the T-arm and acceptor stem building the other stretch being deposed by ~90 to one another (interstem angle of 82 by X-ray refinement and 92 in an electron microscopy study). The molecule is about 6 nm in each direction with the anti-codon to acceptor 3'-term ends being 7.6 nm apart. The diameter of both arms is about 2.0 to 2.5 nm.

The structural complexity of tRNA is reminiscent of that of a protein with 71 out of 76 bases participating in stacking interaction (of which 42 in double helical stem structures). 9 bp interactions are cross linking the tertiary structure, i.e., they interact with bases from a different stem and loop region. All of these 9 bp are non-Watson-Crick associations and are highly conserved which makes it likely to predict similar structures for all tRNA molecules (in fact, only few tRNA molecules have been crystallized and their structure determined) (Fig. 5.4).

Table 5.3. Difference between the three types of RNAs

	Ribosomal RNA (rRNA)	Messenger RNA (mRNA)	Transfer RNA (tRNA)
Percentage of total RNA of cell	~ 80%	5 to 10%	10 to 15%
Sedimentation coefficient	28S, 18S, 5.8S, 5S 23S, 16S, 5S	8S	3.8S
Number of nucleotides	5S RNA: 120 nucleotides. 16-18S RNA: 1,600 to 2,500 nucleotides. 23-28S RNA : 3,200 to 5,500 nucleotides	*E. coli:* 900 to 1,500 nucleotides	73 to 93 nucleotides
Unusual bases	Small amount of methylated bases (*E. coli*: 1 per 100-150 nucleotides).	Small amount of unusual bases.	High content of unusual bases. (*E. coli*: 1 per 3040 nucleotides).
Site of synthesis	Derived from nucleolar DNA.	Synthesized in nucleus on DNA template.	Synthesized in nucleus on DNA template.
Beginning of synthesis	Synthesis begins at gastrulation, and increases as development proceeds.	Some mRNAs are found in the ovum. New mRNA is synthesized during early cleavage.	tRNA synthesis occurs at the end of cleavage stages.
Base relationship with DNA	No obvious base relationship to DNA. rRNA is formed from only small sections of of DNA.	mRNA shows base relationship with DNA. It is formed from all sections of DNA.	rRNA also shows base relationship with DNA. It is formed from all sections of DNA.
Function	Unpaired bases may bind mRNA and tRNA to ribosomes.	Conveys genetic information from DNA to the ribosomes, where it takes part in protein synthesis.	Adaptor for attaching amino acids to mRNA template.

5. NEW TYPES OF RNA

Besides mRNA, tRNA and rRNA, the three classically known RNA molecules, new types of RNA molecules are discovered in recent years, which are involved in various activities directly or they modulate the activity.

5.1. Non-coding RNA

RNA genes (sometimes referred to as non-coding RNA or small RNA) are genes that encode RNA that is not translated into a protein. The most prominent examples of RNA genes are transfer RNA (tRNA) and ribosomal RNA (rRNA), both of which are involved in the process of translation. However, since the late 1990s, many new RNA genes have been found, and thus RNA genes may play a much more significant role than previously thought.

In the late 1990s and early 2000, there has been persistent evidence of more complex transcription occurring in mammalian cells (and possibly others). This could point towards a more widespread use of RNA in biology, particularly in gene regulation. A particular class of non-coding RNA, micro RNA, has been found in many metazoans (from *Caenorhabditis elegans to Homo sapiens*) and clearly plays an important role in regulating other genes.

5.2. Small Nuclear RNA (snRNA)

About a dozen genes for snRNAs have been described, each present in multiple copies in the genome. Small nuclear RNAs combine with certain U-proteins to form snRNPs. These executive molecules have roles in editing other classes of RNA. The "U" designation was given to the snRNAs because they were found to be rich in uridylic acid. An important example is the small ribonuclear proteins (snRNPs) that are components of the spliceosomes. Spliceosomes edits introns nitrogen base sequences out of pm RNA forming mRNA. U6 is transcribed in the nucleoplasm by RNA polymerase III, while U1, U2, U4 and U5 are transcribed by RNA polymerase II. The snRNPs are involved in processing of pmRNA.

5.3. Small Nucleolar RNA (snoRNA)

In eukaryotic cells, rRNA and snRNA are extensively modified and processed in the nucleolus. Much of this activity is affected by snoRNAs. It appears that the coding for snoRNAs lies in introns and other intergenic regions (non-protein coding). Small nucleolar RNA can be considered to be a subgroup of the snRNAs but should not be confused with the snRNAs mediating mRNA splicing, i.e. the spliceosomal RNAs. It is interesting to note that snoRNAs have not been found in any prokaryotes. Many snoRNAs and snRNAs have been found to be encoded in introns. It has been suggested that snoRNAs are possibly processed from the out-spliced introns by exonucleases.

5.4. Short Interfering RNAs (siRNA)

Short interfering RNAs and miRNAs were discovered in different works, but their biogenesis and assembly into RNA-protein complexes and their function in down regulating gene expression are closely related. Short interfering RNAs and mi RNAs share common RNAse III processing enzyme, the dicer enzymes and closely related effector complexes for post-transcriptional repression of protein synthesis. On the other hand, siRNAs and miRNAs differ in their molecular origins. In the cytoplasm the dicer enzymes split the dsRNA primer molecules. The finished siRNAs in animals are usually 21-22 nitrogen bases long, similar to miRNAs.

5.5. Micro-RNAs (miRNAs)

Micro-RNAs are a class of small, non-coding RNAs that regulate gene expression in a sequence specific manner as required in embryonic development. Micro-RNAs have been found thorougout diverse eukaryotes genomes including plants. They can inhibit protein expression by shutting off translation or by targeting mRNA for degradation. Micro-RNAs were first discovered in 2001 in the widely studied worm *Caenorhabdtis elegans*. Micro-RNAs genes produce short (~22 nitrogen bases) ss segments that fold over on themselves forming a short section of dsRNA in hairpin like structure. Humans express over 460 genetically encoded miRNAs. These miRNAs make up more than 1% of human genome and may regulate over 30% of all protein coding genes. Micro-RNAs can pair exactly with a mRNA and cause its cleavage and destruction, or it can pair partially with mRNA and produce translational inhibition (block the ribosome). It is presumed that they are involved in regulating development by controlling as transcriptional factor.

5.6. Ribozymes

Ribozymes are catalytic RNA enzymes that act to alter covalent structure in other classes of RNAs and certain molecules. They occur in ribosomes, nucleus and chloroplasts of eukaryotic organisms. Some viruses including several bacteriophages also have ribozymes. An optimum concentration of metal ions such as Mg^{++} and K^+ is associated with their effective functioning. Ribozymes generally act as molecular scissors cutting precursor RNA molecules at specific sites. Surprisingly, they also serve as molecular staplers, which ligate or join two RNA molecules together.

Ribozymes are involved in the transformation of large precursor molecules of tRNA, rRNA and mRNA into smaller final products. In their active form, ribozymes are complexed with protein molecules, e.g., the enzyme ribonuclease-P (RNAse-P) is found in all living organisms. Ribonuclease-P is a heterodimer containing one molecule of protein and one molecule of RNA. This ribozyme cleaves the head 5′ end of the precursors to the tRNAs. These enzymes are involved in autocatalytic splicing of preRNA making contiguous RNA. This cleavage is carried out by RNA part of the heterodimer.

Synthetic ribozymes can be used in genetic engineering applications to stop the expression of any gene in a sequence-specific manner and therefore may be useful in cancer therapy and HIV treatment. Ribozymes can be inserted into the cells has synthetic oligonucleotides with a gene gun or the cell can synthesize them itself with engineered genes.

5.7. XISTRNA

XIST stands for X-inactive-specific transcript. It is a large (~ 17kb) RNA coded by a gene on the X chromosome (8 exons are involved with human xist). xistRNA accumulates in female somatic cells along the X chromosome containing the active xist gene and proceeds to inactivate nearly all the 100s of genes on that X chromosome. The xistRNA does not travel over to any other chromosome in nucleus. The barr body seen in the cell nucleus with light microscopy is inactive X chromosome covered with xistRNA. The entire xistRNA X-blocking mechanism is very complex.

5.8. Double-stranded RNA (dsRNA)

Double-stranded RNA (dsRNA) is RNA with two complementary strands, similar to the DNA found in all 'higher' cells. dsRNA forms the genetic material of some viruses. In eukaryotes, it acts as a trigger to initiate the process of RNA interference and is present as an intermediate step in the formation of siRNAs (small interfering RNAs). siRNAs are often confused with miRNAs; siRNAs are double-stranded, whereas miRNAs are single-stranded. Although initially single stranded, there are regions of intra-molecular association causing hairpin structures in pre-miRNAs. Very recently, dsRNA has been found to induce gene expression at transcriptional level, a phenomenon named "small RNA induced gene activation (RNAa)". Such dsRNA is called "small activating RNA (saRNA)".

5.9. RNA Secondary Structures

The functional form of single stranded RNA molecules (like proteins) frequently requires a specific tertiary structure. The scaffold for this structure is provided by secondary structural elements which are hydrogen bonds within the molecule. This leads to several recognizable "domains" of secondary structure like hairpin loops, bulges and internal loops. The secondary structure of RNA molecules can be predicted computationally by calculating the minimum free energies (MFE) structure for all different combinations of hydrogen bondings and domains. There has been a significant amount of research directed at the RNA structure prediction problem.

QUESTIONS

1. Describe the chemical nature of RNA. How it differs from DNA?
2. Describe the different types of RNA molecules present in a cell.
3. Write short notes on:
   (a) tRNA (b) Clover leaf model
   (c) Ribosomal RNA (d) snRNA
   (e) snoRNA (f) siRNA
   (g) xist RNA (h) miRNA
   (i) Ribozyme
4. Differentiate between the following:
   (a) RNA and DNA
   (b) Prokaryotic and eukaryotic RNA
   (c) mRNA and tRNA

CHAPTER 6

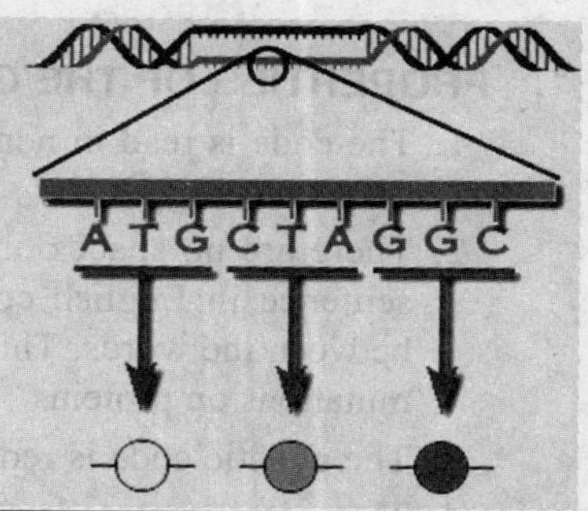

Genetic Code

We need several thousands of different proteins in our body. It is the genetic material, the DNA, in our cells that provides the information needed to produce all these proteins. Though the linear sequence of nucleotides in DNA contains the information for protein sequences, proteins are not made directly from DNA. Instead, a messenger RNA (mRNA) molecule is synthesized from the DNA and directs the formation of the protein. RNA is composed of four nucleotides: adenine (A), guanine (G), cytosine (C), and uracil (U). Three adjacent nucleotides constitute a unit known as the codon, which codes for an amino acid.

The genetic code consists of 64 triplets of nucleotides (Fig. 6.1). These triplets are called codons. With three exceptions, each codon encodes for one of the 20 amino acids used in the synthesis of proteins. Most of the amino acids are encoded by more than one codon. The deciphering of the genetic code was accomplished by the American biochemists Marshall W. Nirenberg, Robert W. Holley, and Har Gobind Khorana in the early 1960s.

The codons for each amino acid have been deciphered by using a variety of synthetic polyribonucleotides, which were added to the protein synthesizing system isolated from *E. coli.* Radioactive amino acid were added to the system and the protein synthesized was monitored, e.g., poly A i.e., A-A-A-A-A-A-A led to poly lysine being formed, e.g., poly C i.e., C-C-C-C-C-C-C led to poly proline being formed, so AAA codes for lys and CCC codes for proline. Development of this technique has enabled the full genetic code to be deciphered.

First base	Second base U	C	A	G	Third base
U	UUU } Phe UUC UUA } Leu UUG	UCU UCC } Ser UCA UCG	UAU } Tyr UAC UAA } NSC UAG	UGU } Cys UGC UGA } NSC UG G Trp	U C A G
C	CUU CUC } Leu CUA CUG	CCU CCC } Pro CCA CCG	CAU } His CAC CAA } Gln CAG	CGU CGC } Arg CGA CGG	U C A G
A	AUU AUC } Ile AUA AUG* Met	ACU ACC } Thr ACA ACG	AAU } Asn AAC AAA } Lys AAG	AGU } Ser AGC AGA } Arg AGG	U C A G
G	GUU GUC } Val GUA GUG	GCU GCC } Ala GCA GCG	GAU } Asp GAC GAA } Glu GAG	GGU GGC } Gly GGA GGG	U C A G

Fig. 6.1. The Genetic Code (DNA).

1. PROPERTIES OF THE GENETIC CODE

1. The code is read in non-overlapping groups of three mRNA nucleotides. Each group is called a codon.
2. There are no spaces or commas separating neighboring codons. This is like having a sentence in English consisting entirely of 3 letter words where there are no spaces between the words. This property is especially important in understanding the effects of mutations on proteins.
3. The genetic code is redundant. There are 64 possible codons but only 20 amino acids.
4. There is a start codon corresponding to the amino acid methionine. When translation begins the first amino acid is always methionine. After translation this amino acid is removed as part of editing the protein. Note: methionine can be incorporated during peptide chain elongation and can occur in the protein (Fig. 6.2).
5. There are three non-coding stop or nonsense codons. These tell the machinery of translation that the end of the protein has been reached.
6. Not all amino acids have an equal number of codons coding for it. Observe that tryptophan has one codon while arginine has six codons.
7. The code is almost universal. However, certain bacteria, mitochondria and protista have minor variations in their codes. The near universality of the code suggests that the code arose very early in the evolution of life.

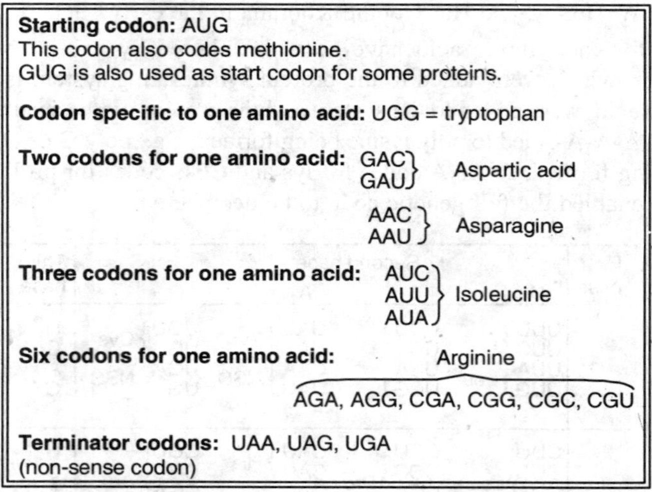

Fig. 6.2. Start and stop codons.

1.1. A Non-overlapping Code

The genetic code is read in groups (or "words") of three nucleotides. After reading one triplet, the "reading frame" shifts over three letters, not just one or two. In the following example, the code would not be read CAT, ATG. Rather, the code would be read ACA, TGA (Fig. 6.3).

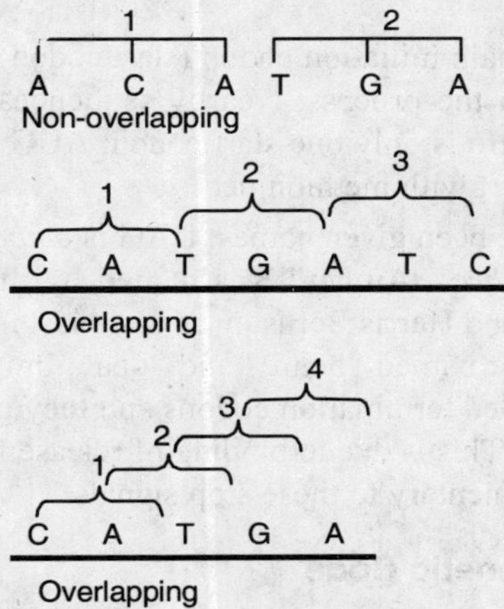

Fig. 6.3. Overlapping and non-overlapping codons.

1.2. Exceptions to the Code

The genetic code is almost universal. The same codons are assigned to the same amino acids and to the same START and STOP signals in the vast majority of genes in animals, plants, and microorganisms. However, some exceptions have been found. Most of these involve assigning one or two of the three STOP codons to an amino acid instead.

1.3. Transfer of Information via the Genetic Code

The genome of an organism is inscribed in DNA, or in some viruses RNA. The portion of the genome that codes for a protein or an RNA is referred to as a gene. Those genes that code for proteins are composed of tri-nucleotide units called codons, each coding for a single amino acid. Each protein-coding gene is transcribed into a template molecule of the related polymer RNA, known as messenger RNA or mRNA. This in turn is translated on the ribosome into an amino acid chain or polypeptide. The process of translation requires transfer RNAs specific for individual amino acids with the amino acids covalently attached to them, guanosine triphosphate as an energy source, and a number of translation factors. tRNAs have anticodons complementary to the codons in mRNA and can be "charged" covalently with amino acids at their 3′ terminal CCA ends. Individual tRNAs are charged with specific amino acids by enzymes known as aminoacyl tRNA synthetases which have high specificity for both their cognate amino acids and tRNAs. The high specificity of these enzymes is a major reason why the fidelity of protein translation is maintained. The standard genetic code is shown in the figure 6.1.

1.4. Reading Frame of a Sequence

Note that a codon is defined by the initial nucleotide from which translation starts. For example, the string GGGAAACCC, if read from the first position, contains the codons GGG, AAA and CCC; and if read from the second position, it contains the codons GGA and AAC; if read starting from the third position, GAA and ACC. Partial codons have been ignored in this example. Every sequence can thus be read in three reading frames, each of which will produce a different amino acid sequence (in the given example, Gly-Lys-Pro, Gly-Asp, or Glu-Thr, respectively). With double-stranded DNA there are six possible reading frames, three in the forward orientation on one strand and three reverse (on the opposite strand). The actual frame a protein sequence is translated is determined by a start codon, usually the first AUG codon in the mRNA sequence.

1.5. Start/stop Codons

Translation starts with a chain initiation codon (start codon). Unlike stop codons, the codon alone is not sufficient to begin the process. Nearby sequences and initiation factors are also required to start translation. There is only one start codon: AUG, which codes for methionine, so every amino acid chain must start with methionine.

The three stop codons have been given names: UAG is *amber*, UGA is *opal* (sometimes also called *umber*), and UAA is *ochre*. "Amber" was named by discoverers Richard Epstein and Charles Steinberg after their friend Harris Bernstein, whose last name means "amber" in German. The other two stop codons were named "ochre" and "opal" in order to keep the "color names" theme. Stop codons are also called termination codons and they give signal to release the nascent polypeptide from the ribosome. This is due to binding of release factors in the absence of cognate tRNAs with anticodons complementary to these stop signals.

1.6. Degeneracy of the Genetic Code

The genetic code has redundancy but no ambiguity. For example, although codons GAA and GAG both specify glutamic acid (redundancy), neither of them specifies any other amino acid (no ambiguity). Degenerate codons may differ in their third positions; e.g., both GAA and GAG code for the amino acid glutamic acid. A codon is said to be fourfold degenerate if any nucleotide at its third position specifies the same amino acid; it is said to be twofold degenerate if only two of four possible nucleotides at its third position specify the same amino acid. In twofold degenerate codons, the equivalent third position nucleotides are always either two purines (A/G) or two pyrimidines (C/T). Only two amino acids are specified by a single codon; one of these is the amino-acid methionine, specified by the codon AUG, which also specifies the start of translation; the other is tryptophan, specified by the codon UGG. Degeneracy results because a triplet code designates 20 amino acids and a stop codon. Because there are four bases, triplet codons are required to produce at least 21 different codes. For example, if there were two bases per codon, then only 16 amino acids could be coded for ($4^2=16$). Because at least 21 codes are required, then 4^3 gives 64 possible codons, meaning that some degeneracy must exist. These properties of the genetic code make it more fault-tolerant for point mutations.

Wobble hypothesis

There are 64 different triplet codons, and only 20 amino acids. Unless some amino acids are specified by more than one codon, some codons would be completely meaningless. Therefore, some redundancy is built into the system: some amino acids are coded for by multiple codons. In some cases, the redundant codons are related to each other by sequence; for example, leucine is specified by the codons CUU, CUA, CUC, and CUG. Note how the codons are the same except for the third nucleotide position. This third position is known as the **"wobble"** position of the codon (Fig. 6.4). This is because in a number of cases, the identity of the base at the third position

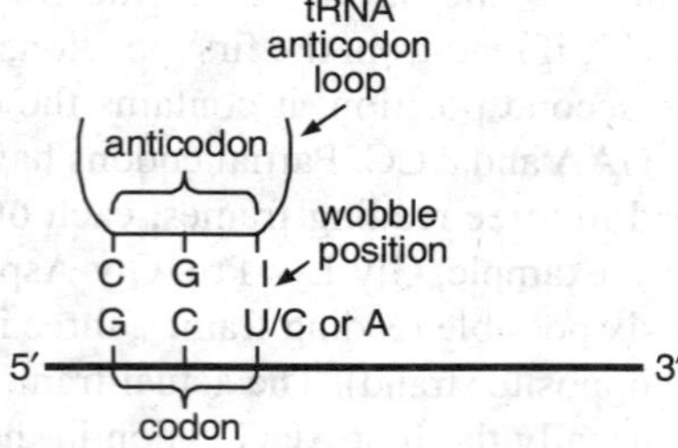

Fig. 6.4. Wobble hypothesis showing insignificant role of third nucleotide.

can wobble, and the same amino acid will still be specified (Table 6.1). This property allows some protection against mutation - if a mutation occurs at the third position of a codon, there is a good chance that the amino acid specified in the encoded protein won't change. In 1966, Francis Crick proposed the wobble concept to explain this phenomenon The wobble rules do not permit any single tRNA molecule to recognize four different codons. These codons can be recognized only when inosine occupies the first (5′) position of the anticodon.

Table 6.1. Base pairing according to Wobble hypothesis.

First base of anticodon 5′ end	Third base of codon 3′ end
C	G
A	U
U	A or G
G	UC or A
I	UC or A

1.7. Variations to the Standard Genetic Code

Slight variations in the standard code were observed by researchers while studying human mitochondrial genes. They discovered that mitochondrial genes use some alternative codes. Such small variants were also seen in organisms such as *Mycoplasma* translating the codon UGA as tryptophan. In bacteria and Archaea, GUG and UUG are common start codons. However, in rare cases, certain specific proteins may use alternative initiation (start) codons not normally used by that species.

In certain proteins, non-standard amino acids are substituted for standard stop codons, depending upon associated signal sequences in the messenger RNA: UGA can code for selenocysteine and UAG can code for pyrrolysine. Selenocysteine is now viewed as the 21st amino acid, and pyrrolysine is viewed as the 22nd.

QUESTIONS

1. Describe the genetic code. How codons determine the amino acids for protein?
2. Describe the properties of genetic code.
3. Write short notes on:
   (a) Wobble hypothesis
   (b) Codon and anticodon
   (c) Start and stop codon
   (d) Chemical nature of codons

CHAPTER 7

Gene Expression-Transcription

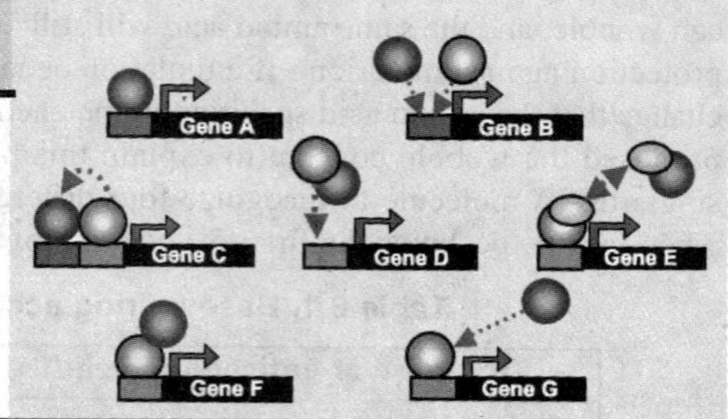

1. INTRODUCTION

Transcription is the process by which genetic information from DNA is transferred into RNA. DNA sequence is enzymatically copied by RNA polymerase to produce a complementary nucleotide RNA strand. Transcription produces different types of RNA molecules such as, mRNA, tRNA, rRNA and microRNA as described in chapter 5. tRNA and rRNA are involved in the synthesis of protein. RNA molecules are of different sizes depending upon the gene. The amount of RNA produced from different genes also varies depending upon the requirement of the gene product as described in Chapter 9. RNA transcription in prokaryotes and eukaryotes is similar but several fundamental differences exist. The most striking difference is that prokaryotic cells lack a nucleus whereas it is present in eukaryotes. Therefore, in eukaryotes mRNA is transported from nucleus to the cytoplasm before the information within it can be converted into a protein product.

In the case of protein-encoding DNA, transcription is the first step that ultimately leads to the translation of the genetic code, via RNA intermediate, into a functional peptide or protein. The stretch of DNA that is transcribed into an RNA molecule is called a transcription unit. A transcription unit that is translated into protein contains sequence that directs and regulates protein synthesis in addition to coding sequence that is translated into protein.

Regulatory sequence that is before, or 5′, of the coding sequence is called 5′ untranslated (5′UTR) sequence, and sequence found following, or 3′, of the coding sequence is called 3′ untranslated (3′UTR) sequence. Transcription has some proofreading mechanisms, but they are fewer and less effective than the controls for copying DNA; therefore, transcription has a lower copying fidelity than DNA replication. Only one stand of DNA, called the template strand (antisense strand), acts as a template. The other strand is called as coding strand (sense strand).

SENSE AND ANTISENSE STRAND

In any gene, one of the two DNA strands is transcribed. The transcribed strand is called 'template' or 'antisense strand'. The other strand that is not transcribed is called sense strand for two reasons: 1. The sequence of base in the non-transcribed strand is the same as that of mRNA (except for T in DNA and U in RNA) so that the sequence of codons in mRNA is reflected in the base sequence of the non-transcribed strand; 2. The 5′ → 3′ polarity of the non-transcribed strand is the same as that of mRNA. All the genes that reside in the same chromosomal DNA may not be transcribed from the same DNA strand. For some genes, one strand may serve as the template, while for the other genes the other strand may serve as the template. Because transcription always proceeds in the 5′ → 3′ direction, and because the template DNA strand and the RNA that is synthesized on it are antiparallel, the location of the promoter automatically determines which of the two DNA strands can serve as template for transcription.

As in DNA replication, transcription proceeds in the 5′ → 3′ direction. The DNA template strand is read 3′ → 5′ by RNA polymerase and the new RNA strand is synthesized in the 5′ → 3′

direction (Fig. 7.1). RNA polymerase binds to the 3′ end of a gene (promoter) on the DNA template strand and travels toward the 5′ end. Except for the fact that thymines in DNA are represented as uracils in RNA, the newly synthesized RNA strand will have the same sequence as the coding (non-template) strand of the DNA. For this reason, scientists usually refer to the DNA coding strand that has the same sequence as the resulting RNA when referring to the directionality of genes on DNA, not the template strand. Transcription is divided into 3 stages: *initiation*, *elongation* and *termination*.

```
5′ ... A T G G C C T G G A C T T C A ... 3′  Sense strand of DNA
3′ ... T A C C G G A C C T G A A G T ... 5′  Antisense strand of DNA
                  ↓ Transcription of antisense strand
5′ ... A U G G C C U G G A C U U C A ... 3′  mRNA
                  ↓ Transcription of mRNA
          Met – Ala – Trp – Thr – Ser –      Peptide
```

Fig. 7.1. One strand of DNA forms RNA. RNA formed is complementary to antisense or template strand and identical to sense or coding strand.

A molecule which allows the genetic material to be realized as a protein was first hypothesized by Jacob and Monod. RNA synthesis by RNA polymerase was established in vitro by several laboratories by 1965; however, the RNA synthesized by these enzymes had properties that suggested the existence of an additional factor needed to terminate transcription correctly. Recently, Roger D. Kornberg got the 2006 Nobel Prize in Chemistry "for his studies of the molecular basis of eukaryotic transcription".

2. RNA POLYMERASE OF PROKARYOTES

RNA polymerase is the enzyme that is responsible for synthesis of RNA. It acts like DNA polymerase requiring a template, nucleotide tri phosphate (ATP, CTP, GTP, UTP) and Mg^{2+} ions. The special characteristics of RNA polymerase is its ability to initiate chain growth without the need of a primer (DNA polymerase requires a primer to initiate DNA synthesis). RNA polymerase from *E. coli* is composed of 6 polypeptide subunits-two alpha (α) subunits, one beta (β) subunit, one beta prime (β') subunit, one omega (ω) subunit, one sigma (σ) subunit (Fig. 7.2). These five subunits are tightly bound together and constitute the core enzyme. The core enzyme in itself is capable of synthesizing RNA from DNA, but it is not capable of identifying the initiation site of a gene. The sigma subunit is essential for initiating transcription at the beginning of a gene and binds to a core enzyme to form the holoenzyme. Core enzyme has 400kD molecular weight and four types of sigma units have 30-90 kD molecular weight.

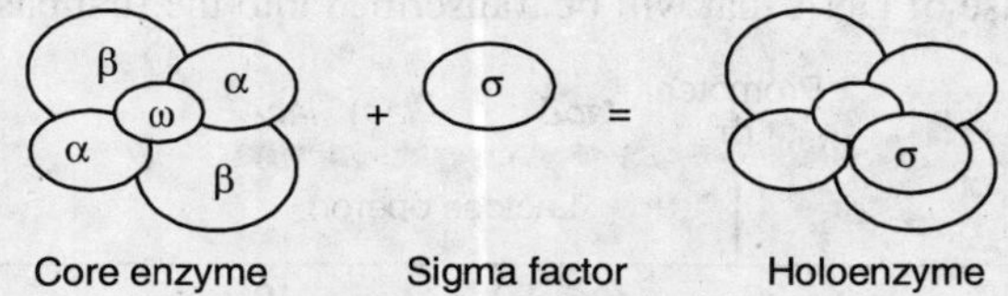

Fig. 7.2. RNA polymerase holoenzyme.

Bacterial cells have several different sigma subunits and each is responsible for initiating RNA synthesis for a specific group of genes (Table 7.1). Each bacterial cell has ~4-7 thousands RNA polymerase molecules and 80-90% are engaged in active RNA synthesis in a rapidly dividing cells. A specific sigma subunit (in association with core enzyme) is able to recognize and binds to a specific DNA sequence called as promoter. The promoter site defines the beginning of a gene for transcription. Consensus sequences located at –35 and –10 base pair upstream are recognized by sigma subunits. These sequences are quite similar in nucleotide arrangement.

Table 7.1 Characteristics and functions of bacterial sigma subunits

Subunit	MW kD	Function	Positions at consensus sequence –35	–10	Gene symbol
σ 70	70	House keeping gene	TTGACA	TATAAT	rpo D
σ 54	54	Nitrogen metabolism	CTGGPyAPyPu Actually at –25	TTGCA	nNtr A
σ 32	32	Heat shock genes	CTTGAA	CCCCATTA	rpo H
σ ?	?	Flagellar synthesis and chemotaxis	TAAA	GCCGATAA	flb B

Py- Pyrimidine; Pu- Purine

In bacteria, promoter along with sigma subunit decides how frequently a gene is transcribed. For example, sigma^{70} serves to initiate transcription of house keeping genes (those genes that are needed all the times to maintain the metabolism and structure of the cell) and thus initiates the synthesis of mRNAs from the largest class of genes. However, there is variation in the promoter sequence among genes recognized by sigma^{70} subunit and this variation affects the frequency of transcription initiation. As a result, one gene may be transcribed with one initiation event every second, whereas, another may be transcribed every 5 minutes. This is a fundamental system of regulating the level of gene expression in prokaryotes.

3. PROCESS OF TRANSCRIPTION IN PROKARYOTES

3.1. Initiation

Unlike DNA replication, transcription does not need a primer to start. RNA polymerase simply binds to the DNA and, along with other cofactors, unwinds the DNA to create an initiation bubble so that the RNA polymerase has access to the single-stranded DNA template. Transcription initiation is far more complex in eukaryotes and archaea, the main difference being that eukaryotic polymerases do not recognize directly their core promoter sequences. Transcription factors must first mediate the binding of RNA polymerase and the initiation of transcription. The completed assembly of transcription factors and RNA polymerase bound to the promoter is called the transcription initiation complex.

All promoters recognized by a single sigma subunit have similar sequences, and nucleotides that occur most frequently in the promoter are known as a "consensus sequence". The promoter consensus sequence is composed of two short sequences separated by about 20bp. These two groups of sequences are labeled the –35 and –10 sequences because they are situated 35 and 10 bases before the first base of DNA that will be transcribed into the first base of the RNA (Fig. 7.3).

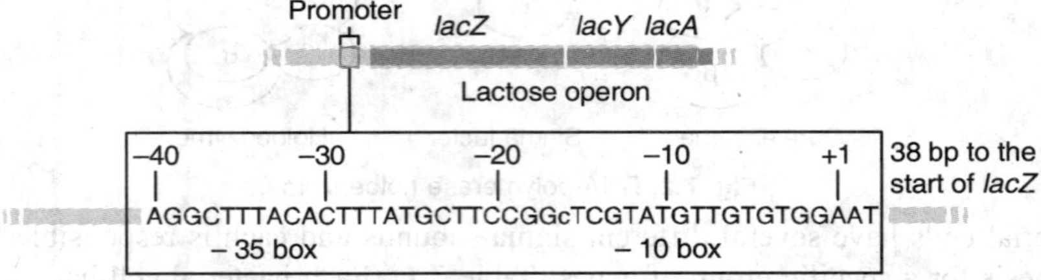

Fig. 7.3. Details of promoter region showing consensus sequences.

DNA bases to the left (negative numbers) of the 'start' site are said to be 'upstream' and those on right side (positive numbers) are called as 'downstream'.

RNA polymerase binds to DNA in two steps. In first step, the enzyme binds to promoter. In 2nd step, the helix unwinds to allow base-pairs recognition to the template-the DNA sequence for the RNA polymerization reaction. DNA unwinding starts at the –10 sequence and proceeds to the right of the start point i.e., downstream. The sigma subunit is required only for recognition of the consensus sequence and thus detaches from the core enzyme.

The product of transcription is always RNA (tRNA, mRNA, rRNA, snRNA). The precursors for RNA synthesis are nucleotide triphosphates. The greatest varieties of RNAs are found in the mRNAs. To start RNA synthesis a ribonucleotide triphosphate is base paired at the RNA start site on the DNA. This represents the 5′ end of the new molecule and it grows when ribonucleotides are added to the 3′ end. The 5′ end retains the triphosphate precursor.

3.2. Elongation

One strand of DNA, the *template strand* (or non-coding strand), is used as a template for RNA synthesis. As transcription proceeds, RNA polymerase traverses the template strand and uses base pairing complementarity with the DNA template to create an RNA copy. Although RNA polymerase traverses the template strand from 3′ → 5′, the coding (non-template) strand is usually used as the reference point, so transcription is said to go from 5′ → 3′. This produces an RNA molecule from 5′ → 3′, an exact copy of the coding strand (except that thymines are replaced with uracils, and the nucleotides are composed of a ribose (5-carbon) sugar where DNA has deoxyribose (one less Oxygen atom) in its sugar-phosphate backbone). In *E.coli*, the polymerization reaction occurs at a rate of 40 nucleotides per second at 37°C; thus a gene for 1000bp is transcribed in about 30 seconds. Finally, RNA polymerase will reach a terminator sequence that tells the enzyme to stop polymerization of the RNA and to dissociate from the DNA. Unlike DNA replication, mRNA transcription can involve multiple RNA polymerases on a single DNA template and multiple rounds of replication, so many mRNA molecules can be produced from a single copy of a gene. This step also involves a proofreading mechanism that can replace incorrectly incorporated bases (Fig. 7.4).

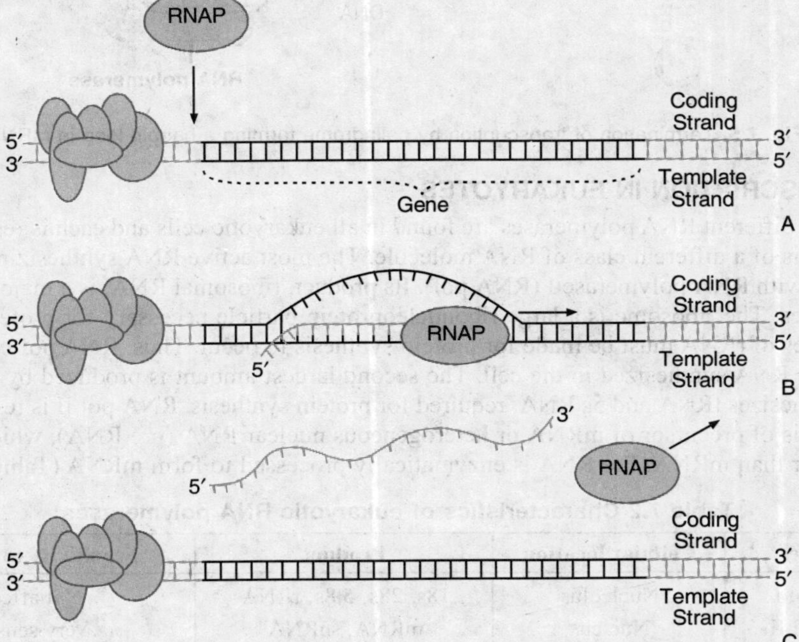

Fig. 7.4. Process of transcription showing attachment of RNA polymerase- Initiation (A), Separation of DNA strands and formation of transcription bubble- Elongation (B) RNA is synthesized by complementary base pairs and terminated with the release of RNA polymerase- Termination (C).

1.3.3. Termination

RNA synthesis is terminated by two mechanisms;

1. One type of termination sequence is composed of a stretch of AT nucleotides preceded by two symmetrical 8bp GC sequences (a palindrome) followed by short sequence of AT pairs (Fig. 7.5).
2. In other mechanism rho (ρ) protein terminates the synthesis.

In bacteria, a gene can be defined as physical entity-having a start site (promoter), the coding information and a terminator sequence. Another term for a gene is a cistron, which is the information required to make one polypeptide. Unfortunately, in bacteria organelles of eukaryotic cells, more than one polypeptide is encoded in a single message. These are called polycistronic mRNAs, because more than one cistron is present in mRNA. This is illustrated in detail in chapter 9 describing *lac* Operon and *trp* Operon.

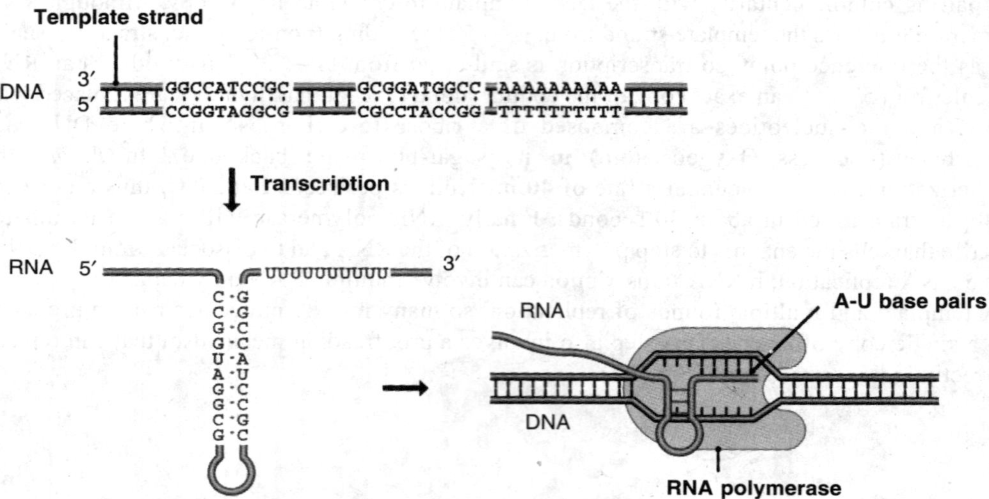

Fig. 7.5. Termination of transcription by palindrome forming a hairpin loop in mRNA.

4. TRANSCRIPTION IN EUKARYOTES

Three different RNA polymerases are found in all eukaryotic cells and each is responsible for the synthesis of a different class of RNA molecule. The most active RNA synthesizing abilities is associated with RNA polymerase I (RNA pol). Its product, ribosomal RNA, is a major component of ribosomes. The ribosome is a large ribonucleoprotein particle necessary for protein synthesis. Many copies of rRNA must be made for protein synthesis to occur. Thus, RNA pol I accounts for most of the RNA synthesized in the cell. The second largest amount is produced by RNA pol III which synthesizes tRNA and 5s RNA, required for protein synthesis. RNA pol II is responsible for the synthesis of precursor of mRNA or heterogeneous nuclear RNA (pre RNA), which is usually much larger than mRNA. Pre RNA is enzymatically processed to form mRNA (Table 7.2).

Table 7.2 Characteristics of eukaryotic RNA polymerases*

Enzymes	Cellular location	Product	Sensitivity to amanitin
RNA pol I	Nucleolus	18s, 28s, 5.8s, rRNA	Not affected
RNA pol II	Nucleus	mRNA, snRNA	Very sensitive
RNA pol III	Nucleus	tRNA, 5sRNA, snRNA	Lesser inhibition than pol II

* All the RNA polymerases are constructed of multiple subunits and the holoenzymes have molecular weight up to half a million daltons. Amanitin is a transcription inhibitor

Like prokaryotic genes eukaryotic genes have promoter and terminator sequences that direct the transcriptional machinery to start and stop. However, the signal sequences in the DNA for starting and stopping RNA synthesis are variable in eukaryotes. It is thus, impossible to specify a single sequence responsible for stop and start sequences for all genes. Some common features are as follows:

1. In eukaryotes most promoters have a sequence called TATA box, which has the consensus sequence 5′-TATAAA-3′.
2. The TATA box is located about 25bp upstream (–25) from the starting point and determines the start site for transcription. It is surrounded by GC rich regions.
3. Two other sequences upstream of the start point also effect gene expression and are present in many eukaryotic promoters. These are called CAAT box and GC box. The CAAT box is named for its consensus sequence (GGCCAATCT) and located about 80bp upstream (–80).
4. Mutation in these sequences affects transcription.

RNA polymerase cannot bind itself and requires a group of proteins known as transcription factor. The eukaryotic transcription requires multiple protein binding sites of initiation of transcription and a number of transcription factors. In prokaryotes, RNA pol requires only sigma factor to bind to the core enzyme, which can then bind to promoter region (table 7.3).

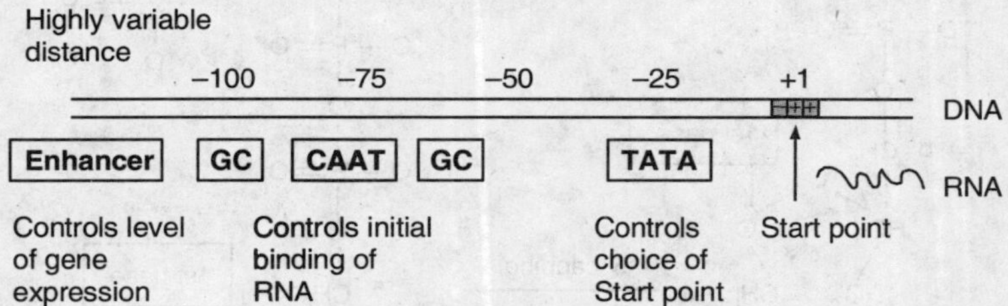

Fig. 7.6. Controlling region of an average eukaryotic gene. Enhancer sequences are located at a very far distance and can control upstream and downstream.

In eukaryotes, promoters do not act alone in regulating the amount of RNA transcript synthesized. Another very important class of DNA sequences that control RNA transcription is the enhancer sequences or enhancer elements. Enhancer sequences are binding site for a wide variety of transcription factors. The enhancer sequences in conjunction with the proper transcription factors can have an enormous effect on the level of synthesis of a RNA transcript, increasing it by as much as thousand-folds. Enhancer sequences are situated at highly variable distances from the gene that they control, sometimes several thousands base pairs from the promoter. These sequences can regulate the genes located upstream or down stream positions.

Table 7.3. Comparison of eukaryotic and prokaryotic RNA polymerase.

Prokaryotic	Eukaryotic
Only one RNA polymerase	Three different RNA polymerases
Synthesize polycistronic mRNAs	Do not synthesize polycistronic mRNA
mRNA translated into protein	mRNA undergoes modifications before translated into protein

4.1. RNA processing

RNA processing is to generate a mature mRNA (for protein genes) or a functional tRNA or rRNA from the primary transcript. In this section, we discuss first the processing of pre-mRNA and then processing of pre-rRNA and pre-tRNA. In some cases, RNA editing is also involved. Processing of pre-mRNA involves the following steps:

1. Capping - add 7-methylguanylate (m^7G) to the 5′ end.
2. Polyadenylation - add a poly-A tail to the 3′ end.
3. Splicing - remove introns and join exons.

5′-Capping

Capping occurs shortly after transcription begins. The chemical structure of the "cap" is shown in the following figure, where m^7G is linked to the first nucleotide by a special 5′-5′ triphosphate linkage. In most organisms, the first nucleotide is methylated at the 2′-hydroxyl of the ribose. In vertebrates, the second nucleotide is also methylated (Fig. 7.7).

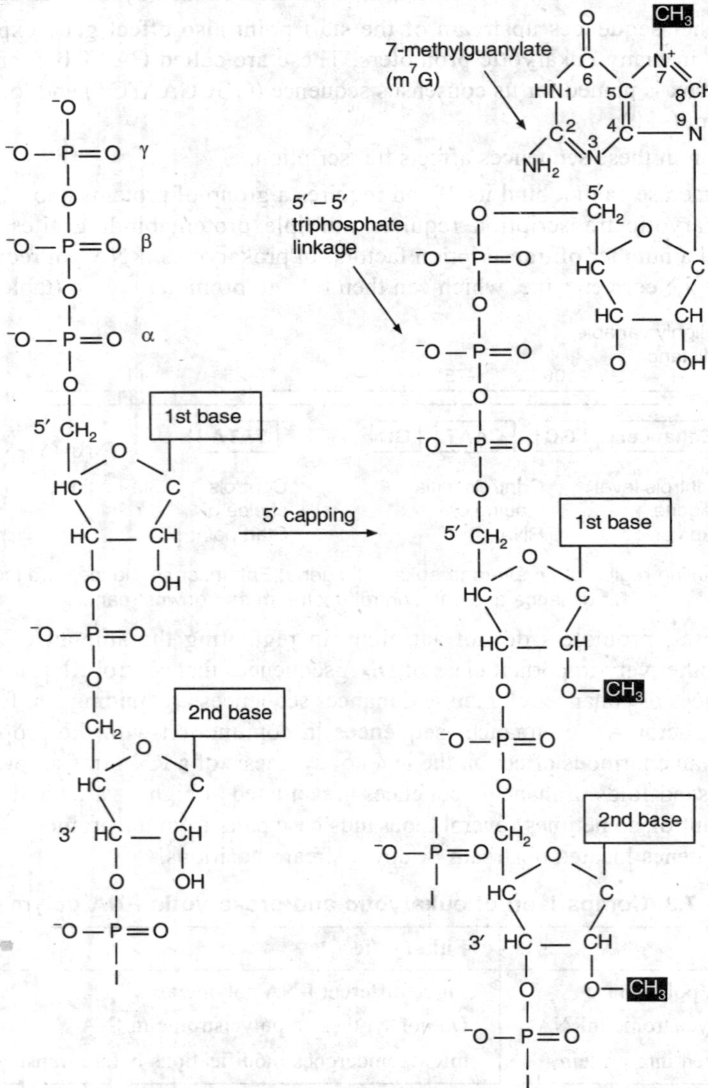

Fig. 7.7. Modifications at the 5′ end of newly formed RNA.

3′-Polyadenylation

A stretch of adenylate residues are added to the 3′ end. The poly-A tail contains ~ 250 A residues in mammals, and ~ 100 in yeasts (Fig. 7.8).

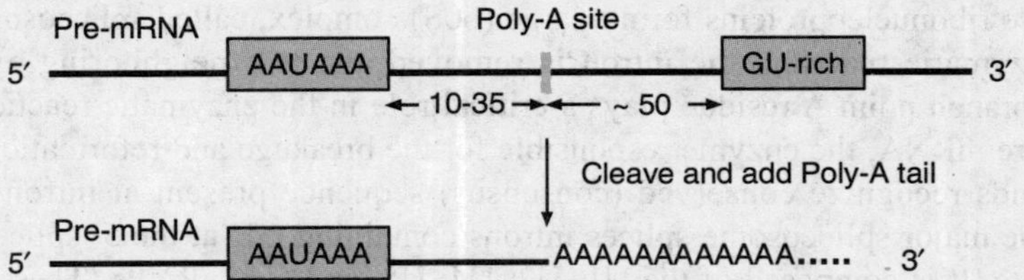

Fig. 7.8. Polyadenylation at the 3′ end. The major signal for the 3′ cleavage is the sequence AAUAAA. Cleavage occurs at 10-35 nucleotides downstream from the specific sequence. A second signal is located about 50 nucleotides downstream from the cleavage site. This signal is a GU-rich or U-rich region.

In general, introns tend to be much longer than exons. An average eukaryotic exon is only 140 nucleotides long, but one human intron stretches for 480,000 nucleotides. Removal of the introns - and splicing the exons together - are among the essential steps in synthesizing mRNA (Fig. 7.9).

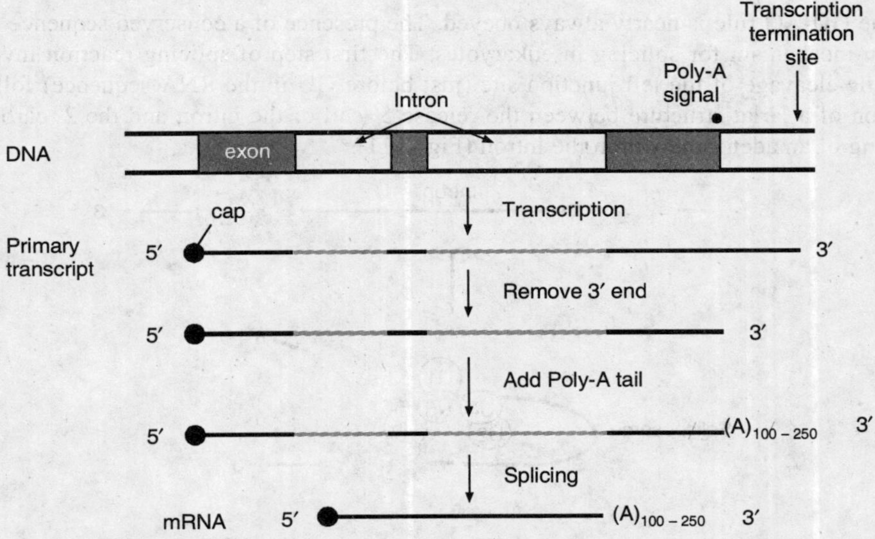

Fig. 7.9. The procedure of RNA processing for forming mRNA which will produce a protein.

Splicing

In genetics, splicing is a modification of genetic information after transcription, in which introns of precursor messenger RNA (pre-mRNA) are removed and exons of it are joined. Since in prokaryotic genomes introns do not exist, splicing naturally only occurs in eukaryotes. The splicing prepares the pre-mRNA to produce the mature messenger RNA (mRNA), which then undergoes translation as part of the protein synthesis to produce proteins. Splicing includes a series of biochemical reactions, which are catalyzed by the spliceosome, a complex of small nuclear ribonucleo-proteins (snRNPs).

Spliceosomal introns often reside in eukaryotic protein-coding genes. Within the intron, a 3′ splice site, 5′ splice site, and branch site are required for splicing. Splicing is catalyzed by the spliceosome which is a large RNA-protein complex composed of five small nuclear ribonucleoproteins (snRNPs, pronounced 'snurps'). The RNA components of snRNPs interact with the intron and may be involved in catalysis. Two types of spliceosomes have been identified (the major and minor) which contain different snRNPs. It involves five snRNAs and their associated proteins. These ribonucleoproteins form a large (60S) complex, called spliceosome. Then, after a two-step enzymatic reaction, the intron is removed and two neighboring exons are joined together. The branch point A residue plays a critical role in the enzymatic reaction. In removing introns from pre mRNA, the enzyme responsible for the breakage and reformation of nucleotide-nucleotide bonds recognize conserved (consensus) sequence present at intron-exon junctions (Fig. 7.10). The major spliceosome splices introns containing GU at the 5' splice site and AG at the 3′ splice site. It is composed of the U1, U2, U4, U5, and U6 snRNPs (These named U1, U2 etc because of abundance of uridines).

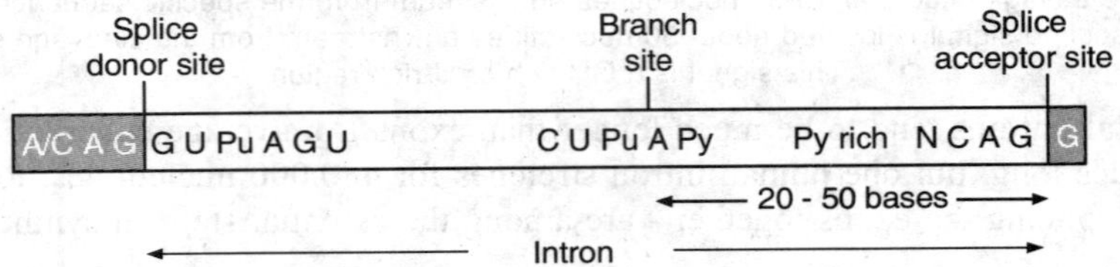

Fig. 7.10. The consensus sequence for splicing. Pu = A or G; Py = C or U.

The GU-AG rule is nearly always obeyed. The presence of a conserved sequence implies a common mechanism for splicing in eukaryotes. The first step of splicing reaction involves the enzymatic cleavage of the left junction site (just before GU in the RNA sequence) followed by formation of a lariat structure between the release 5′ end of the intron and the 2′ carbon in the sugar ring of an adenosine within the intron (Fig. 7.11).

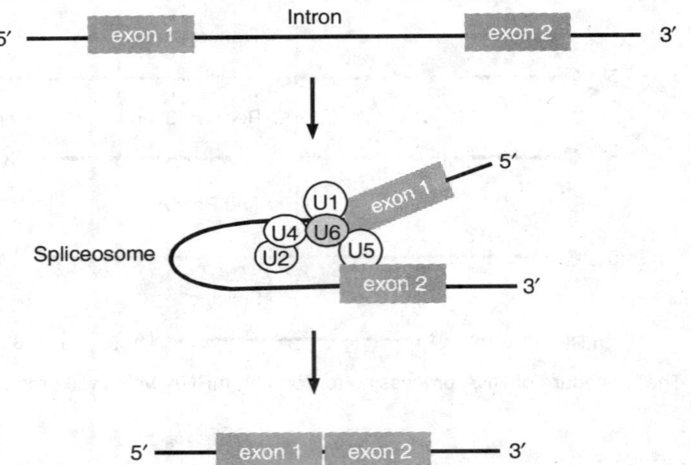

Fig. 7.11. Schematic diagram for the formation of the spliceosome during RNA splicing. U1, U2, U4, U5 and U6 denote snRNAs and their associated proteins. The U3 snRNA is not involved in the RNA splicing, but is involved in the processing of pre-rRNA.

The right junction is then cleaved, releasing the branched lariat and simultaneously joining the two exons. The released intron RNA fragment is degraded and the nucleotides are recycled into new RNA. The splicing reaction is catalyzed by a large complex called spliceosome, which consists of nucleic acids and 8-10 enzymes.

4.2. Self-splicing or autocatalytic splicing

Self-splicing occurs for rare introns that form a ribozyme, performing the functions of the spliceosome by RNA alone. There are two kinds of self-splicing introns, Group I and Group II. Group I and II introns perform splicing similar to the spliceosome without requiring any protein. This similarity suggests that Group I and II introns may be evolutionarily related to the spliceosome. Self-splicing may also be very ancient, and may have existed in an RNA world that was present before protein. The splicing mechanism requires 5 additional RNA molecules and over 50 proteins are used and hydrolyzes many ATP molecules. The splicing mechanisms use ATP in order to accurately splice mRNAs. If the cell were to not use any ATPs, the process would be highly inaccurate and many mistakes would occur.

Two transesterfications characterize the mechanism in which group I introns are sliced: 1) 3′OH of a free guanine nucleoside (or one located in the intron) or a nucleotide cofactor (GMP, GDP, GTP) attacks phosphate at the 5′ intron splice site. These consensus sequences can be located at a distance from intron-exon junctions. Double starnded regions form between intron consensus sequences. The folded secondary structure allows an attacking guanosine residue to release side of intron, joining and guiding the two intron-exon junctions in close proximity. The guanosine becomes linked to the intron, and the free 3′-hydroxyl on the left exon then attaches to the right 3′ exon-intron junction and releasing the intron (Fig. 7.12).

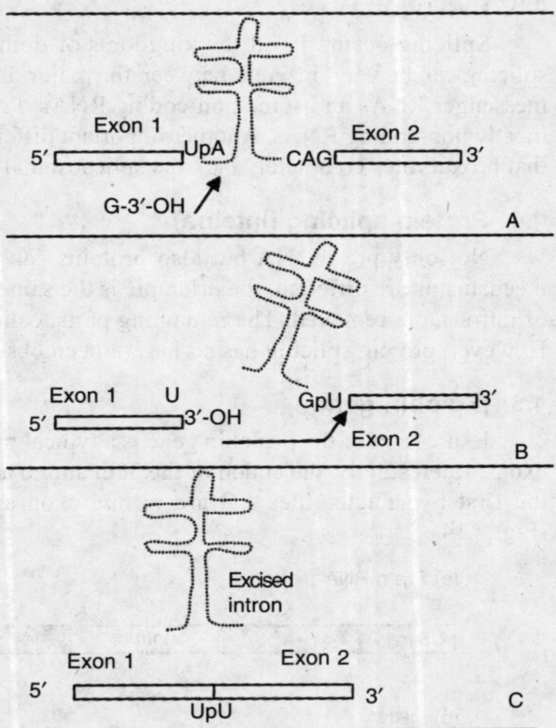

Fig. 7.12. Autocatlytic splicing. (A). The reaction is initiated by 3'-OH from a free guanosine, (B). The G attacks the phosphate linking the intron and the exon, freeing the 5'end of the exon, and attaching G to it, (C). The free 3'-OH of exon 1then attacks the phosphate of the right border junction joining exon 1 to exon 2 and releasing the intron.

3′OH of the 5′ end (exon1) and 5′OH of exon1 2 becomes a nucleophile and the second transesterfication results in the joining of the two exons. Autocatalytic introns splicing is found in the nuclear genes coding for rRNA in *Tetrahymena* and *Physarum* (a ciliate and a slime mold), in the mitochondria of fungi, and in mitochondria and chloroplasts of many plants. There are RNAs capable of self splicing. Catalysis by RNA may be quite wide spread and may have significant evolutionary implications. Such RNA molecules are called ribozymes. A number of ribozymes are able to cleave a phosphodiester bond linking nucleotide in an RNA strand. RNA-cleaving ribozymes provide on strategy to disrupt the function of specific mRNA. These ribozyme cleave at specific sequences and make double stranded structures by base pair complementation. Such strategy can be used to destroy RNA produced by a pathogen, e.g., AIDS virus (Fig. 7.13).

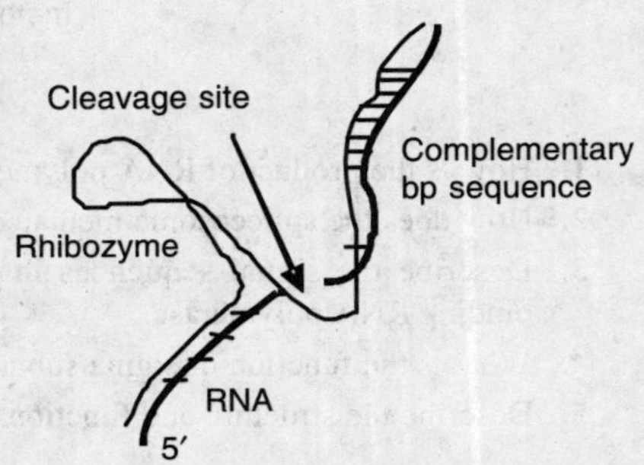

Fig. 7.13. A ribozyme cleaving the RNA molecule.

4.3. Evolution

Splicing occurs in all the kingdoms or domains of life, however, the extent and types of splicing can be very different between the major divisions. Eukaryotes splice many protein-coding messenger RNAs and some non-coding RNAs. Prokaryotes, on the other hand, splice rarely, but mostly non-coding RNAs. Another important difference between these two groups of organisms is that prokaryotes completely lack the spliceosomal pathway.

4.4. Protein splicing (Inteins)

Not only pre-mRNA but also proteins can undergo splicing. Although the biomolecular mechanisms are different, the principle is the same, that parts of the protein, called inteins instead of introns, are removed. The remaining parts, called exteins instead of exons, are fused together. However, protein splicing has so far not been observed in humans, but in yeast.

4.5. β-globin gene

Expression of the β-globin gene is a typical process. This gene contains two introns and three exons. Interestingly, the codon of the 30th amino acid, AGG, is separated by an intron. As a result, the first two nucleotides AG are in one exon and the third nucleotide G is in another exon (Fig. 7.14).

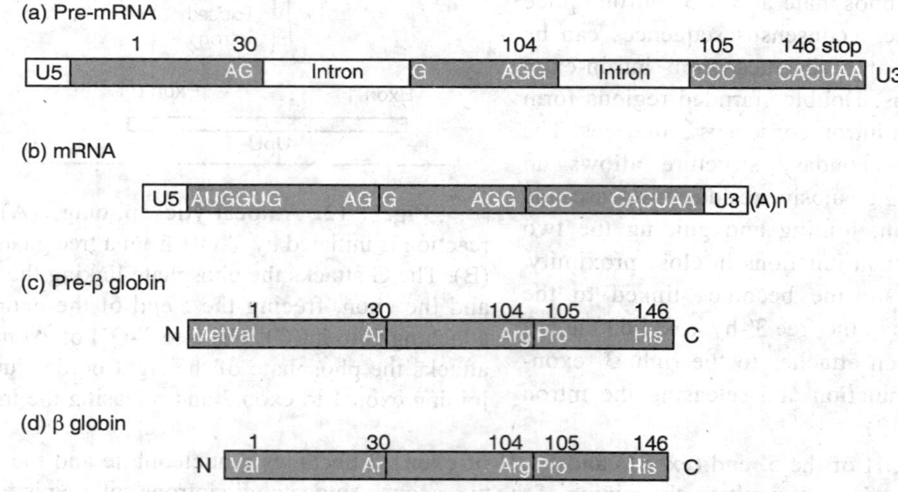

Fig. 7.14. Expression of the human β-globin gene. U5 and U3 represent untranslated regions at the 5′ and 3′ end, respectively. Note that the mature β-globin protein does not contain the initiating methionine for protein synthesis.

QUESTIONS

1. How is the product of RNA polymerase II altered after synthesis?
2. How does the spliceosome mediated splicing differ from autocatalytic splicing?
3. Describe the signal sequences in promoter region of prokaryotes present for recognisation and binding RNA polymerase.
4. What is the function of sigma subunit of RNA polymerase holoenzyme?
5. Describe the structure and function of RNA polymerase.

CHAPTER 8

Gene Expression – Translation: Molecular Basis

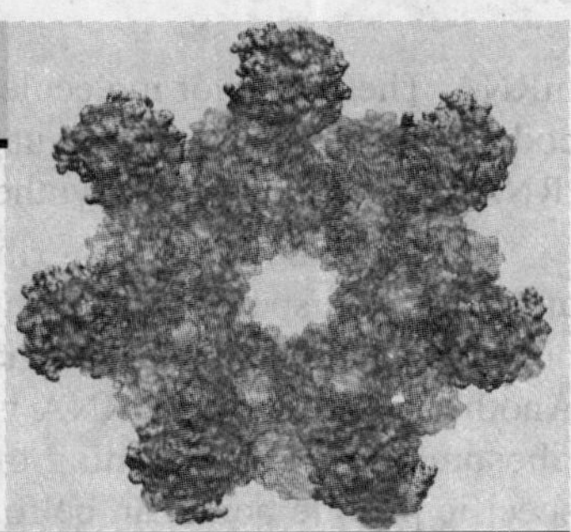

1. INTRODUCTION

There is a genetic code that stores information in the form of triplet nucleotides in DNA, and this information is expressed through the process of transcription into a mRNA that is complementary to one of the two strands of the DNA helix. However, the final product of gene expression, in most cases, is a polypeptide chain consisting of a linear series of amino acids, whose sequence has been prescribed by the genetic code. In this chapter we will explain how the information present in mRNA is translated to create a polypeptide. This information provides understanding about transfer of information from DNA to protein and how DNA can control a function or phenotypic expression.

The order of nucleotides in mRNA is used to generate the linear sequences of amino acids in proteins. This process is known as translation. Translation is among the most highly conserved process in all the organisms and among the most energetically costly for the cell. In rapidly growing bacterial cells, upto 80% of cell's energy and 50% of the cell's dry weight are involved in protein synthesis. In fact, the synthesis of a single protein requires the coordinated action of over 100 proteins and RNAs. In this process, ribosomes are actively involved providing polymerization site. Transfer RNA (tRNA) is actively involved in recognition of specific amino acids for their arrangement into a polypeptide chains. In 1955, Francis Crick proposed that prior to their incorporation into polypeptides, amino acids must attached to a special adaptor molecule that is capable of directly interacting with and recognizing the three-nucleotides-long coding units of the

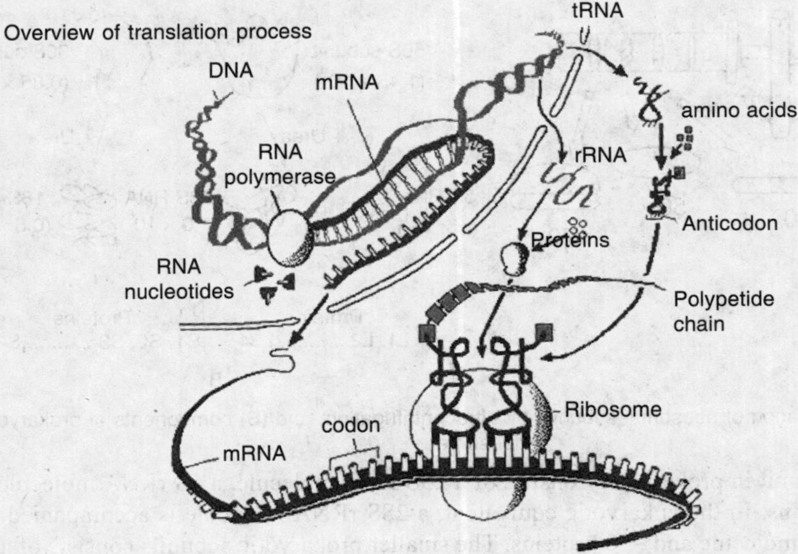

Fig. 8.1. Overview of translation process demonstrating assembly of ribosome, tRNA and polypeptide synthesis.

mRNA. These adaptor molecules are now known as tRNA. The process involved in translating the code of mRNAs into the language of proteins is composed of four primary components: mRNAs, tRNAs, aminoacyl tRNA synthetases and the ribosomes (Fig. 8.1).

In combination with a ribosome, mRNA presents a triplet codon that calls for a specific amino acid. A specific tRNA molecules contain within its nucleotide sequence three consecutive nucleotide complementary to the codon, called the anticodon, which can base pair with the codon. Another region of this tRNA is covalently bonded to its corresponding amino acid. Inside the ribosome, hydrogen bonding of tRNAs to mRNA holds the amino acids in proximity so that a specific peptide bond can be formed. The processes occur over and over as mRNA runs through the ribosomes, and amino acids are polymerized into a peptide.

2. STRUCTURE OF RIBOSOME

Ribosome is a key component in the process of translation therefore studied extensively. One bacterial cell contains ~10,000 such structures and a eukaryotic cell contains many times more. A bacterial ribosome is about 250 nm in diameter and consists of two subunits, one large and one small. Both subunits consist of one or more molecules of rRNA and an array of ribosomal proteins. Association of two subunits is called monosome. The structure of prokaryotic ribosome is given in the figure 8.2 B. The subunit and rRNA components are most easily isolated and characterize on the basis of their sedimentation behavior in sucrose gradient (there rate of migration is called Svedberg's coefficient 'S'. It is a unit of sedimentation velocity; sedimentation in ultracentrifuge depends on both the mass and shape of molecule, and is not a simply a measure of molecular mass) (Fig. 8.2A). The monosome of prokaryotes is a 70S particle and in eukaryotes it is 80S. The prokaryotic 70S monosome consists of a 50S and 30S subunit and the eukaryotic 80S monosome consists of a 60S and 40S subunit. The sum value of 'S' is not a simple arithmetic addition.

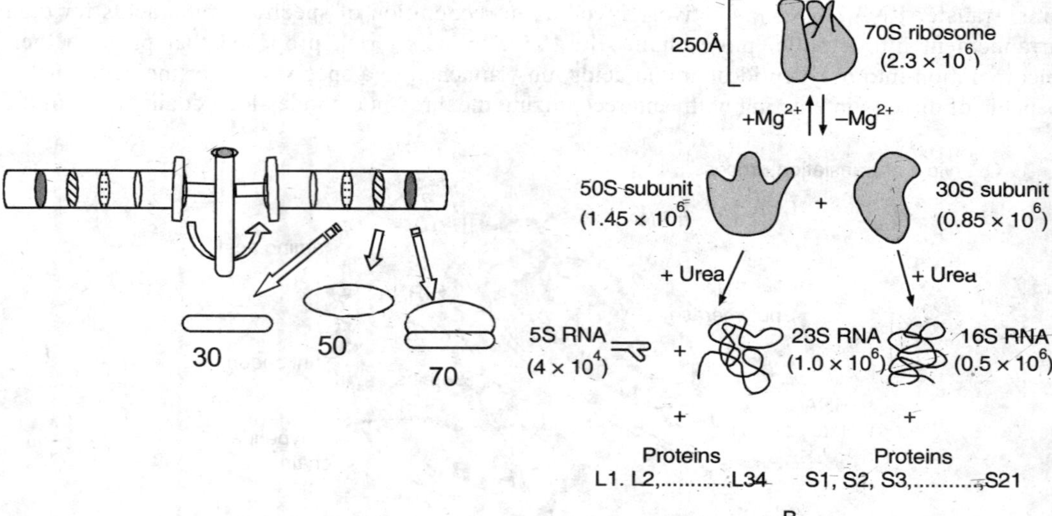

Fig. 8.2. (A) Separation of ribosomal subunits by ultracentrifugation, and (B) components of prokaryotic ribosome.

The large subunit in prokaryotes consists of a 23S RNA molecule, a 5S rRNA molecule and 31 ribosomal proteins. In the eukaryotic equivalent, a 28S rRNA molecule is accompanied by a 5.8S and 5S rRNA molecule and ~50 proteins. The smaller prokaryotic subunits consists of a 16S rRNA component and 21 proteins. Similarly, in eukaryotes smaller subunits consist of a 18S rRNA component and ~33 proteins. These proteins are involved in binding of various molecules involved in translation and precise control of process. In eukaryotes, many more copies of a sequence encoding the 28S and 18S components are present. The exact functions of various components are still not very much clear. In bacteria, the monosome has a combined molecular weight of 2.5 million dalton.

Structure of tRNA is described in Chapter 5. Transfer RNA has an anticodon loop and a amino acid binding site besides other loops and stems as described earlier. Before translation can proceed, the tRNA molecule must be chemically linked to their respective amino acids. This activation process, called charging (Fig. 8.2), governs by the enzymes called aminoacyl tRNA synthetases. Because there are 20 different amino acids, there must be 20 different tRNA molecules and as many different enzymes. In theory, because there are 61 triplet codes, there could be the same number of specific tRNAs and enzymes. However, because of the ability of the third member of a codon to wobble, it is now thought that there are at least 32 different tRNAs. It is also believed that there are only 20 synthetases, one for each amino acid.

During the initial step, the amino acid is converted to an activated form, reacting with ATP to create an aminoacylicadenylic acid. A covalent linkage is formed between the 5′-phosphate group of ATP and the carboxyl end of the amino acid. This enzyme-amino acid complex then reacts with a specific tRNA molecule. In the next step, the amino acid is transferred to the appropriate tRNA and bonded covalently to the adenine residue at the 3′ end. The charged tRNA may participate directly in protein synthesis. Aminoacyl tRNA synthetases are highly specific enzymes because they recognize only one amino acids and only a subset of corresponding tRNAs, called isoaccepting tRNAs. This is a crucial point in maintaining the correctness of translation.

3. PROCESS OF TRANSLATION

Similar to transcription the process of translation can be divided into three steps. This is a continuous process like cell division but for the sake of study divided into different steps.

3.1. Initiation

Initiation of translation in *E.coli* involves the small ribosome subunit, a mRNA molecule, a specific charge initiator tRNA, GTP, Mg++ and number of proteinaceous initiation factors (IFs). These are initially part of the small subunit and are required to enhance binding affinity of the various translational components (Table 8.1). Unlike ribosomal proteins, IFs are released from the ribosome once initiation is completed. In prokaryotes, the initiation codon of mRNA-AUG-requires modified amino acid, formylmethionine (fmet), in which a formyl group has been added

Table 8.1. Various protein factors involved in translation.

Process	Factors	Role
Initiation of translation	IF_1	Stabilizes 30S subunit
	IF_2	Binds fmet tRNA to 30S mRNA complex; binds to GTP and hydrolyse
	IF_3	Binds 30S subunit to mRNA; dissociates monosomes after termination
Elongation of polypeptide	EF-Tu	Binds GTP, bring amino acyl-tRNA to site 'A' of ribosome
	EF-Ts	Generates active EF-Tu
	EF-G	Stimulates translocation, GTP dependent
Termination of translation and release of polypeptide	RF_1	Catalyses release of polypeptide chain from tRNA, dissociation of translocation complex, specific for UAA, UAG
	RF_2	Behaves like RF_1, specific for UGA and UAA
	RF_3	Stimulates RF_1 and RF_2

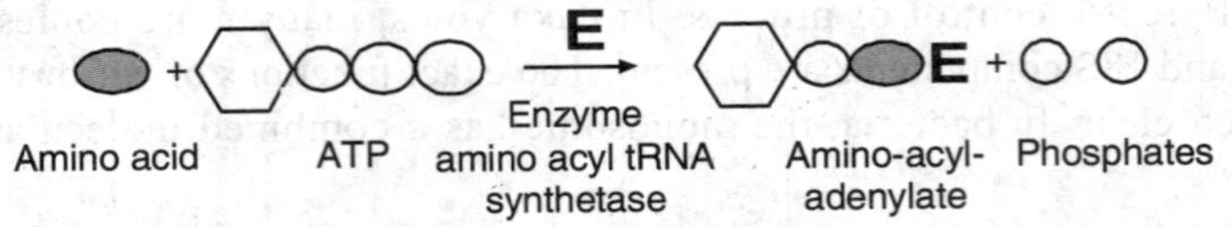

Fig. 8.3. Formation of amino-acyl-adenylate complex by the enzyme aminoacyl tRNA synthetase.

to the methionine's amino group. The fmet is brought to the ribosome attached to a special tRNA called fmet.tRNA, which has the anticodon 5′-CAU-3′ to bind to the AUG start codon. This tRNA is special, since it is involved specifically with initiation process of protein synthesis. The aminoacylation of tRNA.fmet occurs as follows; Methionyl- tRNA synthetase catalyses the addition of methionine to tRNA (Fig. 8.3). Then an enzyme called transformylase catalyses addition of the formyl group to the methionine. The resulting molecule is designated fmet-tRNA.fmet.

Small ribosomal subunits bind to several initiation factors, and this complex in turn binds to mRNA (step I). Protein synthesis is regulated by the sequence and structure of the 5′ untranslated region (UTR) of the mRNA transcript. In prokaryotes, the ribosome binding site (RBS), which promotes efficient and accurate translation of mRNA, is called the Shine-Dalgarno sequence after the scientists that first described it. This purine-rich sequence of 5′ UTR is complementary to the UCCU core sequence of the 3′-end of 16S rRNA (located within the 30S small ribosomal subunit). Various Shine-Dalgarno sequences have been found in prokaryotic mRNAs (*see* Figure 8.4 for the consensus sequence). These sequences lie about 10 nucleotides upstream from the AUG start codon. Activity of a RBS can be influenced by the length and nucleotide composition of the spacer separating the RBS and the initiator AUG. The characteristics of RBS sequences are the following:

1. RBS sequences are rich in purine bases, i.e., rich in Adenine (A) and Guanine (G);
2. They are localized from three to 14 base pairs upstream from the beginning of a gene;
3. Their size vary from three to nine base pairs;
4. Their consensus sequence is "A G G A G";
5. The RBS sequences are complementary to the pyrimidine-rich sequence found in the rRNA in the 6S unit of the ribosome (end 3′ - HO-AUUCCUCCACUAG-5′).

Initiation of protein synthesis in eukaryotic cytoplasm resembles the process in bacteria, but the order of events is different and the number of accessory factors is greater. Some of the differences in initiation are related to a difference in the way the bacterial 30s and eukaryotic 40S subunits find their binding sites for initiating protein synthesis on mRNA. In eukaryotes, small subunits first recognize the 5′ end of the mRNA and then move to the initiation site, where they are joined by large subunits (in prokaryotes, small subunits bind directly to the initiation site). In

Consensus RBS sequences

Prokaryotic (Shine –Dalgarno sequence)

+1
5'-AGGAGGACAGCUAUG-3'
RBS Spacer initiator

Eukaryotic (Kozak sequence)

+1
5'-A/GCCACCAUGG –3'
RBS |
Initiator

Fig. 8.4. Consensus RBS Sequences. The +1 A is the first base of the AUG initiator codon (shaded) responsible for binding of fmet-tRNA.fmet. The underline indicates the ribosomal binding site sequence, which is required for efficient translation.

eukaryotes, the Kozak sequence A/GCCACCAUGG, which lies within a short 5′ untranslated region, directs translation of mRNA. An mRNA lacking the Kozak consensus sequence may be translated efficiently in Ambion's in vitro systems if it possesses a moderately long 5′ UTR that lacks stable secondary structure. Our data demonstrate that in contrast to the *E. coli* ribosome, which preferentially recognizes the Shine-Dalgarno sequence, eukaryotic ribosomes (such as those found in retic lysate) can efficiently use either the Shine-Dalgarno or the Kozak ribosomal binding sites. Shine-dalgarno sequence is a short stretch of nucleotides on a prokaryotic mRNA molecule, upstream of the translational start site, that serves to bind to ribosomal RNA and thereby bring the ribosome to the initiation codon on the mRNA. Eukaryotic protein synthesis requires many initiation factors for all stages of initiation, including binding the initiator tRNA, 40S subunit attachment to mRNA, movement along mRNA and joining of 60S subunit. The overall process involves many regulatory factors and is not described in detail in this chapter.

Another initiation protein then enhances the binding of charged formyl methionyl tRNA to the small subunit in response to the AUG triplet (step II). This step establishes the reading frame so that all subsequent groups of three ribonucleotides are translated accurately. The aggregate represents the initiation complex, which then combines with the large ribosomal subunit. In this process, a molecule of GTP is hydrolysed providing required energy and the initiation factors are released (step III).

3.2. Elongation

Once both subunits of the ribosome are assembled with the mRNA, binding site for two charged tRNA molecules are formed. These are designated as the 'P' or peptidyl and the 'A' or aminoacyl sites. The charged initiator tRNA binds to the P site, provided that the AUG triplet of mRNA is in the corresponding position of the small subunit. The increase of the growing polypeptide chain by one amino acid is called elongation. The sequence of the second triplet in mRNA dictates which charged tRNA molecule will become positioned at the A site (step I). Once it is present, peptidyl transferase catalyses the formation of the peptide bond that links the two amino acids together (stepII). The catalytic activity of peptidyl transferase is a function of rRNA of the large subunit, not one of the ribosomal proteins. At the same time, the covalent bond between the amino acid and the tRNA occupying the P site is hydrolysed (broken). The product of this reaction is a dipeptide, which is attached to the 3′ end of tRNA still present at the A site (Fig. 8.5 A, B).

Before addition of another amino acid, the tRNA attached to the P site which is now uncharged, must be released from the large subunit. The uncharged tRNA moves transiently through a third site on the ribosome called 'E' site (E stands for exit). The entire mRNA-tRNA-aa2-aa1 complex then shifts in the direction of P site by a distance of three nucleotides (step III). This event requires several protein elongation factors as well as the energy derived from hydrolysis of GTP. The result is that the third triplet of mRNA is now in a position to accept another specific charged tRNA into the A site (step IV). The sequence of elongation is repeated over and over (step V and VI). An additional amino acid is added to the growing polypeptide chain each time the mRNA advances through the ribosome. Once a polypeptide chain is of reasonable size is assembled (~30 amino acids), it began to emerge from the base of large subunits. A tunnel exists within the large subunit, through which the elongating polypeptide emerges.

The principle role of small subunit during elongation is to decode the triplet present in mRNA while the role of large subunit is peptide bond synthesis. The efficiency of the process is very high with one error in 10-4 amino acids added. An incorrect amino acids may present in one of the 20 polypeptides (500 amino acids in length) synthesized. In *E. coli*, elongation occurs at a rate of ~15 amino acids per second at 37°C.

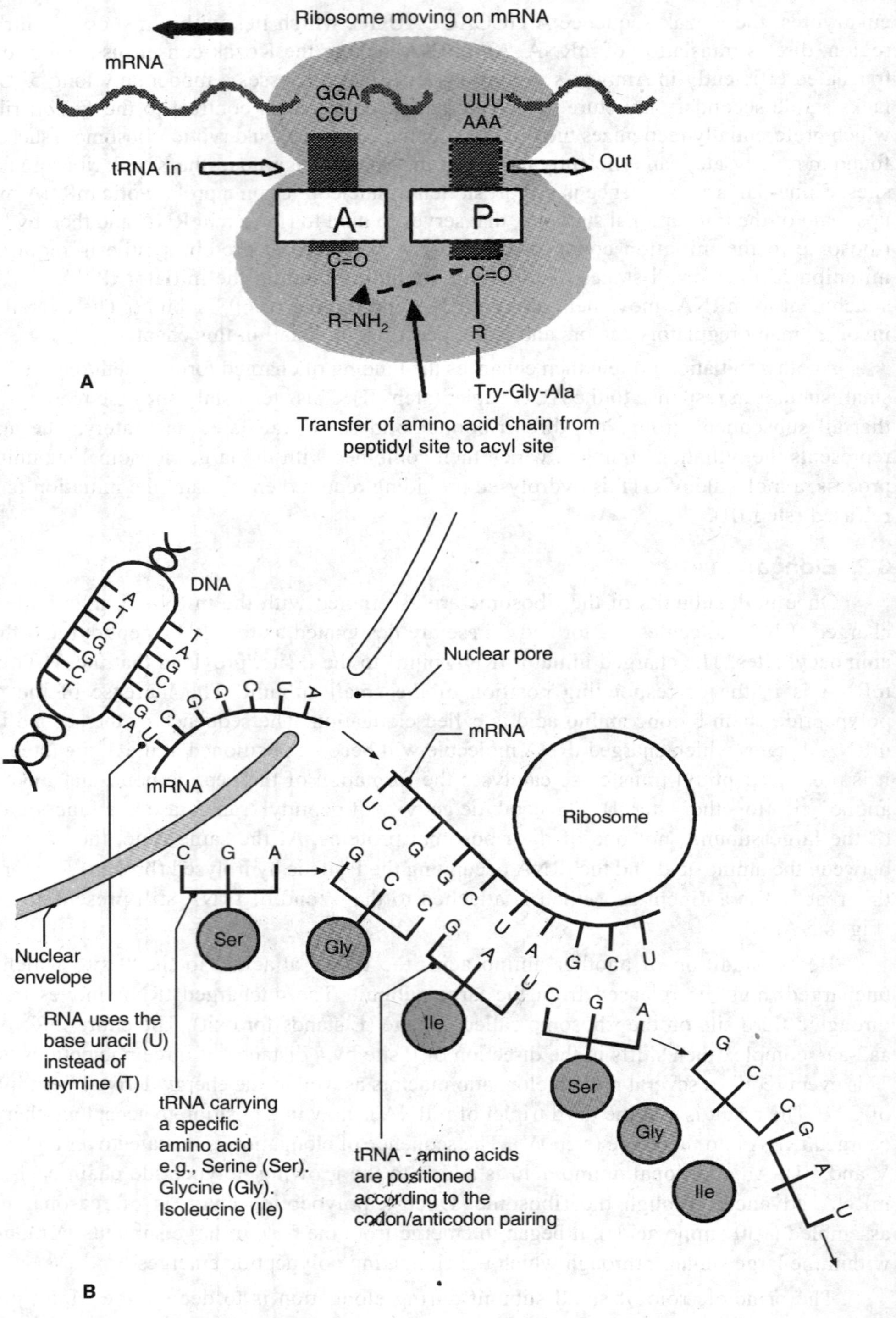

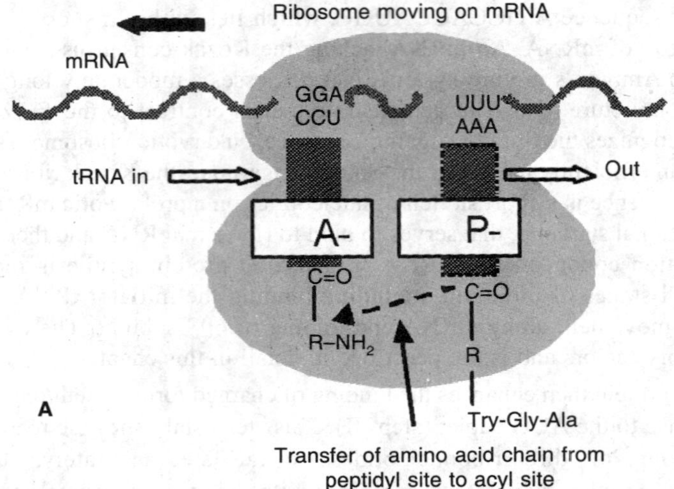

Fig. 8.5. (A) Addition of aminoacids in the peptide chain, and (B) elongation of polypeptide chain.

3.3. Termination

Termination of protein synthesis is carried out by triplet codes (UAG, UAA, UGA; stop codons) present at site A. These codons do not specify an amino acid, nor do they call for a tRNA in the A site. These codons are called stop codons, termination codons or nonsense codons. The finished polypeptide is still attached to the terminal tRNA at the P site, and the A site is empty. The termination codon signals the action of GTP-dependent release factor, which cleave the polypeptide chain from the terminal tRNA, releasing it from the translation complex (step I). Once this cleavage occurs, the tRNA is released from the ribosome, which then dissociates into its subunits (step II). If a termination codon appears in the middle of an mRNA molecule as a result of mutation, premature termination of polypeptide occurs.

3.4. Polyribosomes

With start of elongation process and passing of the initial portion of mRNA through ribosome, the mRNA is free to associate with another small subunit to form a second initiation complex. This process can be repeated several times with a single mRNA and results in attachment of several ribosomes on an mRNA called as polysomes or polyribosomes.

Table 8.2. Difference between prokaryotic and eukaryotic translation.

S.No.	Prokaryotic	Eukaryotic
1.	Initiating amino acid (methionine) needs to be formylated, therefore two types of tRNA are present.	Initiating amino acid (methionine) is not formylated, therefore one type of tRNA is present.
2.	Ribosomes enter the mRNA at AUG codon.	Ribosomes enter at the capped 5′ end of mRNA and then advanced to AUG codon by linear scanning.
3.	No initiation factor or co-factors are require for initial contact between ribosomes and mRNA.	ATP and a number of protein factors are required for ribosomes to engage mRNA.
4.	Small ribosomal subunit (30S) can engage mRNA before binding of initiator fmet-tRNA.	Small ribosomal subunit (40S) binds stably to mRNA only after initiator met-tRNA is bound.

4. PROTEIN STRUCTURE

The primary structure of a segment of a polypeptide chain or of a protein is the amino-acid sequence of the polypeptide chain(s), without regard to spatial arrangement (apart from configuration at the alpha-carbon atom). The "R" in the amino acid generic structure stands for the term "radical" and represents one of twenty or so different possibilities. So with 20 different "R" groups, there are twenty different amino acids in nature. All amino acids have attached to the same carbon both an amino group (NH_2) and a carboxyl group (COOH). The hydrogen from the carboxyl group actually exists in a cloud around the amino group and itself, thus creating the zwitterion. This renders the amino acid active and proteins formed from them capable of great activity. Buffers and enzymes are examples of active proteins.

The peptide bond is formed when the amino group of one amino acid and the carboxyl group of another amino acid unite with the loss of one water molecule per bond.

The commonly occurring amino acids are of 20 different kinds (a protein may contain a chain of 100 to 1000 amino acids; 26 alphabets in English making several thousands words, 20 amino acids forms different proteins) which contain the same dipolar ion group $H_3N^+.CH.COO^-$. They all have in common a central carbon atom to which are attached a hydrogen atom, an amino group (NH_2) and a carboxyl group (COOH). The central carbon atom is called the C_{alpha}-atom and is a

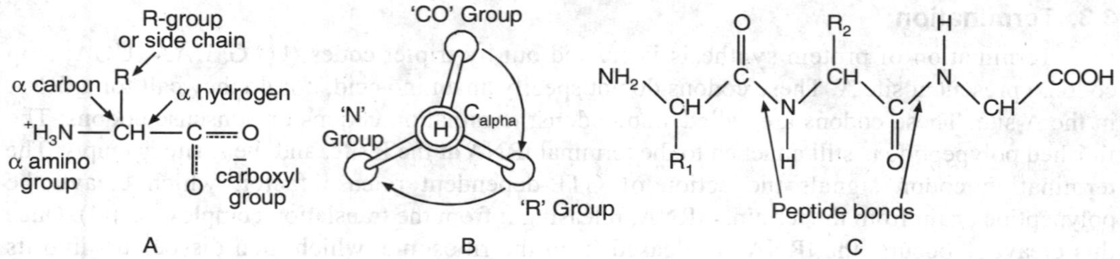

Fig. 8.6. (A) Carboxylic and amino group of an amino acid, (B) CORN orientation of groups, (C) Peptide bonds.

chiral centre. All amino acids found in proteins encoded by the genome have the L-configuration at this chiral centre. This configuration can be remembered as the **CORN** law (Fig. 8.6). Imagine looking along the H-C_{alpha} bond with the H atom closest to you.

When read clockwise, the groups attached to the C_{alpha} spell the word CORN. There are 20 side chains found in proteins encoded by the genetic machinery of the cell. The side chains confer important properties on a protein such as the ability to bind ligands and catalyse biochemical reactions. They also direct the folding of the nascent polypeptide and stabilise its final conformation. Amino acids in proteins (or polypeptides) are joined together by peptide bonds. The sequence of R-groups along the chain is called the **primary structure**. Proteins can occur as primary, secondary, tertiary or quaternary structures.

Secondary Structure: The secondary structure of a segment of polypeptide chain is the local spatial arrangement of its main-chain atoms without regard to the conformation of its side chains or to its relationship with other segments. There are three common secondary structures in proteins, namely **alpha helices**, **beta sheets** and **turns**.

The alpha-helix and beta-structure conformations for polypeptide chains are generally the most thermodynamically stable of the regular secondary structures. However, particular amino acid sequences of a primary structure in a protein may support regular conformations of the polypeptide chain other than alpha-helical or beta-structure.

Tertiary structure: The tertiary structure of a protein molecule, or of a subunit of a protein molecule, is the arrangement of all its atoms in space, without regard to its relationship with neighbouring molecules or subunits.

Quaternary structure: The quaternary structure of a protein molecule is the arrangement of its subunits in space and the ensemble of its intersubunit contacts and interactions, without regard to the internal geometry of the subunits. The subunits in a quaternary structure must be in noncovalent association. Haemoglobin contains four polypeptide chains (alph a_2b_2) held together noncovalently in a specific conformation as required for its function.

5. POST–TRANSCRIPTIONAL MODIFICATIONS

Polypeptide chains like RNA transcripts are also modified after their synthesis. This additional processing is termed as post transcriptional modification. These types of post translational modifications are important in achieving the functional status specific to any given protein. Because the final 3D structure of the molecule is closely elated to its specific function, the folding of protein is also important. These complex biochemical processes are briefly described here for an understanding about overall process.

1. The N-terminus and C-terminus amino acids are usually removed or modified. The initial and terminal formylmethionine residue in bacterial polypeptide is usually removed enzymatically. In eukaryotes, initial methionine residue is removed and the amino group of the N-terminal residue is chemically modified.

2. Individual amino acid residues are sometimes modified, e.g., phosphate may be added to the hydroxyl groups of certain amino acids such as tyrosine. The process of phosphorylation is extremely important in regulating several cellular activities and is a result of the action of enzymes called kinases. In other proteins, methyl group may be added enzymatically.
3. In some proteins, carbohydrate side chains are sometimes added. Covalently added carbohydrates form a class of molecules called glycoproteins having antigenic properties.
4. Sometimes polypeptide chains are trimmed to make active protein molecules, e.g., insulin is produced as a large molecule and then trimmed to 51 amino acid molecules.
5. At the end terminal end of some proteins, a sequence of up to 30 amino acids is found that plays an important role in directing the protein to the location in the cell where it becomes functional. This is called a signal sequence and it determines the final destination of protein in the cell.
6. Polypeptide chains often complexes with metals. The quaternary and tertiary levels of protein structure often include and are dependent on metal atoms. The function of the protein is thus dependent on the molecular complex that includes both polypeptide chains and metal atoms, e.g., haemoglobin containing four iron atoms.

6. INTEINS

Inteins are selfish DNA elements inserted in-frame and translated together with their host proteins. This precursor protein undergoes an autocatalytic protein splicing reaction resulting in two products: the host protein and the intein. Protein splicing was shown to occur in heterologous organisms, in different *in vitro* systems and with various natural and engineered host flanks. Hence, protein splicing is probably independent of specific host cell factors. Most inteins have an endonuclease domain inserted in the protein splicing domain. The endonuclease activity of these inteins can mediate the specific transposition of their gene into unoccupied integration sites of intein-less homologs (homing) (Fig. 8.7).

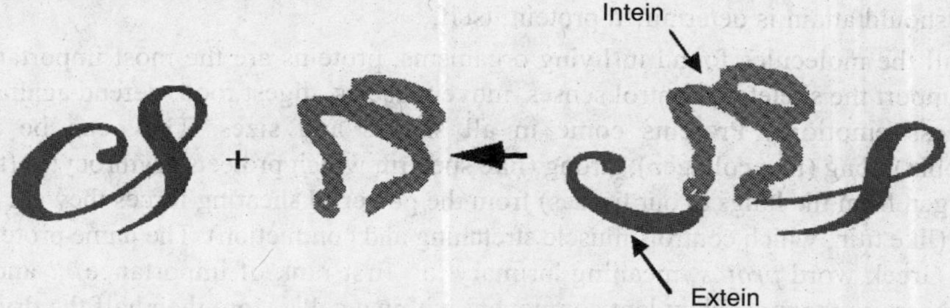

Fig. 8.7. Protein processing and formation of inteins.

Protein-splicing scheme. A precursor protein is shown on the left, with intein protein-splicing domain shown in dotted and the host protein flanks (exteins) shown in black. The intein protein-splicing domain autocatalyzes its excision and the ligation of its two flanks. Inteins are currently known in more than 50 types of proteins with diverse in functions. These proteins include metabolic enzymes, DNA and RNA polymerases, proteases, ribonucleotide reductases, and vacuolar-type ATPases. Inteins integration points also vary in structure and in function. Their only apparent common feature is being in highly conserved protein motifs. Intein proteins contain a number of conserved sequence motifs (blocks). The motifs can be grouped in three domains according to their location and inferred function. Intein structures show that the inteins protein-splicing and endonuclease active sites are formed from conserved motifs. The intein's domain organization, deduced by sequence analysis, exactly corresponds to the structural domains.

The protein-splicing N-terminal (N) domain spans about 100-150 aa. Mutations in this domain affect the first step in protein splicing, an N-S/O acyl shift in the peptide bond connecting the N terminus of the intein and the N-terminal flank (intein flanks are termed exteins). The N domain motifs are similar in sequence and function to motifs found at the Hint domain of Hog regions found in the Hedgehog animal developmental proteins and a few related nematode protein families. Hog Hint domains self-process their precursor proteins, cleaving themselves off the N-terminal parts of the proteins. Cleavage was shown to utilize a cholesterol molecule that is consequently covalently attached to the cleavage site in the N-terminal part modulating the activity of that part. The relation between inteins and Hog Hint domains was first determined by their similarity in sequence and function. Determination of protein structures from both families provided final proof for the common origin of inteins and Hog Hint domains. The intein protein-splicing C-terminal (C) domain is composed of the two adjacent motifs in the C-terminal 25-40 aa (including the conserved aa immediately C′ to the intein). Residues in these motifs are necessary for the catalyzing the next steps of protein splicing: the branch formation and its resolution. Most (but not all) inteins also include a central endonuclease (EN) domain. Functional inteins with no EN domain (minimal inteins), the relation of the protein splicing domain to other Hint domains and the presence of different EN domains in inteins all indicate that the primeval inteins had no EN domains. Different EN domains, perhaps from homing endonucleases and DNA binding domains invaded intein genes to form the typical present day intein. Some present day minimal inteins clearly lost their EN domain (such as Mxe_gyrA) and some maybe never acquired one. Some protein families, such as ribonucleotide reductases and archaeal DNA polymerase type B, are more prone to contain inteins. These proteins contain inteins in different organisms and in different integration sites. Some of the ribonucleotide reductases and most of the DNA polymerases with inteins contain more than one intein.

7. PROTEIN FOLDING

Proteins are the biological workhorses that carry out vital functions in every cell. To carry out their task, proteins must fold into a complex three-dimensional structure, but how and which shape a protein should attain is determined protein itself.

Of all the molecules found in living organisms, proteins are the most important. They are used to support the skeleton, control senses, move muscles, digest food, defend against infections and process emotions. Proteins come in all shapes and sizes. They can be round (like haemoglobin), long (like collagen), strong (like spectrin which protects erythrocytes (the cells that carry oxygen from the lungs to our tissues) from the powerful shearing forces they are exposed to), or elastic (like titin, which controls muscle stretching and contraction). The name protein is derived from the Greek word *prôtos*, meaning 'primary' or 'first rank of importance? ? and with good reason. These are the most abundant component within a cell'. More than half the dry weight of a cell is made up of proteins and they have a range of indispensable roles; for example, enzymes, the biocatalysts that carry out crucial biochemical reactions in every cell that would otherwise be too slow to sustain life.

What is remarkable is that the more than 100,000 proteins in our bodies are produced from a set of only 20 building blocks, known as amino acids. All amino acids have the same basic structure - an amino group, a carboxyl group and a hydrogen atom, but differ due to the presence of a side-chain. This side-chain varies dramatically between amino acids, from a simple hydrogen atom in the amino acid glycine to a complex structure found in tryptophan. Depending on the nature of the side-chain, an amino acid can be hydrophilic (water-attracting) or hydrophobic (water-repelling), acidic or basic; and it is this diversity in side-chain properties that gives each protein its specific character.

7.1. Creating a Functional Protein

The sequence of amino acids in a protein defines its primary structure. The blueprint for each amino acid is laid down by sets of three letters known as base triplets that are found in the coding regions of genes. These base triplets are recognized by ribosomes, the protein building sites of the cell, which create and successively join the amino acids together. This is a remarkably quick process: a protein of 300 amino acids will be made in little more than a minute. The result is a linear chain of amino acids, but this only becomes a functional protein when it folds into its three-dimensional (tertiary structure) form. This occurs through an intermediate form, known as secondary structure, the most common of which are the rod-like a-helix and the plate-like β-plated sheet. These secondary structures are formed by a small number of amino acids that are close together, which then, in turn, interact, fold and coil to produce the tertiary structure that contains its functional regions (called domains).

Although it is possible to deduce the primary structure of a protein from a gene/s sequence, its tertiary structure cannot be determined (although it should become possible to make predictions when more tertiary sequences are submitted to databases). It can only be determined by complex experimental analyses and, at present, this information is only known for about 10% of proteins. It is therefore not yet known how an amino-acid chain folds into its tertiary structure in the short time scale (fractions of a second) that occurs in the cell. So, there is a huge gap in our knowledge of how we move from protein sequence to function in living organisms: the line of sight from the genetic blueprint for a protein to its biological function is blocked by the impenetrable jungle of protein folding.

7.2. The Quest to Understand Protein Folding

One of the most important results in understanding the process of protein folding was a thought-provoking experiment that was carried out by Christian Anfinsen and colleagues in the early 1960s. They investigated a protein called ribonuclease, which they isolated from the pancreatic tissue of cattle. This enzyme, made up of 124 amino acids, cleaves any ribonucleic acid (RNA) that could be harmful to the cell, such as truncated RNA that would not make a fully operational protein. It briefly binds RNA in a binding site and requires several sulphur-containing amino-acid cysteine residues in the protein, which form bonds with each other (called disulphide bridges) and hold the protein structure together. Ribonuclease can be denatured by adding certain chemicals or by heat. The disulphide bridges break and other forces of attraction between amino acids disappear, which makes the enzyme collapse into a tangled, useless ball. The ribonuclease then folds back to its natural functional state on its own. So, Anfinsen concluded that the amino-acid sequence determines the shape of a protein, a finding for which Anfinsen received the Nobel Prize in Chemistry in 1972.

7.3. Finding the Energy to Fold

As with all processes in nature, protein folding also needs energy, the process has to obey the laws of thermodynamics. A protein always folds so that it achieves the lowest possible energy just as we always try to adopt the most comfortable position, in which we need to move about least, when going to sleep. It is thought that this is achieved by using an energy gradient or funnel along the path from the random tangle to the folded protein. Alan Fersht of Cambridge University used the following analogy to illustrate this model: if you blindfold a golfer and let him hit the ball in any direction he likes, the probability that he will hole the ball is almost infinitesimal. The same is true of a protein, finding the right form by chance. However, if all parts of the golf course slope toward the hole, which is at the lowest point in the area, even a blindfolded golfer has a good chance of finding the hole. So, fixed reaction pathways are not necessary, as each protein seeks out

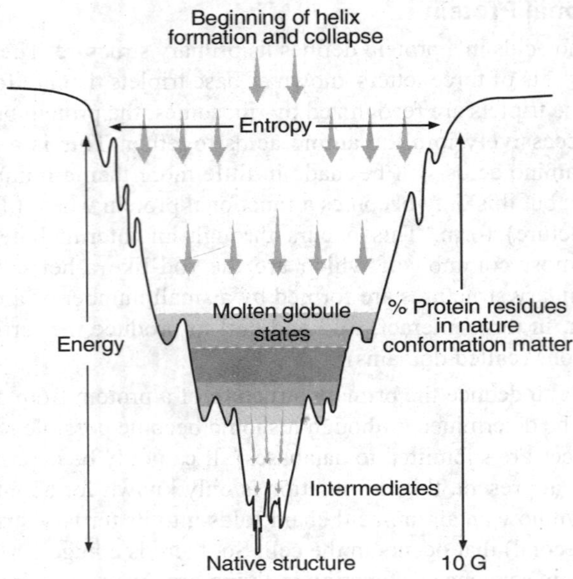

Fig. 8.8. Schematic presentation of energy involvement in protein folding.

its natural shape through a funnel of declining energy; it can take many folding routes and still reach its target of the completed tertiary structure (Fig. 8.8).

7.4. The Process of Folding

The folding pathway of a large polypeptide chain is very complicated, and not all the principles that guide the process have been worked out. However, many plausible models have attempted to describe protein folding. One model views folding as a hierarchical process, where local secondary structures form first. Under this model, a helices and β sheets form first, with longer range interactions between helices and sheets forming super-secondary structures later. This process continues until the entire polypeptide folds. An alternative model describes folding as a spontaneous collapse of the polypeptide into a compact state. This collapsed state is known as a molten globule. It may be that the actual folding process of proteins incorporates features of both models. Instead of following a single pathway, a population of peptide molecules may take a variety of routes. Thermodynamically, the folding process can be viewed as a kind of free-energy funnel, where the unfolded states are characterized by a high degree of conformational entropy and relatively high free energy. In a trivialized definition, entropy is a measure of chaos, a measure of all different conformational states that the protein can be in. Obviously, there is more chaos in the protein in its unfolded state. On the other hand, high free energy is a measure of unstableness, which is higher in a protein's unfolded state. Therefore, as folding proceeds, the narrowing of the funnel represents a decrease in the number of conformational states present. At the bottom of the funnel, also known as the global minimum, an assembly of folding intermediates are reduced to a single conformation. It is important to realize that although we often describe the free energy funnel as having *one* global minimum - that is, one native conformation - a protein can have a small set of native conformations, each one important for its biological function(s).

It has been experimentally confirmed that not all proteins fold spontaneously in the cell. For many proteins the folding process is facilitated by the action of specialized proteins known as chaperones. Molecular chaperones are proteins that interact with partially folded or improperly folded polypeptides to facilitate correct folding pathways of provide microenvironments so that folding can occur. Chaperones are not the only proteins to facilitate protein folding. Two enzymes, protein disulfide isomerase (PDI) and peptide prolyl *cis-trans* isomerase (PPI), catalyze isomerization reactions and are required for the folding pathways of a number of proteins.

7.5. Fold Recognition (Threading)

Threading uses a database of known three-dimensional structures to match sequences without known structure with protein folds. This is accomplished through a scoring function that assesses the fit of a sequence to a given fold. These scoring functions are usually derived from a database of known structures and generally include a pair-wise atom contact and solvation terms. Threading methods are very similar to comparative modelling in that threading compares a target sequence against a library of structural templates, producing a list of scores. The scores are then ranked and the fold with the best score is assumed to be the one adopted by the sequence. The methods to fit a sequence against a library of folds can be extremely elaborate computationally, such as those involving double dynamic programming.

8 PROTEIN TARGETING

Proteins that must be targeted within the cell must be intercepted early during synthesis so that this can happen correctly. As a protein is being synthesized, decisions must be taken about sending it to the correct location in the cell where it will be required. The information for doing this resides in the nascent protein sequence itself. Once the protein has reached its final destination, this information may be removed by proteolytic processing.

8.1. Targeting in Bacteria

In bacterial cells, the targeting decision is relatively straightforward: is the protein destined to be an intracellular protein or an extracellular one? However, even here gradations are possible. In a gram negative bacterium, such as *E. coli*, there must be some way of knowing whether a protein is destined to go to the cell membrane, the periplasmic space, or the outer membrane.

Secreted proteins contain a **signal sequence**. This is a short (6 - 30) stretch of hydrophobic amino acids, flanked on the N-terminal side by one or more positively charged amino acids such as lysine or arginine, and containing neutral amino acids with short side-chains (such as glycine or alanine) at the cleavage site. As proteins with signal sequences are synthesized, they are bound by the SecB protein. This prevents the protein from folding. SecB delivers the protein to the cell membrane where is is secreted through a pore formed by the SecE and SecY proteins. Secretion is driven by the SecA ATPase. After the protein has been secreted, the signal sequence is removed by a membrane bound leader peptidase.

8.2. Targeting in Eukaryotes

In eukaryotic cells, the situation is more complex. Extracellular proteins can be targeted for secretion or to the cell membrane, or to one of the many internal organelles such as the lysosome. Intracellular proteins can be targeted for the cyoplasm, to the nucleus or to special organelles such as the mitochondrion or the chloroplast. The Signal Sequence hypothesis was first enunciated by Gunther Blöbel who was awarded the Nobel Prize in Medicine in 1999 for his work.

The following diagram summarizes the choices/fates available to newly synthesized proteins in a eukaryotic cell. Protein secretion in eukaryotic cells also involves a signal sequence. As in bacteria, the signal sequence is a short (6 - 30) stretch of hydrophobic amino acids, flanked on the N-terminal side by one or more positively charged amino acids such as lysine or arginine, and

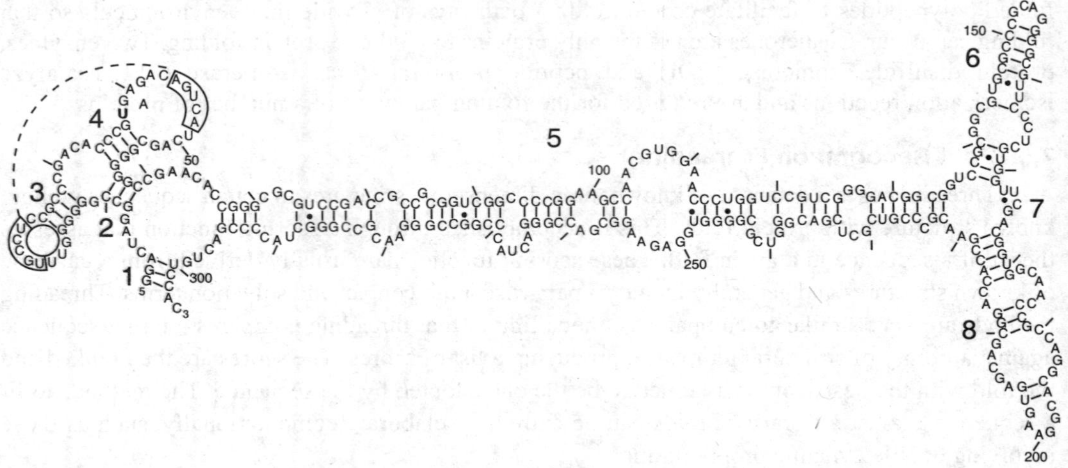

Fig. 8.9. Image of *Halobacterium halobium* SRP RNA.

containing neutral amino acids with short side-chains (such as glycine or alanine) at the cleavage site. Signal sequences are recognized by Signal Recognition Particles (SRPs) which are ribonucleoprotein particles containing a stable 305 nt 7S RNA and 6 polypeptides (Fig. 8.9). The 7S RNA has 2 domains - the Alu domain (so called because it is related to the Alu family of repeating sequences) and the S domain. The 7S RNA has 3 activities - a signal sequence recognition activity, an SRP receptor binding activity, and an elongation arrest activity.

As proteins are being synthesized, the N-terminal signal sequence is bound by the SRP particle. Protein synthesis is temporarily halted (elongation arrest) until the particle- with the nascent polypeptide and the ribosome -attaches to an SRP receptor complex in the endoplasmic reticulum membrane. The SRP receptor is a heterodimer formed by a larger 68 kD alpha subunit and a smaller 30 kD subunit. The subunit is a GTP-binding protein. Attachment of the ribosome triggers the release of GDP and binding of GTP. The GTP form of the SRP receptor binds the ribosome very tightly. After the transfer takes place, GTP is hydrolyzed and the SRP receptor is free to interact with another SRP particle-ribosome complex. Protein synthesis then resumes but now the nascent polypeptide is secreted into the lumen of the ER as it is being synthesized.

Shortly after the polypeptide enters the ER, signal peptidase cleaves the signal peptide. The finished polypeptide is then targeted through the ER and Golgi apparatus to the cell surface or to the lysosome. During this process, the proteins are carried in coated vesicles which are enclosed either by a coat protein skeleton (ER → Golgi) or by a clathrin skeleton (Golgi → exterior).

- Proteins that are targeted for organelles have their own N-terminal uptake-targeting sequence(s) that determines whether the protein must cross one or two membranes. In the former case, proteins destined for the intermembrane space of the chloroplast or mitochondrion are first transported into the stroma (chloroplast) or matrix (mitochondrion) and then re-transported back through the inner chloroplast or mitochondrial membrane to the intermembrane space.
- Proteins that must be targeted to the nucleus have a nuclear localization signal (NLS). One common type of signal is a series of five or so closely spaced positively charged amino acids.

Table 8.3 Different types of signal associated with protein sorting.

Organelle	Location of signal in the protein (Usually)	Nature of signal
Endoplasmic Reticulum	C-terminus	KDEL or HDEL (in yeast)
Mitochondrion	N-terminal	3-5 nonconsecutive Arg or Lys; often contains Ser and Thr; never has acidic amino acids; no consensus sequence.
Chloroplast	N-terminal	Generally rich in Ser, Thr, and small hydrophobic amino acids; no acidic amino acids
Nucleus	Internal	Single cluster of 5 basic amino acids (e.g. KKKRK in the SV40 T antigen), or 2 smaller clusters separated by 10 amino acids
Lysosome	Internal	Covalent attachment of Mannose-6-phosphate
Peroxisome	C-terminus	SKF tripeptide at C-terminus

QUESTIONS

1. Describe the process of translation in prokaryotes giving suitable diagrams.
2. Give different modes of post translational modifications in proteins.
3. Describe the ribosome binding sites in prokaryotes and eukaryotes.
4. Discuss phenomenon and importance of protein folding.
5. Write short note;
   (a) Protein folding
   (b) Protein targeting
   (c) Transport across membrane
   (d) Signal sequences
   (e) Ribosomes
   (f) Translation
   (g) Polysomes
   (h) Protein structure
   (i) Inteins

CHAPTER 9

Regulation of Gene Expression in Bacteria

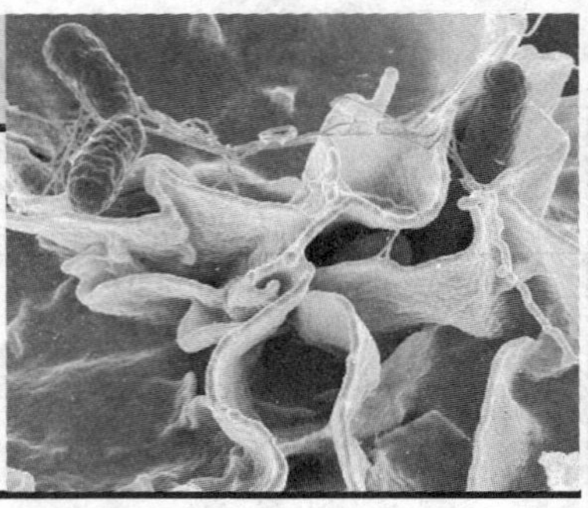

1. CONTROL POINTS IN GENE REGULATION

Gene expression is the process by which the inheritable information in a gene, such as the DNA sequence, is made into a functional gene product, such as protein or RNA. Several steps in the gene expression process may be modulated, including the transcription step and the post-translational modification of a protein. Gene regulation gives the cell control over structure and function, and is the basis for cellular differentiation, morphogenesis, and the versatility and adaptability of any organism. Gene regulation may also serve as a substrate for evolutionary change, since control of the timing, location, and amount of gene expression can have a profound effect on the functions (actions) of the gene in the organism. All genes are not expressed all the time, e.g., genes for flowering in plants are expressed once in a year. The DNA in the human genome contains about 100,000 genes and if all genes are expressed equally, then all cells would be the same. It is estimated that a typical higher eukaryotic cells expresses between 10,000 and 20,000 genes, which is about 10-20% of the total gene complement of the cell. Thus, cells present in different organs have differential expression though they have the same genetic makeup. This is known as differential gene expression. It has been observed that cells in different organs differ slightly in their protein content and about 95% of the proteins are identical and the remaining 5%

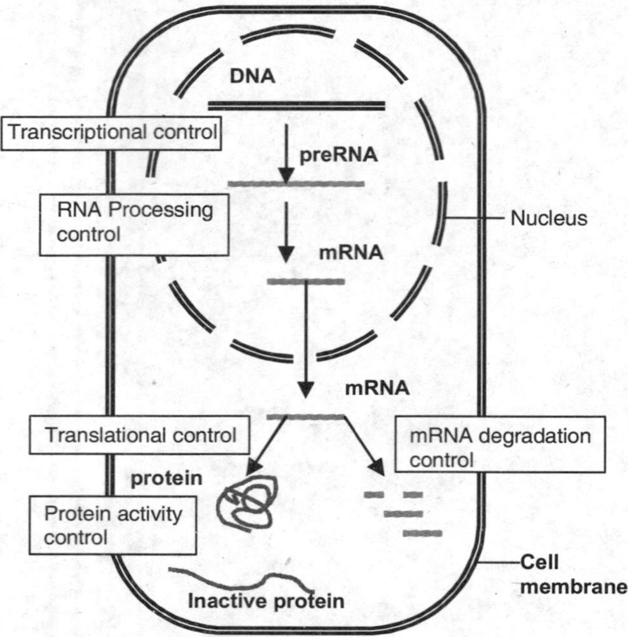

Fig. 9.1. Control points in gene expression.

proteins are responsible for the difference between two types of cell. This can be seen in the protein contents of pollen grains and root cells of a plant. These common proteins are produced by genes called as 'house-keeping genes'. Non-protein coding genes (e.g., rRNA genes, tRNA genes) are not translated into protein. The expression of many genes is regulated after transcription (i.e., by microRNAs or ubiquitin ligases), so an increase in mRNA concentration need not always increase expression. Nevertheless, mRNA levels can be quantitatively measured by Northern blotting, a process in which a sample of RNA is separated on an agarose gel and hybridized to a radio-labelled RNA probe that is complementary to the target sequence. Northern blotting requires the use of radioactive reagents and can have lower data quality than more modern methods (due to the fact that quantification is done by location, and amount of gene expression can have a profound effect on the functions (actions) of the gene in the organism.

In the previous chapters we have seen that transcription results in mRNA formation and translation results in protein synthesis. There are differences in prokaryotic and eukaryotic transcription and translation particularly processing of RNA molecules. There are multiple points in the steps between gene expression and protein synthesis at which gene expression can be controlled in cells (Fig. 9.1). These points of control can be separated in two areas; transcriptional control and post transcriptional control. In translational control sequence specific, DNA binding proteins are involved which can 'switch on' and 'switch off' the genes. Posttranscriptional control are secondary mechanisms for controlling gene expression after transcription which includes, (*i*) RNA processing control (*ii*) Translational control (*iii*) mRNA degradation control and (*iv*) protein activity control. RNA processing control (in eukaryotes) determines how and when the primary transcripts are spliced to form an active mRNA. Further, translation can be controlled by specific proteins binding to mRNA or control of degradation of mRNA. A protein that is needed in large quantities over an extended time period may have very stable mRNA, *e.g.,* β-globin mRNA which has half life time (time required for 50% of molecules to be degraded) of over 10 hours. Protein activity control selectively activates, inactivates or compartmentalizes specific protein molecules within the cell.

2. OPERON CONCEPT

Jacob and Monod (1961) gave the concept of Operon model to explain the gene regulation in prokaryotes using simple experiments to grow *E. coli* in petriplates containing histidine, lactose or lactose and glucose in the medium (Fig. 9.2). When histidine is present in the medium, activity of histidine synthesizing enzymes decreases rapidly showing switching off the genes. Similarly, when concentration of lactose increases in the medium, activity of galactosidase increases. In a medium if both glucose and lactose are present, bacteria will utilize glucose first and there will be no β-galactosidase activity. Activity of β-galactosidase increases with the depletion of glucose and availability of lactose. Activity of β-galactosidase shut off if glucose is added in the medium. A molecule of lactose forms one molecule of galactose and one molecule of glucose.

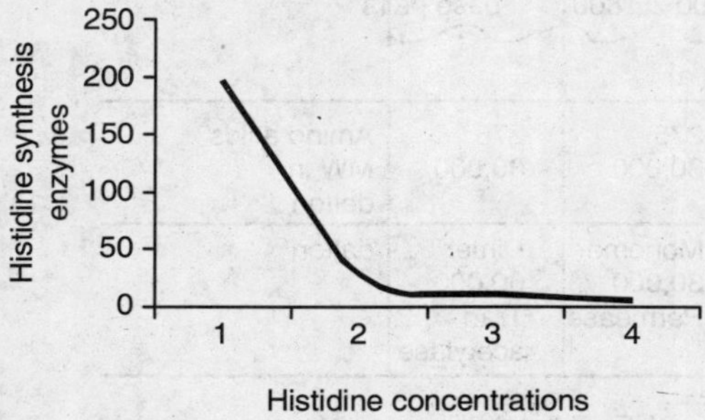

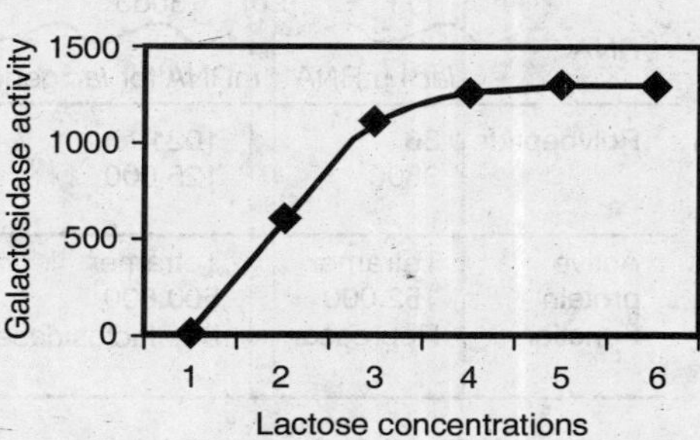

Fig. 9.2. Substrate dependent switch off (histidine) and switch on (lactose) of Operon.

$$\text{Lactose} \xrightarrow{\beta\text{- galactosidase}} \text{Lactose + Glucose}$$

These experiments clearly demonstrate substrate induced gene expression or inhibition. Therefore, a precise control mechanism must exist to switch on and switch off the genes. During 60s there was no clear vision about physical nature of gene even then they proposed gene regulation which holds true till today. Jacob and Monod were awarded Nobel Prize in 1965.

3. STRUCTURE OF OPERON

An Operon consists of a group of coordinately controlled genes that are all related to a particular cellular function. Operons occur primarily in prokaryotic organisms. Typically, the genes are physically adjacent to one another and all controlled from a single upstream promoter. An operator locus immediately adjacent to the promoter determines when the Operon is actually actively transcribed. In bacteria, genes are clustered into Operons: gene clusters that encode the proteins necessary to perform coordinated function, such as biosynthesis of a given amino acid. RNA that is transcribed from prokaryotic Operons is polycistronic a term meaning that multiple proteins are encoded in a single transcript.

In bacteria, control of the rate of transcriptional initiation is the predominant site for control of gene expression. As with the majority of prokaryotic genes, initiation is controlled by two DNA sequence elements that are approximately 35 bases and 10 bases, respectively, upstream of the site of transcriptional initiation and as such are identified as the –35 and –10 positions. These 2

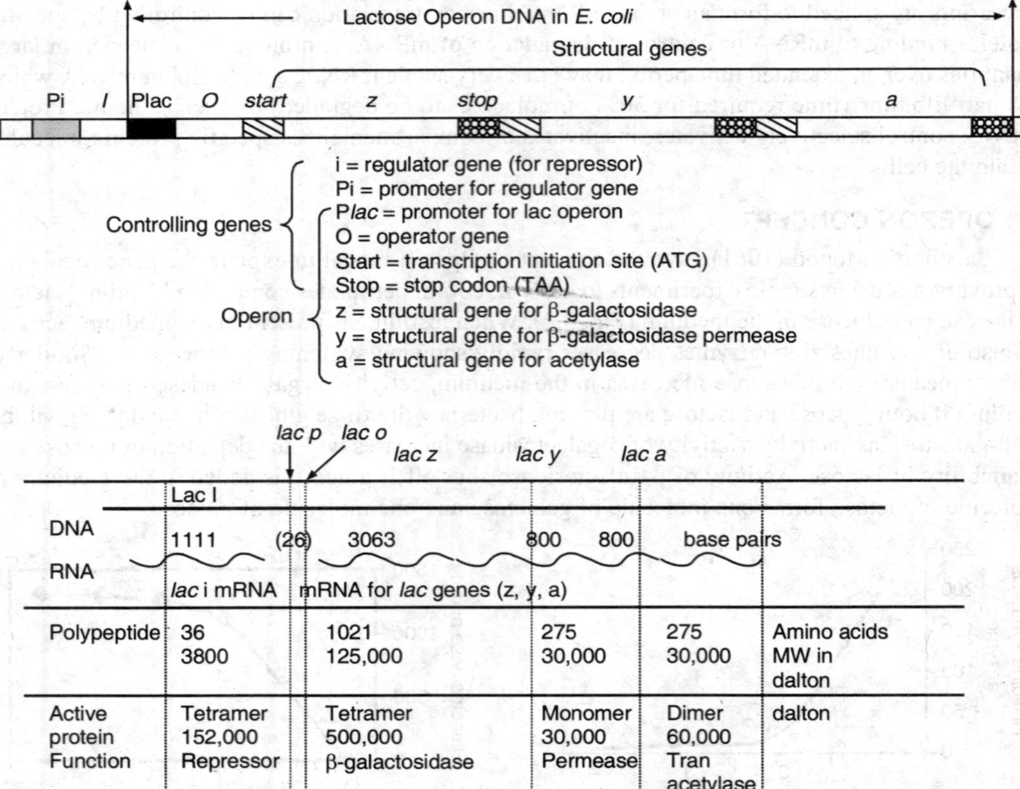

Fig. 9.3. Structural details of *lac* Operon of *E. coli* including size of genes, gene product and their molecular weight.

sequence elements are termed promoter sequences, because they promote recognition of transcriptional start sites by RNA polymerase. The consensus sequence for the –35 position is TTGACA, and for the –10 position, TATAAT. (The –10 position is also known as the Pribnow-box). These promoter sequences are recognized and contacted by RNA polymerase (Fig. 9.3).

The *lac* Operon consists of one regulatory gene (the *i* gene) and three structural genes (*z*, *y*, and *a*). The *i* gene codes for the repressor of the *lac* operon. The *z* gene codes for β-galactosidase (β-gal), which is primarily responsible for the hydrolysis of the disaccharide, lactose into its monomeric units, galactose and glucose. The *y* gene codes for permease, which increases permeability of the cell to β-galactosides. The *a* gene encodes a transacetylase.

The activity of RNA polymerase at a given promoter is in turn regulated by interaction with accessory proteins, which affect its ability to recognize start sites. These regulatory proteins can act both positively (activators) and negatively (repressors) (Fig. 9.4). The accessibility of promoter regions of prokaryotic DNA is in many cases regulated by the interaction of proteins with sequences termed operators. The operator region is adjacent to the promoter elements in most Operons and in most cases the sequences of the operator bind a repressor protein.

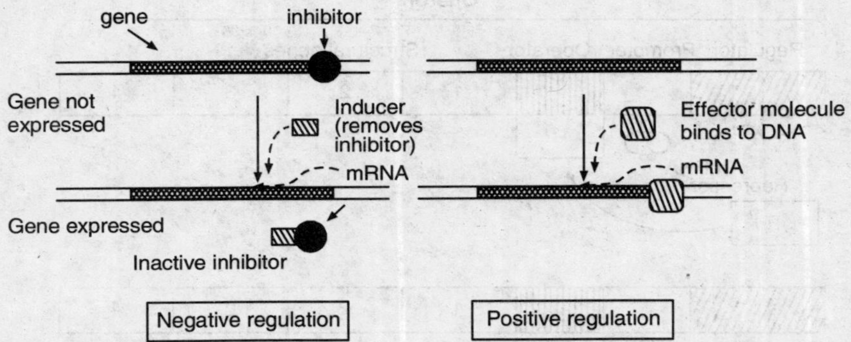

Fig. 9.4. Negative and positive regulation of Operon. Presence of inducer is necessary to remove inhibitor in negative regulation while in positive effector binds directly to DNA to initiate transcription.

4. REGULATION OF GENE EXPRESSION IN PROKARYOTES

The expression of a gene in response to a substance in its environment is called induction and such genes are called inducible genes, e.g., lactose dependent genes or *lac* Operon. Other genes that are always expressed (such as tRNA, rRNA) are called constitutive genes. Some genes are

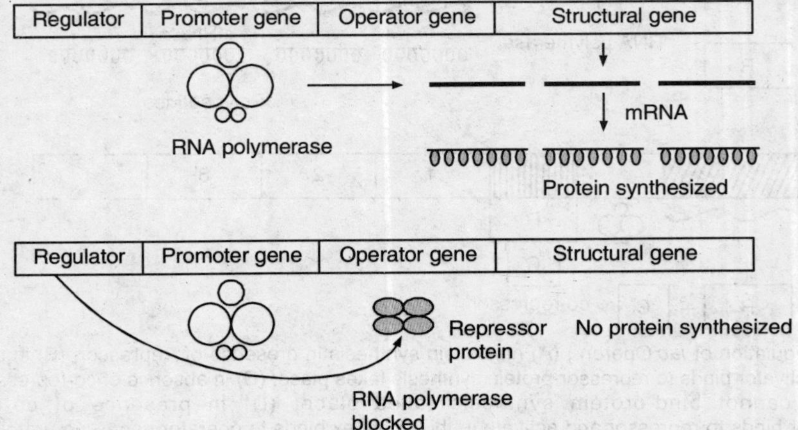

Fig. 9.5. Normal process of translation (above) and repression of translation by repressor.

turned off when a particular substance is present in the environment and such genes are called repressible genes, e.g., presence of histidine in the medium as described above. The *lac* Operon controls three genes responsible for synthesis of β-galactosidase (*lac* Z gene), β-galactosidase permease (*lac* Y gene) and transacetylase (*lac* A gene) enzymes. Therefore, mRNA produced by this Operon is known as polycistronic mRNA. As shown in Figure 9.3. *Lac* Operon contains a regulator gene called as *lac I*, which is responsible for synthesis of a protein referred to as repressor protein. The repressor protein binds to as specific DNA sequence and inhibits β-galactosidase synthesis. Normally β-galactosidase is not synthesized in a bacterial cell, thus inhibition of β-galactosidase synthesis and binding of the repressor protein is the normal condition. Gene encoding the repressor protein does not need to be located adjacent to the protein's site of binding because the repressor protein can diffuse throughout the cytoplasm. A protein that can diffuse and act at a distant site is known as a trans-acting protein and the gene that produces it is known as trans-acting gene. The repressor protein binds to operator site (*lac O*) on DNA which is 21 nucleotides long. In presence of lactose (activator), repressor does not bind to *lac O* and transcription of the three adjacent structural genes proceeds (Fig. 9.5).

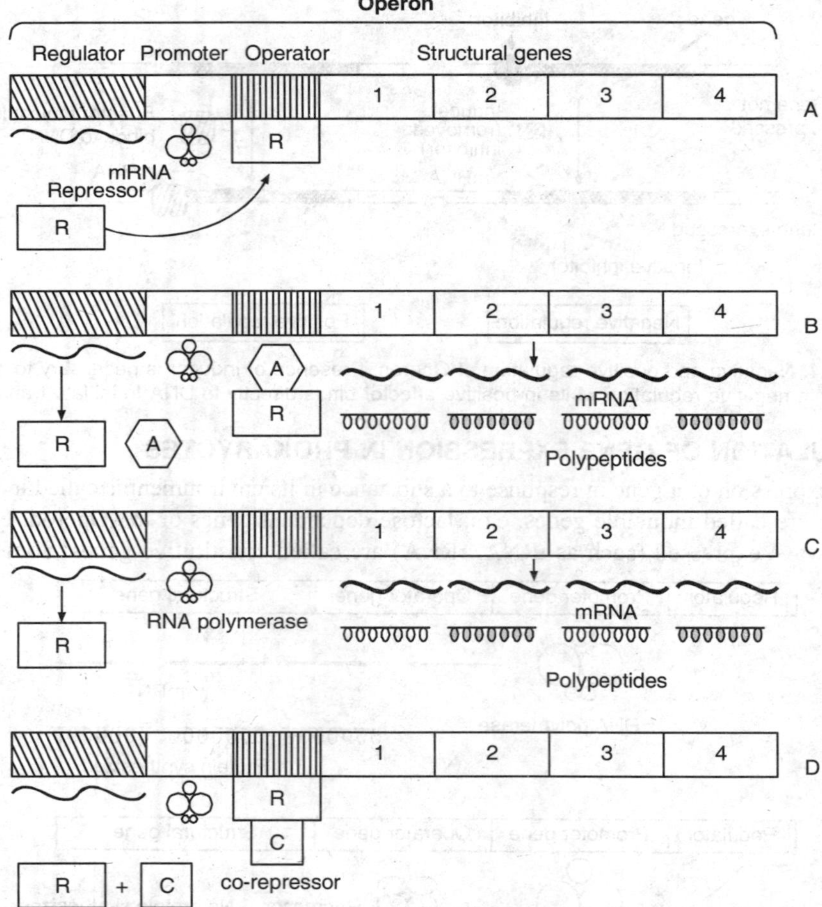

Fig. 9.6. Regulation of *lac* Operon ; (A) no protein synthesis in presence of repressor; (B) in presence of activator, activator binds to repressor-protein synthesis takes place; (C) in absence of co-repressor, inactive repressor cannot bind-protein synthesis takes place; (D) in presence of co-repressor, co-repressor binds to repressor and activate it, this complex binds to operator gene - no protein synthesis takes place.

In contrast to regulation by repressor, in the reversible system repressor protein remains inactive and fails to bind to the operator. Consequently, proteins are synthesized by all the three structural genes. However, the repression protein can be activated in the presence of co-repressor. The co-repressor together with repressor protein forms the repressor - co-repressor complex. This complex binds to operator gene and blocks protein synthesis. This repressor protein was isolated by Gilbert and Muller-Hill in 1966 (Fig. 9.6).

5. CATABOLITE REPRESSION

In presence of glucose and lactose preferential utilization of glucose is explained by catabolite repression. An additional control system is superimposed upon the repressor-operator system regulating the expression of *lac* Operon. This system is known as catabolite repression and allows the cell to use glucose first. It is general, mechanism that controls Operons that catabolize carbon sources (galactose, arabinose) and allows the cell to first absorb glucose as a carbon source. Another protein, the cyclic AMP receptor protein (CRP) binds to cyclic AMP as shown in the Figure 9.7. The CRP protein is a dimer composed of two identical subunits of 22.5 KD each and is activated by single molecule of cAMP. The RNA polymerase will not initiate RNA synthesis unless the CRP protein-cAMP complex is bound to the CRP site on the lactose Operon.

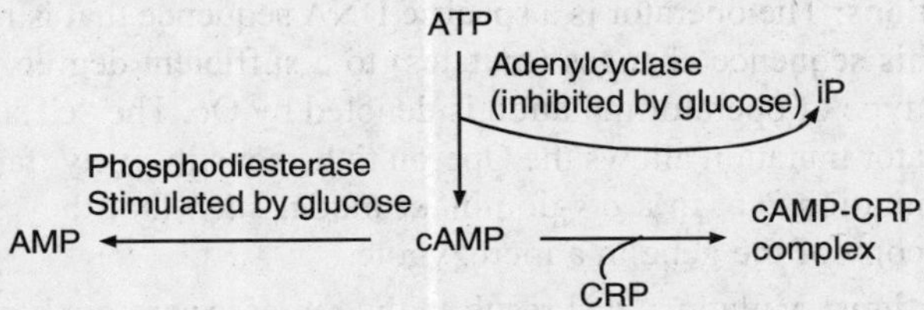

Fig. 9.7. Catabolite repression of *lac* Operon.

In *E. coli* cAMP is converted into AMP by the enzyme phosphodiestrase. Thus, this enzyme will degrade all the cyclic AMP, if synthesis does not occur simultaneously. In the absence of glucose, cAMP is synthesized by adenylcyclase and the CRP-cAMP complex forms, activating transcription of the Operon. The presence of glucose indirectly inactivates adenylcyclase, which results in a rapid reduction of the amount of cAMP in the cell. If cAMP is absent, the cAMP-CRP-protein complex does not form and does not activate the lac Operon or other catabolite repressible Operons.

6. PROKARYOTES VS. EUKARYOTES

In prokaryotes regulation of transcription is needed for the cell to quickly adapt to the ever changing outer environment. The presence of the quantity and type of nutrients determines which genes are expressed. In order to do that, genes must be regulated in some fashion. In prokaryotes, repressors bind to regions called operators that are generally located downstream from the promoter (normally part of the transcript). Activators bind to the upstream portion of the promoter, such as the CAP region (completely upstream from the transcript). A combination of activators, repressors and rarely enhancers (in prokaryotes) determines whether a gene is transcribed. The controls that act on gene expression (i.e., the ability of a gene to produce a biologically active protein) are much more complex in eukaryotes than in prokaryotes. A major difference is the presence in eukaryotes of a nuclear membrane, which prevents the simultaneous transcription and translation that occurs in prokaryotes. Whereas, in prokaryotes, control of transcriptional initiation is the major point of regulation, in eukaryotes the regulation of gene expression is controlled nearly equivalently from many different points (Table 9.1).

Table 9.1. Difference between eukaryotic and prokaryotic gene regulation.

Prokaryotic	Eukaryotic
Prokaryotes have proteins that bind to DNA, DNA is free	Eukaryotes have histones and DNA is tightly folded into chromatin; requires defolding for expression
Prokaryotes have operons	In eukaryotes each gene has its own regulatory region.
Transcriptional factor and enhancer sequences absent	Transcriptional factor and enhancer sequences present
Prokaryotes do not have introns and no processing of mRNA	Introns are present and RNA is processed before expression

7. MUTATIONS IN OPERON

A mutation in any sequence of promoter or structural gene results in changed gene expression, e.g., altered protein, non-transcription etc. Some of the mutations are expressed below. These mutations are helpful in determining the role of a particular gene.

Operator mutations: The operator is a specific DNA sequence that is recognized by the lac-repressor protein. If this sequence changes (mutates) to a sufficient degree, the repressor can no longer bind to it. This type of operator mutation is denoted by Oc. The "c" stands for constitutive, since this type of operator mutation allows the Operon to be constitutively transcribed (always 'on' position). Operator mutations act in a *cis*-dominant mode, such that they are not overcome by introducing a normal copy of the gene in a merozygote.

Repressor mutations: Mutations that result in the repressor protein being unable to bind the operator locus (denoted I-) will always be recessive when in trans to a wild type copy of the gene, whose product will bind to both operators. Repressor protein mutations that are unable to bind allolactose (denoted IS) will remain bound to the operator locus, and will be dominant in trans.

Structural gene mutations: A mutation that blocks the activity of β-galactosidase will prevent the conversion of lactose to allolactose. Induction of the Operon by lactose will be blocked, but externally supplied allolactose will still serve as an inducer. The induced Operon will not generate a functional β-galactosidase, but the other two enzymes will be induced normally. A mutation that blocks the activity of the permease will block (or greatly impair) all induction unless artificial means are employed to get lactose into the cell.

8. THE *trp* OPERON

The *trp* Operon encodes the genes for the synthesis of tryptophan. This cluster of genes, like the *lac* Operon, is regulated by a repressor that binds to the operator sequences. However, differs significantly from *lac* Operon. The activity of the *trp* repressor for binding the operator region is enhanced when it binds to tryptophan; in this capacity, tryptophan is known as a corepressor. Since the activity of the *trp* repressor is enhanced in the presence of tryptophan, the rate of expression of the *trp* Operon is graded in response to the level of tryptophan in the cell. The *trp* Operon was found to possess secondary mechanism of gene – expression regulation known as attenuation.

In the *lac* Operon the level of gene expression (i.e., the concentration of the products of the Operon) varies from nearly zero without induction to several hundred times in presence of lactose. The operator and repressor control the level of gene expression. In the *trp* Operon, however, the attenuation mechanism adds another level of regulation of gene expression (~10 folds). Combined, induction and attenuation can vary the level of gene expression over 1000 fold range.

The *trp* Operon of *E. coli* was characterized (Fig. 9.8) by Charles Yanofsky and his colleagues. This is a large Operon (7kb transcripts) having five structural genes (*trp E*, *trp D*, *trp C*, *trp B*, *trp A*), a repressor gene (*trp R*) and an operator site (*trp O*). It is responsible for synthesizing amino acid tryptophan from chorismate. *E. coli* produce a polycistronic mRNA that has a life time of 3 minutes, which enables bacteria to respond quickly to the environment induced changes in the need for tryptophan.

The first level of control in the *trp* Operon is achieved by interaction between a 58 kD regulatory protein, encoded by the *trp R* cistron and *trp O* operator site. The interaction between the *trp R* protein and the *trp O* site regulates the production of tryptophan. When tryptophan is in excess in the medium the Operon is turned off by the repressor protein- tryptophan complex binding to the operator site. In this complex, tryptophan is said to be corepressor. Resultantly, there would be no tryptophan synthesis. When there is low level of tryptophan in the medium, the repressor tryptophan complex does not form and the *trp* Operon forms mRNA by transcription and synthesizes tryptophan.

The attenuator region, which is composed of sequences found within the transcribed RNA, is involved in controlling transcription from the Operon after RNA polymerase has initiated synthesis. The attenuator of sequences of the RNA is found near the 5′ end of the RNA termed the leader region of the RNA. The leader sequences are located prior to the start of the coding region for the first gene of the Operon (the *trp E* gene). The attenuator region contains codons for a small leader polypeptide that contains tandem tryptophan codons. This region of the RNA is also capable of forming several different stable stem-loop structures.

Coordinated transcription of these five genes is controlled in two very different ways: (*i*) through the use of an end-product activated repressor protein; and (*ii*) through the use of a novel leader-attenuator mechanism.

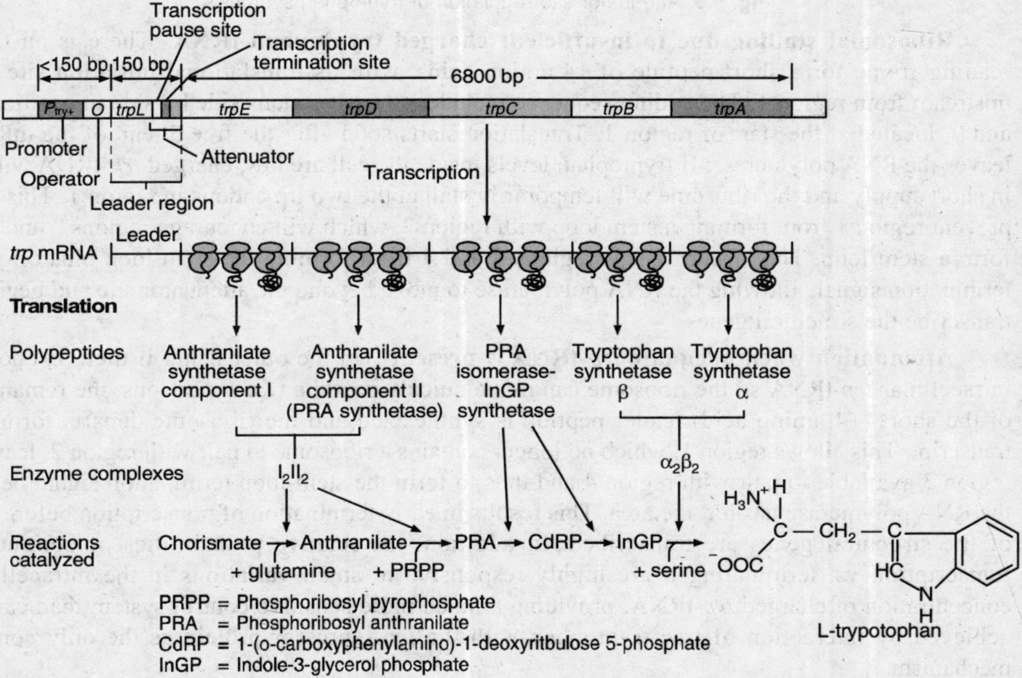

Fig. 9.8. Structure of tryptophan Operon and synthesis of tryptophan.

Attenuation of transcription in the *trp* operon- The first part of the *trp* transcript includes four regions (1-4) that can form two stem loops by pairing 1:2 and 3:4, or a single stem loop by pairing 2:3, with regions 1 and 4 unpaired. Immediately following region 4, there is a sequence of seven uracils (U). If the 3:4 stem loop is allowed to form as regions 3 and 4 are being transcribed, that loop and the oligo-U sequence that is synthesized immediately afterward generate a prokaryotic transcription termination signal. This causes transcription to stop before it reaches the structural genes for the five enzymes coded by the *trp* operon, which begin just downstream from the oligo-U sequence. The leader-attenuator system operates by controlling the extent of formation of the 3:4 loop, and thus of the termination signal, as shown in the Figure 9.9.

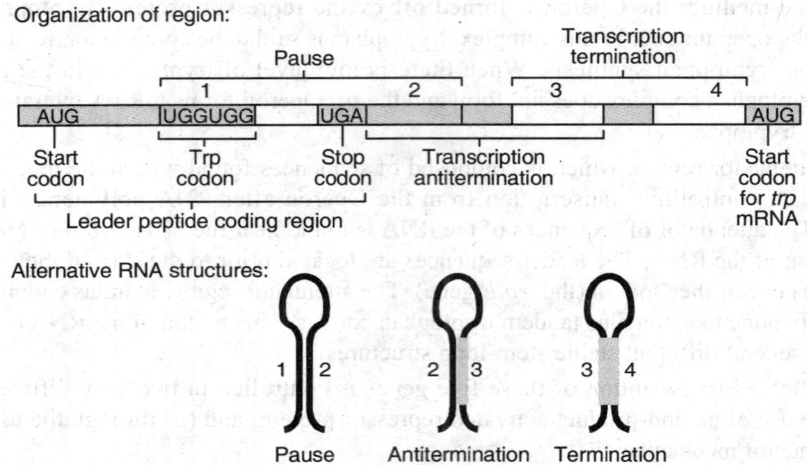

Fig. 9.9. Attenuator site regulation of tryptophan synthesis.

Ribosomal stalling due to insufficient charged tryptophan-tRNA- There is an open reading frame for a short peptide of 14 amino acids, with its translational initiation site just upstream from region 1. The coding sequence includes two *trp* codons side by side in positions 8 and 9, located at the start of region 1. Translation starts soon after the free 5′-end of the mRNA leaves the RNA polymerase. If tryptophan levels inside the cell are low, charged *trp*-tRNA will be in short supply and the ribosome will temporarily stall at the two trp codons in region 1. This will prevent region 1 from forming a stem loop with region 2, which will encourage regions 2 and 3 to form a stem loop. This in turn keeps regions 3 and 4 from forming the stem-loop plus oligo-U termination signal, allowing the RNA polymerase to move beyond the attenuator site and begin to transcribe the structural genes.

Attenuation when adequate *trp*-tRNA is present- On the other hand, if there is enough intracellular *trp*-tRNA so the ribosome can move quickly past the two *trp* codons, the remainder of the short (14 amino acid) leader peptide is synthesized and the ribosome departs form the transcript. This allows region 1, which no longer contains a ribosome to pair with region 2, leaving region 3 available to pair with region 4 and thus to form the stem-loop termination signal before the RNA polymerase has left the area. This results in early termination of transcription before any of the structural genes are transcribed. In addition, the relative probabilities of continuing transcription vs. terminating it are highly responsive to small variations in the intracellular concentration of charged *trp*-tRNA, providing a much more sensitive control system than can be achieved by interaction of free tryptophan with the *trp* repressor protein as the only control mechanism.

As described at the beginning of the chapter three dimensional structures of some proteins is changed by binding to some small molecules. This is a conformational change of the 3-D structure of the protein called an allosteric transition and such proteins are known as allosteric proteins. This is a post-translational regulation control mechanism.

QUESTIONS

1. What is an Operon? Describe the characteristic features of Operon which coordinately control the gene expression.
2. How does an inducible Operon differs from a constitutive Operon?
3. Describe the role of repressor in the lac Operon.
4. Describe the structure of lac Operon or trp Operon.
5. What is cis and trans acting gene?
6. Write short notes on:
   (a) Catabolic repression
   (b) Attenuation
   (c) Control points in gene expression

CHAPTER 10

DNA Damage and Repair Mechanisms

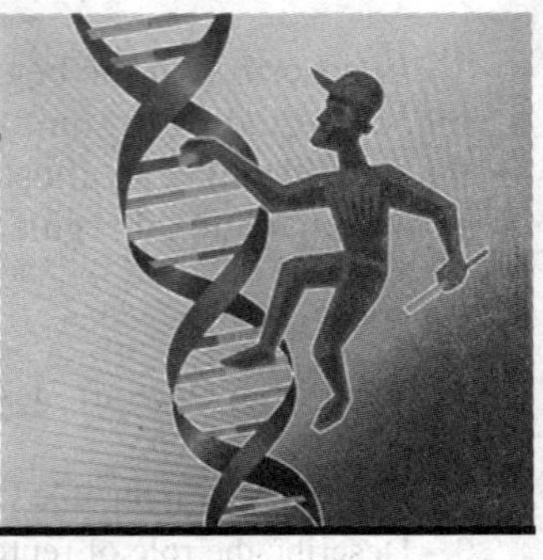

1. INTRODUCTION

As a major defense against environmental damage to cells, DNA repair is present in all organisms including bacteria, yeast, drosophila, fish, amphibians, rodents and humans. DNA repair is involved in processes that minimize cell killing, mutations, replication errors, persistence of DNA damage and genomic instability. Abnormalities in these processes have been implicated in cancer and aging (Fig. 10.1).

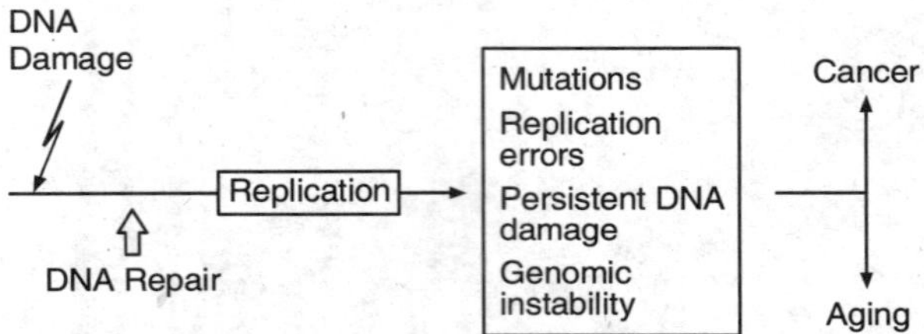

Fig. 10.1. DNA Repair functions.

Although genetic variation is important for evolution, the survival of the individual demands genetic stability. Maintaining genetic stability requires not only an extremely accurate mechanism for replicating DNA, but also mechanisms for repairing the many accidental lesions that occur continually in DNA. These mechanisms are collectively called DNA repair. Of the thousands of random changes created every day in the DNA of a human cell by heat, metabolic accidents, radiation of various sorts, and exposure to substances in the environment, only a few accumulate as mutations in the DNA sequence. We now know that less than one in 1000 accidental base changes in DNA results in a permanent mutation; the rest are eliminated with remarkable efficiency by DNA repair.

The importance of DNA repair is evident from the large investment that cells make in DNA repair enzymes. Analysis of the genomes of bacteria and yeasts has revealed that several percent of the coding capacity of these organisms is devoted solely to DNA repair functions. The importance of DNA repair is also demonstrated by the increased rate of mutation that follows the inactivation of a DNA repair gene.

Although DNA is a highly stable material, as required for the storage of genetic information, it is a complex organic molecule that is susceptible, even under normal cellular conditions, to spontaneous changes that would lead to mutations if left unrepaired. The DNA sequence can be changed as the result of copying errors introduced by DNA polymerases during replication and by environmental agents such as mutagenic chemicals and certain types of radiation.

DNA undergoes major changes as a result of thermal fluctuations: for example, about 5000 purine bases (adenine and guanine) are lost every day from the DNA of each human cell because their N-glycosyl linkages to deoxyribose hydrolyze, a spontaneous reaction called depurination.

Similarly, a spontaneous deamination of cytosine to uracil in DNA occurs at a rate of about 100 bases per cell per day. DNA bases are also occasionally damaged by an encounter with reactive metabolites (including reactive forms of oxygen) or environmental chemicals. Likewise, ultraviolet radiation from the sun can produce a covalent linkage between two adjacent pyrimidine bases in DNA to form, for example, thymine dimmers. If left uncorrected when the DNA is replicated, most of these changes would be expected to lead either to the deletion of one or more base pairs or to a base-pair substitution in the daughter DNA chain. The mutations would then be propagated throughout subsequent cell generations as the DNA is replicated. Such a high rate of random changes in the DNA sequence would have disastrous consequences for an organism. Moreover, the DNA in germ cells might acquire too many damages for viable offspring to be formed. Thus, the correction of DNA sequence errors in all types of cells is important for survival.

2. TYPES OF DNA DAMAGE

Damage to DNA can result from several different types of processes. Hydrolysis, deamination, alkylation, and oxidation are all capable of causing a modification in one or more bases in a DNA sequence.

2.1. Hydrolysis

DNA consists of long strands of sugar molecules called deoxyribose that are linked together by phosphate groups. Each sugar molecule carries one of the four natural DNA bases: adenine, guanine, cytosine, or thymine (A, G, C, or T). The chemical bond between a DNA base and its respective deoxyribose, although relatively stable, is nonetheless subject to chance cleavage by a water molecule in a process known as spontaneous hydrolysis. Loss of the "purine" bases (guanine and adenine) is referred to as depurination, whereas loss of the "pyrimidine" bases (cytosine and thymine) is called depyrimidination. In mammalian cells, it is estimated that depurination occurs at the rate of about 10,000 purine bases lost per cell generation. The rate of depyrimidination is considerably slower, resulting in the loss of about 500 pyrimidine bases per cell generation.

The baseless sugars that result from these processes are commonly referred to as AP-sites (apurinic/apyrimidinic). They are potentially lethal to the cell, as they act to block the progress of DNA replication, but are efficiently repaired in a series of enzyme-catalyzed reactions collectively referred to as the base excision repair (BER) pathway. In fact, AP-sites are intentionally created during the course of BER.

2.2. Deamination

The bases that make up DNA are also vulnerable to modification of their chemical structure. One form of modification, called spontaneous deamination, is the loss of an amino group ($-NH_2$). For example, cytosine (C), which is paired with guanine (G) in normal, double-stranded DNA, has an amino group attached to the fourth carbon (C4) of the base. When that amino group is lost, either through spontaneous, chemical, or enzymatic hydrolysis, a uracil (U) base is formed, and a normal C-G DNA base pair is changed to a premutagenic U-G base pair (uracil is not a normal part of DNA).

The U-G base pair is called premutagenic because if it is not repaired before DNA replication, a mutation will result. During DNA replication, the DNA strands separate, and each strand is copied by a DNA polymerase protein complex. On one strand, the uracil (U) will pair with a new adenine (A), while on the other strand the guanine (G) will pair with a new cytosine(C). Thus, one DNA double-strand contains a normal C-G base pair, but the other double-strand has a mutant U-A base pair. This process is called mutation fixation, and the mutation of the G to an A is said to be fixed (meaning "fixed in place," not "repaired"). In other words, the cell now accepts the new mutant base pair as normal. It is estimated that approximately 400 cytosine deamination events per genome occur every day. Clearly, it is very important for the cell to repair DNA damage before DNA replication commences, in order to avoid mutation fixation. One cause of normal human aging is the gradual accumulation over time of mutations in our cellular DNA.

2.3. Alkylation

Another type of base modification is alkylation. Alkylation occurs when a reactive mutagen transfers an alkyl group (typically a small hydrocarbon side chain such as a methyl or ethyl group, denoted as $-CH_3$ and $-C_2H_5$, respectively) to a DNA base. The nitrogen atoms of the purine bases (N_3 of adenine and N_7 of guanine) and the oxygen atom of guanine (O_6) are particularly susceptible to alkylation in the form of methylation. Methylation of DNA bases can occur through the action of exogenous (environmental) and endogenous (intracellular) agents. For example, exogenous chemicals such as dimethylsulfate, used in many industrial processes and formed during the combustion of sulfur-containing fossil and N-methyl-N-nitrosoamine, a component of tobacco smoke, are powerful alkylating agents. These chemicals are known to greatly elevate mutation rates in cultured cells and cause cancer in rodents.

Inside every cell is a small molecule known as S-adenosylmethionine or "SAM" SAM, which is required for normal cellular metabolism, is an endogenous methyl donor. The function of SAM is to provide an activated methyl group for virtually every normal biological methylation reaction. SAM helps to make important molecules such as adrenaline, a hormone secreted in times of stress; creatine, which provides energy for muscle contraction; and phosphatidylcholine, an important component of cell membranes. However, SAM can also methylate inappropriate targets, such as adenine and guanine. Such endogenous DNA-alkylation damage must be continually repaired; otherwise, mutation fixation can occur.

2.4. Oxidation

Oxidative damage to DNA bases occurs when an oxygen atom binds to a carbon atom in the DNA base. High-energy radiation, like X rays and gamma radiation, causes exogenous oxidative DNA base damage by interacting with water molecules to create highly reactive oxygen species, which then attack DNA bases at susceptible carbon atoms. Oxidative base damage is also endogenously produced by reactive oxygen species released during normal respiration in mitochondria, the cell's "energy factories."

Table 10.1. DNA Lesions and their causes.

DNA damage	Example/Cause
Missing base	Removal of purines by acid and heat (under physiological conditions $\approx 10^4$ purines/day/cell in a mammalian genome); removal of altered bases (e.g., uracil) by DNA glycosylases
Altered base	Ionizing radiation; alkylating agents (e.g., ethylmethane sulfonate)
Incorrect base	Mutations affecting $3' \rightarrow 5'$ exonuclease proofreading of incorrectly incorporated bases
Bulge due to deletion or insertion of a nucleotide	Intercalating agents (e.g., acridines) that cause addition or loss of a nucleotide during recombination or replication
Linked pyrimidines	Cyclotubyl dimers (usually thymine dimers) resulting from UV irradiation
Single- or double-strand breaks	Breakage of phosphodiester bonds by ionizing radiation or chemical agents (e.g., bleomycin)
Cross-linked strands	Covalent linkage of two strands by bifunctional alkylating agents (e.g., mitomycin C)
3′-deoxyribose fragments	Disruption of deoxyribose structure by free radicals leading to strand breaks

Humans enjoy a long life span; thus, it would seem that healthy, DNA repair-proficient cells could correct most of the naturally occurring endogenous DNA damage. Unfortunately, when levels of endogenous DNA damage are high, which might occur as the result of an inactivating mutation in a DNA repair gene, or when we are exposed to harmful exogenous agents like radiation or dangerous chemicals, the cell's DNA repair systems become overwhelmed. Lack of DNA repair results in a high mutation rate, which in turn may lead to cell death, cancer, and other diseases. Also, if the level of DNA repair activity declines with age, then the mutational burden of the cell will increase as we grow older (Table 10.1).

3. DNA DAMAGE REMOVAL BY DIFFERENT MECHANISM

3.1. Proofreading by DNA Polymerase

Because the specificity of nucleotide addition by DNA polymerases is determined by Watson-Crick base pairing, a wrong base (e.g., A instead of G) occasionally is inserted during DNA synthesis. Indeed, a subunit of *E. coli* DNA polymerase III introduces about 1 incorrect base in 104 internucleotide linkages during replication in vitro. Since an average *E. coli* gene is about 103 bases long, an error frequency of 1 in 10^4 base pairs would cause a potentially harmful mutation in every tenth gene during each replication, or 10^{-1} mutations per gene per generation. However, the measured mutation rate in bacterial cells is much less, about 1 mistake in 10^9 nucleotide polymerization events or, equivalently, 10^{-5} to 10^{-6} mutations per gene per generation (assuming ˜1000 base pairs per gene).

This increased accuracy *in vivo* is largely due to the proofreading function of *E. coli* DNA polymerases. An experiment demonstrated that the 3′ → 5′ exonuclease activity of *E. coli* DNA polymerase I can remove a mismatched base at the 3′ growing end of a synthetic primer-template complex. In DNA polymerase III, this function resides in the e subunit of the core polymerase. When an incorrect base is incorporated during DNA synthesis, the polymerase pauses, then transfers the 3′ end of the growing chain to the exonuclease site where the mispaired base is removed. Then the 3′ end is transferred back to the polymerase site, where this region is copied correctly. Proofreading is a property of almost all bacterial DNA polymerases. Both the δ and ε DNA polymerases of animal cells, but not the polymerase, also have proofreading activity. It seems likely that this function is indispensable for all cells to avoid excessive genetic damage.

3.2. Excision Repairs

There are multiple pathways for DNA repair, using different enzymes that act upon different kinds of lesions. Two of the most common pathways are shown in figure 10.3 A,B. In both, the damage is excised, the original DNA sequence is restored by a DNA polymerase that uses the undamaged strand as its template, and the remaining break in the double helix is sealed by DNA ligase (Fig. 10.2). As shown, DNA ligase uses a molecule of ATP to activate the 5′ end at the nick (step 1) before forming the new bond (step 2). In this way, the energetically unfavorable nick-sealing reaction is driven by being coupled to the energetically favorable process of ATP hydrolysis.

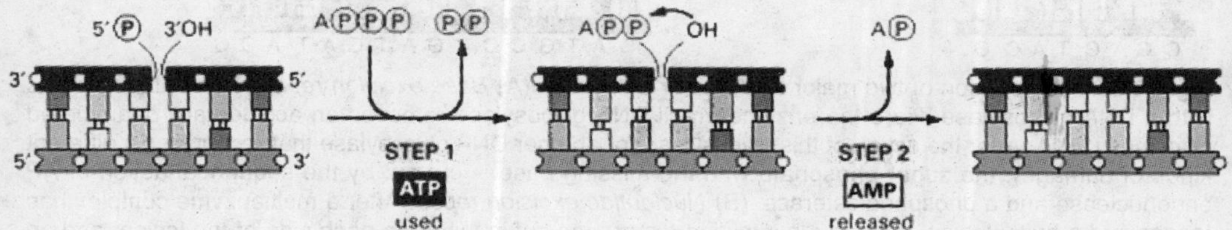

Fig. 10.2. Repair by DNA ligase to seal the nick making a phosphodiester bond.

The two pathways differ in the way in which the damage is removed from DNA. The first pathway, called **base excision repair**, involves a battery of enzymes called *DNA glycosylases*, each of which can recognize a specific type of altered base in DNA and catalyze its hydrolytic removal. There are at least six types of these enzymes, including those that remove deaminated Cs, deaminated As, different types of alkylated or oxidized bases, bases with opened rings, and bases in which a carbon-carbon double bond has been accidentally converted to a carbon-carbon single bond.

As an example of the general mechanism of base excision repair, the removal of a deaminated C by uracil DNA glycosylase is shown in figure. 10.3 A. It is thought that an altered base is detected when DNA glycosylases travel along DNA using base-flipping to evaluate the status of each base pair. Once a damaged base is recognized, the DNA glycosylase reaction creates a deoxyribose sugar that lacks its base. This "missing tooth" is recognized by an enzyme called AP endonuclease, which cuts the phosphodiester backbone, and the damage is then removed and repaired. Depurination, which is by far the most frequent type of damage suffered by DNA, also leaves a deoxyribose sugar with a missing base. Depurinations are directly repaired beginning with AP endonuclease, following the bottom half of the pathway.

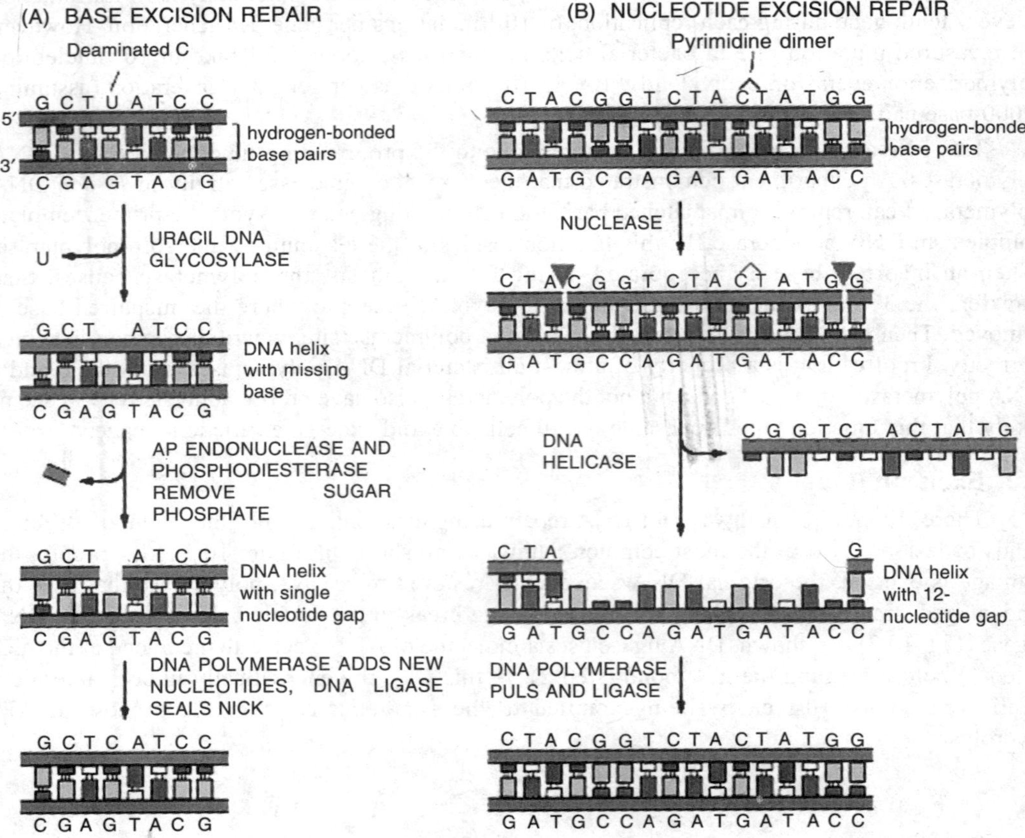

Fig. 10.3. A comparison of two major DNA repair pathways. (A) *Base excision repair*. This pathway starts with a DNA glycosylase. Here the enzyme uracil DNA glycosylase removes an accidentally deaminated cytosine in DNA. After the action of this glycosylase (or another DNA glycosylase that recognizes a different kinds of damage), the sugar phosphate with the missing base is cut out by the sequential action of AP endonuclease and a phosphodiesterase. (B) *Nucleotide excision repair*. After a multienzyme complex has recognized a bulky lesion such as a pyrimidine dimer, one cut is made on each side of the lesion, and an associated DNA helicase then removes the entire portion of the damaged strand.

The second major repair pathway is called **nucleotide excision repair**. This mechanism can repair the damage caused by almost any large change in the structure of the DNA double helix. Such "bulky lesions" include those created by the covalent reaction of DNA bases with large hydrocarbons (such as the carcinogen benzopyrene), as well as the various pyrimidine dimers (T-T, T-C, and C-C) caused by sunlight. The DNA helix distortions are recognized by the UvrABC endonuclease, a multisubunit enzyme encoded by the three genes uvrA, uvrB, and uvrC. This enzyme makes one cut in the damaged DNA strand, the phosphodiester backbone of the abnormal strand is cleaved on both sides of the distortion, and an oligonucleotide containing the lesion is peeled away from the DNA double helix by a DNA helicase enzyme. The large gap produced in the DNA helix is then repaired by DNA polymerase I and sealed by DNA ligase (Fig. 10.3 B).

3.3. Spontaneous Repair by DNA Double Helix

The double-helical structure of DNA is ideally suited for repair because it carries two separate copies of all the genetic information-one in each of its two strands. Thus, when one strand is damaged, the complementary strand retains an intact copy of the same information, and this copy is generally used to restore the correct nucleotide sequences to the damaged strand. The nature of the bases also facilitates the distinction between undamaged and damaged bases. Thus, every possible deamination event in DNA yields an unnatural base, which can therefore be directly recognized and removed by a specific DNA glycosylase. An indication of the importance of a double-stranded helix to the safe storage of genetic information is that all cells use it; only a few small viruses use single-stranded DNA or RNA as their genetic material. The chance of a permanent nucleotide change occurring in these single-stranded genomes of viruses is thus very high.

3.4. Homologous and Non-homologous DNA Repair

A potentially dangerous type of DNA damage occurs when both strands of the double helix are broken, leaving no intact template strand for repair. Breaks of this type are caused by ionizing radiation, oxidizing agents, replication errors, and certain metabolic products in the cell. If these lesions were left unrepaired, they would quickly lead to the breakdown of chromosomes into smaller fragments. However, two distinct mechanisms have evolved to restructure the potential damage. The simplest to understand is nonhomologous end-joining, in which the broken ends are juxtaposed and rejoined by DNA ligation, generally with the loss of one or more nucleotides at the site of joining (Fig. 10.4) This end-joining mechanism, which can be viewed as an emergency solution to the repair of double-strand breaks, is a common outcome in mammalian cells. Although a change in the DNA sequence (a mutation) results at the site of breakage, so little of the mammalian genome codes for proteins that this mechanism is apparently an acceptable solution to the problem of keeping chromosomes intact. The specialized structure of telomeres prevents the ends of chromosomes from being mistaken for broken DNA, thereby preserving natural DNA ends.

An even more effective type of double-strand break repair exploits the fact that cells that are diploid contain two copies of each double helix. In this second repair pathway, called homologous end-joining, general recombination mechanisms come into play that transfer nucleotide sequence information from the intact DNA double helix to the site of the double-strand break in the broken helix. This type of reaction requires special recombination proteins that recognize areas of DNA sequence matching between the two chromosomes and bring them together. A DNA replication process then uses the undamaged chromosome as the template for transferring genetic information to the broken chromosome, repairing it with no change in the DNA sequence. In cells that have replicated their DNA but not yet divided, this type of DNA repair can readily take place between the two sister DNA molecules in each chromosome; in this case, there is no need for the broken

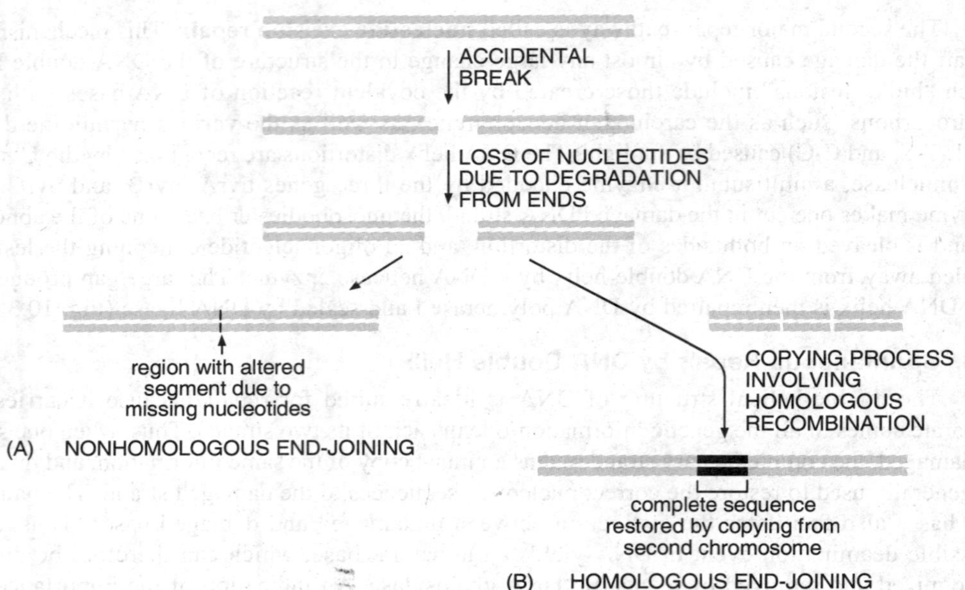

Fig. 10.4. Two different types of end-joining for repairing double-strand breaks. (A) Nonhomologous end-joining alters the original DNA sequence when repairing broken chromosomes. These alterations can be either deletions (as shown) or short insertions. (B) Homologous end-joining is more difficult to accomplish, but is much more precise.

ends to find the matching DNA sequence in the homologous chromosome. Such type of repair is called as **recombination repair**.

3.5. SOS Repair

Cells have evolved many mechanisms that help them survive in an unpredictably hazardous world. Often an extreme change in a cell's environment activates the expression of a set of genes whose protein products protect the cell from the deleterious effects of this change. One such mechanism shared by all cells is the heat-shock response, which is evoked by the exposure of cells to unusually high temperatures. The induced "heat-shock proteins" include some that help stabilize and repair partly denatured cell proteins.

Cells also have mechanisms that elevate the levels of DNA repair enzymes, as an emergency response to severe DNA damage. The best-studied example is the so-called SOS response in *E. coli*. The SOS response is a post replication DNA repair system that allows DNA replication to bypass lesions or errors in the DNA. The SOS uses the *Rec*A protein. The signal (thought to be an excess of single-stranded DNA) first activates the RecA protein. The *Rec*A protein, stimulated by single-stranded DNA, is involved in the activation of the *Lex*A thereby inducing the response. It is an error-prone repair system.

During normal growth, the SOS genes are negatively regulated by LexA repressor protein dimers. Under normal conditions, LexA binds to a 20 bp consensus sequence (the SOS box) in the operator region for those genes. Activation of the SOS genes occurs after DNA damage by the accumulation of single stranded (ssDNA) regions generated at replication forks, where DNA polymerase is blocked. RecA forms a filament around these ssDNA regions in an ATP-dependent fashion, and becomes activated. The activated form of *Rec*A interacts with the *Lex*A to facilitate the LexA self-cleavage from the operator (Fig. 10.5).

Once the LexA protein are cleaved then no repression of the SOS genes occurs. Operators that bind *Lex*A weakly are the first to be fully expressed. In this way *Lex*A can sequentially activate different mechanisms of repair. Genes having a weak SOS box (such as *lex*A, *rec*A, *uvr*A, *uvr*B, and *uvr*D) are fully induced in response to even weak SOS-inducing treatments. Thus the first SOS repair mechanism to be induced is nucleotide excision repair (NER), whose aim is to fix DNA damage without commitment to a full-fledged SOS response.

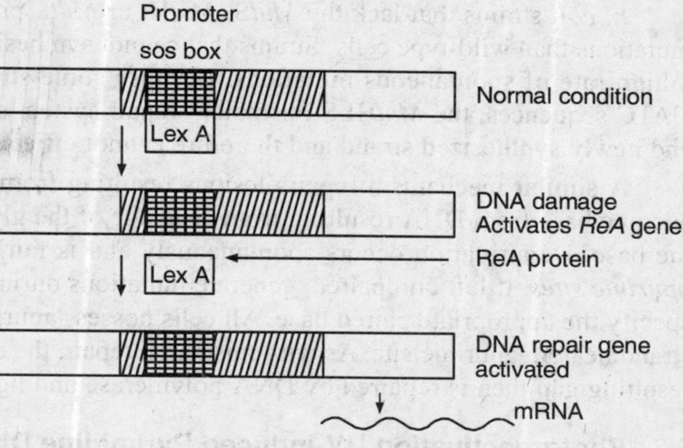

Fig. 10.5. Mechanism of SOS repair.

Cells have an additional mechanism that helps them respond to DNA damage: they delay progression of the cell cycle until DNA repair is complete. For example, one of the genes expressed in response to the *E. coli* SOS signal is *sul*A, which encodes an inhibitor of cell division. Thus, when the SOS functions are turned on in response to DNA damage, a block to cell division extends the time for repair. When DNA repair is complete, the expression of the SOS genes is repressed, the cell cycle resumes, and the undamaged DNA is segregated to the daughter cells. While this "error-prone" DNA repair can be harmful to individual bacterial cells, it is presumed to be advantageous in the long term because it produces a burst of genetic variability in the bacterial population that increases the likelihood of a mutant cell arising that is better able to survive in the altered environment.

3.6. Mismatch Repair of Single-Base Mispairs

Many spontaneous mutations are point mutations, which involve a change in a single base pair in the DNA sequence. These can arise from errors in replication, during genetic recombination, and, particularly, by base deamination whereby a C residue is converted into a U residue. The conceptual problem with mismatch repair is determining which is the normal and which is the mutant DNA strand, and repairing the latter so that it is properly base-paired with the normal strand. How this is accomplished has been elucidated in considerable detail for the *E. coli* methyldirected mismatch repair system, often referred to as the *Mut*HLS system.

In *E. coli* DNA, adenine residues in a GATC sequence are methylated at the 6 position. Since DNA polymerases incorporate adenine, not methyl-adenine, into DNA, adenine residues in newly replicated DNA are methylated only on the parental strand. The adenines in GATC sequences on the daughter strands are methylated by a specific enzyme, called *Dam methyltransferase*, only after a lag of several minutes. During this lag period, the newly replicated DNA contains hemimethylated GATC sequences:

$$\begin{array}{ll} \quad\quad\quad\;\; CH_3 & \\ \quad\quad\quad\;\;\; | & \\ 5' - G - A - T - C - 3' & \text{Parental strand} \\ 3' - C - T - A - G - 5' & \text{Daughter strand} \end{array}$$

An *E. coli* protein designated *MutH*, which binds specifically to hemimethylated sequences, is able to distinguish the methylated parental strand from the unmethylated daughter strand. If an error occurs during DNA replication, resulting in a mismatched base pair near a GATC sequence, another protein, *Mut*S, binds to this abnormally paired segment. Binding of *Mut*S triggers binding of *Mut*L, a linking protein that connects *Mut*S with a nearby *Mut*H. This cross-linking activates a latent endonuclease activity of *Mut*H, which then cleaves specifically the unmethylated daughter strand. Following this initial incision, the segment of the daughter strand containing the misincorporated base is excised and replaced with the correct DNA sequence.

E. coli strains that lack the *Mut*S, *Mut*H, or *Mut*L protein have a higher rate of spontaneous mutations than wild-type cells. Strains that cannot synthesize the Dam methyltransferase also have a high rate of spontaneous mutations. Because some strains cannot methylate adenines within GATC sequences, the *Mut*HLS mismatch- repair system cannot distinguish between the template and newly synthesized strand and therefore cannot efficiently repair mismatched bases.

A similar mechanism repairs lesions resulting from *depurination*, the loss of a guanine or adenine base from DNA resulting from cleavage of the glycosidic bond between deoxyribose and the base. Depurination occurs spontaneously and is fairly common in mammals. The resulting *apurinic sites*, if left unrepaired, generate mutations during DNA replication because they cannot specify the appropriate paired base. All cells possess apurinic (AP) endonucleases that cut a DNA strand near an apurinic site. As with mismatch repair, the cut is extended by exonucleases, and the resulting gap then is repaired by DNA polymerase and ligase.

3.7. Photoreactivation UV-induced Pyrimidine Dimers

Direct correction of damaged part can occur also in the repair of UV light induced thymine dimers. By this process of photoreactivation or light repair, the dimers are reverted directly to the original form by exposure to visible light in the range of 320-370 nm. Photoreactivation is catalysed by an enzyme called photolyase encoded by *phr* gene. When this dimer is activated by a photon of light, it splits the dimer apart. Bacterial strains with mutations in the *phr* gene are defective in light repair. Photolyase has been reported in prokaryotes and in lower eukaryotes but not in humans. Experimental bacteria are kept in dark to avoid photoreactivation of mutated DNA (Fig. 10.6).

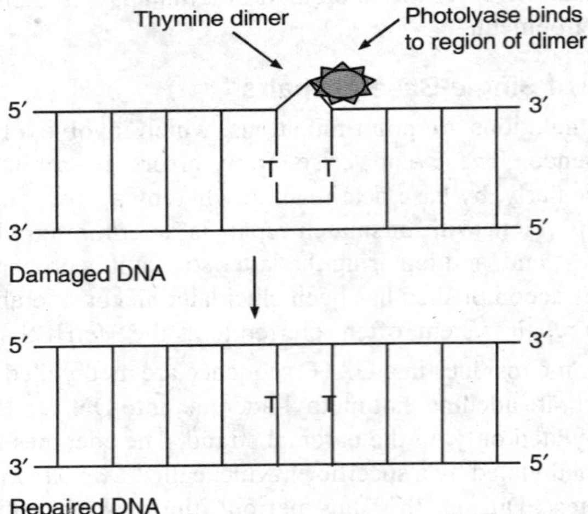

Fig. 10.6. Light activation of enzyme photolyase for repair of damaged DNA.

QUESTIONS

1. What are different factors that cause DNA damage?
2. Explain different DNA repair mechanisms.
3. Compare the base excision repair with nucleotide excision repair.
4. Write short notes on:
   (a) SOS repair
   (b) Photo repair
   (c) Recombination repair
   (d) Excision repair
   (e) Mismatch repair

CHAPTER 11

Biological Nitrogen Fixation and its Genetic Engineering

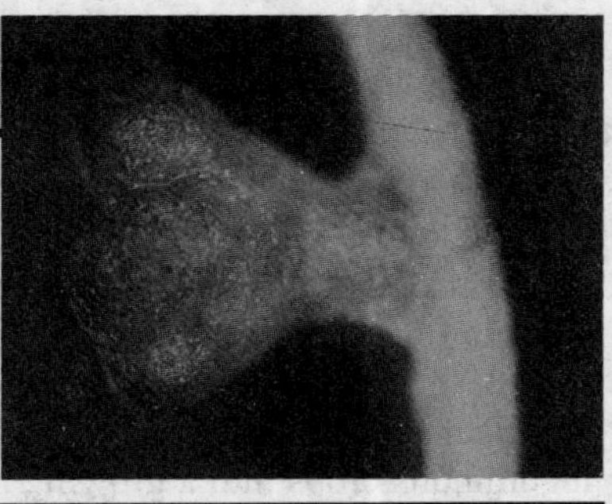

1. INTRODUCTION

There are economic, environmental, and agronomic benefits from using BNF in cropping systems. All living things require nitrogen. It is a key element of amino acids, the building blocks of proteins. There is a huge reservoir of N in the atmosphere. Approximately 80% of the air consists of nitrogen gas (N_2), but plants cannot use atmospheric N directly to make protein. The gaseous N must first be converted, or "fixed," into forms plants can use. Unless fertilizer N is applied, most plants obtain their nitrogen from natural sources in the soil. Natural reserves of soil N are normally low, so N fertilizers must be added to increase plant growth. Nature has an alternative method of providing N to plants and enriching soil N resources-biological nitrogen fixation. Many members of the legume plant family, such as beans, peas, alfalfa, and leucaena, have the special ability to use BNF to meet their N needs.

2. NITROGEN FIXING BACTERIA

The free-living bacteria having the ability to fix molecular nitrogen (Table 11.1) can be distinguished into obligate aerobic, facultative aerobic and anaerobic organisms. Obligate aerobic bacteria belong to the genera *Azotobacter*, *Beijerinckia*, *Derxia*, *Achromobacter*, *Mycobacterium*, *Arthrobacter* and *Bacillus*.

Table 11.1. Well-studied free-living nitrogen-fixing bacteria.

Species	Super family	Aerobic Anaerobic Ae-Ana	Characteristics
Clostridium pasteurianum	Firmibacterial (Low GC Gram positive)	Ana	Isolated first (1893) A cell free-extract nitrogenase was first made (1962)
Klebsiella pneumoniae	Proteobacteria-gamma	Ana	Close to *E. coli*. Genetics was first studied
Klebsiella oxytoca	Proteobacteria-gamma	Ana	Isolated from rice root in Japan
Azobacter chroococcum	Proteobacteria-gamma	Ae	Azotobacter was isolated next (1901). Widely used for research
Azospirilium brazilense	Proteobacteria-alfa	Ae	Isolated from rhizosphere of C4-plant. Widely studied as rhizosphere bacteria
Rhodospirillum rubrum	Proteobacteria-alfa	Ana	Non-S-photosynthetic bacteria. Active in H_2 production
Rhodobacter capsulatus	Proteobacteria-alfa	Ana	Non-S-photosynthetic bacteria. Active in H_2 production
Anabaena sp. 7120	Cyanobacteria	Ae	Heterocyst-forming. Most well studied among cyanobacteria

Among the facultative anaerobic bacteria are the genera *Aerobacter*, *Klebsiella* and *Pseudomonas*. Anaerobic nitrogen-fixing bacteria are represented by the genera *Clostridium*, *Chlorobium*, *Chromatium*, *Rhodomicrobium*, *Rhodopseudomonas*, *Rhodospirillum*, *Desulfovibrio* and *Methanobacterium*.

Nitrogen is an essential plant nutrient. It is the nutrient that is most commonly deficient, contributing to reduced agricultural yields throughout the world. Developing countries used more than 85 million metric tones of nitrogenous fertilizer in 2003, worth billions of US dollars. Such fertilizer expenditure can be significantly reduced by incorporating biological nitrogen fixed leguminous crops into a growing rotation.

For example, 1.4 millions tonnes of N, worth approximately US $400 million/year is fixed by 14 million ha of soybean in Brazil (Table 11.2). This means a substantial saving for farmers in terms of reducing N fertiliser use. Many agriculturally important plants in the legume family can use nitrogen (N) from the atmosphere for growth through biological nitrogen fixation (BNF). Legume BNF involves a symbiosis between legume plants and the rhizobia that live in nodules on their roots.

Table 11.2. Estimates of the amount of nitrogen fixed by various legumes.

Plant	Scientific name	Nitrogen fixed (kg N/ha/yr)
Horse bean	*Vicia faba*	45-552
Pigeon pea	*Cajanus cajan*	168-280
Cowpea	*Vigna unguiculata*	73-354
Mung bean	*Vigna mungo*	63-342
Soybean	*Glycine max*	60-168
Chickpea	*Cicer arietinum*	103
Pea	*Pisum sativum*	55-77
Alfalfa	*Medicago sativa*	229-290

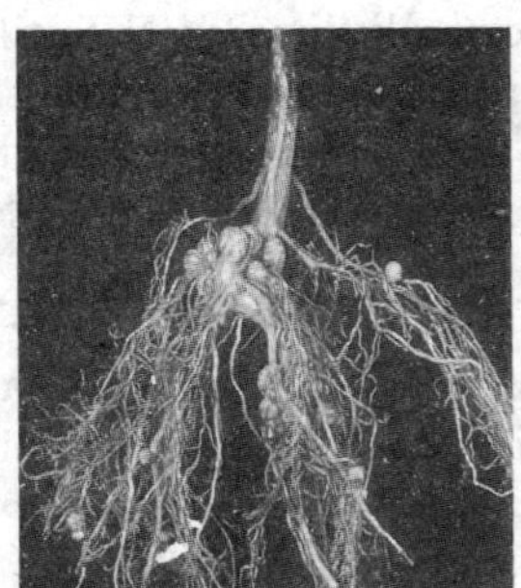

Fig. 11.1. Root nodule on roots of a legume plant.

Legume BNF involves a remarkable symbiosis, or mutually beneficial relationship, between the plant and N-fixing soil bacteria called rhizobia. The rhizobia invade the host plant's roots and cause the formation of structures called nodules. Within these nodules (Fig. 11.1) the rhizobia use enzymes to biologically convert ("fix") N_2 gas from the air into a form of N that can be used by its plant host to make proteins. In turn, the plant provides the rhizobia with products of photosynthesis: sugars and carbohydrates that fuel the bacteria and the BNF process.

3. NITROGEN CYCLE

Free atmospheric nitrogen is involved in nitrogen cycle, where it forms inorganic nitrogen to be used by plants and again released in atmosphere after decomposition. Nitrogen content of different systems is as follows: Atmosphere:79/100 (as N_2), Plants:5-2/100(mostly organic), soil: 0.5-5/1000(mostly organic).

- N gain process: From N_2 to microorganisms, and eventually to plant
- N oxidation: ammonia is oxidized to nitrate (nitrification)
- N loss process: Nitrate is ultimately lost as N_2 (return to atmosphere)

Comparison of Industrial and biological nitrogen fixation - Microorganism and some plants with the help of microorganisms fix atmospheric nitrogen. In 1910-16, Habar-Bosch process to synthesize ammonia from N_2 and H_2 was established, and nitrogen fertilizer industry started. This is how large scale nitrogen is fixed and known as Industrial Nitrogen Fixation.

Both processes have common characters:

$$N_2 + (6H) = 2NH_3$$

Conditions for biological N_2 fixation: 30 degree, 1 atmosphere pressure, enzyme catalysts-nitrogenase and reducing agents are organic substances.

Conditions for industrial N_2 fixation: 300-400 degree, 500 atm, chemical catalysts-Fe, Al oxides. Reducing agent is hydrogen.

4. SYMBIOTIC NITROGEN FIXATION

The relationship between the plant and the bacteria is known as a symbiotic relationship since both the plant and the bacteria benefit from their relationship. Legumes including peas, lentils and alfalfa can form symbiotic associations for nitrogen fixation with a soil bacterium called *Rhizobium*. The *Rhizobium* enters into the roots and forms nodules, which the bacteria then use as their home (Fig. 11.2 A, B).

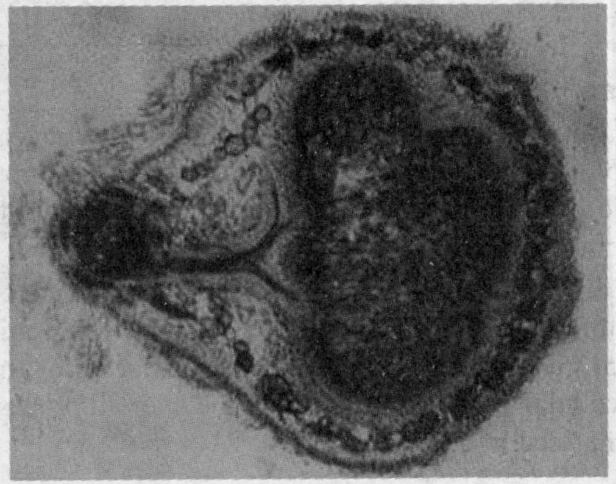

Fig. 11.2 A. T.S. of root showing nodule structure

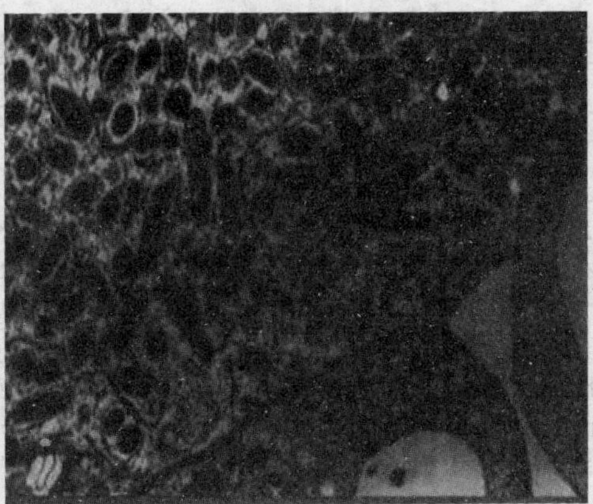

Fig. 11.2 B. *Rhizobium bacterioids in the cells of nodule*

In the nodules, the *Rhizobium* fixes N_2 into a form that the plant can use. Nitrogen is fixed by binding it to hydrogen and making it into ammonia, which the legume plant can use. Rhizobia benefit as the plant provides carbohydrates to the rhizobia. Carbohydrates are required by rhizobia as a source of energy. Also, the carbohydrates produced by the legume plant are transported to the nodules where they are used by the rhizobia as a source of hydrogen in the conversion of nitrogen to ammonia. This interaction is quite specific. Specific *Rhizobium* species fix nitrogen only with a specific type of legume.

This symbiotic relationship is very specific and this specificity requires a mechanism for host recognition (Table 11.3). This is mediated by plant compounds, usually flavonoids, which are produced by the roots of the host plant. These flavonoids influence a series of genes in *Rhizobium* known as the *nod* genes, the genes involved in infection and nodule formation. The rhizobia then bind to the root surface. In response to factors produced by the rhizobia, the root hair begins a curling growth, and the rhizobia continue to divide within the coils. An infection thread, which allows the rhizobia to move into the root hair and eventually to other plant root cells, is subsequently formed. Eventually the bacteria stop dividing and form bacteroids, which fix nitrogen. The *nod* genes are expressed at specific stages of development. The *nod* genes code for proteins called nodulins. Nodulins, including leghemoglobin, are expressed at specific stages of nodule development. There are a series of genes involved in nitrogen fixation, referred to as the nif genes or *fix* genes. The *nif* genes are found in both free-living and symbiotic nitrogen fixing bacteria and include the structural genes for nitrogenase and other regulatory genes.

Table 11.3. Host specificity of certain Rhizobium strains.

Genus	Species	hosts (Genus or species)
Rhizobium		
	R. meliloti	*Medicago* (Alfalfa), *Melilotus*, *Trigonella* spp.
	R. fredii	*Glycine max*, (Soybean), *Glycine soja*
	R. leguminosarum bv. viciae	*Vicia faba* (Faba bean), *Pisium sativa* (pea), *Lathyrus* spp.
	R. leguminosarum bv. trifolii	*Trifolium* spp. (clovers)
	R. leguminosarum bv. phaseoli	*Phaseolus vulgaris* (common bean)
Bradyrhizobium		
	B. japonicum	*Glycine max* (soybean), *Glycine soja* (wild soybean)
Azorhizobium		
	A. caulinodans	*Sesbania rostratastem nodules*

4.1. Structure and Operation of Nitrogenase

Active nitrogenase can be reconstituted by the addition of purified Mo-Fe and Fe proteins of different microorganisms. For examples, proteins of *Klebsiella pneumoniae* and *Bacillus polymyxa* and those of blue-green algae and photosynthetic bacteria have been combined to reconstitute active nitrogenases, capable of reducing acetylene to ethylene.

During catalysis by nitrogenase, protons and nitrogen compete for electrons. Therefore, in an atmosphere containing nitrogen, hydrogen evolution occurs simultaneously with ammonia formation. This evolution of hydrogen diverts 25-35% of the total reductants available for the nitrogenase reaction, which is regarded as an intracellular wastage of energy in the over all process of nitrogen fixation, and the reaction can be summarized in Figure 11.3.

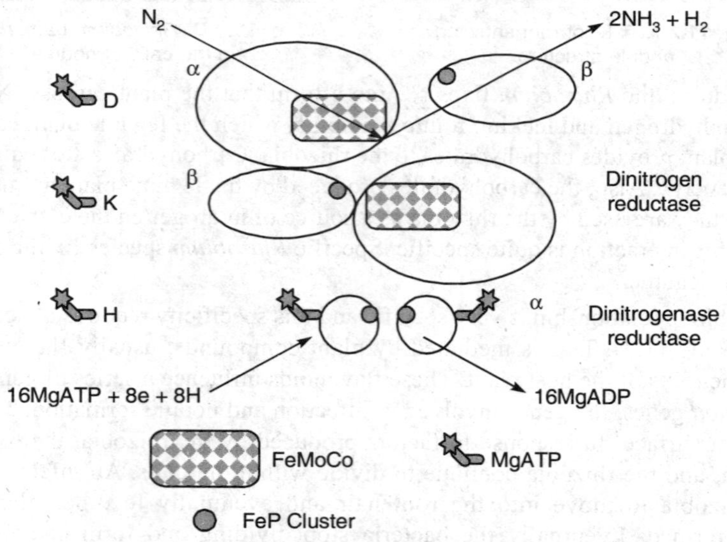

Fig. 11.3. Nitrogenase structure and reaction.

Nitrogenase contains the two proteins molybdoferredoxin and azoferredoxin (Fig. 11.3). This MoFe cofactor is unique to nitrogen fixation and distinct from the Mo-pterin cofactor of other Mo proteins (e.g., nitrate reductase, xanthine oxidase).

4.2. Organization of *Nif* Genes

Nitrogen fixation is carried out by three groups of genes. These are; Nod gene (responsible for nodule formation), *Nif* gene (responsible for nitrogen fixation) and *Hup* gene (responsible for nitrogen uptake). All these three types of genes are present in a group on a single chromosome. This makes their copying and transfer mechanism simple for genetic engineering purposes. Though the mechanism of nodule formation is complex nod gene is responsible for nodule formation as well as host recognization and specificity. However, a few genes located on plasmids can produce nodules. Plasmid of *R. leguminosarum* is less than 10kb even then it has property to recognize host and nodule formation.

4.2.1. *Nod* gene

Most of the biological nitrogen fixing bacteria contains a large plasmid called megaplasmid. In several functions it is similar to Ti plasmid and contains genes responsible for auxin and cytokinin production. Excess production of these plant growth regulators helps in nodule formation. According to Rosenberg (1981) several special genes are present along with nod genes. Such plasmids are absent in non-symbiotic bacteria. A nod gene is a group of genes containing Nod A, B, C, D genes having 8.5 kb length. These genes form polypeptides of different lengths (196, 197, 402, 211 amino acid). Nod genes of different rhizobium species have almost 70% homology which are called common Nod genes.

4.2.2. *Nif* genes

This gene is responsible for nitrogen fixation and present in the genome of symbiotic and non symbiotic nitrogen fixing bacteria. In symbiotic bacteria *Rhizobium*, it is present near nod genes on the megaplasmid, while in non-symbiotic cyanobacteria it is present on the main DNA. Initially *Nif* gene has been transferred in *E. coli*. In higher plants, chloroplast is a cell organelle which might have been originated from prokaryotes, therefore attempt are made to transfer *Nif* gene into chloroplast. Easy availability of ATP and $NADPH_2$ in chloroplast also makes them ideal recipient for this gene transfer. Most of the cereal plants are monocots and any such effort to transfer such *Nif* gene will revolutionize the yield, economics and environmental pollution. However, there are many difficulties in transferring, integration and expression of a prokaryotic gene into a monocot.

4.2.3. Hup gene

Gene responsible for nitrogen uptake is *Hup* gene. In symbiotic bacteria this gene recycles the hydrogen produced during nitrogen fixation as shown in the Figure 11.4. Hydrogen produced at different steps is assimilated in the reduction of nitrogen. In most of the legumes 30-50% energy (in the form of ATP) is spent on hydrogen liberation. This results in loss in capacity of nitrogen fixation. If this hydrogen can be recycled by nitrogenase enzyme we can save a lot of energy, and this can be carried out by improved Hup gene.

Klebsiella pneumoniae strain M5 a1(Enterobacteriaceae) is a free living bacteria which has been studied extensively for genetics of nitrogen fixation. This bacterial genome is quite similar to that of *E. coli* and *Salmonella typhimurium*. Therefore most of the techniques of genetic engineering can be applied to *Klebsiella*.

4.4. *Nif* Gene Organization in *Klebsiella*

Several mutants of *Klebsiella* were developed by growing the bacteria on medium containing a mutagen, methylnitro nitrosoguanidine. Different mutants obtained were used in transformation,

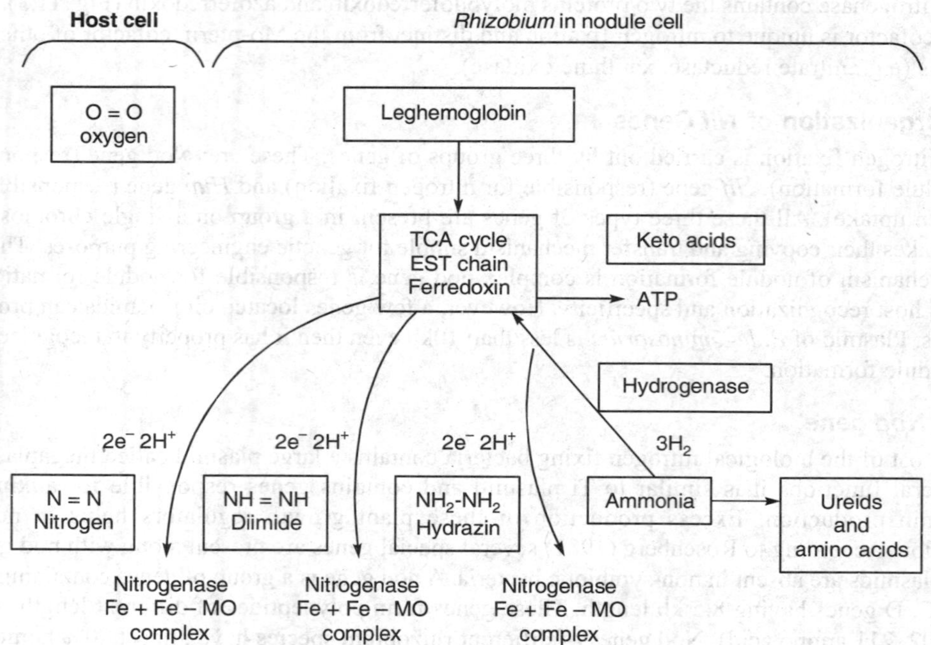

Fig. 11.4. Hydrogen evolution and utilization.

transduction to map the *Nif* gene (Table 11.4). This provided the information that *Nif* gene is downstream to histidine operator (Fig. 11.4). On the basis of this *Nif* gene of *Klebsiella* was transferred in *E. coli* on the basis of homology in the plasmid and not in genomic DNA. Nif gene organization of *Klebsiella* is similar to that of *Azotobacter*, *Asospirillum*, *Clostridium* and prokaryotic blue-green algae.

Table 11.4. Description and product of *Nif* genes of *Klebsiella*.

Nif H	Structural gene 35kd protein, a subunit of nitrogenase reductase.
Nif D	Structural gene 56kd, β subunit of nitrogenase.
Nif K	Structural gene 60kd, a subunit of nitrogenase, helps in transcription of Nif H-D and -K.
Nif E and N	46 and 50 kd protein formation; helps in formation of Fe-Mo co-factor and protein for Nif B.
Nif M and S	Formation of 18 kd protein which activates nitrogenase reductase.
Nif F	17 kd protein, function of electron transport factor.
Nif Q and V	Function unknown.
Nif A and L	Important role in expression of other *Nif* genes, product unknown.
Nif J	120 kd structural gene, important for activity of nitrogenase.

4.5. Regulation of Nitrogen Fixation

All the *nif* genes in *Klebsiella* are clustered and coordinately regulated (Fig. 11.5). *E. coli* to which the nif genes of *Klebsiella* have been transferred can fix N_2. In both the original *Klebsiella* and the *E. coli* nitrogenase is expressed only in the absence of both O_2 and NH_3 in the growth medium. The *nif* genes are regulated by the *nifLA* operon. The nitrogen regulators *Ntr*C (= GlnG), and *Ntr*B determine whether or not the *nifLA* operon is expressed (depending on the presence of ammonia or organic nitrogen). In the absence of ammonia or organic nitrogen the *Ntr*C protein is phosphorylated by the *Ntr*B protein. *Ntr*C-P then binds to the upstream region of the *nifLA* operon and activates transcription.

NtrA (= GlnF = RpoN = σ54) is the nitrogen sigma factor, which is needed for expression of the *nifLA* operon and the *nif* structural genes. NtrA is an alternative sigma factor used by RNA polymerase to recognize many genes involved in nitrogen metabolism which are not recognized by the standard sigma factor. The *nifA* gene encodes a protein required for switching on all of the *nif* genes except the regulatory genes *nifLA* themselves. If *nif*A protein is made, its function is to activate the other *nif* genes. The *nifL* gene is required for O_2 repression. In the absence of NifL protein, nitrogenase is made in the presence of O_2 (but is inactivated by O_2). When oxygen is present, the *nif*L protein binds to *nif*A and prevents it from activating the other *nif* genes.

The greater study of the infectious process and the genes involved and the physiological aspects will lead to better nitrogen use by plants grown on nitrogen-poor soils in agriculture and can reduce the requirement for chemical fertilizers.

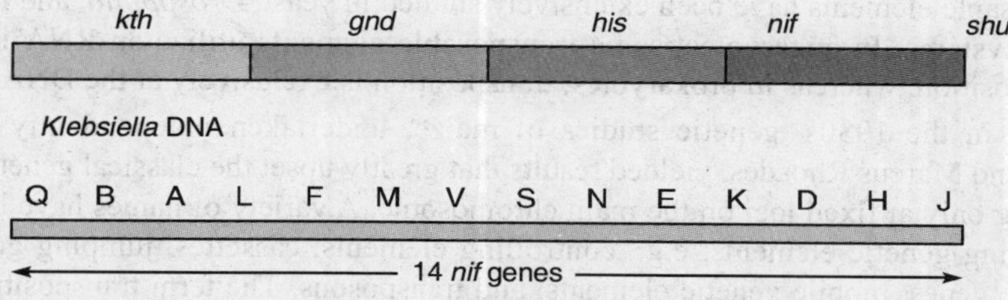

Fig. 11.5. Nif genes on *Klebsiella* DNA and details of Nif genes.

4.6. Transgenes with *Nif* Genes

It is important to develop transgenic plants containing *Nif* genes to solve the problem of nitrogen feriliser supplement to crop plants. This will have beneficial effects of economics and environment also. For this purpose, Ti based plasmid and Cauliflower mosaic virus based promoter (CAM promoter) was used to transfer *Nif* genes in to non-legume plants. To test the efficacy of this system, phaseolin gene from legume (pulses) has been transferred to sunflower where it was expressed and produced phaseolin. Protoplasts isolation from root nodules and preparations of rhizobia were used to develop hybrids by protoplasts fusion and organelles uptake. Major contributions were made by groups headed by Prof. E.C. Cocking, Prof. M.R. Davey, Dr. I. Portykus (inventor of Golden rice) and Prof. I.K.Vasil. However, true nitrogen fixing hybrids are yet to be obtained by this method.

QUESTIONS

1. What is biological nitrogen fixation? Compare it with chemical nitrogen fixation.
2. Write short notes:
   (a) Free living bacteria and nitrogen fixation. (b) Rhizobia and legumes.
   (c) Root nodule formation. (d) Biological nitrogen fixation.
   (e) Nif gene (f) Hup gene
   (g) Nod gene (h) Nitrogenase.
   (i) Klebsiella. (j) Nif genes in Klebsiella.

CHAPTER 12

Transposable Elements in Prokaryotes and Eukaryotes

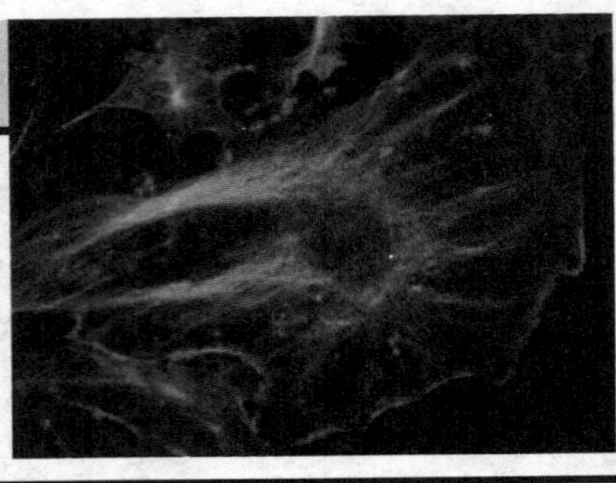

1. INTRODUCTION

A series of genetic elements can occasionally move, or transpose, from one position on a chromosome to another position on the same chromosome or on a different chromosome.

In bacteria, insertion sequences, transposons, and phage mu are examples of transposable genetic elements. Transposable elements can mediate chromosomal rearrangements. In higher cells, transposable elements have been extensively studied in yeast, *Drosophila,* and maize and in mammalian systems. In eukaryotes, some transposable elements utilize an RNA intermediate during transposition, whereas in prokaryotes, transposition is exclusively at the DNA level.

Starting in the 1930s, genetic studies of maize, undertaken independently by Barbara McClintock and Marcus Rhoades, yielded results that greatly upset the classical genetic picture of genes residing only at fixed loci on the main chromosome. A variety of names have been applied to these moving genetic elements, e.g., controlling elements, cassettes, jumping genes, roving genes, mobile genes, mobile genetic elements, and transposons. The term transposition has long been used in genetics to describe transfer of chromosomal segments from one position to another in major structural rearrangements. In the present context, what is being transposed can be a gene or a small number of linked genes or a gene-sized fragment.

Transposable genetic elements can move to new positions within the same chromosome or even to a different chromosome. The normal genetic role of these elements is not known with certainty. They have been detected genetically through the abnormalities that they produce in the activities and structures of the genes near the sites to which they move. A variety of physical techniques have been used to detect them as well, including DNA sequencing. Transposable genetic elements have been found in most organisms in which they have been sought. Today, transposable elements provide valuable tools both in prokaryotes and in eukaryotes for genetic mapping, creating mutants, cloning genes, and even producing transgenic organisms.

2. TRANSPOSABLE ELEMENTS IN PROKARYOTES

2.1. Bacterial Insertion Sequences

2.1.1. Insertion sequences, or **insertion-sequence (IS) elements**, are now known to be segments of bacterial DNA that can move from one position on a chromosome to a different position on the same chromosome or on a different chromosome. An IS element contains only genes required for mobilizing the element and inserting the element into a chromosome at a new location. Is elements are normal constituents of bacterial chromosome and plasmids. When IS elements appear in the middle of genes, they interrupt the coding sequence and inactivate the expression of that gene. Owing to their size and in some cases the presence of transcription and translation termination signals, IS elements can also block the expression of other genes in the same operon if those genes are downstream from the promoter of the operon.

IS elements were first found in *E. coli* as a result of their affects on the expression of a set of three genes whose products are needed to metabolize the sugar galactose as a carbon source. Careful investigations showed that the mutant phenotypes resulted from the insertion of an approximately 800 base pairs (bp) DNA segment into a gene. This particular DNA segment is now called insertion sequence1 (IS1).

2.1.2. Properties of IS elements

Is1 is the genetic element capable of moving around the genome. It integrates into the chromosome at locations with which it has no homology, thereby distinguishing it from recombination. This event is an example of transposition event. There are number of IS elements that have been identified in *E. coli*, including IS1, IS2, and IS10, each present in 0 to 30 copies per genome, and each with a characteristic length and unique nucleotide sequence. IS1 is 768 bp long, and is present in 4 to 19 copies on the *E. coli* chromosomes. IS2 is present in 0 to 12 copies on the *E. coli* chromosome and in one copy on the F plasmid, and IS10 is found in a class of plasmids called R plasmid that can replicate in *E. coli* (Fig. 12.1). Among prokaryotes, the IS elements are normal cell constituents, that is, they are found in most cells. Altogether, IS elements constitute approx. 0.3% of the cell's genome. All IS elements that have been sequenced, end with perfect or nearly perfect inverted terminal repeats (IRs) of between 9 and 41 bp. This means that essentially the same sequence is found at each end of an IS but in opposite orientations.

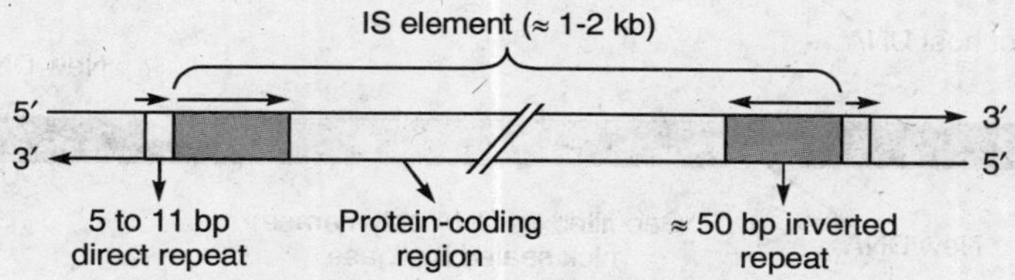

Fig. 12.1. Structure of bacterial IS elements.

2.1.3. IS transposition

When transposition of an IS element takes place, a copy of the IS element inserts into a new chromosome location while the original IS elements remains in place. That is, transposition requires the precise replication of the original IS element, using the replication enzymes of the host cell. The actual transposition also requires an enzyme encoded by the Is element called transposase. The IR sequences are essential for the transposition process, that is, those sequences are recognized by transposase to initiate transposition. Is elements insert into the chromosomes at sites with which they have no sequence homology. Genetic recombination between nonhomologous sequences is called illegitimate recombination. The sites into which IS elements insert are called target sites. The process of IS insertion into a chromosome is shown in Figure 12.2. Firstly, a staggered cut is made in the target site and the IS element is then inserted, becoming joined to the single-stranded ends. The gaps are filled in by DNA polymerase and DNA ligase, producing an integrated IS element with two direct repeats of the target site sequence flanking the IS element. 'Direct' in this case means that the two sequences are repeated in the same orientation. The direct repeats are called target site duplications. The sizes of target site duplication vary with the IS elements, but tend to be small. Intergration of some IS elements show preference for certain regions, while others integrate only at particular sequences.

All copies of a given IS element have the same sequence, including that of the inverted terminal repeats. Mutations that affect the inverted terminal repeat sequence of IS elements affect transposition, indicating that the inverted terminal repeat sequences are the key sequences recognized by transposase during a transposition event.

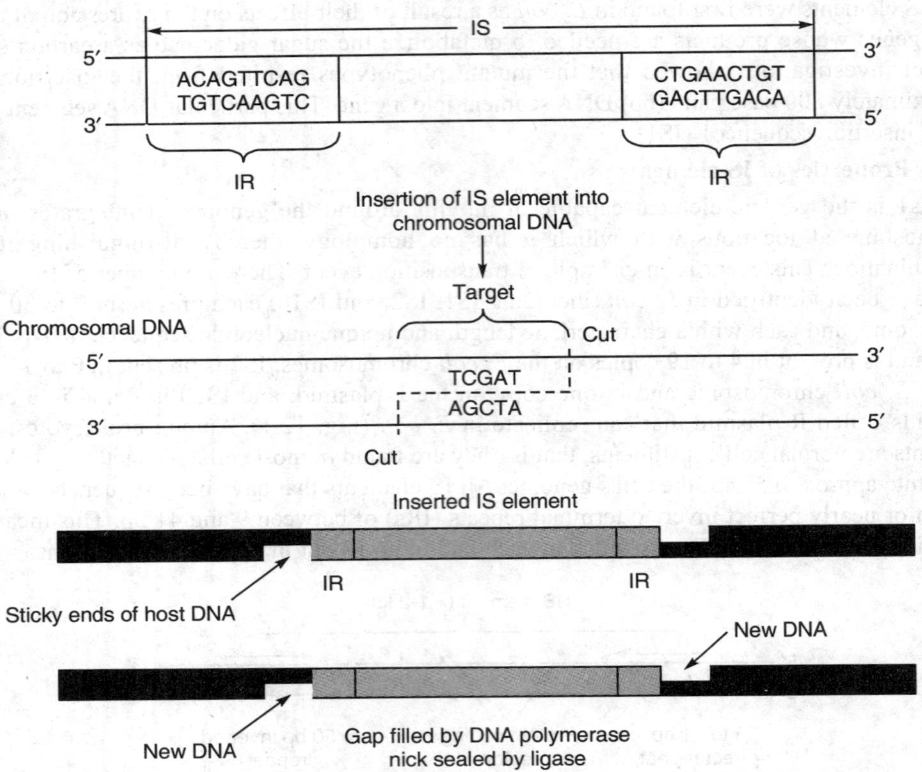

Fig. 12.2. Schematic presentation of the integration of an IS elements into chromosomal DNA.

2.2. Prokaryotic Transposons

A transposon (Tn) is more complex than an IS elements. A transposon is a mobile DNA segment that, like an IS element, contains genes for the insertion of the DNA segment into the chromosome and for the mobilization of the element to other locations on the chromosome. There are two types of prokaryotic transposons: composite transposons and noncomposite transposons.

2.2.1. Composite transposons- They are complex transposons with a central region containing genes, e.g., drug resistance genes, flanked on both sides by IS elements (also called IS modules). Composite transposons may be thousands of base pairs long. The IS elements are both of the same types and are called IS-L (for "left") and IS-R (for "right"). Depending upon the transposon, IS-L and IS-R may be in the same or inverted orientation relative to each other. Because the ISs themselves have terminal inverted repeats, the composite transposons also have terminal inverted repeats.

Figure 12.3 shows the structure of the composite transposon Tn10 to illustrate the general features of such transposons. The Tn 10 transposon is 9,300 bp long and consists of 6,500 bp of central, nonrepeating DNA containing the tetracycline resistance gene flanked at each end with a 1,400-bp IS element. These IS elements are designated IS10L and IS10R and are arranged in an inverted orientation. Cells containing Tn10 are resistant to tetracycline resistance gene contained within the central DNA sequence.

Transposition of composite transposon occurs because of the function of the IS elements they contain. One or both IS element supplies the transposase. The inverted repeats of the IS elements at the two ends of the transposon are recognized by transposase to initiate transposition (as with

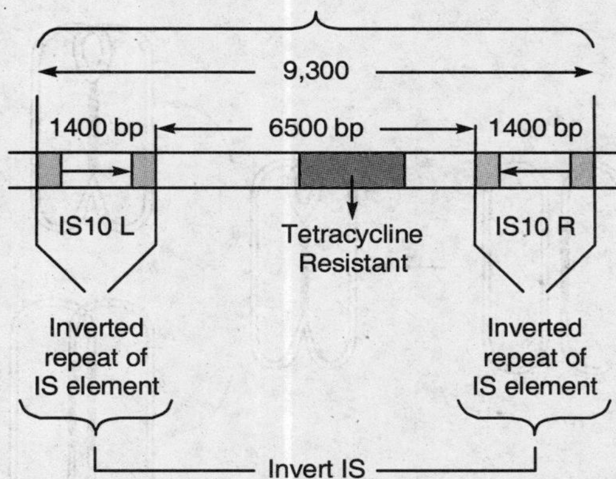

Fig. 12.3. Detailed structure of Tn10 transposon.

transposition of IS elements). Transposition of Tn10 is rare, occurring once in 10^7 cell generations. This is the case because less than one transposase molecule per cell generation is made by Tn10. Like IS elements, composite transposons produce target site duplications after transposition.

2.2.2. Noncomposite transposons- They like composite transposons, contain genes such as those for drug resistance. Unlike composite transposons, they do not terminate with IS elements. However, they do have the repeated sequences at their ends that are required for transposition. Tn3 is a noncomposite transposon. Tn3 has 38 bp inverted terminal repeats and contains three genes in its central region. One of those genes, bla, encodes β-lactamase which breaks down ampicillin and therefore makes cells containing Tn3 resistant to ampicillin. The other two genes, tnpA and tnpB, encode the enzymes transposase and resolvase that are needed for transposition of Tn3 (Fig. 12.4). Transposase catalyzes insertion of the Tn into new sites, and resolvase is an enzyme involved in the particular recombinational events associated with transposition. Resolvase is not found in all transposons. The genes for transposition are in the central region for noncomposite transposons, while they are in the terminal IS elements for composite transposons. Non composite transposons also cause target site duplications when they move.

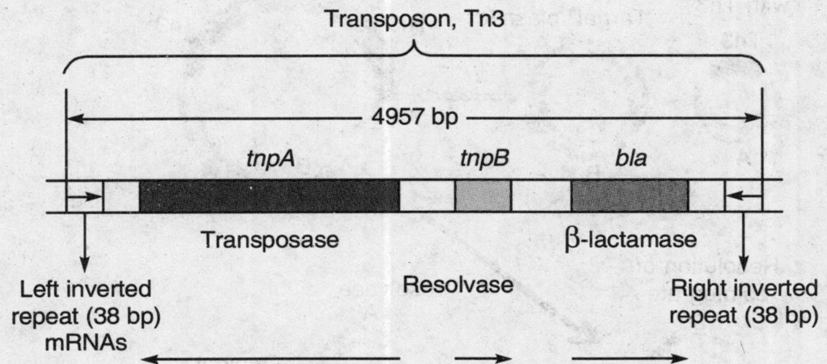

Fig. 12.4. Detailed structure of Tn3 transposon.

2.2.3. Mechanism of transposition in prokaryotes

Several different mechanisms of transposition are employed by prokaryotic transposable elements. And, as we shall see later, eukaryotic elements exhibit still additional mechanisms of transposition.

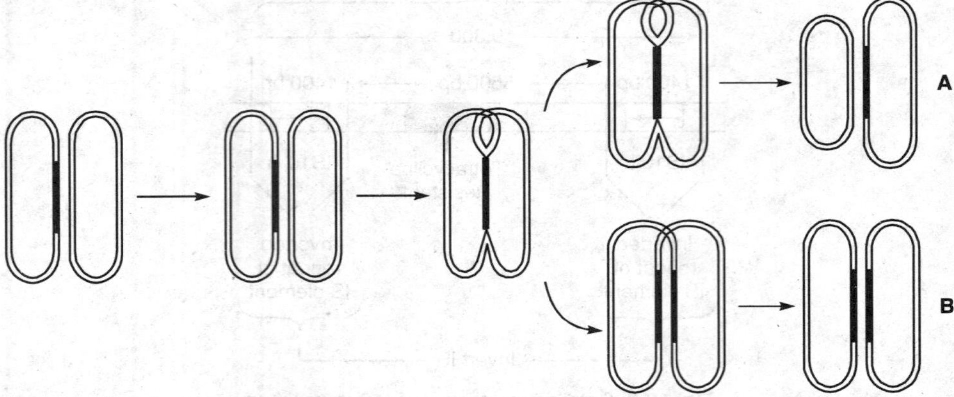

Fig.12.5. A. Conservative mode of transposition **B.** Replicative mode of transposition.

In *E. coli*, we can identify **replicative** and **conservative** (nonreplicative) modes of transposition. In the replicative pathway, a new copy of the transposable element is generated in the transposition event. The results of the transposition are that one copy appears at the new site and one copy remains at the old site. In the conservative pathway, there is no replication. Instead, the element is excised from the chromosome or plasmid and is integrated into the new site (Fig. 12.5).

Replicative transposition- The transposition of Tn3 occurs in two stages. Firstly, the transposase mediates the fusion of two molecules, forming a structure called a **cointegrate**. During this process, the transposon is replicated, and one copy is inserted at each junction in the cointegrate. The two Tn3 are oriented in the same direction. In the second stage of transposition, the tnpR-encoded resolvase mediates a site-specific recombination event between the two Tn3 elements. This event occurs at a sequence in Tn3 called *res*, the resolution site, and generates two molecules, each with a copy of the transposon.

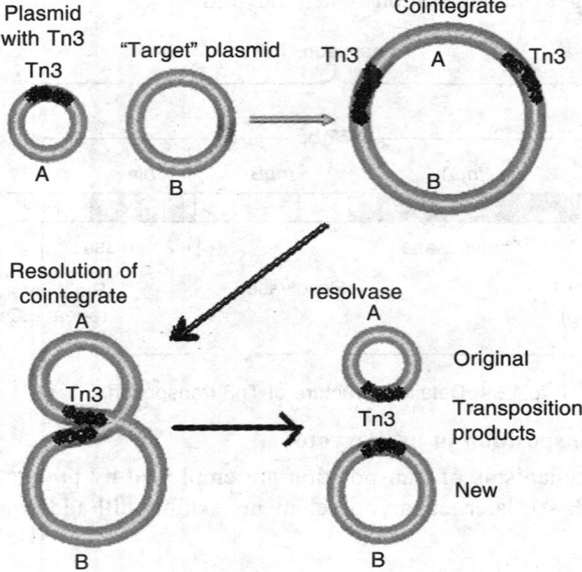

Fig. 12.6. Process of Tn3 transposition through cointegrate intermediate.

The tnpR gene-product also has another function, namely, to repress the synthesis of both the transposase and resolvase proteins. This repression occurs because the res site is located in between the tnpA and tnpR genes. By binding to this site, the tnpR protein interferes with the synthesis of both gene-products, leaving them in chronic short supply. Consequently, the Tn3 element tends to remain immobile (Fig. 12.6).

Conservative transposition- Some transposons, such as Tn10, excise from the chromosome and integrate into the target DNA. In these cases, DNA replication of the element does not occur, and the element is lost from the site of the original chromosome. This mechanism is called conservative (non replicative) transposition or simple insertion. Tn10, e.g., transposes by conservative transposition.

Insertion of a transposon into the reading frame of a gene will disrupt it, causing a loss of function of that gene. Insertion into gene's controlling region can cause changes in the level of expression of the gene. Deletion and insertion events also occur as a result of activities of the transposons, and from crossing-over between duplicated transposons in the genome.

2.3. IS elements and Transposons in Plasmids

The transfer of genetic material between conjugating *E. coli* is the result of the function of the fertility factor F. The F factor, a circular double stranded DNA molecule, is one of the example of bacterial plasmid. Plasmids such as F that are also capable of integrating into the bacterial chromosomes are called episomes. F factor consists of 94,500 bp of DNA that code for a variety of proteins. The important elements are (i). transfer gene (tra) required for the conjugation transfer of the DNA. (ii). genes that encode proteins required for the plasmid's replication. (iii). Four IS elements, two copies of IS3, one of IS2, and one of an insertion sequence element called gamma-delta.

It is because the *E. coli* chromosome has copies of these four insertion sequence at various positions, that the F factor can integrate into the *E. coli* chromosome at different sites and in different orientations with homologous sequence of the insertion elements.

Another class of plasmids that has medical significance is the R plasmid group, which was discovered in Japan in the 1950s, during the cure for dysentery. The disease is the result of infection by the pathogenic bacterium *Shigella*. *Shigella* was found to be resistant to most of the commonly used antibiotics. Subsequently, they found that the genes responsible for the drug resistances were carried on R plasmids, which can promote the transfer of genes between bacteria by conjugation, just as the F factor.

One segment of an R plasmid that is homologous to a segment in the F factor is the part needed for the conjugal transfer of genes. That segment and the plasmid-specific genes for DNA replication constitute what is called the RTF (resistance transfer factor) region (Fig. 12.7). The rest

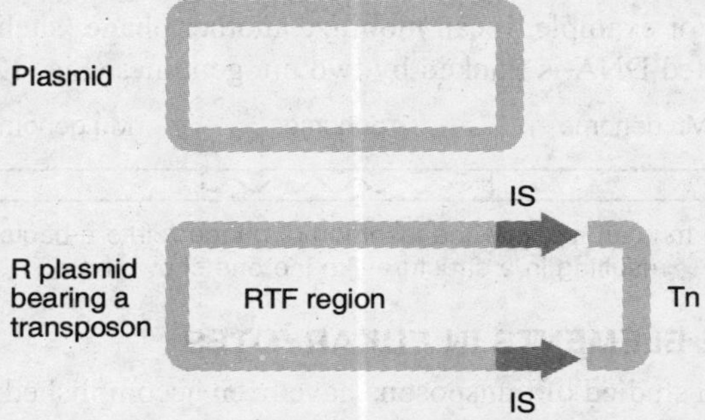

Fig. 12.7. The insertion of a transposon (Tn) into a plasmid. RTF represents the resistance-transfer functional genes of the plasmid.

of the R plasmid differs from type to type and includes the antibiotic-resistance genes or other types of genes of medical significance, such as resistance to heavy metal ions.

The resistance genes in R plasmid are, in fact, transposons, that is each resistance gene is located between flanking, directly repeated segments such as one of the IS modules (Fig. 12.8). Thus, each transposon with its resistance gene in the R plasmid can be inserted into new location on other plasmids or on the bacterial chromosome, while at the same time leaving behind a copy of itself in the original position.

Fig. 12.8. Two different transposons having different inverted repeat (IR) regions and carrying different drug-resistance genes. (a) Tn9 has a short IR region, because the two IS1 elements are in the same orientation and each element has a short inverted repeat. (b) Tn10 has a large IR region because the two IS10 components have opposite orientations, and the entire IS10 sequence constitutes the inverted repeat.

2.4. Phage mu

Phage mu is a normal-appearing phage. We consider it here because, although it is a true virus, it has many features in common with IS elements. The DNA double helix of this phage is 36,000 nucleotides long-much larger than an IS element. However, it does appear to be able to insert itself anywhere in a bacterial or plasmid genome in either orientation. Once inserted, it causes mutation at the locus of insertion-again like an IS element. (The phage was named for this ability: mu stands for "mutator.") Normally, these mutations cannot be reverted, but reversion can be produced by certain kinds of genetic manipulation. When this reversion is produced, the phages that can be recovered showing no deletion, proving that excision is exact and that the insertion of the phage therefore does not involve any loss of phage material either.

Each mature phage particle has on each end a piece of flanking DNA from its previous host (Fig. 12.9). However, this DNA is not inserted anew into the next host. Its function is unclear. Phage mu also has an IR sequence, but neither of the repeated elements is at a terminus.

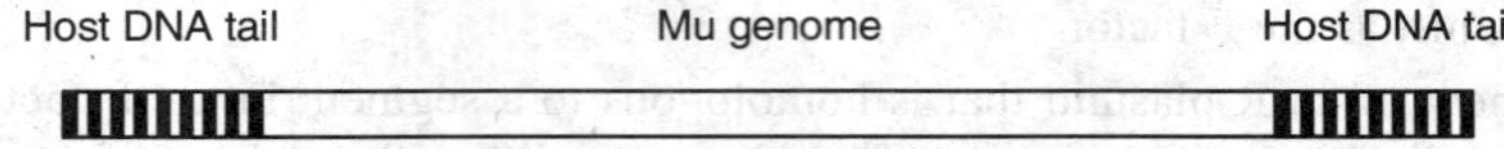

Fig. 12.9. The DNA of a free mu phage has tails derived from its previous host.

Mu can also act like a genetic snap fastener, mobilizing any kind of DNA and transposing it anywhere in a genome. For example, it can mobilize another phage (such as λ) or the F factor. In such situations, the inserted DNA is flanked by two mu genomes (Fig. 12.10).

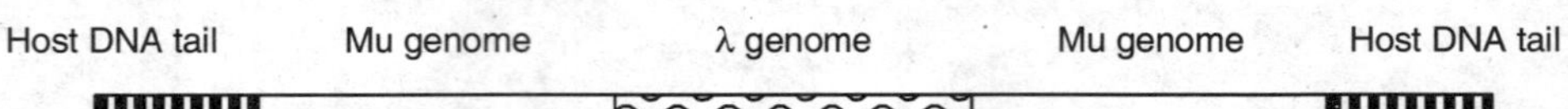

Fig. 12.10. Phage mu can mediate the insertion of phage λ into a bacterial chromosome, resulting in a structure like the one shown here.

3. TRANSPOSABLE ELEMENTS IN EUKARYOTES

Most of the detailed studied of transposons have been accomplished in bacteria. However, in eukaryotes beginning with the classic work of McClintock, some of the recent studies on yeast and *Drosophila* have been discussed here.

12.3.1. Yeast Ty Elements

The yeast carries about 35 copies of a transposable element called Ty in its haploid genome. These transposons are about 5900 nucleotide-pairs long and are bounded at each end by a DNA segment called the δ sequence, which is ~340bp long. Each δ sequence is oriented in the same direction, forming what are known as direct long terminal repeats of LTRs. Sometimes an LTR becomes detached from a Ty element, creating a so called solo δ. It is thought that these solo δ are generated by recombination between the LTRs of a complete Ty element. Ty elements are flanked by five nucleotide-pair direct repeats created by the duplication of DNA at the site of the Ty insertion. These target site duplications do not have a standard sequence, but they tend to contain AT bp. This may indicate that Ty elements preferentially insert into A-T-rich regions of the genome.

The genetic organization of the Ty elements resembles that of the eukaryotic retroviruses. These single stranded RNA viruses synthesizes DNA from their RNA after entering a cell. The DNA then inserts into a site in the genome, creating a target site duplication. This inserted material has the same overall structure as a yeast Ty element-a DNA bounded by LTRs and is called a provirus. Ty elements have only two genes, A and B which are analogous to the *gag* and *pol* gene of the retroviruses. These two genes can form virus like particles inside yeast cells.

One hypothesis is that yeast Ty elements are primitive retroviruses, capable of moving from one site to another inside a cell, but not capable of moving between cells. In this regard, is has been

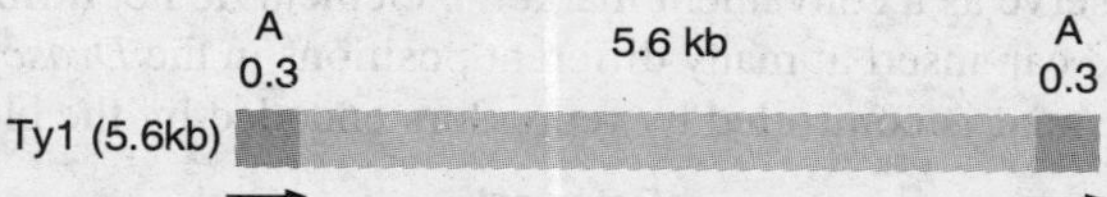

Fig. 12.11. The structure of a yeast transposable element. The Ty1 sequence appears approximately 35 times in the yeast genome. It contains two copies of the delta (d) sequence in direct orientation at each end. Delta appears approximately 100 times in the yeast genome.

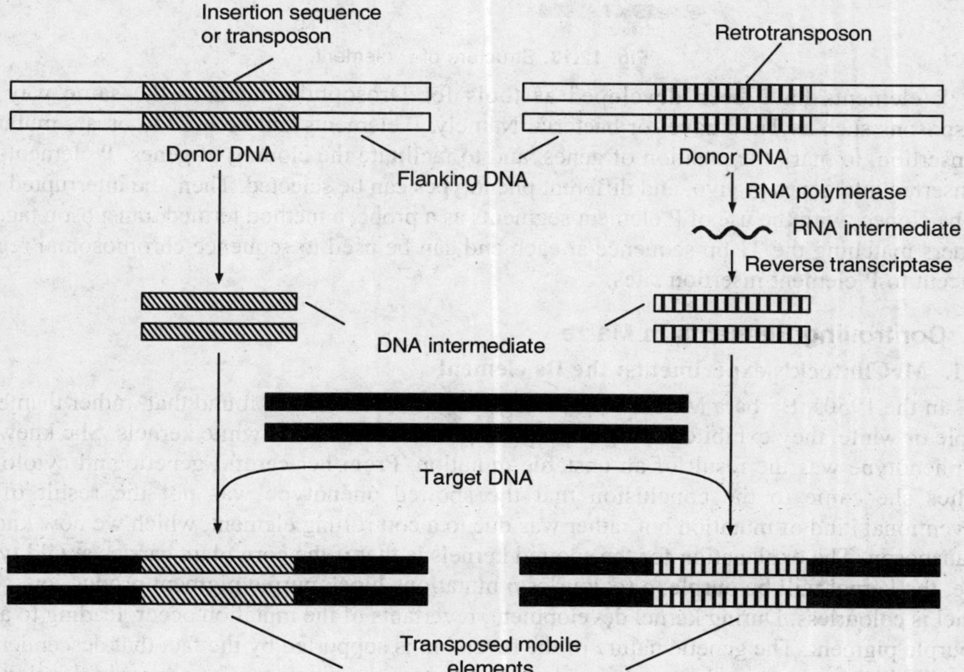

Fig. 12.12. Formation of copy of retrotransposon by reverse transcription.

shown that the transposition of Ty elements involves an RNA intermediate. After the RNA is synthesized from Ty DNA, a product of the TyB gene uses the RNA to make double stranded DNA. The process is reverse transcription. Then the newly synthesized DNA is inserted somewhere in the genome, creating a new Ty element (12.11). Because of their overall similarity to the retroviruses, yeast Ty elements are sometimes called retrotransposons (12.12).

12.3.2. *Drosophila* transposons

Transposable elements have been discovered in many animals, but some of the best information comes from studies with *Drosophila*, in which as much as 15% of the DNA is mobile. The largest group of Drosophila transposons comprises the retroviruslike elements, or retrotransposons. These elements are 5000 to 15,000 nucleotide pair long and resemble the integrated forms of retrotransposon.

P elements-The P element in *Drosophila* is one of the best examples of exploiting the properties of transposable elements in eukaryotes. This element, shown in figure 12.13, is 2907 bp long and features a 31 bp inverted repeat at each end. DNA sequence analysis of the 2.9 kb element reveals a gene, composed of four exons and three introns, that encodes transposase. There is a perfect 31 bp inverted repeat at each terminus. Although the transposase is required for transposition, it can be supplied by a second element. Therefore, P elements with internal deletions can be mobilized and then remain fixed in the new position in the absence of the second element; thus the P element can serve as a convenient marker. P element do not utilize an RNA intermediate during transposition and can insert at many different positions in the *Drosophila* chromosome. The transposition of a P element is controlled by repressors encoded by the element.

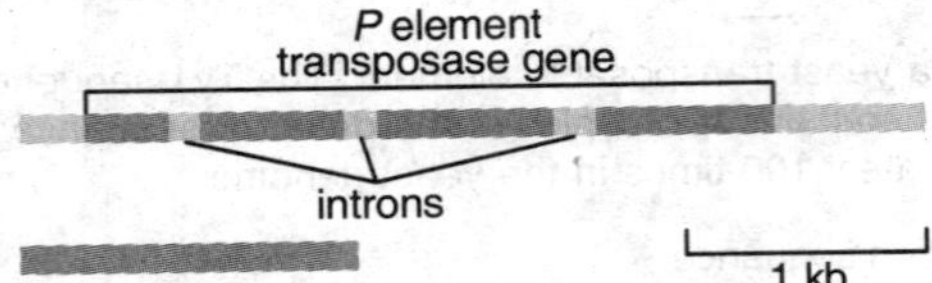

Fig. 12.13. Structure of P element.

P elements have been developed as tools for Drosophila much in the same way that transposons such as Tn10 have for bacteria. Namely, P elements can be used to create mutations by insertion, to mark the position of genes, and to facilitate the cloning of genes. P elements can be inserted into genes in vivo, and different phenotypes can be selected. Then, the interrupted gene can be cloned, with the use of P element segments as a probe, a method termed transposon tagging. Primers matching the 31 bp sequence at each end can be used to sequence chromosomal regions adjacent to P element insertion sites.

3.3. Controlling Elements in Maize

3.3.1. McClintock's experiments: the Ds element

In the 1950s, Barbara McClintock during study of corn kernels found that, rather than being purple or white, they exhibited spots of purple pigment on normally white kernels. She knew that the phenotype was the result of an unstable mutation. From her careful genetic and cytological studies she came to the conclusion that the spotted phenotype was not the result of any conventional kind of mutation but rather was due to a controlling element, which we now know is a transposon. The explanation for the spotted kernels is that if the corn plant carries a wild type C gene, the kernel will be purple, c (colourless) mutations block purple pigment production, so the kernel is colourless. During kernel development, revertants of the mutation occur, leading to a spot of purple pigment. The genetic nature of the reversion is supported by the fact that descendents of the cell which underwent the reversion also can produce the pigment. The earlier in development the reversion occurs, the larger is the purple spot. McClintock determined that the original c (colourless) mutation resulted from a mobile controlling element, a genetic factor called as Ds (Dissociation). This element gets inserted into the C gene. This action of Ds is dependent on the presence of an unlinked gene, Ac (Activator). Ac is required for transposition of Ds into the gene. Ac can also move the Ds out of the C gene, resulting in the wild type revertant, i.e., a purple spot. McClintock found it impossible to map Ac. In some plants, it mapped to one position; in other plants of the same line, it mapped to different positions. Moreover, the Ds locus itself was constantly changing position on the chromosome arm, as indicated by the differing phenotypes of the variegated sections of the seeds. Ac is the autonomous element of the family, and hence mutations caused by Ac are unstable. Ds is the nonautonomous element of the family. Ds mutations are stable if only Ds are present; they are unstable in the presence of an Ac element.

Ac is 4,563 bp long, with 11bp imperfect terminal inverted repeats (IRs). Ac contains a single transcription unit that comprises most of the element's length. Ds does not transpose in the absence of Ac and remains as a stable insertion in the chromosome. When an Ac element is present, it activates Ds, causing it to transpose to a new site or to break the chromosome in which it is located. Ds elements are heterogenous in length and sequence. All Ds elements have the same terminal IRs as Ac elements, and many of these Ds elements have been generated from Ac by deletion of various lengths. Ac exhibits conservative transposition. Ac transposes only during replication (Fig. 12.14).

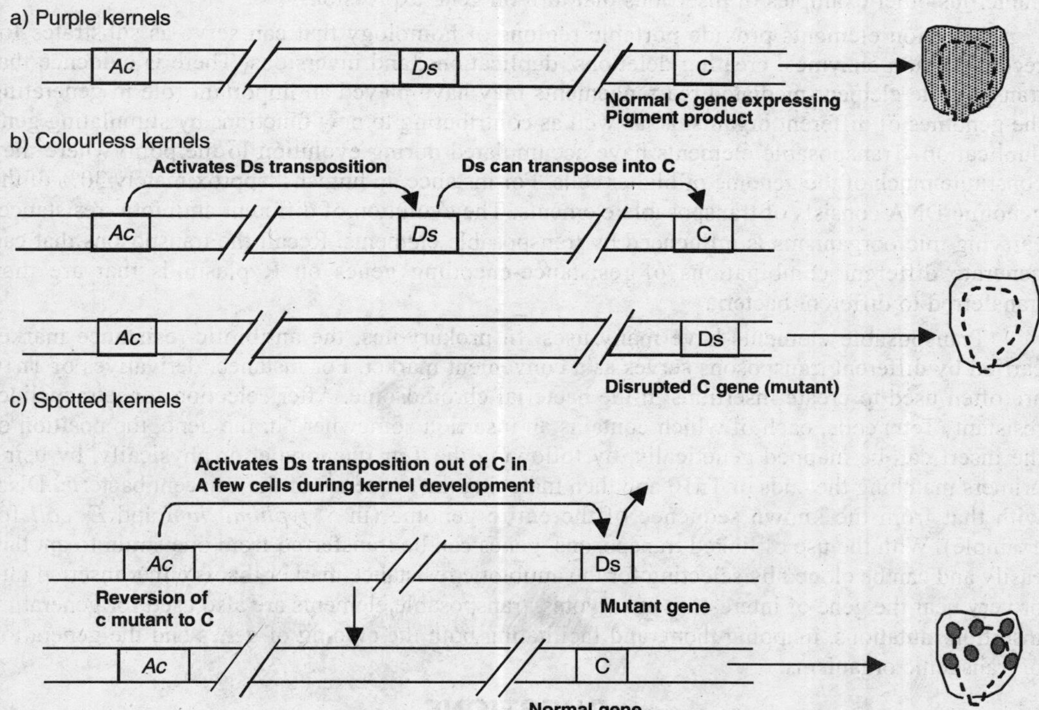

Fig. 12.14. Change in kernel color in corn due to transposon effect.

3.3.2. General characteristics of controlling elements

Several systems like Ds-Ac have now been found in corn. Each shows similar action, having a **target gene** that is inactivated, presumably by the insertion of some receptor element into it, and a distant **regulator gene** that maintains the mutational instability of the locus, presumably through its ability to "unhook" the receptor element from the target locus and return the locus to normal function. The receptor and the regulator are termed **controlling elements**.

In the examples considered so far, the unstable allele is said to be nonautonomous: it can revert only in the presence of the regulator. Sometimes, however, a system such as the Ac-Ds system can produce an unstable allele that is **autonomous**. Such mutants are recognizable because they show Mendelian ratios (such as 3:1 for pigmented to dotted) that apparently are independent of any other element. In fact, such alleles appear to be caused by the insertion of Ac itself into the target gene. An allele of this type can subsequently be transformed into a nonautonomous allele. In such cases, the nonautonomy seems to result from the spontaneous generation of a Ds element from the inserted Ac element. In other words, Ds is in all likelihood an incomplete version of Ac itself.

4. FUNCTION OF TRANSPOSABLE ELEMENTS

Transposable elements play a role in the biology of organisms. They cause mutations by insertion into genes and affect the regulation of genes by inserting near promoters. They also provide substrates for genetic rearrangements and thus act as agents of genome evolution. For instance, in *E. coli* growing in natural environments, a sizable fraction of spontaneous mutations are caused by the IS series of transposable elements. As mentioned earlier, more than half of the spontaneous mutations in *Drosophila* result from transposable-element insertion. Insertions of IS elements near the promoter region of the bgl operon, which encodes proteins participating in β-glucosidase metabolism, activates transcription in this normally "cryptic" operon. There are numerous other examples of insertions that turn on gene expression.

Insertion elements provide portable regions of homology that can serve as substrates for recombination enzymes, creating deletions, duplications, and inversions. There is evidence that transposable element-mediated rearrangements may have played an important role in generating the genomes of different organisms, as well as contributing to new functions by stimulating gene duplication. Transposable elements have accumulated during evolution to the point where they constitute much of the genome of higher cells. For instance, in humans, approximately 30% of the genomic DNA consists of transposable elements. The evolution of different antibiotic-resistance-carrying microorganisms is influenced by transposable elements. Recall the transposons that can generate different combinations of resistance-encoding genes on R plasmids that are then transferred to different bacteria.

Transposable elements have many uses. In prokaryotes, the antibiotic-resistance marker carried by different transposons serves as a convenient marker. For instance, derivatives of Tn10 are often used to create insertions in the bacterial chromosome. After selection for tetracycline-resistant (Tetr) cells, each of which contains an insertion somewhere in the gene, the position of the insert can be mapped genetically, by following the Tetr phenotype, or physically, by using primers matching the ends of Tn10 and then matching the sequence of the adjacent bacterial DNA with that from the known sequence of the entire genome (in *S. typhimurium* and *E. coli* for example). With the use of linked transposons, genes can be transferred from one strain to another easily and can be cloned by selecting for the antibiotic resistance marker that is either inserted into or very near the gene of interest. In eukaryotes, transposable elements are also used for generating insertion mutations, mapping them, and facilitating both the cloning of genes and the generation of transgenic organisms.

QUESTIONS

1. What do you understand by jumping genes? Describe contribution of Barbara McClintok in the discovery of jumping genes.
2. What are trnasposons? Describe various types of their transposition.
3. Write short notes on:
   (a) Ty element
   (b) Retrotransposons
   (c) IS elements
   (d) P element
   (e) Ac-Ds elements

CHAPTER 13

Enzymes involved in DNA Synthesis and Cloning

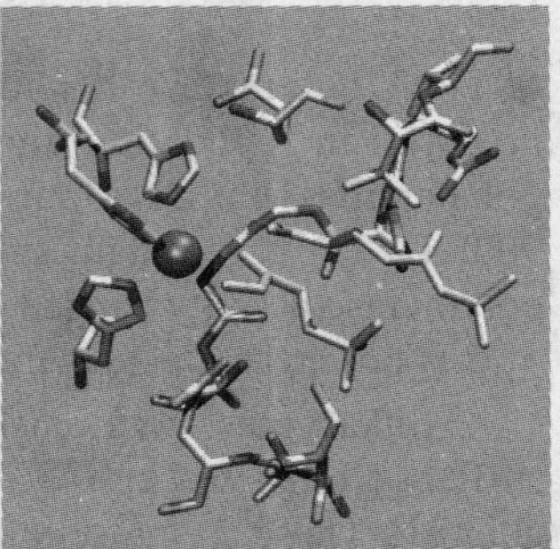

In this chapter we will describe various enzymes used in rDNA technology. In general, DNA polymerases are described with DNA replication where there primary work is to add nucleotides. The isolation and identification of many types of polymerases make them specific for different purposes. The specific role of these enzymes has been described and compared with similar enzymes available in the market.

1. DNA POLYMERASE

The principal chemical reaction catalysed by a DNA polymerase is the $5' \rightarrow 3'$ synthesis of a DNA polynucleotide. DNA polymerases carry out the process of addition of nucleotides and formation of polynucleotide chain. This enzyme is also called as replicase when it replicates the DNA molecules. This enzyme can add as well as remove polynucleotide (exonuclease activity): (a) A $3' \rightarrow 5'$ exonuclease activity is shown by template dependent DNA polymerases. This activity enables the enzyme to remove nucleotides from the $3'$ end of the strands that it has just synthesized. It is called as proofreading activity and enzyme corrects the error by removing the mismatch nucleotide. (b) A $5' \rightarrow 3'$ exonuclease activity is less common. However, this activity is required by the enzyme to remove already attached small DNA fragments on the template (during ligation of Okazaki fragments).

Table 13.1. Characteristics of prokaryotic and eukaryotic DNA polymerases.

Enzyme	Subunits	Mol.wt. (Kilo Dalton)	Exonuclease activity $3' \rightarrow 5'$	Exonuclease activity $5' \rightarrow 3'$	Function
Bacterial DNA polymerases					
DNA polymerase I	1	109	Yes	Yes	Repair, replication
DNA polymerase II	1	90	Yes	No	Repair
DNA polymerase III	10	900	Yes	No	Main replicating enzyme
Eukaryotic DNA polymerase					
DNA polymerase α	4	300	No	No	Priming during replication
DNA polymerase γ	2	180-300	Yes	No	Mitochondrial DNA replication
DNA polymerase δ	2	170-230	Yes	No	Main replicating enzyme
DNA polymerase ε	1	250	Yes	?	Repair

DNA polymerase was isolated from *E. coli* by Arthur Kornberg in 1957, which was also called as Korenberg polymerase. Later on it was named as **DNA polymerase I** after the discovery of DNA polymerase II and III. Now we know that DNA polymerase III, isolated in 1972, is involved in replication along with DNA polymerase I. DNA polymerase I and II are single polypeptides, but DNA polymerase III is a ten subunits protein with a molecular mass of approx. 900KD. The three main subunits which form the core enzyme are called α, ε and θ, with the polymerase activity carried out by α subunit and $3' \rightarrow 5'$ exonuclease activity by ε. The function of θ is not clear. Besides these, there are smaller subunits like τ, γ, δ, δ', χ and ψ (Table 13.1).

Eukaryotes have ten DNA polymerases, which in mammals are distinguished by Greek suffixes [(α, β, γ, ε, τ, δ, δ', θ, χ and ψ) alpha, beta, gamma, epsilon, tau, delta, delta prime, theta, chi, psi]. These are present in the holoenzyme in equal ratio. The α and β polymerases are located in the nucleus. The β polymerase copies a poly (A) or poly (C) template. The γ polymerase copies many poly ribonucleotides such as poly (A) and poly (C). The mitochondrial DNA polymerase is like γ polymerase. There are only 10-15 holoenzyme molecules per cell. The holoenzyme attaches firmly to the template and move along with the leading strand until the synthesis is complete. The holoenzyme is a dimer with two active sites, one for synthesizing the leading strand and one for extending the lagging strand.

1.1. DNA Polymerase I (Pol I)

This is the first DNA polymerase enzyme isolated by Kornberg having molecular weight of 109 KD. It is a single-chain protein that requires magnesium as a cofactor. Each of its three enzymatic activities are located into distinct domains of the holoenzyme, such that proteolytic deletions can be generated that lack one or more of the activities. The so-called Klenow fragment is one such molecule that is widely used in recombinant DNA work. It is the largest single chain of peptide forming globular protein. One atom of zinc per chain is present, therefore it is a metalloenzyme. In *E. coli* approx. 400 molecules of Pol I are present. Experimental evidences obtained by Kornberg by in vitro synthesis of DNA in a test tube suggested that pol I can synthesize only complementary strand. Electron microscopic study of purified pol I reveals its shape as a sphere of 65 Å. It has several attachment sites such as template site (for attachment to the template), a primer site (attachment with RNA primer for elongation of DNA), a primer terminus site and a triphosphate site (for recognizing incoming nucleotide).

The primary function of Pol I is polymerization (synthesis of DNA) and it has subsidiary role as an exonuclease (removal or degradation of nucleotides). It is mainly involved in short fragment synthesis and repair. This enzyme is cleaved by trypsin into two fragments, a large fragment of 75 KD and a small fragment of 36 KD. The large fragment shows $3' \rightarrow 5'$ exonuclease activity and the small fragment shows $5' \rightarrow 3'$ exonuclease activity. Due to $5' \rightarrow 3'$ exonuclease activity it involves in removal of RNA primer and filling the gap between Okazaki fragments by synthesis of DNA fragments.

1.2. DNA Polymerase II (Pol II)

This is a 90KD polypeptide mainly involved in $5' \rightarrow 3'$ repair synthesis. It has $3' \rightarrow 5'$ exonuclease activity which shows its involvement in repair. It has no $5' \rightarrow 3'$ exonuclease activity. A cell has about 40 molecules of pol II. Experimental evidences suggest that it can seal Okazaki fragments and carry out the functions of pol I in its absence.

1.3. DNA Polymerase III (Pol III)

It is heterodimer of two polypeptide chains having molecular weight of 900 KD. It is the main and most active enzyme which can add about 150,000 nucleotides per minute during DNA polymerization. However, it requires auxillary proteins (copolymerase II) for the synthesis of long DNA strand. It has only $3' \rightarrow 5'$ exonuclease activity.

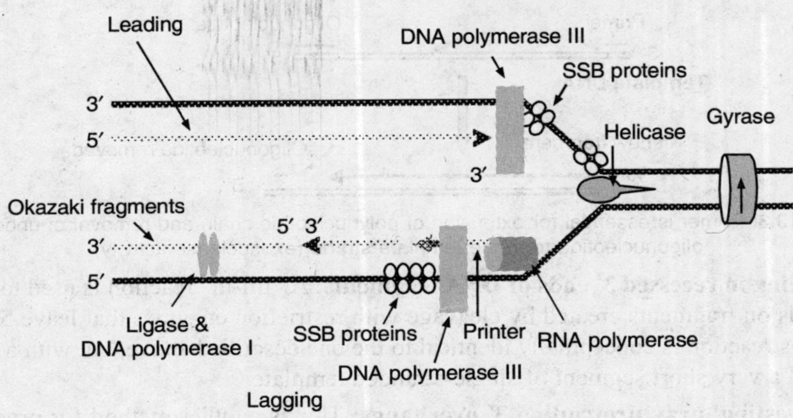

Fig. 13.1. Role of enzymes of DNA replication.

DNA polymerases are obtained by transferring the gene for this protein into *E. coli*. Thus, polymerases obtained from different sources differs slightly in their properties. These properties and specific uses are described here. The *E. coli* DNA polymerase I is a DNA-dependent DNA polymerase that possesses both $3' \rightarrow 5'$ and $5' \rightarrow 3'$ exonuclease activities. DNA polymerase I was used frequently in the early days of recombinant DNA technology for radiolabelling DNA and synthesizing cDNA. However, other enzymes have proven to be more effective for these purposes, including a proteolytic fragment of DNA polymerase I called Klenow fragment and T4 DNA polymerase. The holoenzyme DNA polymerase I is no longer frequently used (Fig. 13.1).

1.4. Large (Klenow) Fragment of *E. coli* DNA Polymerase I

The $5' \rightarrow 3'$ exonuclease activity of E. coli's DNA polymerase I makes it unsuitable for many applications. However, this enzymatic activity can readily be removed from the holoenzyme. Exposure of DNA polymerase I to the protease subtilisin cleaves the molecule into a small fragment, which retains the $5' \rightarrow 3'$ exonuclease activity, and a large piece called Klenow fragment (Fig. 13.2). The large or Klenow fragment of DNA polymerase I has DNA polymerase and $3' \rightarrow 5'$ exonuclease activities, and is widely used in molecular biology.

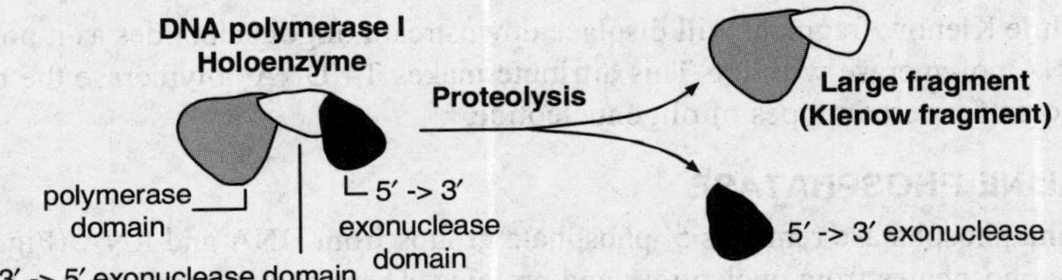

Fig. 13.2. Cleavage of DNA polymerase I to produce Klenow fragment.

In addition to generating Klenow fragment by proteolysis, it can be expressed in bacteria from a truncated form of the DNA polymerase I gene.

Uses of Klenow fragment

i. Synthesis of double-stranded DNA from single-stranded templates: The function of DNA polymerases is to synthesize complementary strands during DNA replication.

DNA polymerases require a primer to provide a free 3′ hydroxyl group for initiation of synthesis. The primers used for most in vitro polymerization reactions are single-stranded DNAs, typically 6 to 20 bases in length, called oligonucleotides. The oligonucleotides must be complementary to some section of template DNA. The reaction proceeds are depicted in the Figure 13.3.

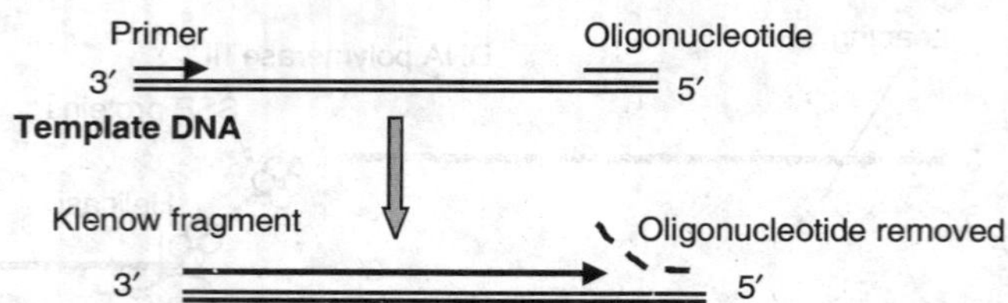

Fig. 13.3. Primer is essential for extension of polynucleotide chain and removal of undesired oligonucleotide from the template strand (exonuclease activity).

ii. **Filling in recessed 3′ ends of DNA fragments:** A "fill-in" reaction is used to create blunt ends on fragments created by cleavage with restriction enzymes that leave 5′ overhangs. This reaction is conceptually identical to the one described above, but with a huge primer and a very short segment of single-stranded template.

iii. **Digesting away protruding 3′ overhangs:** This is another method for producing blunt ends on DNA, in this with ends generated from restriction enzymes that cleave to produce 3′ overhangs. The 3′ → 5′ exonuclease activity of Klenow will digest away the protruding overhang. Removal of nucleotides from the 3′ ends will continue, but, in the presence of nucleotides, the polymerase activity will balance the exonuclease activity, yielding blunt ends. This reaction is more efficiently conducted with T4 DNA polymerase, which has much more potent exonuclease activity.

1.5. T4 DNA Polymerase

T4 is a bacteriophage of *E. coli*. The activities of T4 DNA polymerase are very similar to Klenow fragment of DNA polymerase I ; it functions as a 5′ → 3′ DNA polymerase and a 3′ → 5′ exonuclease, but does not have 5′ → 3′ exonuclease activity.

In general, T4 DNA polymerase is used for the same types of reactions as Klenow fragment, particularly in blunting the ends of DNA with 5′ or 3′ overhangs. There are however, two differences between the two enzymes that have practical signficance:

1. The 3′ → 5′ exonuclease activity of T4 DNA polymerase is roughly 200 times that of Klenow fragment, making it preferred by many investigators for blunting DNAs with 3′ overhangs.
2. While Klenow fragment will displace downstream oligonucleotides as it polymerizes, T4 DNA polymerase will not. This attribute makes T4 DNA polymerase the more efficient choice for certain types of oligonucleotide.

2. ALKALINE PHOSPHATASE

Alkaline phosphatase removes 5′ phosphate groups from DNA and RNA (Fig. 13.4). It will also remove phosphates from nucleotides and proteins. These enzymes are most active at alkaline pH therefore known as alkaline phosphatase.

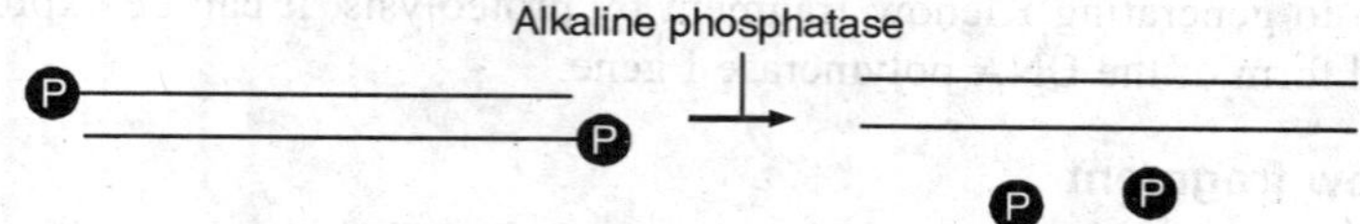

Fig. 13.4. Removal of phosphate by alkaline phosphatase.

There are several sources of alkaline phosphatase that differ in how easily they can be inactivated:

1. Bacterial alkaline phosphatase (BAP) is the most active of the enzymes, but also the most difficult to destroy at the end of the dephosphorylation reaction.

2. **Calf intestinal alkaline phosphatase** (CIP) is purified from bovine intestine. This is phosphatase most widely used in molecular biology labs because, although less active than BAP, it can be effectively destroyed by protease digestion or heat (75°C for 10 minutes in the presence of 5 mM EDTA).
3. Shrimp alkaline phosphatase is derived from a cold-water shrimp and is promoted for being readily destroyed by heat (65°C for 15 minutes).

There are two primary uses for alkaline phosphatase in DNA manipulations:

1. Removing 5′ phosphates from plasmid and bacteriophage vectors that have been cut with a restriction enzyme. In subsequent ligation reactions, this treatment prevents self-ligation of the vector and thereby greatly facilitates ligation of other DNA fragments into the vector (e.g. subcloning).
2. Removing 5′ phosphates from fragments of DNA prior to labeling with radioactive phosphate. Polynucleotide kinase is much more effective in phosphorylating DNA if the 5′ phosphate has previously been removed.

It is usually recommended that dephosphorylation of DNAs with blunt or 5′-recessed ends be conducted using a higher concentration alkaline phosphatase or at higher temperatures than for DNAs with 5′ overhangs.

3. TERMINAL TRANSFERASE

Terminal transferase catalyzes the addition of nucleotides to the 3′ terminus of DNA. Interestingly, it works on single-stranded DNA, including 3′ overhangs of double-stranded DNA, and is thus an example of a DNA polymerase that does not require a primer. It can also add homopolymers of ribonucleotides to the 3′ end of DNA.

The much preferred substrate for this enzyme is protruding 3′ ends, but it will also, less efficiently, add nucleotides to blunt and 3′-recessed ends of DNA fragments. Cobalt is a necessary cofactor for activity of this enzyme. Some of its uses are-

1. **Labelling the 3′ ends of DNA:** Most commonly, the substrate for this reaction is a fragment of DNA generated by digestion with a restriction enzyme that leaves a 3′ overhang, but oligodeoxynucleotides can also be used. When such DNA is incubated with tagged nucleotides and terminal transferase, a string of the tagged nucleotides will be added to the 3′ overhang or to the 3′ end of the oligonucleotide (Fig. 13.5).

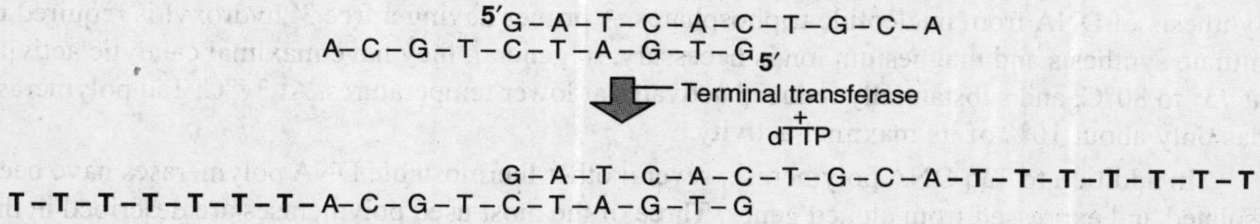

5′G–A–T–C–A–C–T–G–C–A
A–C–G–T–C–T–A–G–T–G 5′

Terminal transferase
+
dTTP

G–A–T–C–A–C–T–G–C–A–T–T–T–T–T–T–T
T–T–T–T–T–T–T–A–C–G–T–C–T–A–G–T–G

Fig. 13.5. Role of terminal transferase in adding poly (T).

2. **Adding complementary homopolymeric tails to DNA:** This clever procedure was commonly used in the past to clone cDNAs into plasmid vectors, but has largely been replaced by other, much more efficient techniques. The principles of this technique are depicted in the Figure 13.6. Basically, terminal transferase is used to tail a linearized plasmid vector with G′s and the cDNA with C′s. When incubated together, the complementary G′s and C′s anneal to "insert" the cDNA into the vector, which is then transformed into *E. coli*.

Terminal transferase is a mammalian enzyme, expressed in lymphocytes. The enzyme purchased commercially is usually produced by expression of the bovine gene in *E. coli*.

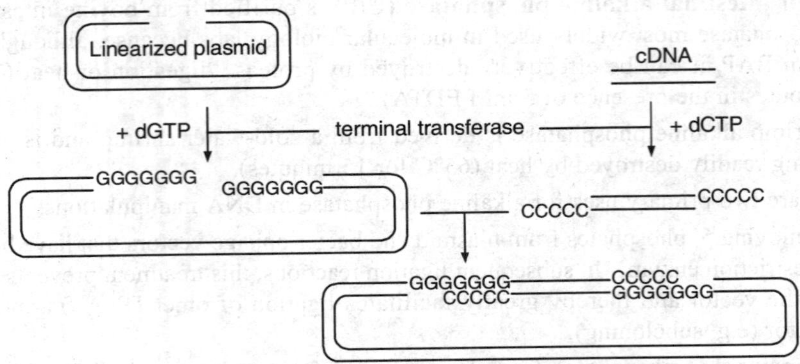

Fig. 13.6. Adding complementary homopolymeric tails to DNA.

4. THERMOSTABLE DNA POLYMERASES (TAQ POLYMERASE)

It is interesting how some unimportant discoveries become something of immense practical importance after some time. Such is the history of Taq DNA polymerase. The original report of this enzyme, purified from the hot springs bacterium *Thermus aquaticus*, was published in 1976 (Fig. 13.7). Roughly 10 years later, the polymerase chain reaction was developed and shortly thereafter "Taq" became a household word in molecular biology circles. Currently, the world market for Taq polymerase is in the hundreds of millions of dollars each year.

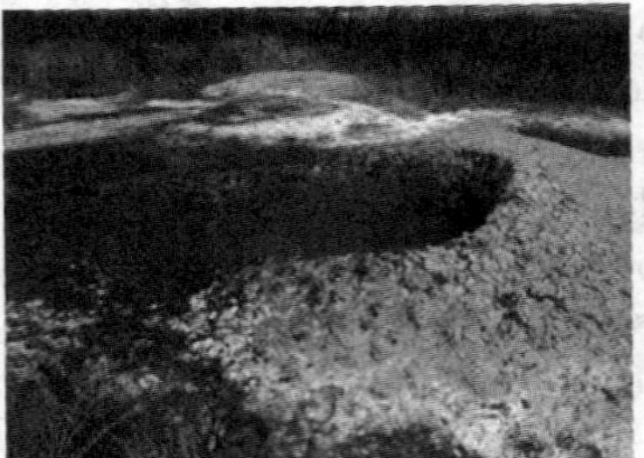

Fig. 13.7. Hot spring, habitat of bacteria.

The thermophilic DNA polymerases, like other DNA polymerases, catalyze template-directed synthesis of DNA from nucleotide triphosphates. A primer having a free 3′ hydroxyl is required to initiate synthesis and magnesium ion is necessary. In general, they have maximal catalytic activity at 75° to 80°C, and substantially reduced activates at lower temperatures. At 37°C, Taq polymerase has only about 10% of its maximal activity.

In addition to Taq DNA polymerase, several other thermostable DNA polymerases have been isolated and expressed from cloned genes. Three of the most used polymerases are described in the Table 13.2.

Table 13.2. Properties of commonly used polymerases.

Polymerase	3′ → 5′ Exonuclease	Source and Properties
Taq	No	From Thermus aquaticus. Half-life at 95°C is 1.6 hours.
Pfu	Yes	From Pyrococcus furiosus. Appears to have the lowest error rate of known thermophilic DNA polymerases.
Vent	Yes	From Thermococcus litoralis; also known as Tli polymerase. Half-life at 95°C is approximately 7 hours.

5. BACTERIOPHAGE RNA POLYMERASES

Phage-encoded DNA-dependent RNA polymerases are used for in vitro transcription to generate defined RNAs. Most commonly, the reaction utilizes ribonucleotides that are labeled with radionuclides or some other tag, and the resulting labeled RNA is used as a probe for hybridization. Other applications of in vitro transcription including making RNAs for in vitro translation or to study RNA struction and function.

Several bacteriophage RNA polymerases are commercially available. They are named after the phage that encodes them, and either purified from phage-infected bacteria or produced as recombinant proteins.

Many of the plasmids used for carrying cloned DNA incorporate promoters for bacteriophage RNA polymerases adjacent to the cloning site. This allows one to readily obtain either mRNA sense or antisense transcripts from the inserted DNA. The process is often called run-off transcription, because the plasmid is cut with a restriction enzyme downstream of the inserted DNA, which causes the polymerase to fall off the template when it reaches that spot.

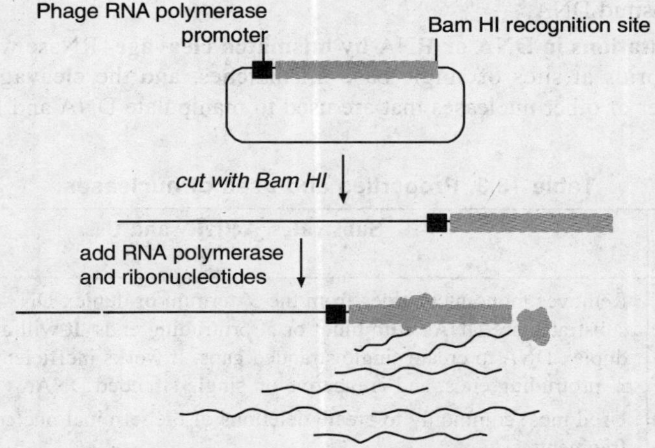

Fig. 13.8. RNA formation from inserted DNA.

If we assume that the RNA transcribed (Fig. 13.8) has the polarity of a mRNA (e.g. sense), it is easy to modify the construct to express an antisense RNA - simply reverse the orientation of the transcribed region. Indeed, most plasmids used for in vitro transcription have two different phage polymerase promoters flanking the insertion site, which allows transcription of sense RNA with one polymerase and antisense with the other.

6. NUCLEASES: DNase AND RNase

Most of the time nucleases are the enemy of the molecular biologist who is trying to preserve the integrity of RNA or DNA samples. However, deoxyribonucleases (DNases) and ribonucleases (RNases) have certain indispensible roles in molecular biology laboratories.

Numerous types of DNase and RNase have been isolated and characterized. They differ among other things in substrate specificity, cofactor requirements, and whether they cleave nucleic acids internally (endonucleases), chew in from the ends (exonucleases) or attack in both of these modes. In many cases, the substrate specificity of a nuclease depends upon the concentration of enzyme used in the reaction, with high concentrations promoting less specific cleavages.

The most widely used nucleases are DNase I and RNase A, both of which are purified from bovine pancreas: **Deoxyribonuclease I** cleaves double-stranded or single stranded DNA. Cleavage preferentially occurs adjacent to pyrimidine (C or T) residues, and the enzyme is therefore an endonuclease. Major products are 5′-phosphorylated di, tri and tetranucleotides.

In the presence of magnesium ions, DNase I hydrolyzes each strand of duplex DNA independently, generating random cleavages. In the presence of manganese ions, the enzyme cleaves both strands of DNA at approximately the same site, producing blunt ends or fragments with 1-2 base overhangs. DNase I does not cleave RNA, but crude preparations of the enzyme are contaminated with RNase A; RNase-free DNase I is readily available.

Applications

1. Eliminating DNA (e.g. plasmid) from preparations of RNA.
2. Analyzing DNA-protein interactions via DNase footprinting.
3. Nicking DNA prior to radiolabelling by nick translation.

Ribonuclease A is an endoribonuclease that cleaves single-stranded RNA at the 3′ end of pyrimidine residues. It degrades the RNA into 3′-phosphorylated mononucleotides and oligonucleotides. The major use of RNase A is eliminating or reducing RNA contamination in preparations of plasmid DNA.

Mapping mutations in DNA or RNA by mismatch cleavage- RNase will cleave the RNA in RNA-DNA hybrids at sites of single base mismatches, and the cleavage products can be analyzed. A number of other nucleases that are used to manipulate DNA and RNA are described in the Table 13.3.

Table 13.3. Properties and uses of nucleases.

Nuclease and Source	Substrates, Activity and Uses
Exonuclease III (*E. coli*)	Removes mononucleotides from the 3′ termini of duplex DNA. The preferred substrates are DNAs with blunt or 5′ protruding ends. It will also extend nicks in duplex DNA to create single stranded gaps. It works inefficiently on DNA with 3′ protruding ends, and is inactive on single-stranded DNA. Used most commonly to create deletions of the terminal nucleotide of linear DNA fragments.
Mung Bean Nuclease (Mung bean sprouts)	Digests single stranded DNA to 5′-phosphorylated mono or oligonucleotides. High concentrations of enzyme will also degrade double stranded nucleic acids. Used to remove single stranded extensions from DNA to produce blunt ends.
Nuclease BAL 31 (Alteromonas)	Functions as an exonuclease to digest both 5′ and 3′ ends of double-stranded DNA. It also acts as a single stranded endonuclease that cleaves DNA at nicks, gaps and single stranded regions. Does not cleave internally in duplex DNA. Used for shortening fragments of DNA at both ends.
Nuclease S1 (Aspergillus)	The substrate depends on the amount of enzyme used. Low concentrations of S1 nuclease digests single stranded DNAs or RNAs, while double-stranded nucleic acids (DNA:DNA, DNA:RNA and RNA:RNA) are degraded by large concentrations of enzyme. Moderate concentrations can be used to digest double stranded DNA at nicks or small gaps. Used commonly to analyze the structure of DNA:RNA hybrids (S1 nuclease mapping), and to remove single stranded extensions from DNA to produce blunt ends.
Ribonuclease T1 (Aspergillus)	An endonuclease that cleaves RNA at 3′ phosphates of guanine residues, producing oligonucleotides terminal guanosine 3′ phosphates. Used to remove unannealed regions of RNA from DNA:RNA hybrids.

a. **Exonucleases** are enzymes (found as individual enzymes, or as parts of larger enzyme complexes) that cleave nucleotides one at a time from an end of a polynucleotide chain. These enzymes hydrolyze phosphodiester bonds from either the 3′ or 5′ terminus of a polynucleotide molecule (Fig.13.9).

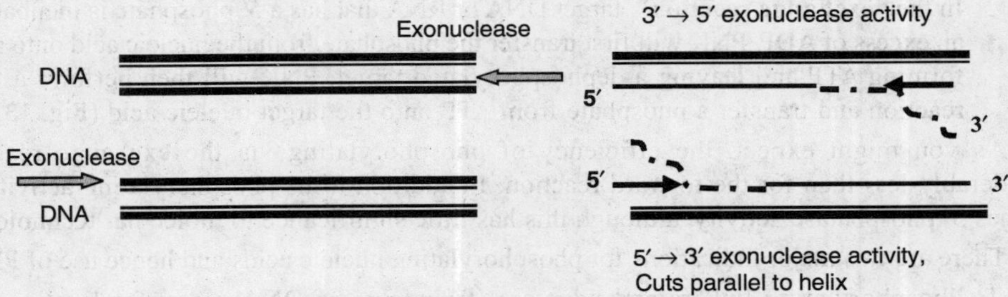

Fig. 13.9. Action of exonuclease.

b. **Endonucleases** are enzymes that cleave the phosphodiester bond within a polynucleotide chain, in contrast to exonucleases, which cleave phosphodiester bonds at the end of a polynucleotide chain. *Restriction endonucleases* (Restriction Enzymes) cleave DNA at specific sites, and are divided into three categories, Type I, Type II, and Type III, according to their mechanism of action. These enzymes are often used in genetic engineering to make recombinant DNA for introduction into bacterial, plant, or animal cells (Fig. 13.10).

Fig. 13.10. Action of endonuclease.

7. POLYNUCLEOTIDE KINASE

Polynucleotide kinase (PNK) is an enzyme that catalyzes the transfer of a phosphate from ATP to the 5′ end of either DNA or RNA. It is a product of the T4 bacteriophage, and commèrcial preparations are usually products of the cloned phage gene expressed in *E. coli.*

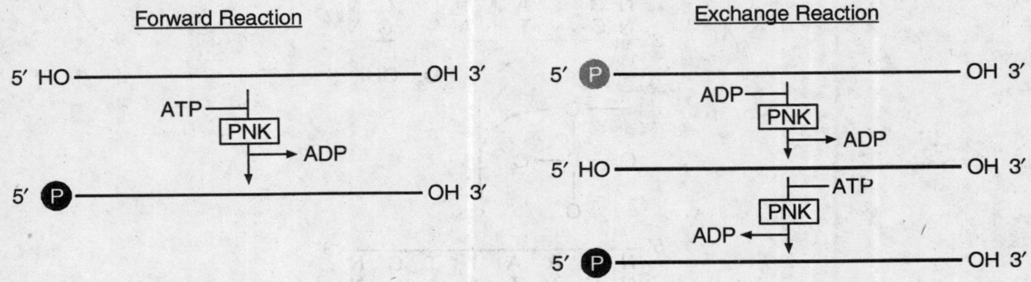

Fig. 13.11. Addition of phosphate at desired position.

The enzymatic activity of PNK is utilized in two types of reactions:

1. In the **"forward reaction"**, PNK transfers the gamma phosphate from ATP to the 5′ end of a polynucleotide (DNA or RNA). The target nucleotide is lacking a 5′ phosphate either because it has been dephorphorylated or has been synthesized chemically.
2. In the **"exchange reaction"**, target DNA or RNA that has a 5′ phosphate is incubated with an excess of ADP, PNK will first transfer the phosphate from the nucleic acid onto an ADP, forming ATP and leaving a dephosphorylated target. PNK will then perform a forward reaction and transfer a phosphate from ATP onto the target nucleic acid (Fig. 13.11).

As you might expect, the efficiency of phosphorylating via the exchange reaction is considerably less than for the forward reaction. In addition to its phosphorylating activity, PNK also has 3′ phosphatase activity, although this has little significance to molecular technologists.

There are two major indications for phosphorylating nucleic acids and hence use of PNK are:

1. Phosphorylating linkers and adaptors (fragments of DNA ready for ligation) which requires a 5′ phosphate. This includes products of polymerase chain reaction, which are typically generated using non-phosphorylated primers.
2. Radiolabelling oligonucleotides, usually with 32P, for use as hybridization probes.

PNK is inhibited by small amounts of ammonium ions, so ammonium acetate should not be used to precipitate nucleic acids prior to phosphorylation. Low concentrations of phosphate ions, or NaCl concentrations greater than about 50 mM, also inhibit this enzyme.

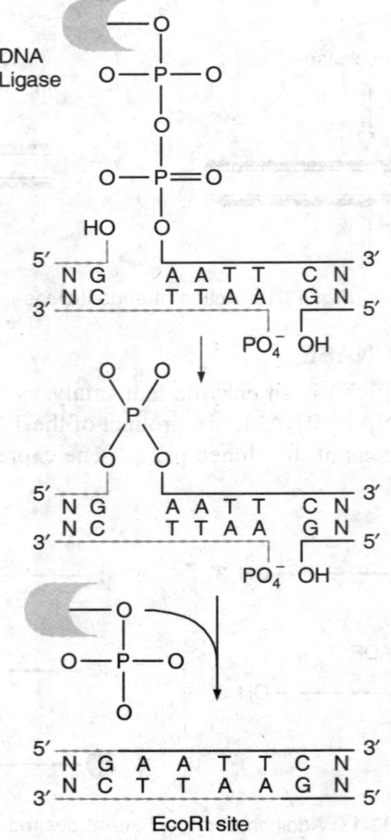

Fig. 13.12. Sealing of nick by ligase

8. DNA LIGASE

The term *recombinant DNA* includes the concept of recombining fragments of DNA from different sources into a new and useful DNA molecule. Joining linear DNA fragments together with covalent bonds is called ligation. More specifically, DNA ligation involves creating a phosphodiester bond between the 3′ hydroxyl of one nucleotide and the 5′ phosphate of another.

The enzyme used to ligate DNA fragments is **T4 DNA ligase**, which originates from the T4 bacteriophage. This enzyme will ligate DNA fragments having overhanging, cohesive ends that are annealed together, as in the EcoRI (Fig. 13.12). This is equivalent to repairing "nicks" in duplex DNA. T4 DNA ligase will also ligate fragments with blunt ends, although higher concentrations of the enzyme are usually recommended for this purpose.

9. REVERSE TRANSCRIPTASE

The initial information came from study of tobacco mosaic virus (TMV) in virology that, only RNA can act as genetic material, can infect, can produce complete virus, and finally isolation of replicase. This was followed by evidence for double stranded RNA by Weismann and August in 1968. **Reverse transcription** is the process of making a double stranded DNA (deoxyribonucleic acid) molecule from a single stranded RNA (ribonucleic acid) template. It is called reverse transcription as it acts in the opposite or reverse direction to transcription. This idea was very unpopular at first as it contradicted the central dogma of molecular biology which states that DNA is transcribed into RNA which is then translated into proteins. However, in 1970 when Mizutani and Howard Temin at Madison and David Baltimore at MIT, USA discovered the enzyme responsible for reverse transcription,

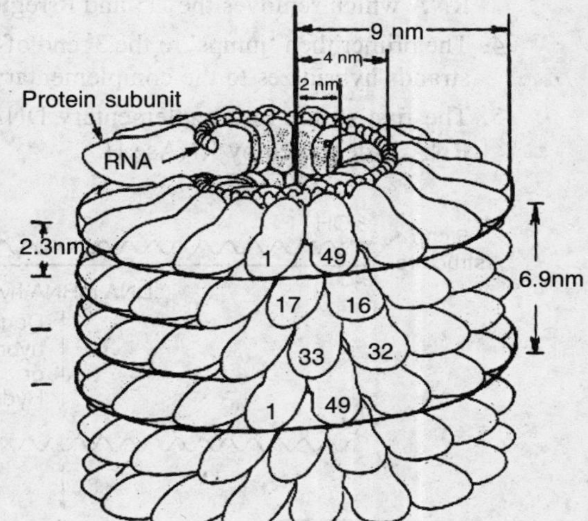

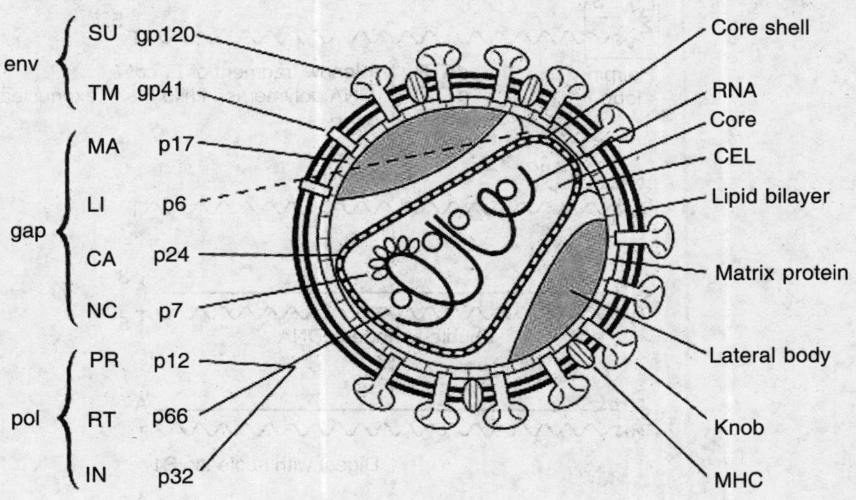

Fig. 13.13. Structure of TMV and HIV virus.

named reverse transcriptase, the possibility that genetic information could be passed on in the reverse manner was finally accepted (Fig. 13.13).

Class VI viruses ssRNA-RT, also called the retroviruses are RNA reverse transcribing viruses with a DNA intermediate. Their genomes consist of two molecules of positive sense single stranded RNA with a 5′ cap and 3′ polyadenylated tail. Examples of retroviruses include *Human Immunodeficiency Virus* (HIV) and *Human T-Lymphotropic virus* (HTLV). Once the viruses have entered the cell and been uncoated the genome is reverse transcribed into double stranded DNA which can be incorporated into the host cell and subsequently expressed. Reverse transcription by the enzyme reverse transcriptase occurs in a series of steps:

1. A specific cellular tRNA acts as a primer and hybridizes to a complementary part of the virus genome called the primer binding site or PBS.
2. Complementary DNA then binds to the U5 (non-coding region) and R region (a direct repeat found at both ends of the RNA molecule) of the viral RNA.
3. A domain on the reverse transcriptase enzyme called RNAse H degrades the 5′ end of the RNA which removes the U5 and R region.
4. The primer then 'jumps' to the 3′ end of the viral genome and the newly synthesised DNA strands hybridizes to the complementary R region on the RNA.
5. The first strand of complementary DNA (cDNA) is extended and the majority of viral RNA is degraded by RNAse H.

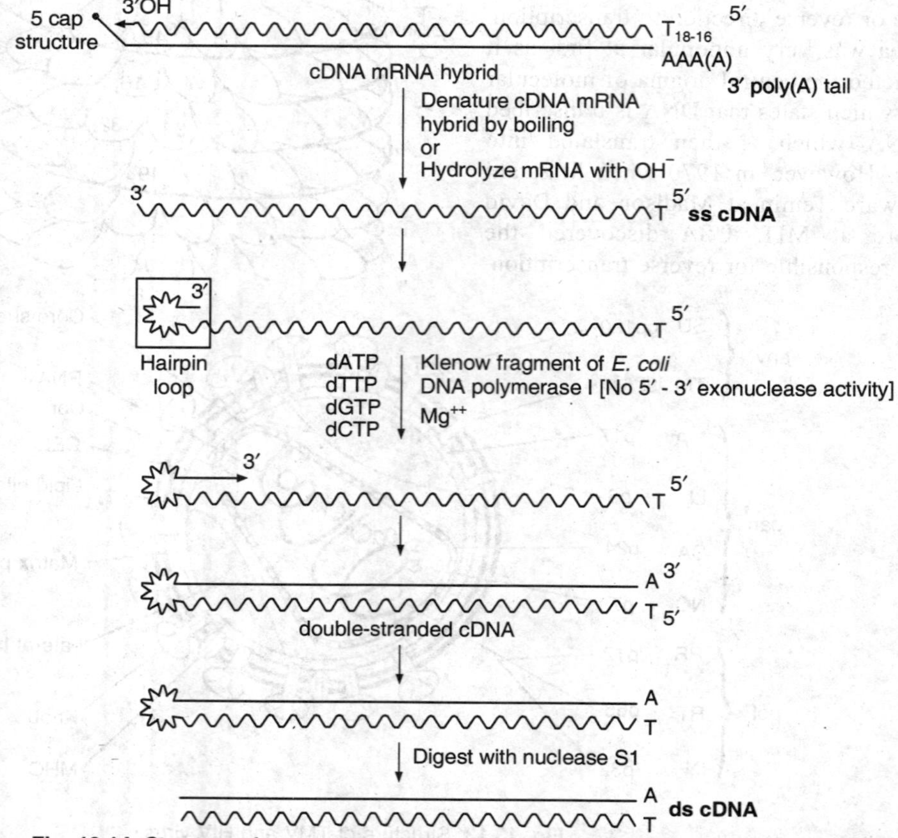

Fig. 13.14. Structure of reverse transcriptase and process of DNA formation from RNA.

6. Once the strand is completed, second strand synthesis is initiated from the viral RNA.
7. There is then another 'jump' where the PBS from the second strand hybridizes with the complementary PBS on the first strand.
8. Both strands are extended further and can be incorporated into the hosts genome by the enzyme integrase. DNA duplex so generated directs the remainder of the viral infection process (immediate lysis or integrate in to host genome and remains dormant) (Fig. 13.14).

Like other DNA and RNA polymerases, RT synthesizes polynucleotides in the 5′ to 3′ direction. Similar to DNA polymerase it requires a primer. Here, primer is a tRNA molecule captured by the virion from the host cell in which it was produced. 3′ end of the tRNA is base-paired with the viral template at the site where DNA synthesis initiates and its free 3′-OH accepts the deoxynucleotides. This enzyme has three types of enzyme activities:

1. RNA-directed DNA polymerase, hence called RT.
2. RNAse H activity, degrade RNA in RNA-DNA hybrid.
3. DNA-directed DNA polymerase.

The enzyme is a heterodimer consisting of 66kDa and 51kDa subunits.The two polypeptides have identical N-terminal sequences, indicating that the 51 kDa unit is cleaved from 66 kDa unit. The 66kDa monomer has identifiable 'finger', 'palm', and 'thumb' components which are functionally analagous to those in the Klenow fragment structure. The interface of the 66 kDa and 51kDa monomers provides a track from the polymerase active site to the RNaseH active site, and allow for the concomitant destruction of the RNA primer template strand directly after it has been copied (Fig. 13.15).

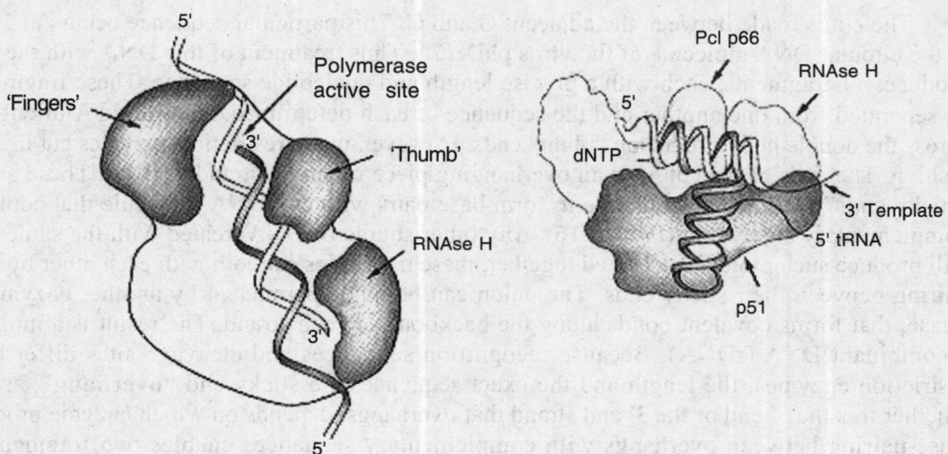

Fig. 13.15. Structure and domains of reverse transcriptase.

In a complex of RT with a double stranded DNA molecule, the DNA close to the polymerase active site is in the A-form, while the rest is in the normal B-form. The 51kDa monomer may also provide a binding site for the lysine-tRNA that primes initial DNA synthesis. HIV RT is of great clinical significance because it is responsible for replication and infection of HIV. HIV RT produces error of incorporation mismatch base at a rate of 1base /2000-4000 nucleotides polymerized. This means virus changes rapidly making vaccine development a difficult task.

10. RESTRICTION ENZYMES

Restriction enzymes are **DNA-cutting enzymes** found in bacteria (and harvested from them for use). Because they cut within the molecule, they are often called **restriction endonucleases**. A restriction enzyme recognizes and cuts DNA only at a particular sequence of nucleotides. For

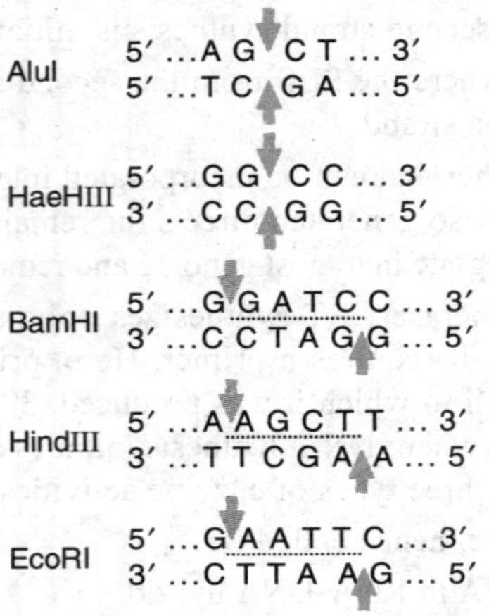

AluI
5′ ...A G C T ... 3′
5′ ...T C G A ... 5′

HaeHIII
5′ ...G G C C ... 3′
3′ ...C C G G ... 5′

BamHI
5′ ...G G A T C C ... 3′
3′ ...C C T A G G ... 5′

HindIII
5′ ...A A G C T T ... 3′
3′ ...T T C G A A ... 5′

EcoRI
5′ ...G A A T T C ... 3′
3′ ...C T T A A G ... 5′

AluI and HaeIII produce blunt ends
BamHI HindIII and EcoRI produce "sticky" ends

Fig. 13.16. Model of action of restriction enzyme.

example, the bacterium **Hemophilus aegypticus** produces an enzyme named **HaeIII** that cuts DNA wherever it encounters the sequence

5′GGCC3′
3′CCGG5′

The cut is made between the adjacent G and C. This particular sequence occurs at 11 places in the circular DNA molecule of the virus phiX174. Thus treatment of this DNA with the enzyme produces 11 fragments, each with a precise length and nucleotide sequence. These fragments can be separated from one another and the sequence of each determined. HaeIII and AluI cut straight across the double helix producing "blunt" ends. However, many restriction enzymes cut in an offset fashion. The ends of the cut have an overhanging piece of single-stranded DNA. These are called "sticky ends" because they are able to form base pairs with any DNA molecule that contains the complementary sticky end (Fig. 13.16). Any other source of DNA treated with the same enzyme will produce such molecules. Mixed together, these molecules can join with each other by the base pairing between their sticky ends. The union can be made permanent by another enzyme, DNA ligase, that forms covalent bonds along the backbone of each strand. The result is a molecule of recombinant DNA (rDNA). Because recognition sequences and cleavage sites differ between restriction enzymes, the length and the exact sequence of a sticky-end **"overhang"**, as well as whether it is the 5′ end or the 3′ end strand that overhangs, depends on which enzyme produced it. Base-pairing between overhangs with complementary sequences enables two fragments to be joined or "spliced" by a DNA ligase.

A sticky-end fragment can be ligated not only to the fragment from which it was originally cleaved, but also to any other fragment with a compatible sticky end. The sticky end is also called a cohesive end or complementary end in some reference. If a restriction enzyme has a non-degenerate palindromic (the sequence on one strand reads the same in the same direction on the complementary strand e.g. GTAATG is not a palindromic DNA sequence, but GTATAC is, GTATAC is complementary to CATATG) cleavage site, all ends that it produces are compatible. Ends produced by different enzymes may also be compatible.

10.1. Naming

Restriction enzymes are named based on the bacteria in which they are isolated in the following manner:

E *Escherichia* (genus)
co *coli* (species)
R RY13 (strain)
I First identified Order ID'd in bacterium

10.2. Patterns of DNA Cutting by Restriction Enzymes

- **5′ overhangs:** The enzyme cuts asymmetrically within the recognition site such that a short single-stranded segment extends from the 5′ ends. BamHI cuts in this manner.

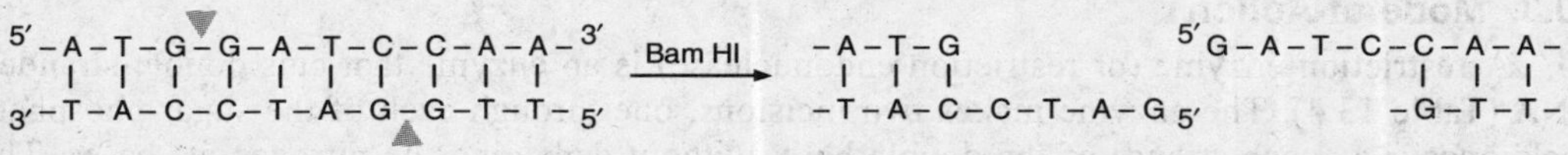

```
    ▼
5′-A-T-G-G-A-T-C-C-A-A-3′   Bam HI    -A-T-G            5′G-A-T-C-C-A-A-
   | | | | | | | | | |      ------>    | | |                    | | |
3′-T-A-C-C-T-A-G-G-T-T-5′              -T-A-C-C-T-A-G 5′        -G-T-T-
                  ▲
```

- 3′ overhangs: Again, we see asymmetrical cutting within the recognition site, but the result is a single-stranded overhang from the two 3′ ends. KpnI cuts in this manner.

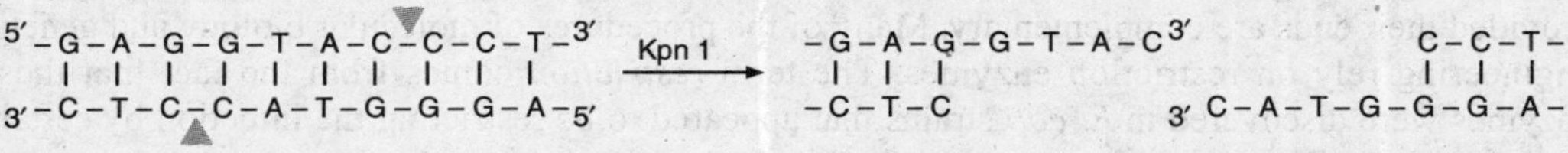

```
                ▼
5′-G-A-G-G-T-A-C-C-C-T-3′   Kpn 1    -G-A-G-G-T-A-C 3′        C-C-T-
   | | | | | | | | | |      ------>   | | |                   | | |
3′-C-T-C-C-A-T-G-G-G-A-5′             -C-T-C            3′ C-A-T-G-G-G-A-
        ▲
```

Table 13.4. Restriction enzymes, their sources and recognition sites.

Enzyme	Source	Recognition Sequence	Cut
*Eco*RI	*Escherichia coli*	5′GAATTC 3′CTTAAG	5′---G AATTC---3′ 3′---CTTAA G---5′
*Bam*HI	*Bacillus amyloliquefaciens*	5′GGATCC 3′CCTAGG	5′---G GATCC---3′ 3′---CCTAG G---5′
*Hind*III	*Haemophilus influenzae*	5′AAGCTT 3′TTCGAA	5′---A AGCTT---3′ 3′---TTCGA A---5′
*Taq*I	*Thermus aquaticus*	5′TCGA 3′AGCT	5′---T CGA---3′ 3′---AGC T---5′
*Hinf*I	*Haemophilus influenzae*	5′GANTC 3′CTNAG	5′---G ANTC---3′ 3′---CTNA G---5′
*Sau*3A	*Staphylococcus aureus*	5′GATC 3′CTAG	5′--- GATC---3′ 3′---CTAG ---3′
*Pov*II	*Proteus vulgaris*	5′CAGCTG 3′GTCGAC	5′---CAG CTG---3′ 3′---GTC GAC---5′
*Hae*III	*Haemophilus egytius*	5′GGCC 3′CCGG	5′---GG CC---3′ 3′---CC GG---5′
*Alu*I	*Arthrobacter luteus*	5′AGCT 3′TCGA	5′---AG CT---3′ 3′---TC GA---5′
*Eco*RV	*Escherichia coli*	5′GATATC 3′CTATAG	5′---GAT ATC---3′ 3′---CTA TAG---5′
*Sal*I	*Streptomyces albue*	5′GTCGAC 3′CAGCTG	5′---G TCGAC---3′ 3′---CAGCT G---5′
*Sca*I	*Streptomyces caespitosus*	5′AGTACT 3′TCATGA	5′---AGT ACT---3′ 3′---TCA TGA---5′

- **Blunts:** Enzymes that cut at precisely opposite sites in the two strands of DNA generate blunt ends without overhangs. SmaI is an example of an enzyme that generates blunt ends.

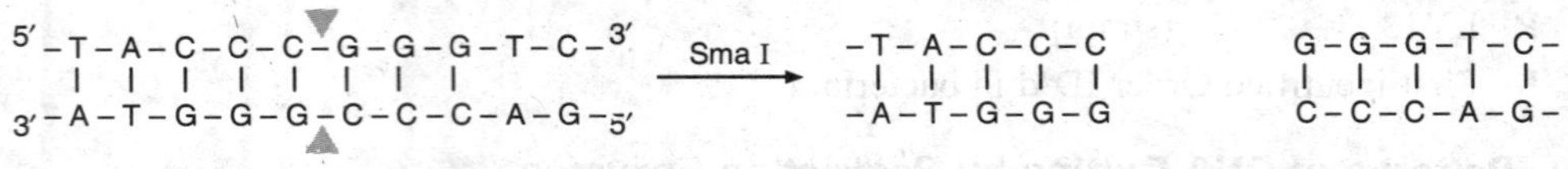

```
         ▼
5′-T-A-C-C-C-G-G-G-T-C-3′           -T-A-C-C-C        G-G-G-T-C-
   | | | | | | | | | |      Sma I      | | | | |        | | | | |
3′-A-T-G-G-G-C-C-C-A-G-5′  ------>  -A-T-G-G-G        C-C-C-A-G-
             ▲
```

The 5′ or 3′ overhangs generated by enzymes that cut asymmetrically are called ***sticky ends*** or ***cohesive ends***, because they will readily stick or anneal with their partner by base pairing.

10.3. Mode of Action

A restriction enzyme (or restriction endonuclease) is an enzyme that cuts double-stranded DNA (Table 13.4). The enzyme makes two incisions, one through each of the sugar-phosphate backbones (i.e., each strand) of the double helix without damaging the nitrogenous bases. The chemical bonds that the enzymes cleave can be reformed by other enzymes known as ligases, so that restriction fragments carved from different chromosomes or genes can be spliced together, provided their ends are complementary. Many of the procedures of molecular biology and genetic engineering rely on restriction enzymes. The term *restriction* comes from the fact that these enzymes were discovered in *E. coli* strains that appeared to be restricting the infection by certain bacteriophages. Restriction enzymes therefore are believed to be a mechanism evolved by bacteria to resist viral attack and to help in the removal of viral sequences. They are part of what is called the restriction modification system.

The 1978 Nobel Prize in Medicine was awarded to Daniel Nathans, Werner Arber and Hamilton Smith for the discovery of restriction endonucleases, leading to the development of recombinant DNA technology. The first practical use of their work was the manipulation of *E. coli* bacteria to produce human insulin for diabetics. The ability to produce recombinant DNA molecules has not only revolutionized the study of genetics, but has laid the foundation for much of the biotechnology industry. The availability of human insulin (for diabetics), human factor VIII (for males with hemophilia A), and other proteins used in human therapy all were made possible by recombinant DNA.

10.4. Types of Restriction Enzymes

Restriction enzymes are classified biochemically into three types. These are designated as Type I, Type II, Type III. A major type of Type II enzymes are sometimes referred to as Type IV enzymes.

Type I and III systems, both the methylase and restriction activities are carried out by a single large enzyme complex. Although these enzymes recognize specific DNA sequences, the sites of actual cleavage are at variable distances from these recognition sites, and can be hundreds of bases away. Both require ATP for their proper function. Type I restriction enzymes produce DNA cleavage following translocation of the DNA, which makes them important molecular motors. The cleavage of DNA appears to occur after blockage of the translocation activity (often following collision with another translocating Type R-M enzyme, but also due to other factors). These enzymes read the methylation status of their recognition sequence, compare the methylation status of two adenines within the recognition sequence, and if both adenines are unmethylated (a signal that the DNA is non-host DNA), the enzyme undergoes a conformational switch that turns the enzyme into a molecular motor and endonuclease. However, if either one of the adenines is methylated (a signal that the DNA is host DNA) then the enzyme acts as a maintenance methylase and methylates the other adenine.

In Type II systems, the restriction enzyme is independent of its methylase, and cleavage occurs at very specific sites that are within or close to the recognition sequence. The vast majority of known restriction enzymes are of type II, and it is these that find the most use as laboratory tools. They produce discrete bands during gel electrophoresis, and are useful for DNA analysis and gene cloning. The first to be discovered and utilized was EcoRI, which is staggered and its recognition sequence is 5′-GAATTC-3′. Type II enzymes are further classified according to their recognition site. Most type II enzymes cut palindromic DNA sequences, while type IIa enzymes recognise non-palindromic sequences and cleave outside of the recognition site, and type IIb enzymes cut sequences twice at both sites outside the recognition sequence. Type IIs enzymes cleave the DNA at a considerable offset from the recognition sequence. The most common type II enzymes are those like HhaI, HindIII and NotI that cleave DNA within their recognition sequences.

Restriction enzymes usually occur in combination with one or two modification enzymes (DNA-methyltransferases) that protect the cell′s own DNA from cleavage by the restriction enzyme. Modification enzymes recognize the same DNA sequence as the restriction enzyme that they accompany, but instead of cleaving the sequence, they methylate one of the bases in each of the DNA strands. The methyl groups protrude into the major groove of DNA at the binding site and prevent the restriction enzyme from acting upon it. Together, a restriction enzyme and its "cognate" modification enzyme(s) form a restriction-modification (R-M) system. In some R-M systems the restriction enzyme and the modification enzyme(s) are separate proteins that act independently of each other. In other systems, the two activities occur as separate subunits, or as separate domains, of a larger, combined, restriction-and-modification enzyme.

10.5. Restriction Enzymes as Tools

Recognition sequences typically are only four to twelve nucleotides long. Because there are only so many ways to arrange the four nucleotides--A,C,G and T--into a four or eight or twelve nucleotide sequence, recognition sequences tend to “crop up” by chance in any long sequence. Furthermore, restriction enzymes specific to hundreds of distinct sequences have been identified and synthesized for sale to laboratories. As a result, potential “**restriction sites**” appear in almost any gene or chromosome. Meanwhile, the sequences of some artificial plasmids include a “**linker**” that contains dozens of restriction enzyme recognition sequences within a very short segment of DNA. So no matter the context in which a gene naturally appears, there is probably a pair of restriction enzymes that can cut it out, and which will produce ends that enable the gene to be spliced into a “plasmid”.

Another use of restriction enzymes can be to find specific SNPs. If a restriction enzyme can be found such that it cuts only one possible allele of a section of DNA (that is, the alternate nucleotide of the SNP causes the restriction site to no longer exist within the section of DNA), this restriction enzyme can be used to genotype the sample without completely sequencing it. The sample is first run in a restriction digest to cut the DNA, then gel electrophoresis is performed on this digest. If the sample is homozygous for the common allele, the result will be two bands of DNA, because the cut will have occurred at the restriction site. If the sample is homozygous for the rarer allele, the sample will show only one band, because it will not have been cut. If the sample is heterozygous at that SNP, there will be three bands of DNA. This is an example of restriction mapping.

QUESTIONS

1. What are DNA polymerase enzymes? Describe different types of eukaryotic polymerases and their role in DNA synthesis.
2. Describe the role of reverse transcriptase in the synthesis of cDNA.
3. What are restriction enzymes and their mode of action? Describe in brief different types of restriction enzymes.
4. Write short notes on:
   (a) DNA ligase
   (b) Alkaline phosphatase
   (c) Klenow fragments
   (d) DNA Taq polymerase
   (e) Terminal transferases
   (f) Nucleases
   (g) Calf intestinal phosphatase

CHAPTER 14

Vectors and Recombinant DNA Technology

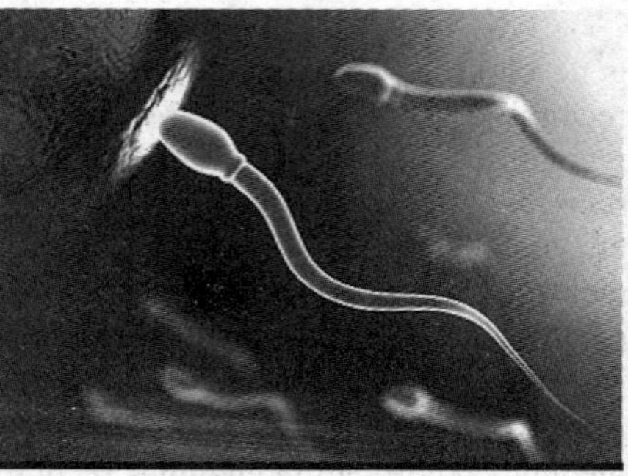

In this chapter, different vectors (carrier DNA) used for gene transfer for bacteria and plants are described. Plasmids are naturally occurring self replicating DNA molecules present in a variety of organisms, which are tailored for carrying foreign gene. For plants, Ti or Ri plasmids from *Agrobacterium* and their suitably modified vectors are used. Basic principle of producing recombinant plasmid, its uptake by bacteria and selection of transformed bacteria are described.

1. PLASMIDS

Plasmids have a wide range of structures; they may be composed of DNA or RNA, of double or single stranded nature in circular or linear form. The size of plasmid vary from 1.5 kb (kilobase) to 1500 kb, with majority of them being circular. Large linear DNA plasmids up to 500 kb are present in streptomyces.

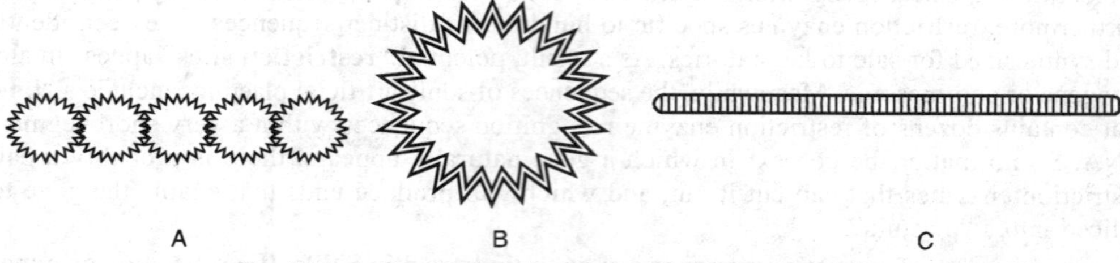

Fig. 14.1. A. Supercoiled; B. Circular; C. Linear plasmid.

DNA plasmids are present in the following forms (Fig. 14.1)

1. Covalently closed circular molecule – both strands of DNA are intact.
2. Open circles – only one of the strands of DNA is intact.
3. Linear DNA molecules.
4. Supercoiled circles.

Because of different sizes and forms, they have different mobility in agarose gel electrophoresis. These are DNA molecules and hence all the enzymes of DNA synthesis and breakdown can act upon to cause changes in their forms. Mainly plasmid in bacteria exists in supercoiled form.

Definition- Plasmids are self replicating, double stranded, extrachromosomal DNAs maintained as independent molecules by most of the bacterial genera. Lederberg in 1952 coined the term 'plasmid' to all extrachromosomal, covalently closed circular (CCC) DNA molecules. Presence or absence of plasmid from a bacterial cell makes no difference to the bacterial cell. Many types of plasmids are found in nature, in bacteria, yeast and fungi.

Classification- Plasmids are of two types:

a. Conjugative or transmissible plasmids
b. Non-conjugative or non-transmissible plasmids.

Conjugative plasmids can mediate their own transfer between bacteria by the process of conjugation. They possess *tra* (transfer) and *mob* (mobilizing) regions to carry on transfer function. Conjugative plasmids are found in many gram negative and some gram positive bacteria. Non-conjugative plasmids are not self-transmissible, but can be mobilized by conjugative plasmids that are present in the same cells.

Plasmids may also be classified on the basis of the number of copies found in host cells.

1. Low copy number (1-4 copies per cell)
2. Moderate copy number (10-100 copies per cell)
3. High copy number (>100 copies per cell).

Copy number is a parameter of bacterial population and may vary around mean values in the individuals of the colony. The copy number of plasmid is generally defined as the ratio of number of moles of plasmid DNA to the number of moles equivalents of chromosomal DNA. Conjugative plasmids are large with stringent control of DNA replication and are present in low copy number. Non-conjugative plasmids are small, high in copy number and have relaxed control of DNA replication. Large number (high copy) of copies is produced because of relaxed control on DNA replication. High copy number plasmids are useful for expression of cloned gene to get higher yields. But when such plasmids are used as vectors, they completely utilize the host's metabolism.

Each plasmid has an origin of replication. Plasmid DNA in a bacterium may be 0.1-5% of the total DNA. Plasmids, in general, may have about 100 genes.

Plasmid Incompatibility- Some micro-organisms may have as many as 8-10 different plasmids. Plasmid uncompatibility is the inability of two different plasmids to co-exist in the same host cell in absence of selection pressure. Currently 30 incompatibility groups are known for *Escherichia coli* plasmids.

Importance as Cloning Vector- Plasmids have basic properties like origin of replication, self transfer capacity and selectable markers. Due to these properties it can act as a potential vector. But plasmids also lack some important characteristics required for high quality cloning vector. The drawbacks are – (a) a small size and efficiency to transfer maximum of 15 kb of foreign DNA; (b) single restriction endonuclease site; (c) one or more selectable genetic markers. These drawbacks have been removed by genetically modifying the natural plasmid into construction of plasmid cloning vectors. The number of copies of such constructed plasmids, which is used as vector, is increased by replication. Inhibitors of protein synthesis are used to inhibit chromosomal transcription (e.g., antibiotics such as chloramphenicol or streptomycin) but plasmid replication occurs without any effect. Natural plasmids are those plasmids, which are not constructed *in-vitro* for the sole purpose of cloning, e.g., pSC 101, Col E1 and RSF 2124.

Plasmid Stability - Stability of plasmid is related to certain loci in plasmid which look after faithful segregation of plasmid copies to daughter cells. These partition loci are apparently similar to the centromere of eukaryotic chromosomes and are sites for attachment of plasmid to cell membrane to ensure distribution at cell division. Some plasmids have alternative mechanism to ensure stability. They have cer loci, which reverses oligomer formation of plasmid (which can cause instability and loss of plasmid).

Stability of plasmid is important when plasmid is used as vector. Genetic modifications have been used to overcome instability problem of plasmid. This is done by introducing *par* locus in high copy number plasmid vector or by inserting cloned gene into chromosome itself.

Examples of Plasmid Types- Genes located on various plasmids can express characteristics such as antibiotic resistance, antibiotic production, sugar fermentation, enterotoxin production, heavy metal resistance, hydrogen sulphide gas production, degradation of aromatic compounds and induction of plant tumours and hairy roots. Different types of plasmid showing various characteristics are presented in Table 14.1.

Table 14.1. Plasmid types and their characteristics.

Plasmid type	Examples	Characteristics
Natural	Col E1, pSC 101 F, RK6	Non-conjugative Conjugative
Low copy number	pBR 322	Moderate copy number plasmids are derived from this plasmid
High copy number	pUC 19, PBR 322	
Conjugative	F, R1, AK6, Ent P307	
Non-conjugative	Col E1, RSF 1030, pSC 101	
Commonly used plasmid vector	pBR 322, pAT153, pUC	Antibiotic resistance

Gene Cloning Vectors- An important requirement in gene cloning is a vector or a vehicle. Plasmids or viral DNA has been used for this purpose. The vehicle DNA is cleaved at a specific site by a restriction endonuclease, preferably one that produces a staggered cut in the two strands. The piece of foreign DNA which is proposed to be introduced into the vector, is also produced by the action of the same restriction endonuclease, thus generating similar ends. Thus foreign DNA or gene is ligated in to vector and then vector is allowed to enter bacteria.

2. pBR322

In the 1980s, one of the best-studied and most often used general-purpose plasmid cloning vectors was pBR322 (p=plasmid, B=F. Bolivar, R=R. Rodriguez, the scientists engineered this plasmid; 322=the numbers used by them to designate the plasmid). Plasmid pBR 322 contains 4361 bp. AS shown in Figure 14.2, pBR322 carries two antibiotic resistance genes; one provides resistance to Ampicillin (Amp^r), and the other confers resistance to tetracycline (Tet^r). This plasmid has also unique Bam HI, HindIII, and SalI recognition sites within Tet^r gene; a unique Eco RI site that is not within any coding DNA; and an origin of replication that functions only in *E. coli.* It is maintained at a high copy number in *E. coli* and can not be readily transferred to other bacteria.

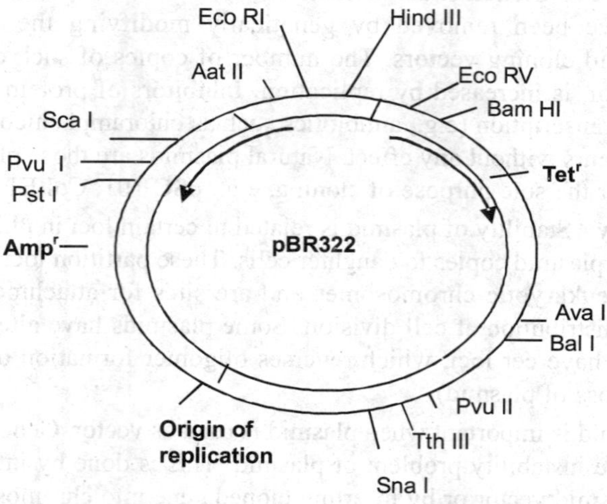

Fig. 14.2. Structure of pBR322.

3. pUC

It is a small plasmid (2.7kb) derived from pBR322. It contains following parts of pBR322: ampicillin resistant gene, and ColE1 origin. The second marker besides ampicillin resistance is *lacZα* of *E. coli* origin encoding for α fragment of β-galactosidase, the enzyme that hydrolyses lactose. The unique restriction sites used for integration of DNA inserts in to pUC vectors interrupt the *lacZα* fragment so that appropriate *E. coli* cells possessing recombinant pUC are β-galactosidase deficient and as result produce white colonies on X-gal medium (compare to blue colonies produced by non-recombinant cells). Thus pUc offers following advantages over pBR322 and similar vectors.

1. Each *E. coli* cell normally produces 500-700 copies of pUC.
2. Cells transformed with recombinant pUC are selected in single step using medium having ampicillin, X-gal and IPTG (isopropylthiogalactoside).
3. The unique restriction sites used for cloning are grouped within the polylinker or multiple cloning site. This allows cloning of a DNA fragment having two different sticky ends, e.g., *Eco*RI at one end and *Bam*HI at the other end. Vector pUC 8 has sites for HindIII, PstI, SalI, AccI, HincII, SmaI, XmaI, and Eco RI within its multiple cloning site.

Several modification of pUC vectors are valuable making them most popular vector.

Fosmids- Recently new vector is created called fosmid, which is similar to cosmid and is based on the bacterial F plasmid. This cloning vector is limited as a host can contain only one copy. Low copy number offers higher stability in comparison to high copy number plasmids. Fosmid system may be useful for constructing a stable library from a complex genome.

4. BACTERIOPHAGE AS CLONING VECTORS

The plasmid based vectors used for cloning DNA molecules generally carry up to 10 kb of inserted DNA. However, for the formation of library, it is often helpful to be able to maintain larger pieces of DNA. For this reason, *E. coli* virus (Bacteriophage, phage) lambda (λ) has been developed as a cloning vehicle. In its life cycle, bacteriophage λ infects *E. coli* and after injection of the viral DNA, two possibilities exist. Bacteriophage λ can enter a lytic cycle, which after 20 minutes lead to the lysis of host cells and the release of about 100 phage particles. Alternatively, the injected bacteriophage λ DNA can be integrated into the *E. coli* chromosome (DNA) as a prophage and can be maintained more or less indefinitely (Lysogeny stage). However, under conditions of nutritional or environmental stress, the integrated bacteriophage λ DNA can be excised and enter a lytic cycle. The bacteriophage λ DNA is about 50 kb in length, of which approximately 20 kb is essential for the integration -excision (I/E) events. For forming genomic libraries, 20 kb of DNA can be replaced with 20kb of cloned DNA.

Cloning Vectors Based on the Bacteriophage Lambda (λ)- Derivatives of the genome of bacterioplage lambda have been constructed to serve as cloning vectors. Transfection or transduction is used to introduce such vectors into *E. coli*. Two properties of the lambda genome make it suitable for use as a vector.

1. Only about 50% of the 50 genes of lambda are essential for its replication and for lysis of the host cell. Most of these non essential genes are located together in a cluster around the middle of the genome.
2. Lambda genome is packaged inside the phage head by what is known as the 'head-full mechanism'. This means that not only there is an upper limit of the amount of DNA that is packaged inside the phage head, but there is a lower limit also. Effective packaging takes place only when a minimum amount of DNA is present, i.e., 35 kb (kilo = thousand base).

An infective bacteriophage λ consists of a tubular protein tail with a few tail fibres and a protein head. The production and assembly of heads and tails, and packaging of DNA are highly coordinated sequence of events. The DNA within head of a λ phage is a 50 kb linear molecule with a 12-base, single stranded extension at the 5′ end. These extensions are called cohesive (*cos*) ends, because they contain sequences that are complementary to each other. In *E. coli*, these cos ends base pair to form a circular DNA. DNA replication from the circular DNA creates a linear form of λ DNA that is composed of several contiguous lengths of 50 kb units. Each new assembly is filled with 50 kb DNA (Fig. 14.3).

Inserting Type Lambda (λ) Vectors- A lambda cloning vector can therefore be constructed by deletion (*in vivo* or *in vitro* by restriction deletion) of a part of the non-essential region such that the remainder is not less than 35kb. Other mutations are also introduced such that restriction sites in the enemies regions are eliminated. A segment of foreign DNA can be cloned in an unique restriction site in the non-essential region, the only condition being that the vector and the insert together would not to be more than 53 kb long. Such vectors are termed "insertion vectors". Some of the Charon vectors are examples of this type of vectors. Bacteriophage λ cloning vector has two *Bam* HI sites that flank the I/E region. When this DNA is cut with *Bam* HI, three segments are produced. The middle segment I/E region, which is replaced by cloned DNA of 20 kb size. The source is cut with *Bam* HI, and DNA pieces that are 15 to 20 kb in length are isolated. The two DNA samples (phage and source) are combined and incubated with T4 DNA ligase. Then empty bacteriophage heads and tail parts are added. Under these conditions 50 kb unit of DNA are packed in to the heads, and infective phages are produced. Other products from ligation reaction cannot be packed, because they are either too large (>52 kb) or too small (>38 kb). Recombinant bacteriophage λ can undergo lytic cycle only in an *E. coli* strain that does not allow reconstituted phage λ (non-recombinant) with intact I/E regions to grow. Recombinant phage is maintained by lytic cycles in fresh *E. coli* cultures. Bacteriophage libraries can be screened by using either DNA probes or immunological assays. For this purpose, individual lytic zones are tested (compared to bacterial colonies in plasmid cloning vectors).

Substitution Type Lambda (λ) Vector- The second type of lambda vector is of the substitution type, the example being the lambda gt vectors and the EMBL vectors. These vectors have two Eco R1 sites or two BamHI sites in the non-essential region. On digestion with Eco R1, at least three piece(s) are produced, two terminal ones containing the essential regions and the central piece(s) containing the non-essential genes. The central piece (s) is separated out by sucrose density gradient sedimentation and replacing by the foreign segment to be cloned. The limits of the size of foreign DNA that can be cloned in the lambda gt vector is 1-14 kb and in the Charon 4 vector is 8.2-22.2 kb. Such a replacement cloning vector has an advantage over the insertion vectors. The terminal pieces by themselves if joined by DNA ligase, do not make up 35 kb and hence cannot be packaged. Packaging occurs only when a segment of foreign DNA gets cloned between the two terminal pieces of the vector and hence no separate selection for recombination molecules to is necessary.

5. PHAGEMIDS AS CLONING VECTORS

A phagemids is a hybrid of a plasmid and a filamentous coliphage that can be propagated in either form. The coliphage could be either of the three virtually identical phages, M13 fd or f1. These are male specific phages that contain single stranded circular DNA as their genome. Upon infection of *E. coli* by the bacteriophage, double stranded DNA is first formed as the replicative intermediate. Finally single stranded DNA is packaged into the virion. Both the replication origins of the plasmid and the coliphage are incorporated in the phagemid. The auxiliary replication functions necessary in trans for the coliphage replication are not, however, incorporate in the phagemid. Hence, replication from the coliphage origin can take place only in the presence of a helper phage. Otherwise, replication takes place from the plasmid replication origin. The pBLUESCRIPT phagemids have both Col E, (pMB 9 like) origin and the filamentous phage f1 origin. The cloning is done in any of the multiple cloning site in the double stranded circular DNA of the plasmid form of the vector. This is introduced in to *E.coli* by transformation and the synthesis of single stranded DNA from the phage f1 origin is induced by superinfection with a helper phage. The single stranded DNA formed is packaged in to the phage rods because of the presence of the phage packaging signals as well in the phagemid. Direct base sequencing can be undertaken using the single stranded DNA isolated from the virions secreted from the *E. coli* cells. Depending on the orientation of the f1 replication origin in the phagemid, either the (+) strand or the (–) strand of the phagemid is replicated in presence of the helper phage. The pEMBL phagemids are similar to the pBLUESCRIPT phagemids. Phagemids that have replication origin of pUC plasmids and of the M13 bacteriophage have also been developed and are available under the trade name of LITMUS vectors. Replication in the circular double stranded plasmid form uses the pUC replication origin. Cloning is done in the multiple cloning site located in the lacZ gene and the usual blue/white selection is available. Here also replication of the single stranded phage DNA form is induced by super infecting with the M13 helper phage.

6. COSMIDS AS CLONING VECTORS

Plasmid vectors are not suitable for cloning DNA fragments very much larger than their own size, as the transformation frequency fall beyond acceptable limits and cloned fragments or their parts very often get deleted. Takagi and co-workers observed as early as 1976 that the presence of the cohesive end site cos λ from the bacteriophage lambda DNA in a plasmid allows it to be packaged *in vivo* into virus particles. The interesting finding was that the *in vivo* packaging mechanism would be select DNA molecules of the full size of the lambda genome (~48.5 kb). Making use of this finding, cosmid vectors were first developed in 1978 by J. Collins and co-workers to facilitate cloning of larger DNA fragments in plasmids. Extracts of lambda lysogens have been successfully used for *in vitro* packaging of the lambda capsids. An example of a commonly used cosmid is pHV79 which is nothing but pBR322 containing the cohesive end site cos λ and which can accommodate up to 45 kb sized inserts. A great advantage of such a cosmid vector is that:

(1) gene libraries consisting of a smaller number of clone members can span the whole genome of an organism. For example, the genome of *Escherichia coli* can be accommodated in just 120 cosmids.

(2) Other advantages are that large gene can be studied intact and genetic linkage studies can be carried out at the molecular level.

(3) An important practical advantage of a cosmid is that background molecules which do not have the intact and genetics linkage studies can be carried out at the molecular level.

(4) An important practical advantage of a cosmid is that background molecules which do not have inserts or have smaller inserts are eliminated during packaging. This is not possible to achieve with plasmid cloning vectors.

(5) Besides, the frequency of transformation of the lambda capsids with an *in vitro* packaging extract is much higher than the transformation frequency of plasmids.

Cosmid cloning vectors can carry 40 kb of cloned DNA and can be maintained as plasmids in *E. coli*. Cosmids combine the properties of plasmids and bacteriophage λ vectors. The commonly used cosmid pLFR-5 (6kb size) has two cos sites (cos ends) from bacteriophage λ separated by a Sca I restriction endonuclease site, a multiple cloning sequence with six unique sites (*Hind* III, PstI, *Sal*I, *Bam*HI, *Sma*I, and *EcoRI*), an origin of DNA replication (*ori*) and a tetracycline resistance (*Tet*) gene. This cosmid carry about about 40 kb of cloned DNA. For this vector, pieces of DNA that are approximately 40 kb in length are purified by sucrose density gradient configuration from a partial digestion of source DNA with Bam HI. The pFLR-5 DNA is cleaved first with *ScaI* and then with *Bam HI*. The two DNA samples are mixed and ligated. Some of the ligand products will have a 40kb DNA piece inserted between the two fragments that are derived from the digestion of the pLFR-5 DNA. The molecules formed by joining will be about 50 kb long, with cos sequences that are about 50 kb apart. Therefore, these DNA constructs can be successfully packaged into bacteriophage λ heads in vitro (as described above, phage λ head accommodate only 50kb DNA). After formation of complete phage, the DNA is delivered by infection into *E. coli*.

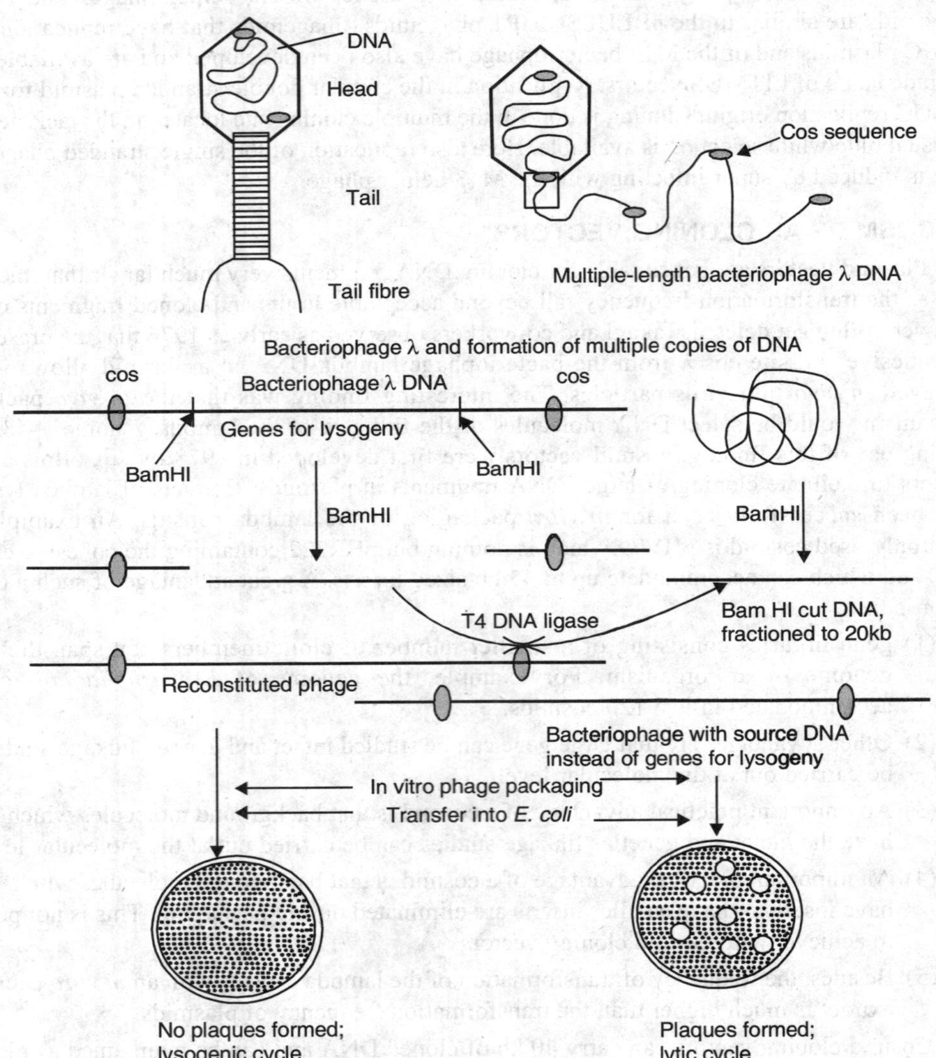

Fig. 14.3. Bacteriophage cloning system.

During phage packaging, cos ends are cleaved. Once inside the bacteria, the cos ends base pair and form a circular DNA molecules (Fig. 14.4). This circular form is stable, so the cloned DNA can be maintained as a plasmid-insert DNA construct because the vector contains a complete set of plasmid functions. Moreover, the tetracycline resistance gene allows colonies that carry the cosmid to grow in presence of tetracycline; non-transformed cells are sensitive to tetracycline and die.

The following are the steps for construction of a cosmid library. (*i*) Cleavage of the genome by partial digestion with restriction endonuclease, (*ii*) Sizing of the fragments by gel electrophoresis or velocity centrifugation; (*iii*) Cleavage of the cosmid vector and treatment with phosphate to minimize polycomid formation; (*iv*) Ligation of the genomic DNA and the cosmid DNA; (*v*) Packaging the ligated DNA into infectious phage particles; (*vi*) Transduction into *E. coli*.

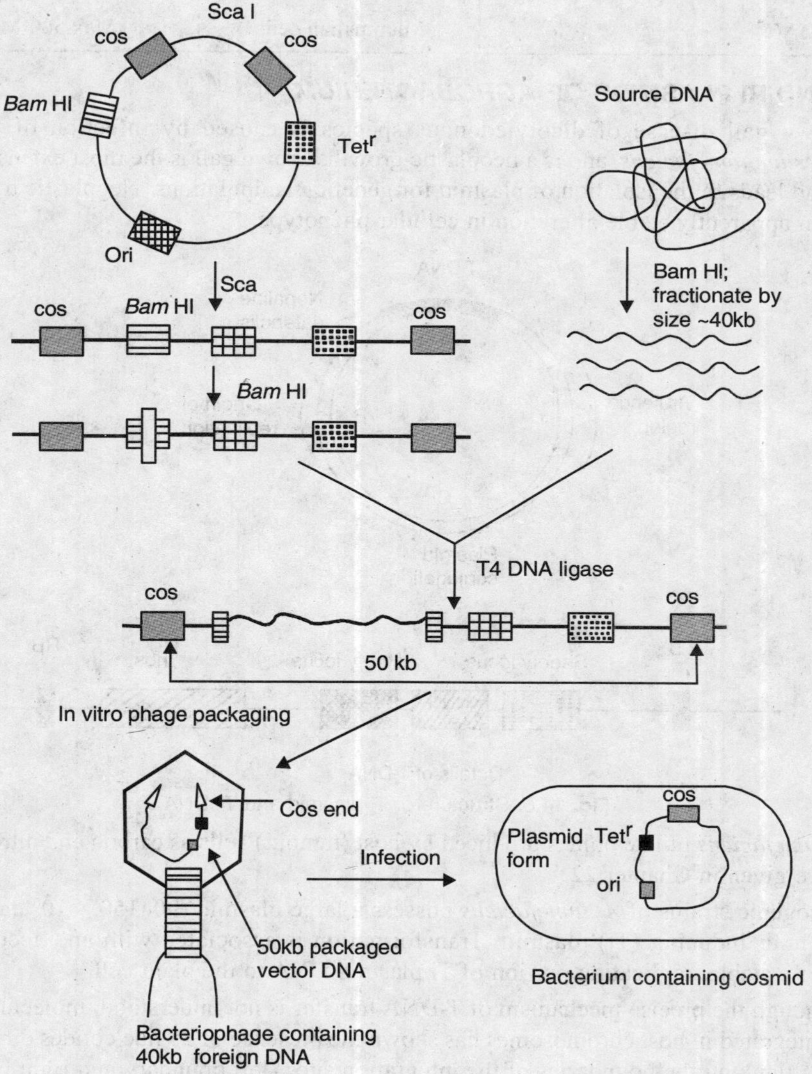

Fig. 14.4. Cosmid vector.

Therefore, it is evident that different vectors have different capacity of carry foreign DNA (Table 14.2).

Table 14.2 Capacity of different cloning vectors and host organisms.

Vector	Host organism	Uptake capacity
Plasmid	*E.coli*	Max.10 kb
Lambda phage	*E.coli*	Max. 25 kb
Cosmids	*E.coli*	35-45 kb
P1 phages (PAC,P1 derived)	*E.coli*	100-300 kb
Artificial chromosomes		
BAC	*E.coli*	Max. 300 kb
YAC	Yeast	100-2000 kb
MAC	Mammalian cells	Max.500 Mb

7. TI AND RI PLASMIDS OF *AGROBACTERIUM*

Crown gall disease of dicotyledonous species is caused by infection of the bacteria, *Agrobacterium tumefaciens*, and is a neoplastic growth. Crown gall is the most extensively studied disease and leads to the isolation of plasmid for genetic manipulations. Neoplastic transformation leads to an apparently stable alteration in cellular phenotype.

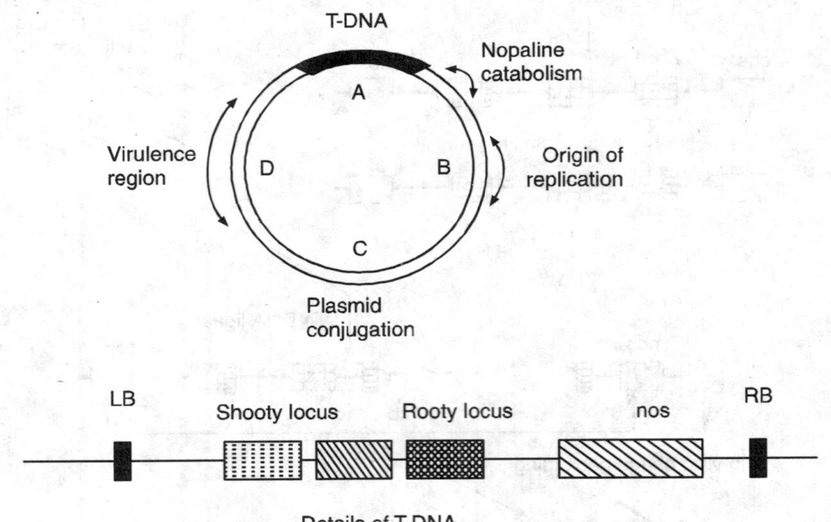

Fig. 14.5. Structure of Ti plasmid and T- DNA.

A. tumefaciens utilize opines produced by host (tumour) cells as carbon and nitrogen sources. Details are given in Chapter 22.

Oncogenic strains of *A. tumefaciens* possess a large plasmid (90–150 × 10^6 daltons) known as the tumour inducing (Ti) plasmid. Transformation is associated with and accomplished by transfer of a stable, replicating portion of Ti plasmid DNA to the plant cell.

Although the precise mechanism of T-DNA transfer is not understood, molecular analysis of T-DNA integrated in host chromosomes has shown the presence of 25 nucleotides directly repeated sequences flanking the boundaries of the integration sites (left boundary and right boundary, LB, RB). Genes that are to be introduced in the plant cells must be inserted between these left and right border sequences, or just adjacent to one border of T-DNA.

(*i*) **Structure of Ti plasmid:** Ti-plasmids have four distinct regions (Fig. 14.5)

A – T-DNA, main transfer DNA responsible for tumour formation.

B – Responsible for replication

C – Responsible for conjugation

D – Responsible for virulence (vir region, vir-genes responsible for virulence; mutation in this region may lead to non-virulence). Important region required for transfer of T-DNA.

(*ii*) **Structure of T-DNA:** Transfer – DNA is transferred and integrated into host genome during the infection. Infection brings about physiological and morphological changes in the tissue due to expression of genes located on T- DNA. These are:

– Onc region - shooty, rooty genes, IAA and cytokinin production.

– OS region - opines - unusual amino acids, octopine and nopaline. Octopine synthase and nopaline synthase (nos) coded by T-DNA. So depending upon synthesis of amino acids plasmid is known as (*a*) Octopine type - Ti plasmid, or (*b*) Nopaline type - Ti plasmid. These amino acids are used only by bacteria as source of carbon and nitrogen and these genes are located outside T-DNA.

– Other important point of interest is 25bp flanking border sequences (Lb,Rb),

Vir region - If vir region and T-DNA are physically separated but present on two plasmids in the same bacteria, transfer of DNA takes place. This is an important property for vector. It consists of 36 kb, and six operons; Vir- A, B, C, D, E & G. Except A & G, others are polycystronic (Fig. 14.6). Operons ABDG are for virulence. Operons C and E are for tumour formation.

Vir-A - chemoreceptor - sense presence of phenolics from wound - acetosyringone, B-hydroxyacetosyringone. Vir A transduces this information, most likely by a mechanism involving protein phosphorylation, to the product Vir-G. Vir-G then acts as transcriptional activator of itself and the other Vir loci. The products of Vir-C and Vir-D loci are involved in the generation and processing of the T-DNA copy. The products of the Vir-B and Vir-E loci are involved in forming most of the structural components that facilitate T-DNA movement.

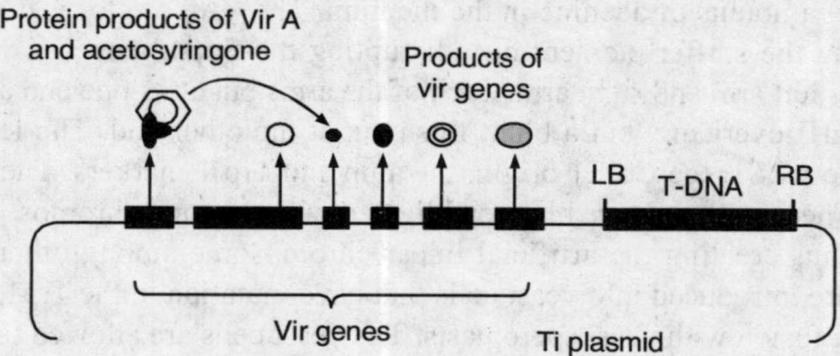

Fig. 14.6. Part of Ti plasmid showing vir region genes and their products.

8. ARTIFICIAL CHROMOSOMES

Efforts to clone and sequence the human genome and genomes of intensively studied organisms such as maize, *Arabidopsis thaliana*, *Drosophila*, and yeast require methods of cloning that gives large sized DNA fragments. Even though cosmids are excellent cloning vectors, a cosmid library of the human genome would require over 500,000 clones. Efforts to develop cloning vectors for large DNA vectors started with the extensively studied eukaryote, yeast. This organism has many of the useful attributes of *E.coli* (such as a rapid growth rate, single cells, and a relatively small genome). Yeast has a doubling time about twice that of *E.coli*, exist as a haploid and diploid, and has a genome only 2½ times (12 mb) that of *E.coli*. Yeast undergoes mitosis and meiosis, has a nucleus, and possesses the enzymes for processing RNA after transcription. This last feature is extremely important because *E.coli* is unable to process RNA in addition, yeast can be transformed easily after treatment with the mixture of snail gut enzymes that degrade the thick, complex yeast cell wall composed of proteins and polysaccharides.

8.1. Yeast Artificial Chromosomes (YAC)

Cloning of DNA fragments much larger than 45 kb became possible in 1987, when D.T. Burke and G.F. Carle developed in the laboratory of M.V. Olson an altogether new type of yeast vector, which they called yeast artificial chromosome (YAC). The development of YAC's were based on the logic that an eukaryotic linear chromosomes needs for its replication and stability, not only replication origins, but also the centromere and the telomere. The centromere sequence would attach to the mitotic spindle during cell division and help in efficient segregation of the chromosomes into the daughter cells. The telomere would preserve the integrity of the ends of the linear chromosomes. Once these elements were provided, the vector could be replicate stably like a chromosome and could accommodate chromosomes sized inserts. Indeed, standard YACs can accommodate around 600 kb DNA inserts, while special type of YACs can accommodate up to 1400 kb DNA inserts. Though an YAC vector is meant to be propagated like a chromosome in yeast, it is a circular double stranded DNA that contains a replication origin (colE 1) compatible with *E. coli* in addition to yeast replication origin or an yeast ARS element. The col E1 replication origin is useful to a yeast replication origin or an yeast ARS elements. The colE1 replication origin is useful for amplification of the vector in *E. coli*. Next to the yeast replication origin is located the centomere of yeast chromosome 4 (Cen 4). The two telomere sequences are from the protozoan Tetrahymenal, which have been found to be functional in yeast. A stuffer element containing the His 3 gene of yeast is present between the two telomere sequences through two BamHI restriction endonuclease sites. Three selectable yeast marker genes, Trp 1, Ura 3 and Sup 4 are also present. The marker genes Trp1 and Ura 3 are on the two sides of the unique restriction endonuclease site SnB1, that produces blunt ends. The SnaB 1 site is used as the cloning site and is located within the Sup4 gene. Sup 4 gene product is a tRNA that suppresses a mutation in the Ade gene of yeast resulting in a change in colour of colonies from red to white in the presence of limiting amounts of adenine in the medium. The YAC vector is digested with BamHI and SnaBI throwing the stuffer element out, disrupting the Sup4 gene and yielding two vector fragments termed as left arm and right arm. Each of the arms but at its one end a telomere sequence followed by a BamHI overhang, but a blunt flush cut at the other end. The left arm also has the Cen 4 sequence, the ARS1, the ColE1 origin, the amp and Trp 1 markers. The right arm contains the Ura 3 marker. These two arms are blunt end ligated with the long chromosomal DNA fragment from any source, thus creating the artificial linear chromosome among other ligation products. Ligation products are introduced into yeast cells that have mutation in the Trp1, Ura3 and Ade loci by lipofection or by fusion with yeast speroplasts. The yeast cells are allowed to regenerate the cell walls and plated on medium lacking trytophan and uracil and containing limiting amount of adenine. Only the yeast cells transformed with artificial chromosomes comprising both the left arm and the right arm would grow. The recombinant artificial chromosomes containing the insert would develop into white colonies. The red colonies would represent cells having the linear vector, but no insert. Cells having other ligation products like two left arms or two right arms would not grow.

8.2. Bacterial Artificial Chromosomes (BAC)

Bacterial artificial chromosomes (BAC) were developed by Mel Simmons and coworkers in the early 1990s and are based on the fertility factor (F factor) of *Escherichia coli*. The F plasmid, a ~ 100 kb circular double stranded DNA, is present is an *E. coli* cell in only 1-2 copies. The synthetic BAC vectors, which are only ~7.5 kb double stranded DNA circles contain the replication origin *oriS* and the gene *repE* of the F plasmid that are responsible for initiation and proper orientation of replication of the BAC vector. The *parA* and *parB* genes of the F plasmid ensure efficient segregation of the F factor into the daughter *E. coli* cells after its replication, are also incorporated in the BAC vector. The BAC vectors also contain multiple cloning sites (mcs), a selectable marker in the form of antibiotic resistance and colour based identification (lac Z complementation system) of recombinants carrying inserts. The naturally occurring F2 factors consist of up to 25% of the *E. coli* genome integrated into the basic F factor and are very stable. This characteristic of the F factor contributes to the ability BACs to accommodate very large amount of external DNA to the extent of 300kb. The recombinant BACs have been found to exhibit a lower level of rearrangement and chimerism of the cloned DNA sequence than exhibited by YACs. The cloning of DNA in BACs is done as is done in a plasmid, by linearising the vector with a restriction endonuclease, treating with phosphatase and then ligating with the DNA fragments to be cloned. *E. coli* has to be transformed by electroporation because of the large size of the recombinant BAC.

8.3. Mammalian Artificial Chromosomes (MACs)

The YACs, the BAC and the PACs (plasmid artificial chromosomes) have found regular use for cloning large genomic DNA fragments in various genome sequencing projects. Another potential application of artificial chromosomes is in gene therapy of human. For gene therapy, we need to have gene in human DNA fragments including their promoters and all the control elements. This would have to be introduced into the target cells efficiently and would have to be stably maintained inside the nucleus, generation after generation through unlimited number of divisions. The DNA would have to be expressed properly without interfering with the function of other resident. Such large artificial human chromosome (aptly termed minichromosomes), if 1-10 mb size have been claimed to be stable for more than 100 cell generations. Satellite DNA based minichromosomes of 20-30 mb have also been reported.

8.4. Artificial Human Chromosomes

Researchers at the School of Medicine and Athersys, Inc. have created the first artificial human chromosome. The synthetic chromosomes represent a breakthrough in medical research and provide scientists with a powerful new tool for the study of human genetics. Artificial chromosomes may also offer a new approach to gene therapy and the treatment of a broad range of genetic diseases. A report of the research was published in the April 2007 issue of *Nature Genetics*.

"This opens the door to a whole new avenue of research in chromosome biology and gene therapy," said Huntington F. Willard, chairman of genetics at the School of Medicine and University Hospitals of Cleveland. "While it's been known since the early years of this century that chromosomes carry genes, until now the complexity and size of normal chromosomes has limited our ability to analyze their structure and function. The synthetic microchromosome system now allows us to perform detailed studies on the nature of chromosomes - essentially the next phase of the Human Genome Project which is to move from just mapping genes to actually understanding how they work and influence human disease."

In this study, the research team created artificial chromosomes from normal human material. The researchers first synthesized arrays of alpha satellite DNA, then introduced the resulting centromeric material into human cells in conjunction with telomeres and genomic DNA. Inside the cells, the independent elements assembled to form miniature chromosomes, or synthetic microchromosomes, that were structurally similar to human chromosomes, but contained less genetic material. Analysis of the newly introduced artificial chromosomes demonstrated normal centromeric activity, genetic stability, and continued gene expression through repeated rounds of the cell cycle.

9. PRINCIPLE OF RECOMBINANT DNA TECHNOLOGY

Knowledge about cell and its functioning has increased to a great magnitude during 20th century. Science and genetics provided basic inputs and now we have gathered information about the structure of master molecule, DNA, its replication and control of gene expression. Developments in last three decades were very rapid and gene sequencing, gene cloning and gene transfer in eukaryotes and prokaryotes (and vice versa) have been achieved. This has become possible because genetic code is assumed to be universal.

Genetic engineering may or may not have recombinant DNA (rDNA) preparation step and with the advancement of gene transfer technology, artificially DNA can be transferred to different hosts (host cells) without any vector and host organism can be engineered to carry desired properties. The genetic manipulations are used to produce individuals having a new combination of inherited properties. Such manipulations may be of two kinds: (1) cellular manipulation involving culturing of cells (e.g., haploid cells) and hybridization of somatic cells (protoplast fusion), and (2) molecular manipulation, involving construction of artificial rDNA molecules, their insertion into a vector and their establishment in a host cell or organism. The latter approach has been called "recombinant DNA (rDNA) technology". Therefore, use of term "genetic engineering" and "rDNA technology" is overlapping. However, all the manipulations involving use of constructed gene or constructed gene transfer are termed as rDNA technology. The different **steps of recombinant technology** include (Fig.14.7):

a. **Gene cloning and development of recombinant DNA**- The foreign DNA (gene of interest) from the source is enzymatically cleaved and ligated (joined) to other DNA molecule i.e. cloning vector (plasmid, phagemid etc) to form recombinant DNA.

b. **Transfer of vector into the host**- This cloning vector with recombinant DNA is transferred into and maintained within a host cell. The introduction of rDNA into a bacterial host cell is called transformation.

c. **Selection of transformed cells (host)**- Those host cells that take up the rDNA are identified and selected from the pool.

d. **Transcription and translation of inserted gene**- If required, a rDNA construct can be prepared to ensure that the protein product that is encoded by the cloned DNA sequence is produced by the host cell.

9.1. Gene Cloning and Development of Recombinant DNA

Any gene to be cloned must be inserted in a cloning vector (plasmid). A foreign gene (DNA fragment) introduced (by transformation) into a bacterium cell will not be replicated with bacterium. The reason for this is that the enzyme DNA polymerase, which is responsible for copying DNA, does not initiate the process at random. It is initiated at selected sites known as "origin of replication". Generally, small fragments of DNA do not possess an origin of replication. Using rDNA technology, it is possible to insert the gene into a 'cloning vector', which in turn will make copies of the fragment (inserted DNA). A **cloning vector** is simply a DNA molecule possessing an 'origin of replication' and which can replicate in the host cell of choice. Most commonly 'plasmids', extra chromosomal, autonomously replicating, circular DNA molecules, are used as vectors. Sometimes, viruses are used as vector for gene insertion into micro-organisms, but they are better vectors for animal cells.

Cutting and insertion of desired foreign gene into the plasmid require special enzymes known as restriction endonucleases or restriction enzymes. These enzymes cut large DNA molecules into shorter fragments by cleavage at specific nucleotide sequences called 'recognition sites'. Therefore, restriction endonucleases are highly specific deoxy-ribonucleases (DNAse). Both vector DNA and foreign DNA to be inserted is cut by the same restriction enzyme, generating complementary ends. Thus, ends of foreign DNA make perfect match with cut ends of vector and join to make again a circular molecule. This process can be understood in detail by taking the example of pBR 322 plasmid. It is commonly used cloning vector (Fig. 14.2). Firstly, purified,

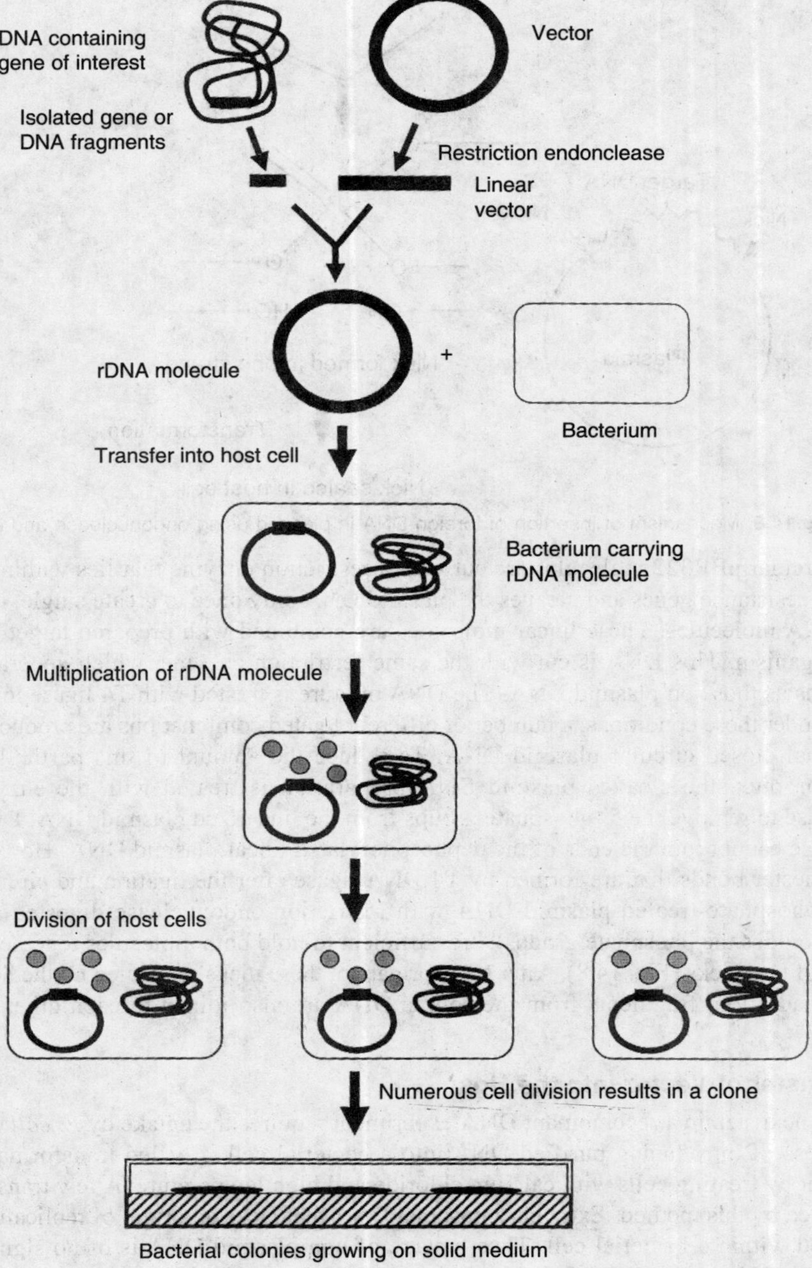

Fig.14.7. Construction of recombinant DNA and recombinant bacteria.

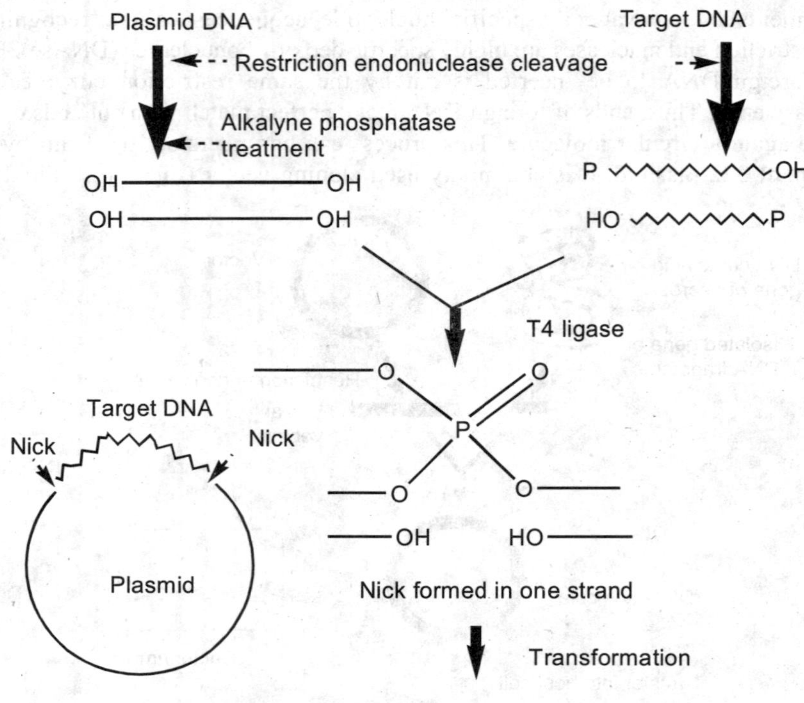

Fig.14.8. Mechanism of insertion of foreign DNA in plasmid using endonuclease and ligase.

closed, circular pBR322 molecules are cut with a restriction enzyme that lies within either of the antibiotic resistance genes and cleaves the plasmid DNA only once to create single, linear, sticky-ended DNA molecules. These linear molecules are combined with prepared target DNA from a source organism. This DNA is cut with the same restriction enzyme, which generates the same sticky ends as those on plasmid DNA. The DNA mixture is treated with T4 ligase in the presence of ATP. Under these conditions, a number of different ligated combinations are produced, including the original closed circular plasmid DNA. To reduce the amount of this particular unwanted ligation product, the cleaved plasmid DNA preparation is treated with the enzyme alkaline phosphatase to remove the 5′-phosphate groups from the linearized plasmid DNA. Due to this, T4 DNA ligase cannot join the ends of the dephosphorylated linear plasmid DNA. However, the two phosphodiester bonds that are formed by T4 DNA ligase after the ligation and circularization of alkaline phosphate-treated plasmid DNA with restriction endonuclease digested source DNA (which provides the phosphate groups), are sufficient to hold both molecules together, despite the presence of two nicks (Fig. 14.8). After transformation, these nicks are sealed by the host cell DNA ligase. In addition, fragments from the source DNA are also joined to each other by T4 DNA ligase.

9.2. Transfer of Vector into the Host

The next step in a recombinant DNA experiment requires the uptake by *E. coli* of the rDNA. The process of introducing purified DNA into a bacterial cell is called transformation. This is carried out by treating cells with calcium chloride and high temperature. A few transformed cells are obtained by this method. Extra chromosomal DNA that lacks an origin of replication cannot be maintained within a bacterial cell. Thus, uptake of non-plasmid DNA is of no significance in a recombinant DNA experiment. Suitable strain of *E. coli* is used which lacks capabilities to destroy plasmid DNA or carrying out exchanges between DNA molecules (Fig.14.9).

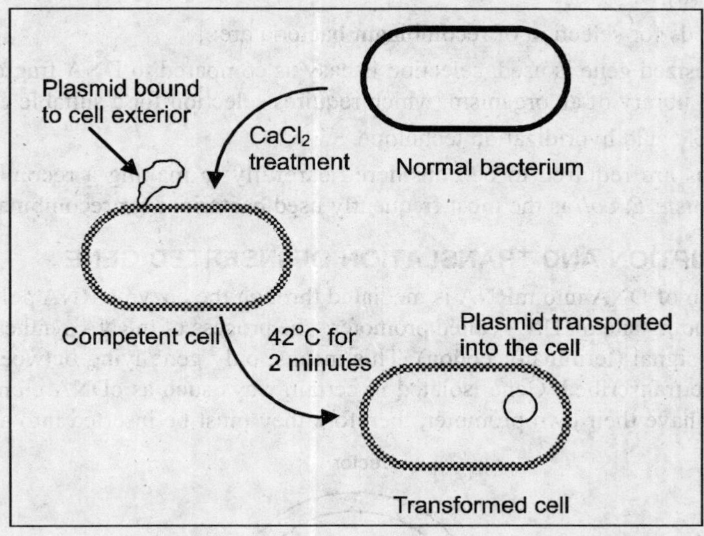

Fig.14.9. Uptake of DNA by bacterial cell.

9.3. Selection of Transformed Cells

After transformation, it is necessary to identify the cells that contain plasmid-cloned DNA constructs. In pBR 322 in which target DNA was inserted onto the BamHI site, recombinant bacteria (bacteria with recombinant plasmid) are selected. All cells are grown successively on media containing antibiotic, ampicillin or tetracycline and cells showing the recombinant DNA (depending upon the restriction enzyme site and loss of particular antibiotic resistance due to

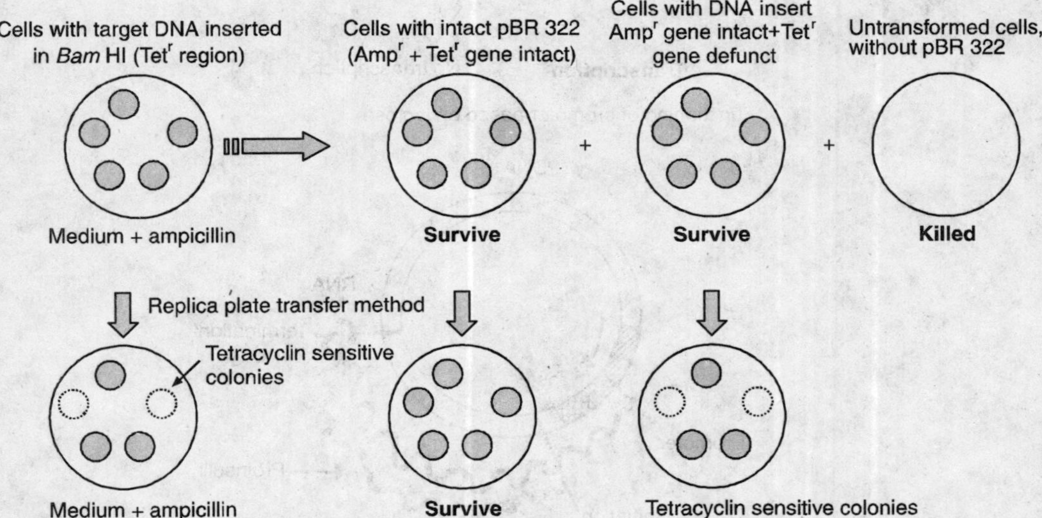

Fig. 14.10. Selection of recombinant bacterial cells containing recombinant plasmid. Colonies corresponding to tetracyclin sensitive colonies (B) are selected from original cultures (A). Growth of three types of cells obtained after transformation is also presented.

disruption of the gene). Selected recombinant bacteria are grown in bioreactor to obtain the gene product (Fig. 14.10).

Other methods for selection of recombinant bacteria are:

1. If synthesized gene is used, selection is easy as compared to DNA fragments used from genomic library of an organism, which requires selection for a suitable characters.
2. By nucleic acid hybridization technique.

All the steps are required to be considered carefully in making a recombinant bacteria. Prokaryotic organism *E. coli* is the most frequently used bacterium for recombinant technology.

9.4. TRANSCRIPTION AND TRANSLATION OF INSERTED GENE

Transcription of DNA into mRNA is mediated through the enzyme RNA polymerase, which recognizes the binding site on DNA called promoter. The process of mRNA synthesis is terminated by a termination signal (terminator codon). This means, only gene lying between promoter and terminator will be transcribed. Gene isolated in certain ways such as cDNA cloning or artificial synthesis, do not have their own promoter, therefore they must be inserted into a vector close to

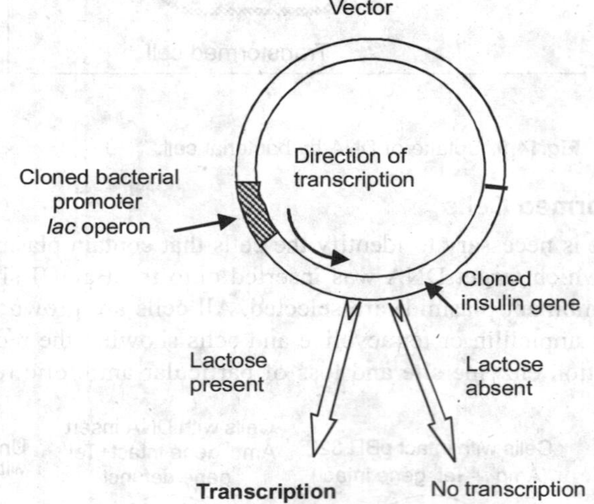

Functioning of promoter based on lactose

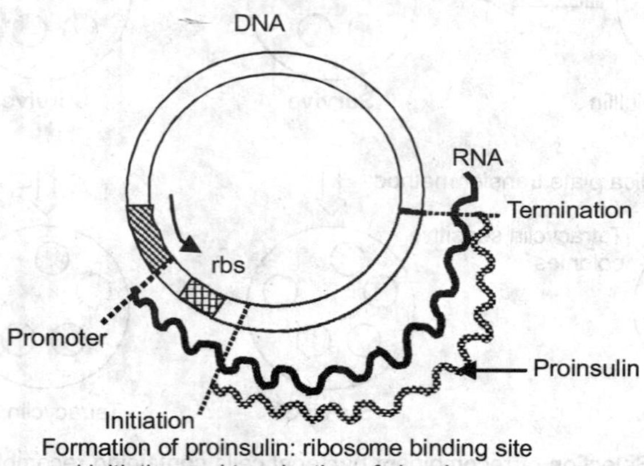

Formation of proinsulin: ribosome binding site and initiation and termination of signal

Fig. 14.11. Transcription of inserted gene and involvement of rbs and termination codon.

promoter site. Even if a cloned gene carries its own promoter, this promoter may not function in the new host cell. In such circumstances, the original promoter has to be replaced. The position of cloned insulin gene with lac promoter in the vector is given in the figure14.11. This gene is transcribed in presence of lactose because lac operon functions in presence of lactose and transcribe the gene attached downstream to attach it. Similarly, positions of ribosome binding site (rbs) and termination codon are shown in the figure 14.11 in respect to the insulin gene in the vector.

In the cell, transcription takes place inside the nucleus or close to the nucleoid in bacteria by the action of the RNA polymerase on DNA. This process can also be mimicked outside the cell, i.e. cloned DNA can be mixed with RNA polymerase and the four nucleotide in a tube and under appropriate conditions, RNA transcripts can be formed as it does inside the cell. This is known as *in vitro* transcription. The cellular RNA polymerase, whether it is bacterial or from higher organisms, is an extremely complicated enzyme, containing several subunits. It is very difficult to purify in an active form. On the other hand, several bacteriophages encode their own RNA polymerases, which are much simpler enzymes, are easy to purify and transcribe genes at a high efficiency. This is because they have evolved to only recognize rate. This phenomenon has been utilized in designing *in vitro* transcription systems. The bacteriophage RNA polymerases commonly used for this purpose include those from bacteriophage T3 and T7 which infect *E. coli* and SP6, which infects *Salmonella typhimurim.* The promoter sequences recognized by each of these RNA polymerases are different, but can be as small as 21bp in length. Thus, these promoters are designed to be part of the special cloning for *in vitro* transcription, just upstream of the cloning sites for the DNA fragments to be transcribed. The cloned DNA, placed downstream of the above promoters, is then incubated with the purified RNA polymerase from the bacteriophage, along with the precursor ribonucleotide triphosphate, which synthesizes transcripts specific for the cloned DNA. *In vitro* derived transcripts are used extensively as probes for the detection of specific nucleic acid fragments both in Southern as well as in Northern hybridization.

Translation: Translation of mRNA into proteins is a complex process which involves interaction of the mRNA with the ribosomes. For translation to take place the mRNA must carry a ribosome binding site in front (upstream) of the gene to be translated. Ribosome binds to this site and move along the mRNA and initiates protein synthesis at the first AUG codon, it encounters. This process is stopped by stop codon (UAA, UAG, UGA). In case the cloned gene lacks a rbs, then it is necessary to use a vector with promoter and rbs and the gene is inserted downstream to both these (promoter and rbs).

Translation in a cell is carried out by the ribosomes, which synthesize polypeptides by decoding the information carried by mRNA. In addition, amino-acyl tRNAs and other proteinaceous accessory factors are also utilized. Biochemically the process of translation is not yet fully characterized, which means that we do not know all the requirements for polypeptide synthesis and very few of the components required for the process of translation have actually been purified. Despite having the above drawbacks, translation of a given mRNA can still be performed outside the cell, if the mRNA is incubated in a mixture of components which supports translation. This is termed *in vitro* translation. The above components are generally isolated from *E. coli* cells (S-30 fraction), plants (wheat germ) or animals (rabbit reticulocytes). Because of the universality of the genetic code a given mRNA gets translated to the same extent and with the same efficiency irrespective of the in vitro translation system used. Since there are already many types of protein molecules in the translation mix, new protein synthesis is usually monitoring by adding radioactive amino acids in the translation mix. Thus, all newly synthesized proteins get radioactively labelled and can be easily detected by autoradiography. In many instances, a combined *in vitro* transcription and translation system is used which can produce the encoded polypeptide directly from a given cloned gene. Such systems are now commercially available.

Splicing mRNA: Bacterial and viral genes have simple structure as all the genetic information in the mRNA between the initiation and stop codons is translated into protein. Many genes of eukaryotic organisms, including the human insulin gene, have a more complex structure. They are consist of coding regions (or exons), which contribute to the final protein sequence, and non-coding regions (or introns), which are not translated into proteins.

In eukaryotes, genes containing introns are transcribed into mRNA in the usual manner, but then the corresponding intron sequences are spliced out. As bacteria can not spliced out introns, they cannot be used directly to express many genes from mammals or other eukaryotes. This is done by ribozymes-RNA molecules having catalytic activity.

Insulin (a dipeptide) formation by processing of mRNA (removal of introns) and translation of exons is well studied. There are two ways to overcome this problem.

1. By use of reverse transcriptase, a cDNA copy of the processed mRNA is prepared and this cDNA (gene) is used for insertion in the vector.
2. The gene for protein may be synthesized in test tube, which lacks introns.

Posttranscriptional modification in the gene

A number of proteins undergo posttranscriptional modifications. Proteins that are destined to be transported out of cell are synthesized with extra 15-30 amino acids at the amino terminal (N-terminus). These extra amino acids are referred to as a signal sequence. The common feature of these sequences is that they have a central core of hydrophobic amino acids flanked by polar or hydrophilic residues. During passage through the membrane the signal sequence is cleaved off, making the protein active.

QUESTIONS

1. What are plasmids? Give characteristics of a vector required for gene transfer.
2. Describe basic principle of recombinant DNA technology in prokaryotes.
3. Describe the factors influencing production recombinant protein by rDNA technology.
4. Write short notes:
   (a) Copy number in plasmid
   (b) Plasmid types
   (c) Phagemids
   (d) YAC
   (e) BAC
   (f) pBR 322
   (g) Bacteriophage
   (h) Cosmid
   (i) Ti plasmid
   (j) T-DNA
   (k) Vir region
   (l) Mammalian artificial chromosomes
   (m) Artificial human chromosomes
   (n) Bacterial transformation
   (o) Recombinant plasmid
   (p) Selection of transformants
   (q) rDNA
   (r) cDNA
   (s) Introns removal
   (t) Translation of recombinant gene
   (u) Restriction endonuclease
   (v) Transgene expression in bacteria
   (w) Transgene expression in yeast
   (x) Transgene expression in plants
   (y) Transgene expression in bacteria

CHAPTER 15

Tools of Genetic Engineering

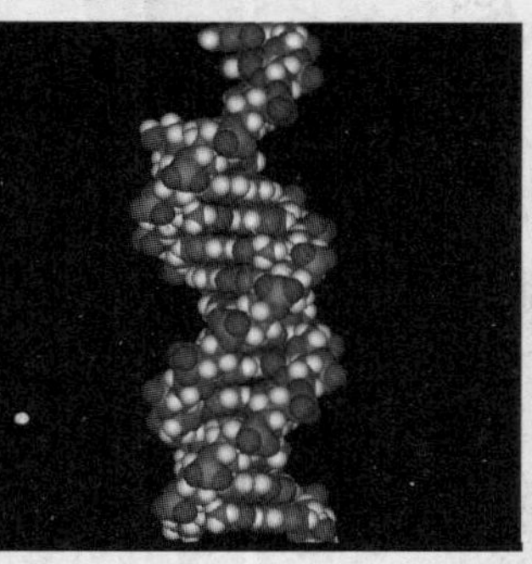

1. GENOMIC LIBRARIES

Creation of genomic library involves cloning of whole genomic DNA. Therefore, a genomic library is a collection of recombinant DNA molecule (plasmids, phages) so that the sum total of DNA inserts in this collection represents the entire genome of the organism. Depending upon the size of genome, prokaryotic or eukaryotic, vector can be selected. This way whole genome can be cut into pieces and cloned in vectors or by PCR technique. This can be referred as **genomic cloning**. Making of genomic library includes the following steps-

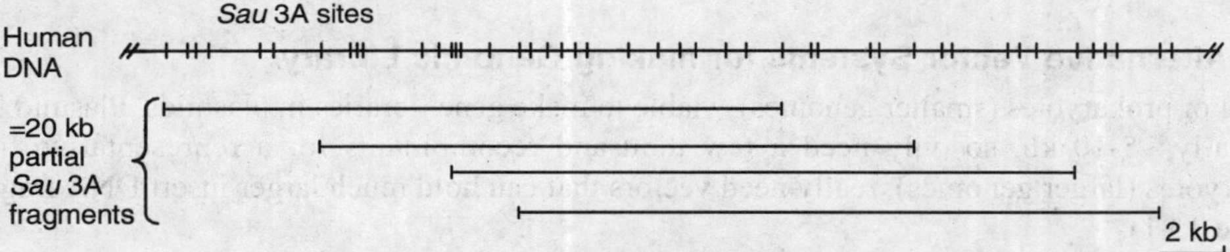

Fig. 15.1. Partial digestion for longer, overlapping DNA fragments.

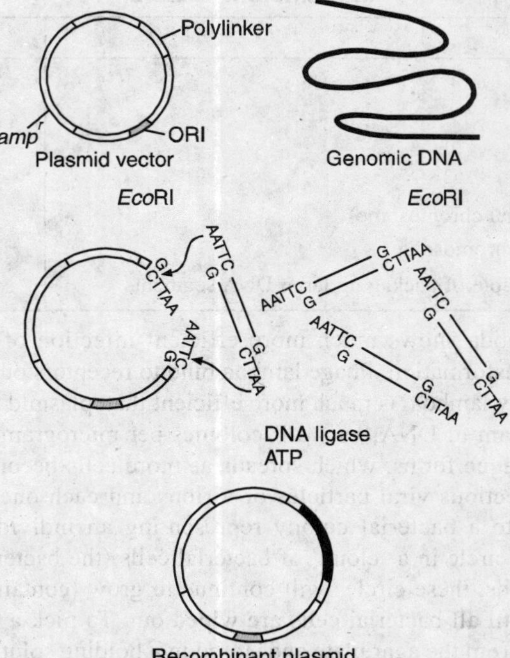

Fig. 15.2. Development of recombinant plasmid for cloning.

1. Isolation of chromosomal DNA; details are described in Chapter 21.
2. Fragmentation of DNA is carried out by mechanical shearing or sonication or by using a suitable endonuclease for partial digestion of DNA (Fig. 15.1). Mechanical shearing generates blunt ends while endonuclease produces cohesive ends. Complete digestion is avoided as it generates fragments too heterogenous in size to be used.
3. Ligation of DNA fragments- The partial digest of genomic DNA are subjected to agarose gel electrophoresis for separation of fragments of required size. The fragments of appropriate size are eluted from gel. These fragments are then inserted into a suitable vector. These vectors are cut with the same restriction enzymes. After that the DNA fragments are ligated into vector using DNA ligase (Fig. 15.2). By this technique about 25kb DNA fragments can be inserted into vectors.
4. Identification of the desired clone- Identification of the bacterial colony containing the desired DNA fragment by employing a suitable hybridization probe in colony hybridization. The probe may be mRNA of the gene, cDNA of its mRNA, homologous gene from another organism, or a synthetic oligonucleotide representing the sequence of a part of the desired gene/DNA fragment. In order to detect hybridization the probe must be labelled usually with a radiolabelled isotope.

1.1. Alternative Vector Systems for making Genomic Library

For prokaryotes (smaller genomes)- viable to make gene libraries in plasmids. Plasmid inserts normally ~5-10 kb, so only need a few thousand recombinants for a representative library. Eukaryotes (larger genomes), really need vectors that can hold much larger insert DNA fragments (Table 15.1).

Table 15.1. Approximate maximum length of DNA that can be cloned into different vectors.

Vector type	Cloned DNA (kb)
Plasmid	20
lambda phage	25
Cosmid	45
P1 phage*	100
BAC (bacterial artificial chromosome)	300
YAC (yeast artificial chromosome	1000
*P1- *E. coli* virus capable of packaging large DNA segments	

Bacteriophage lambda shows much more efficient infection of *E. coli* than that can be achieved by plasmid transformation. Phage lambda bind to receptors on the surface of *E. coli* and injects DNA. Infection by lambda is much more efficient than plasmid transformation - can get ~ 10^9 plaques per microgram of DNA, vs ~ 10^6 colonies per microgram of plasmid DNA. As the bacterial cells lyse, a plaque forms, which spreads as more cells become infected and lyse. The plaque contains the infectious viral particles or virions and each one represents an individual lambda clone (similar to a bacterial colony representing an individual plasmid clone). The appearance is of a clear circle in a 'cloud' of bacterial cells (the bacterial 'lawn') (Fig. 15.3). If incubated for long periods, these circles will continue to grow (containing more and more viral particles) and spread until all bacterial cells are wiped out. To pick a single lambda clone, the plaque is 'punched' out from the agar plate and stored in a holding solution. This can then be used to infect more bacterial cells and to replicate more of the recombinant lambda DNA.

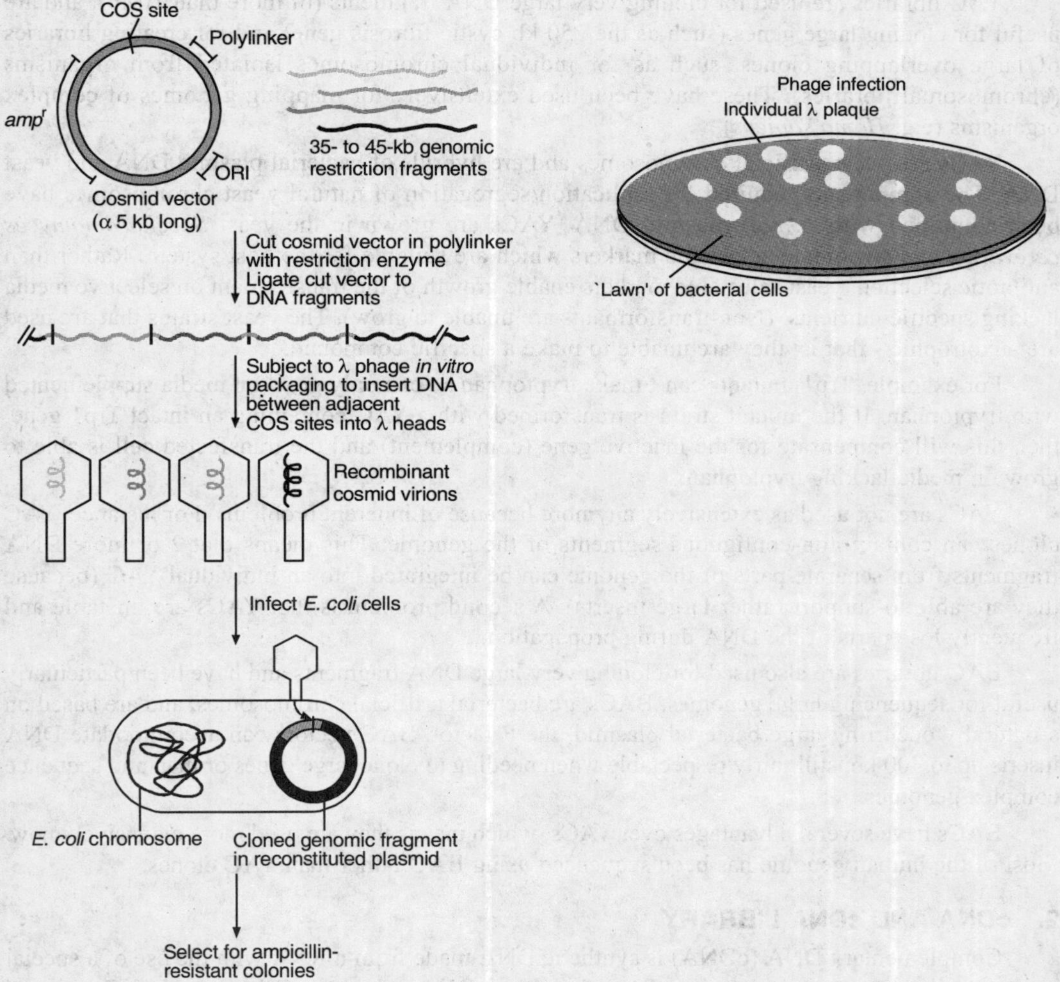

Fig. 15.3. Making a genomic library in bacteriophage lambda vector and plaque formation by transformed bacteriophage in bacterial lawn.

1.2. Vectors for making Libraries with Larger Inserts

Cosmid libraries are used for cloning genes with large introns and for sequencing larger chunks of the genome. Cosmid vectors are hybrids of plasmid and bacteriophage lambda λ DNA (a small ~5 kb plasmid containing the plasmid origin of replication (ori), an antibiotic resistance gene such as amp and a suitable restriction site for cloning along with the COS sequence from phage λ DNA). Because of the COS sequence, cosmid recombinants can be packaged into viral particles (allowing high efficiency transformation). Since most of Genomic libraries in cosmid vector, the λ DNA has been discarded, it can be replaced by the DNA of interest, so long as it doesn't exceed the 50 kb limit for packaging into the viral head. The insert sizes are of the order of 35-45 kb. Since the recombinant DNA does not encode any lambda proteins, cosmids do not form viral particles (or plaques) but rather forms large circular plasmids and the colonies that arise can be selected on antibiotic plates, like other plasmid DNA transformants. Cosmid clones can be manipulated similarly, allowing ease of plasmid isolation (see previous lecture). Since many eukaryotic genes are on the order of 30 - 40 kb, the likelihood of obtaining a DNA clone containing the entire gene sequence is increased significantly when using a cosmid library.

YAC libraries are used for cloning very large DNA fragments (of more than 1 Mb), and are useful for cloning large genes (such as the 250 kb cystic fibrosis gene) and for creating libraries of large overlapping clones, such as for individual chromosomes isolated from organisms (chromosomal libraries). These have been used extensively for mapping genomes of complex organisms (e.g. *Homo sapiens*).

YACs are yeast artificial chromosomes and are hybrids of bacterial plasmid DNA and yeast DNA. The components required for replication/segregation of natural yeast chromosomes have been combined with *E. coli* plasmid DNA. YACs are grown in the yeast *Saccharomomyces cerevesiae* and so contain selectable markers which are suitable for the host system. Rather than antibiotic selection, yeast selectable markers enable growth of the transformant on selective media lacking specific nutrients. (Non-transformants are unable to grow). The yeast strains that are used are auxotrophic - that is, they are unable to make a specific compound.

For example, Trp1 mutants can't make tryptophan so can only grow on media supplemented with tryptophan. If the mutant strain is transformed with a YAC containing an intact Trp1 gene, then this will compensate for the inactive gene (complement) and the transfected cell is able to grow on media lacking tryptophan.

YACs are not used as extensively anymore because of inherent problems. For instance, YAC clones can contain non-contiguous segments of the genome. This means that 2 or more DNA fragments from separate parts of the genome can be integrated into an individual YAC (because they are able to support rather large inserts). A second problem is that YACs are unstable and frequently lose parts of the DNA during propagation.

BAC libraries are also used for cloning very large DNA fragments and have been particularly useful for sequencing large genomes. BACs are bacterial artificial chromosomes, and are based on a naturally occurring large bacterial plasmid, the F factor. BAC vectors can accommodate DNA inserts up to 300 kb, still fairly respectable when needing to clone large genes or map and sequence complex genomes.

BACs have several advantages over YACs, which means they are used more extensively now. Most of the human genome has been sequenced using BAC rather than YAC clones.

2. cDNA AND cDNA LIBRARY

Complementary DNA (cDNA) is synthetic DNA made from mRNA with the use of a special enzyme called reverse transcriptase (RNA dependent DNA polymerase discovered by Temin and Baltimore in 1970). Originally this enzyme was isolated from reteroviruses as described in chapter 13. This enzyme performs similar reactions as DNA polymerase and requires a primer with a free 3′ OH terminal. With the use of a mRNA as a template, reverse transcriptase synthesizes a single stranded DNA molecule that can then be used as template for double-stranded DNA (Fig. 15.4). Because it is made from mRNA, cDNA is devoid of both upstream and downstream regulatory sequences and of introns. This means that cDNA from eukaryotes can be translated into functional protein in bacteria, an important feature when expressing eukaryotic genes in bacterial host.

A short oligo (dT) chain is hybridized to the poly (A) tail of mRNA strand. The oligo (dT) segment serves as a primer for the action of reverse transcriptase, which uses the mRNA as template for the synthesis of cDNA strand. The resulting cDNA ends in a hairpin loop. When the mRNA, from RNA-DNA hybrid, has been degraded by treatment with NaOH or RNase H, a short piece of RNA is left behind. This hair pin loop becomes a primer for DNA polymerase I, which completes the paired DNA strand. The loop is then cleaved by S1 nuclease (which acts only on the single stranded loop) to produce a double stranded cDNA molecule. This double stranded DNA can be inserted into plasmid, cosmid, phage lambda, cloning vectors by blunt-ended ligation.

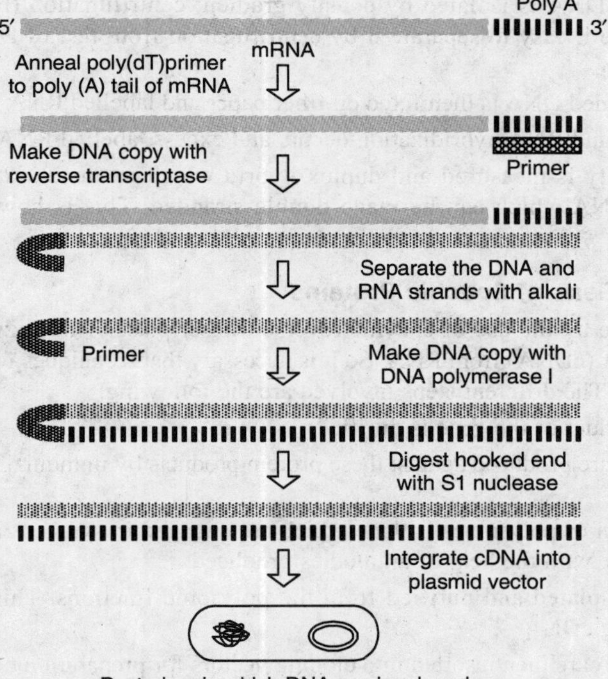

Fig. 15.4. Synthesis of double stranded cDNA from mRNA with the use of reverse transcriptase.

A cDNA library is a population of bacterial transformants or phage lysates in which each mRNA isolated from an organism or tissue is represented as its cDNA insertion in a plasmid or a phage vector. The frequency of specific cDNA in such a library would depend on the frequency of the concerned mRNA in that tissue.

3. ANALYSIS OF GENES AND GENE TRANSCRIPTS

Recombinant DNA technology involves localization of gene of interest. This gene can be either isolated or synthesized before they are manipulated and used for transformation leading to the production of transgenic plants and animals. Different techniques have been used for isolation of different varieties of genes like ribosomal RNA gene, gene for phenotypic traits with unknown gene product, for specific protein products, and genes involve in regulatory functions, e.g. promoter genes etc.

3.1. ISOLATION OF RIBOSOMAL RNA GENES

The ribosomal RNA makes the 80% of total RNA and is synthesized on ribosomal gene which could be isolated. Isolation of this gene was easy due to some characteristics features of rRNA (a) ribomsomal genes are present in multiple copies (b) difference between ribosomal genes and other genes, due to their relatively high G + C content in rRNA. The ribosomal gene was isolated for the first time in 1965 by H. Wallace and M.L. Birnstiel in an amphibian named *Xenopus*. The different steps involved in isolation of rRNA genes are the following:

1. rRNA is isolated from the ribosomes and are made radiolabelled by allowing them to replicate in tritiated uridine containing medium.

2. Ribosomal DNA is isolated by density gradient centrifugation (high G+C content of rDNA make it easy to separate it by centrifugation from rest of DNA) followed by its denaturation.
3. Single stranded DNA is then fixed on filter paper and labelled RNA is added to the paper.
4. Now DNA and RNA hybridization occurs and excess labelled RNA is washed off.
5. Radioactivity is measured and duplex hybrid which on denaturation will give single stranded DNA which can be made double stranded. This is how rRNA gene can be isolated.

3.2. Isolation of Gene of Specific Proteins

This is possible by the use of enzyme reverse transcriptase. This enzyme can make copy/complementary DNA (cDNA) from RNA. So it is necessary that techniques for isolation of mRNA should be available. The different steps involved are the following:

1. Protein product of the gene is purified.
2. Antibodies are produced against these protein products by immunizing animals like rabbit and mouse.
3. Precipitation of polysomes is done which are engaged in synthesizing specific proteins. This is done with the help of antibodies produced.
4. mRNA is isolated and purified from the polysome fractions. This mRNA is used for synthesizing cDNA.
5. These cDNA are then inserted into cloning vectors for preparation of cDNA library. After it, immunological and electrophoretic analysis of the translation products of cDNA clones is done, to identify the specific cDNA clone, having gene for specific protein of interest.
6. Specific cDNA probes are then selected for identification and isolation of the gene from genomic DNA through screening a complete or partial genomic library.

3.3. Isolation of Gene of unknown Products (with Tissue Specific Expression)

It is easy to isolate the genes products which are tissue specific. For example, genes for storage proteins are expressed only in developing seeds, globin gene is expressed in erythrocytes. Such genes can be easily isolated because mRNA extracted from such tissue will be largely of gene of interest. Other mRNA can be eliminated, which can be identified by isolation and comparison of mRNA from the tissue where this gene is not expressed. Then this isolated mRNA is used for synthesis of cDNA using enzyme reverse transcriptase. Then the process is same as described in isolation of genes coding for known specific proteins.

3.4. Isolation of Genes using DNA and RNA Probes

If the specific molecular probes (DNA and RNA probes) are available then they can be used for isolation of specific genes. These probes can be obtained either from another plant species or may be artificially synthesized using a part of the amino acid sequence of the protein product of the gene of interest. The probes obtained from one plant species and used for another plant species are called heterologous probes. These heterologous probes have been found to be effective in identifying gene clones during colony hybridization or plaque hybridization or on southern blot. For example, the gene for chalcone synthase has been isolated from *Antirrhinum majus* and *Petunia hybrida* using heterologous probe from parsely. Heterologous probes are generally used with the cDNA library.

Now with our knowledge of synthesizing cDNA probes, these probes can be helpful in synthesis of synthetic probes. In this procedure, the purified protein using the two dimensional gel electrophoresis is used for microsequencing of 5-15 consecutive amino acids, this information can be used for the synthesis of corresponding oligonucleotides using automated DNA synthesizers. These oligonucleotides may then be directly utilized for screening of cDNA or genomic library for isolation of gene of interest.

4. COLONY HYBRIDIZATION

Colony hybridization is the screening of a library with a labelled probe (radioactive, bioluminescent, etc.) to identify a specific sequence of DNA, RNA, enzyme, protein, or antibody (Fig. 15.5).

Hybridization has two important features:

1. Hybridization reactions are specific, probes will bind only to sites that have complimentary sequences.
2. Hybridization reactions occur in the presence of large quantities of molecules that are similar but not identical to the site. This means a probe can find one molecule of target in a mixture of millions of related but non-complementary molecules.

This specificity allows one to find a specific sequence in a complex mixture full of similar sequences. Hybridization techniques also allows scientists to pick out the molecule of interest from very complex mixtures and study the molecule on its own.

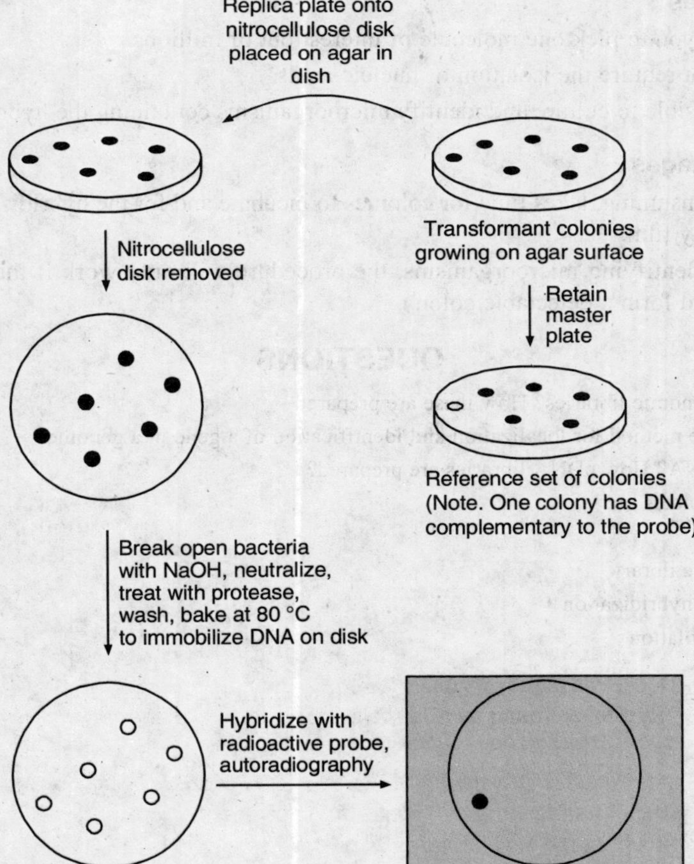

Fig. 15.5. Replica plate method for colony hybridization and localization of hybrid colony.

4.1. Method of Colony Hybridization

The following steps are required:

1. Isolate and grow cultures in a suitable medium (agar).
2. Transfer a sample of the colonies onto a solid matrix such as a nitrocellulose or nylon membrane.
3. The cells on the membrane are lysed and the DNA is then denatured.
4. A labelled probe is added to the matrix and hybridization takes place.
5. The matrix is rinsed to remove the non-hybridized probe molecules.
6. For radioactive probes, one uses autoradiography and the matrix is placed on X-ray film.
7. The film is observed for black spots that correspond to colonies that hybridized with the probe.
8. Compare the X-ray film with the master plate to see which colonies had probe hybridization. These are the colonies that contained the specific sequence that actually hybridized with the probe.
9. Colonies on the master plate that have the desired sequence can then be subcultured if desired.

4.2. Advantages

1. Allows you to pick one molecule of interest out of millions.
2. Does not require the isolation of nucleic acids.
3. It is possible to culture and identify microorganisms containing the hybridized sequence.

4.3. Disadvantages

1. Time consuming, takes time for colonies to incubate and for the hybridization to show on the X-ray film.
2. When identifying microorganisms, the procedure will only work if the organisms will grow and form a detectable colony.

QUESTIONS

1. What are genomic libraries? How these are prepared?
2. Describe the method for localization and identification of a gene in a genome?
3. What is cDNA? How cDNA libraries are prepared?
4. Write short notes on:
   (a) cDNA
   (b) Genomic library
   (c) Colony hybridization
   (d) Gene isolation

CHAPTER 16

Methods of Gene Transfer

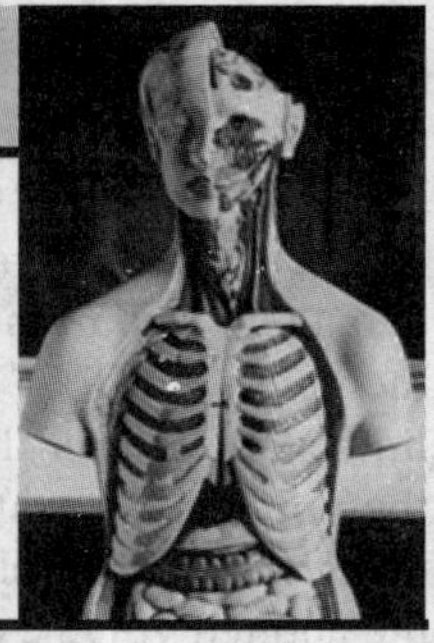

The first transgenic plant was produced via *Agrobacterium* mediated modified transformation of *Nicotiana tabacum* protoplasts by Horsch and co-workers in 1984. Since then several dozen plant species have been genetically engineered using different techniques. Simultaneous development of other techniques such as selectable markers facilitated the development in genetic engineering for obtaining transformed plants. But this technique is not suitable for monocotyledon plants as they are not natural host of *Agrobacterium* (There is evidence that limited gene transfer is possible in monocots by this system). Therefore, other methods of direct gene transfer have been developed for use with monocots and other species. These can be categorized on the basis of the use of protoplasts or cell and tissue as the target materials. The isolation and purification of protoplasts have been presented in Chapter-30. Freshly isolated protoplasts are used for genetic transformation. Protoplast stage is for a short duration, before it regenerates cell wall, and DNA transfer is performed during this period.

I. DNA Transfer in Protoplasts

(i) Electroporation
(ii) Chemically stimulated DNA uptake by protoplasts
(iii) Liposomes
(iv) Microinjection
(v) Sonication

II. DNA Transfer in Plant Tissues

(i) Acceleration of DNA coated microparticles
(ii) Laser microbeam
(iii) Silicon carbide fibres

Direct uptake of DNA by isolated protoplasts is a genotype dependent response. Regeneration from protoplasts is not common in all the species. Due to this, production of fertile transgenic plants has remained difficult in most of the cereal species. However, transgenic fertile plants have been produced in *Sorghum vulgare*, *Oryza sativa* and *Hordeum vulgare*. Direct delivery of free DNA molecules into plant protoplasts by physical (electroporation and micro injection) and chemical (polyethylene glycol) methods have been developed to facilitate DNA delivery across the plasma membrane.

1. DNA TRANSFER IN PROTOPLASTS

1.1. Electroporation

This method is based on the use of the short electrical pulses of high field strength. Electroporation causes the uptake of DNA into protoplasts by temporary permeabilization of the plasma membrane to macromolecules. Protoplasts and foreign DNA are placed in a buffer between two electrodes and a high intensity electric current is passed, the alternating current of about 1 MHz is applied to align the protoplast by dielectrophoresis. Once aligned, fusion is induced by applying one or more direct current pulses (1-3kV /cm, 10-100 μs), then the alternating field is reapplied briefly to maintain close membrane contact for fusion (Fig. 16.1). Electric field damages membranes and creates pores in membranes. DNA diffuses through these pores immediately after the electric field is applied, until the pores are resealed. Technique is optimized by using appropriate electric field strength (defined as the applied voltage divides by the distance between two electrodes). The optimum field strength is dependent on the followings -

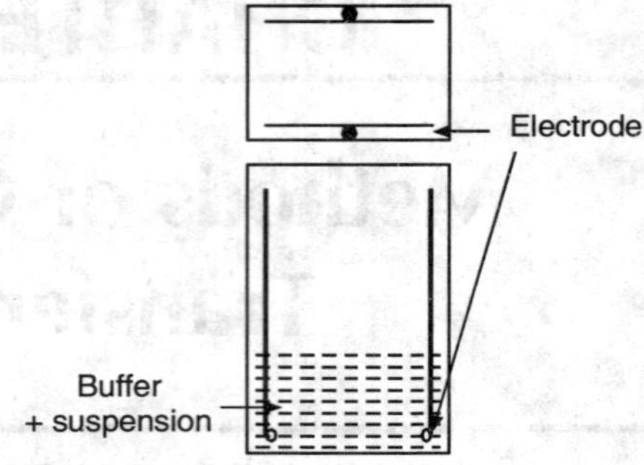

Fig. 16.1. Top (above) and side view (below) of glass cell with electrodes used for electroporation.

1. The pulse length of electric current
2. Composition and temperature of the buffer solution
3. Concentration of foreign DNA in the suspension
4. Protoplasts density, and
5. Size of the protoplasts.

It has been demonstrated that the removal of pectin from the plant wall increases the amount of DNA which can be introduced by electroporation. Tobacco mosaic virus was introduced in tobacco protoplasts by this method. Electroporation has been used successfully for transient (when foreign gene which is present in cell but not integrated in the chromosome, shows expression in cytoplasm) and stable transformation (foreign gene integration in host chromosome and is expressed) of protoplasts from a wide range of species. Plating efficiency (i.e., number of colonies recovered out of number of cells transferred on plates) of electroporated protoplasts grown on selection medium (containing selective marker) can be as high as 0.5%. The highest plant transformation efficiencies have been reported for tobacco, with 0.2% of electroporated leaf mesophyll protoplasts giving rise to transgenic calli. Low transformation efficiency is common in cereals, e.g., in rice 0.002% efficiency was recorded.

1.2. Chemically Stimulated DNA Uptake

Direct uptake of DNA by protoplasts is stimulated by polyethylene glycol (PEG) and PEG is the most widely used chemical for this purpose. PEG mediated transformation involves mixing of freshly isolated protoplasts with DNA and immediately adding 15-20% PEG dissolved in a buffer containing divalent cations. This mixture is incubated for 30 minutes, protoplasts are washed and then plated in Petri plates for culture and growth. The optimization of transformation frequencies by this method include factors that follows.

1. PEG concentration in the mixture.
2. Composition and concentration of salts used.
3. The pH of the solution.
4. Concentration of the foreign DNA.
5. Size and form (linear, supercoiled) of the DNA molecules used.
6. Culture and selection techniques used for protoplasts.

PEG mediated transformation is generally preferred over electroporation for stable transformation of monocot protoplasts due to relatively higher survival rates after treatment.

PEG also stimulates the uptake of liposomes and improves the efficiency of electroporation. PEG causes precipitation of ionic macromolecules like DNA and stimulates their uptake by endocytosis. PEG mediated DNA uptake typically transforms 0.1 to 0.4% of the total protoplasts treated. Production of transgenic plants depends upon the regeneration competance of the transformed protoplasts. In case of Petunia, 40% transformed calli derived from mesophyll protoplasts could be induced to form fertile plants. This is equal to about 0.1% transformation efficiency of the treated protoplasts. In different species different transformation efficiency has been observed, e.g., in embryogenic protoplasts suspension of rice was 0.0004% and, soyabean and tobacco was 0.7-1%.

1.3. Liposomes

Liposomes have also been used as a carrier for the introduction of nucleic acid into plant protoplasts. Liposomes are small lipid sacs containing plasmids and are prepared artificially. The fusion of liposomes with plant protoplasts is stimulated by chemicals such as PEG (endocytosis). Liposomes mediated transformation has been achieved by including positively charged agents such as cations in the transformation mixture or using the cationic liposome preparation (Fig. 16.2).

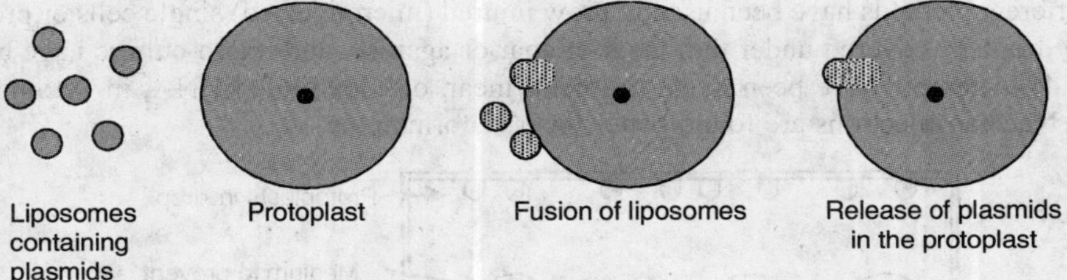

Fig. 16.2. Liposomes mediated gene transfer.

Other chemical agents like polycation Polybrene® or lipofectin have also been used for both transient and stable transformation for maize protoplast. Cationic liposome and polycation mediated DNA delivery are new protoplasts transformation methods and are considered better than other methods of transformation. There are several advantages for the use of this technique.

1. Protection of DNA/RNA from nuclease digestion.
2. They have low cell toxicity.
3. Encapsulation of nucleic acids makes them more stable during storage.
4. High degree of reproducibility.
5. This method is applicable to wide range of plant cell types.

1.4. Microinjection

Delivery of nucleic acids to protoplasts or intact cells via microinjection is a labour intensive procedure that requires special capillary needles, pumps, micromanipulators, inverted microscope and other equipment. However, injection into the nucleus or cytoplasm is possible and cells can be cultured individually to produce callus or plants. In this way selection of transformants by drug resistance or marker genes may be avoided. This method involves skill of the worker to insert needle into the cytoplasm or in the nucleus. The basic technique is similar to that used for animal cell microinjection. In order to microinject protoplasts or other plant cells, the cells need to be immobilized (Fig. 16.3). The cells are immobilized by-

1. The use of a holding pipette which holds the cells by vacuum.
2. Attachment of cells to poly-L-lysin coated cover slips.
3. Embedding the cells in agarose, agar or sodium alginate.

Glass micropipette are prepared to have openings of about 0.3 μM in diameter and are inserted into plant cell cytoplasm and nuclei with the aid of a micromanipulators device. A syringe

like device is used for the controlled delivery of volume (10^{-11} – 10^{-4} l) into the plant cell. Most plant cells are injected while keeping inside microdroplets (2-50 μl) of medium using a chamber which is sterile, vibration free and permits temperature and humidity regulation. A maximum of 100-200 cells per hour can be microinjected by this method.

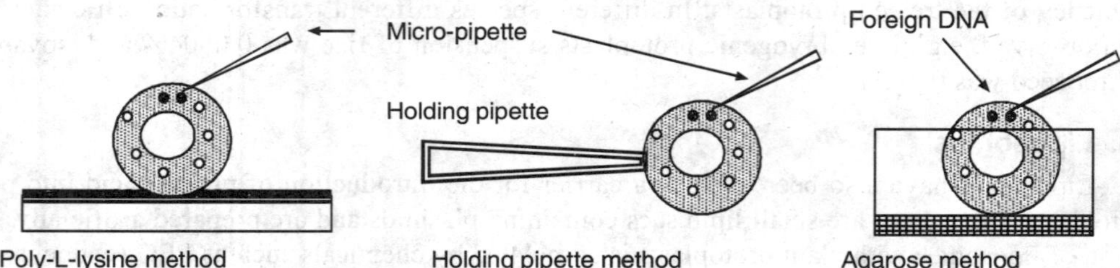

Fig. 16.3. Various methods of immobilizing the cell and microinjection.

The recovery of transformants is dependent upon the regeneration ability of the microinjected cells. Different methods have been used to grow injured (microinjected) single cells or protoplasts. Hanging droplets, covered under thin layer of agar or agarose, and micro-culture have been used (Fig. 16.4). Attempts have been made to inject linear, or supercoiled DNA, in cytoplasm or in nucleus. Nuclear injections are found better for transformations.

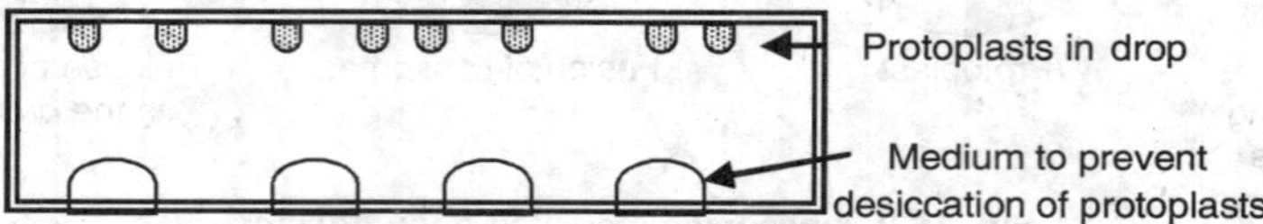

Fig. 16.4. Hanging drop culture method.

1.5. Sonication

Mild sonication (20 KHz ultrasound) has been used to facilitate the uptake and transient expression of a chloramphenicol acetyltransferase (CAT) gene in protoplasts of sugar beet (*Beta vulgaris*) and tobacco. This method was found superior than electroporation method used for the same material. Plating efficiency was also similar to untreated cells. However, transgenic plant production using this technique has not been reported so far.

2. FREE DNA TRANSFER TO INTACT TISSUE

2.1. Acceleration of DNA Loaded Microparticles (Particle Gun or Biolistic® Method)

This is latest technology to transfer DNA into intact tissues. Several devices are developed using different methods. All to achieve the transfer of micro-sized particles (microprojectiles) coated with DNA to penetrate the cells. In this procedure micron size tungsten or gold particles are accelerated in a gun barrel to velocities sufficient for non-lethal penetration of cell walls and membranes. Klein and co-workers in 1987 developed and used for the first time the particle gun to transfer chimeric DNA and viral RNA molecules into intact onion cells. Tungsten acted as a carrier of nucleic acids because it was available in micro-size balls, non-toxic to cells, and dense enough (high density) for rapid penetration of target material.

Microprojectile mediated transformation is a mechanical method of introducing DNA in to any plant species. This method can be successfully used where plasmids or protoplasts mediated transformation can not be used. An acceleration device is used to propel particles (micro projectiles) carrying plasmid DNA is called by various names based on machine or technique used to accelerate the particles such as 'particle gun technology', 'biolistic method', 'DNA bombardment', 'particle acceleration of DNA method' and 'electric discharge particle acceleration method'.

This is a quick method of stable transformation and testing a gene for cell and organelle specific expression, this technique has three components -

1. The basic equipment to generate particle acceleration.
2. Metal particles coated with precipitated DNA (desired gene).
3. Plant tissues to be used for particle penetration. The method for regeneration should be previously standardized and proper tissues be selected for bombardment.

(a) **Instrument:** The instrument is commercially available. Prototype was designed by Klein and co-workers. It uses the explosive force of gun powder (0.22 caliber gun cartridge) to accelerate a polypropylene cylindrical macroprojectile. Thin piece of polypropylene macroprojectile is loaded with microprojectiles coated with DNA. Gun powder explosion forces thin macroprojectile to move with high speed toward another end of barrel, where it is blocked by a polycarbonate disc having an aperture. Macroprojectile is stopped but microprojectiles move fast through the aperture towards tissue placed in the same direction. For each transfer 50 μg tungsten is accelerated up to 2000 ft per second in a partial vacuum. With this speed, particles reach up to lower layers of cells in target tissues (Fig. 16.5).

The other devices are similar in basic design concept but use different methods to accelerate particles like use of compressed air or gas. Compressed air (130 kg/cm^2 pressure) has been used to accelerate microprojectiles at velocities (approximately 440 m/sec) necessary to achieve DNA delivery to plant cells.

An electric discharge particle acceleration device differs in basic design from the above described devices. In this device, a high voltage discharge (14 KV current) delivered to a small water droplet which quickly vaporizes and releases energy to propel DNA coated gold spheres into target cells (Fig. 16.6). In similar way to above devices, a DNA carrier is attracted (accelerated)

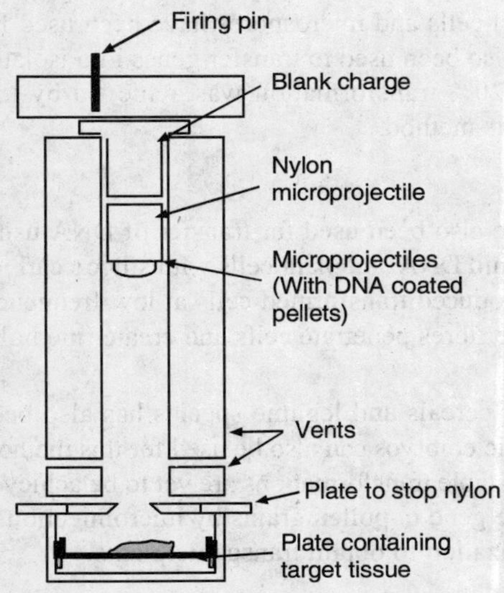

Fig. 16.5. Particle gun or shotgun for delivering DNA coated microprojectiles into plant cells.

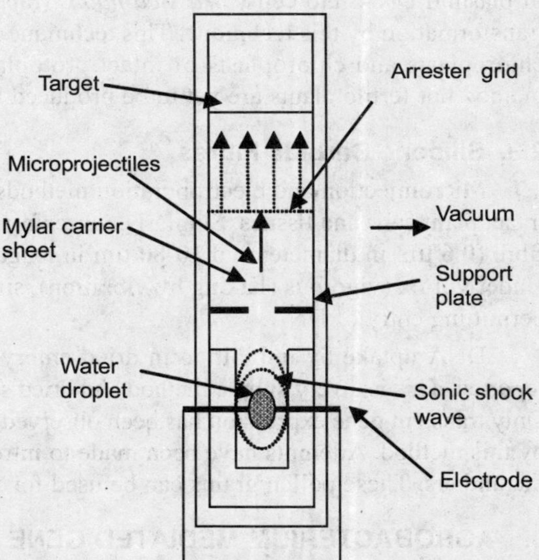

Fig. 16.6. Particle acceleration device based on high voltage charge for delivering DNA coated microprojectiles into plant cells.

due to potential differences, stopped in-between by a screen, DNA coated particles cross the screen and fly towards target tissue and deliver the DNA into cells.

(b) **DNA coating:** This is a sophisticated technology and requires precise preparation of DNA coated gold or tungsten particles. The particles should have following properties.

1. High density (19 g/cm^2 or greater) to ensure proper acceleration and penetration through cell walls.
2. Size (0.5-5 μm) should match with size of the cells. Large sized spheres can be used with large cell size.
3. Gold is costlier than tungsten, but it does not oxidize like tungsten.

DNA is precipitated onto the particles prior to bombardment. Most commonly, $CaCl_2$ and spermidine are used to precipitate plasmid DNA (the desired gene attached to plasmid is suitable for integration into plant genome) onto tungsten particles. Ethanol is also used for precipitation of the DNA on gold particles.

(c) **Plant tissue:** Plant tissue used for transformation should be competent to regenerate. Mostly embryogenic tissue is an ideal material for transformation (e.g., in corn, cotton, soyabean, papaya and wheat) but transformed plants have been obtained after leaf bombardment in tobacco, and stem bombardment in cranberry. Normally, reporter gene and selectable marker genes are used to isolate and select transformed cells/plantlets.

2.2. Laser Microbeam

Weber and co-workers (1988) demonstrated use of laser beam for transformation of plant cells. An ultraviolet (UV) laser microbeam has been used to introduce DNA into plant cells and chloroplasts. A 343 nm beam (wavelength of UV is 200 to 400 nm) is directed through an adjustable attenuator into the optical path of an inverted microscope. The focus of the laser beam is adjusted so that it is identical with that of the objective lens. The laser beam is targeted by focusing on a specimen in the microscope. This laser beam can then make holes in any part of cell which is in focus. Laser micropuncture of the cell wall and plasma membrane allows uptake (entry) of plasmid DNA into cells. *Brassica napus* (rapeseed) cells and microspores have been used for transformation by this technique. This technique has also been used to transfer genes into isolated chloroplasts and chloroplasts of intact protoplasts. 20% transformation was achieved by this method but fertile plants are yet to be produced by this method.

2.3. Silicone Carbide Fibres

Microinjection and electroporation methods have also been used for transfer of DNA using intact plant cells and tissues. Similarly, vortexing plasmid DNA and plant cells with silicon carbide fibre (0.6 μm in diameter and 10-80 μm in length) produced transformed cells at low frequency. Under vortex (vigorous shaking by vibration), silicone fibres penetrate cells and create fine holes permitting entry of DNA.

DNA uptake by imbibition in dried embryos of cereals and legume species has also been reported. This is a very simple method and dried somatic embryos can also be used for this method. Only transient gene expression has been observed and stable transformations are yet to be achieved by this method. Attempts have been made to introduce gene in pollen grains by microinjection in the anthers. These pollen grains can be used for fertilization to obtain transgenic plants.

3. AGROBACTERIUM MEDIATED GENE TRANSFER METHOD

Transgenic plants are produced by two methods – (*i*) Indirect gene transfer using plasmid as a vector and (*ii*) direct gene transfer. Indirect gene transfer requires construction of a vector for carrying foreign genes based on Ti or Ri plasmids.

Vectors are the carrier DNAs into which 'foreign' DNAs or genes of interest are inserted to make a recombinant DNA (rDNA). Vectors along with this 'foreign' DNA (i.e., rDNA) are then introduced into appropriate host cell. Vectors are of two types-cloning vectors [used for obtaining millions of copies (cloning) of DNA segment] and expression vectors [used for expression of cloned gene to produce the product (protein)].

Plasmids are most commonly used vectors in gene cloning work. Isolation and purification of plasmids is a very routine experimental procedure. Several methods are available for isolation and purification of plasmid. The most critical stage in the process is lysis of the cell wall which is just sufficient for isolation of plasmid DNA without contamination by chromosomal DNA. Clear lysate contains plasmid. Alkali-SDS lysis and rapid boiling procedures are commonly used methods for plasmid isolation.

Most vectors carry marker genes which allow recognition e.g., antibiotic resistance - selectable markers, e.g., npt II (kanamycin resistance). Other features include (i) multiple unique restriction sites - a synthetic poly linker and (ii) bacterial origin of replication e.g. (Col E1). The problem is that a vector having these properties may not be easy to transfer to plant system. Therefore, *Agrobacterium* Ti plasmid is preferred because of (1) wide host range and (2) presence of T-DNA border sequence. Plant cells do not have any endogenous plasmid. The plasmid vectors used for gene transfer in plants are based on *p*Ti or *p*Ri (tumour inducing or root inducing plasmid) present in *Agrobacterium tumefaciens* or *A. rhizogenes*. Structure of T-DNA is described in the chapter -14. Presence of auxin and cytokinin producing genes caused tumour formation in plants, when T-DNA present in *p*Ti is transferred in plant cell. Therefore, these tumour forming genes are removed from T-DNA (called as disarming the plasmid) and gene of interest is introduced in that place (between left and right border of T-DNA). Thus this recombinant T-DNA is ready to transfer a gene by natural mechanism of gene transfer in a dicot host cell. Since *p*Ti or *p*Ri are large plasmids, modified plasmids (co-integrative and binary) are prepared from Ti plasmid and used as described later on in this chapter. Thus by placing foreign genes into T-DNA region of Ti-plasmid, it is possible to clone (make copies) the introduced genes with the multiplication of plasmid residing inside the bacteria (self replication of plasmid makes millions of copies) which is grown on a medium and with the multiplication of bacterial population, residing plasmid is also multiplied by this method. It is possible to exploit the natural ability of *Agrobacterium* to transfer new DNA into the plant genome.

3.1. T-DNA: Transfer Mechanism

Though exact mechanism of T-DNA transfer is not clearly known, effective role of vir region is known in the process of transfer. Genes in the vir region is activated by acetosyringone, a phenolic substance secreted by wounded cells of the host. The phenolic signal molecules binds to the vir A gene product, the vir A proten (Fig. 16.7). As consequence of this, several vir rgion genes are activated and these gene products (proteins) help in copy of T-DNA and its insertion into host cell.

- Nicking between 3rd and 4th base of 25 bp repeats (bottom strand). Vir-D operon encodes for an endonuclease that cause nick formation.
- Initiation of DNA synthesis in 5′-3′ direction.
- Involvement of bacterial genome - synthesis and secretion of glucose, cellulose, fibrils, and cell surface proteins. This is common physiological response in all soil bacteria and is involved with pathogenic characters.

The generation of the T-strand is the first step in the complex process of *Agrobacterium*-mediated plant cell transformation. Following its formation, this DNA must pass through the bacterial cell membrane, the bacterial cell wall, the plant cell wall and plant cell and nuclear membranes. Once inside the nucleus, the T-DNA must finally integrate stably into the plant cell genome (Fig. 16.8). During this entire transit process, the T-DNA strand also must avoid degradation by nuclease. The T-DNA exists as a DNA protein complex. The T-DNA complex protects it and mediates its travel. The final step in the genetic transformation of plant cell is integration of T-DNA copy, presumably T-strand, into plant cell DNA. It has been argued that the T-strand might be converted to a double stranded (ds) DNA prior to integration.

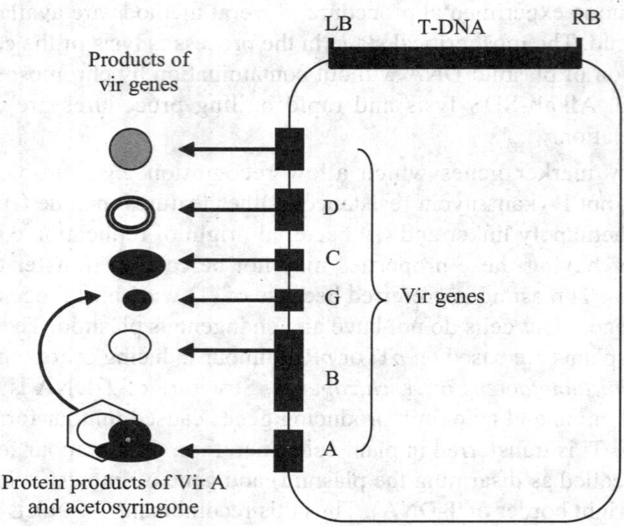

Fig. 16.7. Vir genes and products of vir genes.

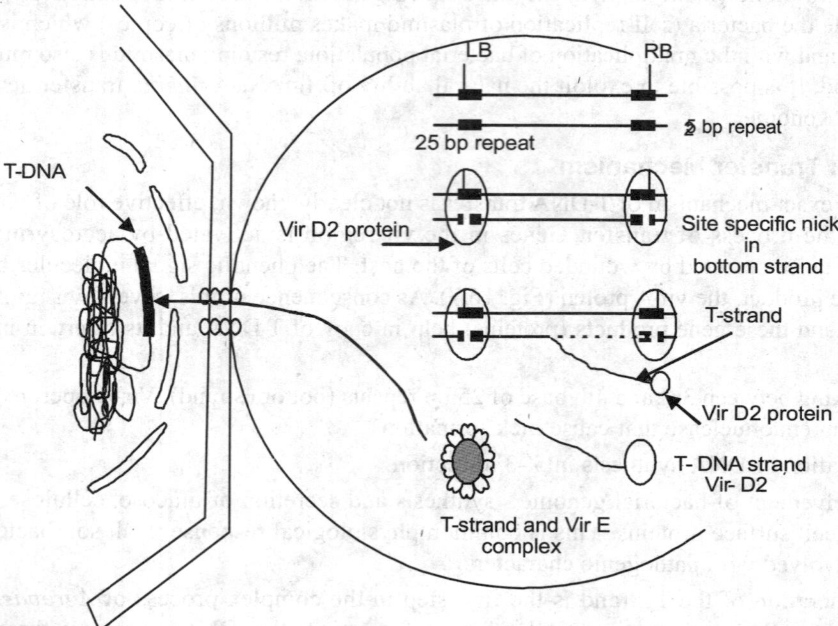

Fig. 16.8. T-DNA transfer process.

3.2. Vectors Based on Ti and Ri Plasmids

The Ti or Ri-plasmid can not be used directly. There are limitations for direct use of these plasmids. These are (i) Large size of vector make it difficult to manipulate (ii) Absence of unique restriction enzymes sites and (iii) Tumour induction. Therefore, vectors are designed with useful characteristics. This involves-removal of tumour induction property or disarming the plasmid. This is achieved by replacing of tumour induction genes in T-DNA by selectable markers such as npt-II (kanamycine). Promoters and polyadenylation signal isolated from octopine and nopaline synthase genes were used for expression of selectable markers. As there is no excess production of plant hormones, whole plants transformed with such disarmed *Agrobacterium* strains can be produced and detected by the production of opines. When a selectable marker gene (kanamycin resistant) is introduced, transformed cells can be selected by their ability to grow on media containing the selective antibiotic. Untransformed cells will not survive on this medium.

Other promoters - *onc* - CaMV35S, CaMV19S isolated from cauliflower mosaic virus have also been used. Therefore, it is concluded that T-DNA and Vir genes are two essential components of a vector.

3.3. Cointegrative Vectors

Cointegrative vectors recombine, via DNA homology, with an intermediate cloning vector, which is used for manipulation and cloning of the gene in *E. coli*. *Agrobacterium* containing cointegrative vector and *E. coli* containing intermediate cloning vector are allowed to undergo conjugation, but the intermediate vector can not replicate in *Agrobacterium* so it has to transfer the marker genes as well as the DNA segment to the resident Ti plasmid (cointegrative vector) through recombination in the region of DNA homology.

Example of such vector - pGV3850 from nopaline type Ti plasmid, where almost all T-DNA has been replaced by pBR322, a small *E. coli* cloning vector (Bolivar and Rodriguez prepared and hence name - plasmid BR, followed by experiment number). The intermediate vector (pGV1103) based on pBR322 is conjugated into pGV3850 at the region of pBR322 homology.

3.4. Binary Vector

A significant advance that bypass the problem of Ti-plasmid size was the discovery that the T-DNA and the vir region could be separated on two different plasmids without loss of the T-DNA transfer capacity, i.e., they worked in a *trans* as well as a *cis* configuration. This discovery led to the development of binary T-DNA vectors that involve two plasmids. The small binary T-DNA plasmid has a wide host range that can replicate both in *E. coli* and *Agrobacterium* cells. The desired foreign gene is inserted into the binary T-DNA plasmid between the left and right border sequences. A selectable plant marker gene is also inserted and (along with desired foreign gene) allow selection of transformed plant material. Several plant species have been transformed by this method.

3.5. Transformation Technique

The critical information that made *Agrobacterium* mediated gene transfer systems possible came from the elegant work by Chilton et al. (1977) who showed that in crown gall disease a region of bacterial plasmid DNA (the T- DNA) is transferred to chromosomal DNA in the plant nucleus, stably maintained there, and expressed in the absence of the bacterium. The first transgenic *Nicotiana tabacum* plant was produced by Horsh and co-workers in 1984 using *Agrobacterium*.

Agrobacterium mediated gene transfer methods are being developed for a wide range of dicotyledonous plants and gymnosperm species. The gene tagging approach is used to demonstrate trnansformation.

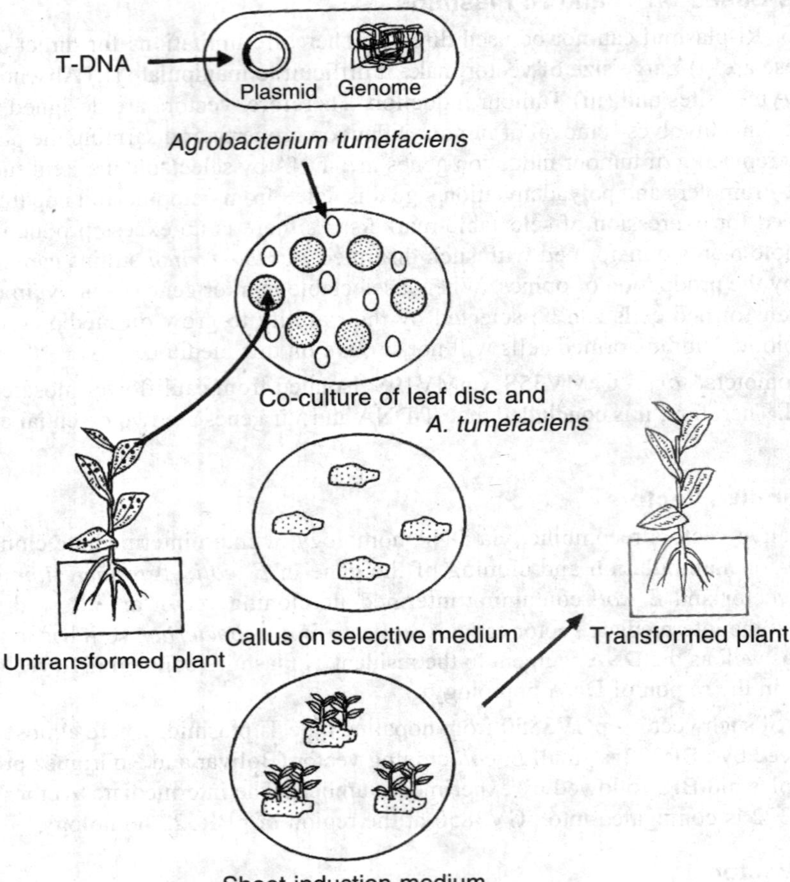

Fig. 16.9. Transformation procedure in laboratory.

T-DNA containing either a reporter or strong transcriptional enhancer transform a large number of cells. Normally bacteria are incubated with plant cells (few hours to few days) during that period T-DNA transfer takes place. The cells are then washed and treated with antibiotics to remove the bacteria. The cells are then cultured in the presence of the selectable agent, and transformed shoots are regenerated and characterised. (Fig. 16.9)

Plasmids of *Agrobacterium* have been used as vector for transfer of foreign DNA into a number of dicot species. However, seed legumes are still not amenable, exception is *Glycine max*. Monocotyledons (cereals) can not be used as *Agrobacterium* is host specific and do not infect monocots but with an exception of *Asparagus* (which has been transformed with this technique). There is no cambial activity and wound healing process in monocots, therefore, acetosyringone is not produced; this results in failure of vir genes to recognise by chemoreception.

(*i*) Pre-requisites for agroinfection-

1. Production of acetosyringone by the host plant cells.
2. Bacteria have access to actively dividing cells (DNA replication) such as meristems, fresh protoplasts, dedifferentiated tissues.
3. Regeneration in transformed tissues should be possible, otherwise transformation will be of no use.

(ii) Explants for co-cultivation- following materials can be used:

(a) Protoplasts

(b) Cell suspension cultures

(c) Callus\thin cell layers - epidermis, tissue slice, organ section (leaf disc, section of roots, stem or flower).

(d) Wounded and inoculated whole plant.

3.6. Selectable Markers

Selectable markers are usually required for efficient recovery of transgenic cells and plants. After gene transfer, transformed cells are few in number compared to untransformed cells. It is not possible to separate transformed and non-transformed cells by any physical method. A selectable marker gene incorporated with the desired gene helps the growth of transformed cells on a nutrient medium containing corresponding selective agent. The availability of multiple selectable markers is useful in developing efficient transformation methods for diverse species as well as for the introduction of multiple novel traits (characters) through successful transformation and regeneration of transformed plants from such cells (Table 16.1).

Selective agents differ in their toxicity to different plant species. The different developmental states of the plant cells or tissues give different response to the selectable marker. The cells will react differently than whole plant or organ. Therefore, it is necessary to use correct concentration of selective agent for the transformed cells of a given species to select the cells and to inhibit the growth of untransformed cells. This can be illustrated by following example. Transgenic plants of *Lycopersicon esculentum, Brassica napus* and *Lactuca sativa* can be selected on low kanamycin concentration (15-100 mg/l), while other plants like *Beta vulgaris* requires high kanamycin concentration (400 mg/l) as selection agent.

Table 16.1. Selectable markers used in transformation.

Marker gene	Enzyme encoded	Resistance against
Antibiotics		
Npt-II	Neomycin phosphotransferase	Kanamycin, G418 neomycin
Npt or *aph IV*	Hygromycin phosphotransferase	Hygromycin
Aad A	Aminoglycoside-3′-adenyl transferase	Streptomycin
Herbicides		
Bar	Phosphinothricine acetyltransferase	Phosphinothricin
Aro A	5-enolpyruvyl shikimate-3-phosphate synthase	Glyphosphate
Other modified *als*	Acetohydroxy acid synthase (or acetolactate synthase)	Chlorsulfuron imidazolanones
DHFR	Dihydrofolate reductase	Methotrexate

Herbicides are generally more toxic to plant cells than antibiotics. This is due to their specific mode of action in plant cells as well as their efficient uptake and translocation within the plant tissues. Herbicides and other highly toxic compounds may require delayed application in order to ensure that the transformed cells have produced sufficient quantities of enzyme responsible for the protection to such compound.

Callus and cell cultures are better systems than organized explants to achieve transformations and selection of transformed cells. Explants may give rise to organs from untransformed cells which may escape selective agent. This is not possible with isolated cells and only transformed cells will be able to grow on the selective agent.

3.7. Reporter Genes

In all transformation experiments, regeneration of non-transformed plants is also observed alongwith transformed plants. If these non-transformed plants can be detected at an early stage, a lot of time, labour and analyses can be saved. There are several methods to screen and identify the transformed plants, but usually these methods are laborious and time consuming, e.g. southern blotting technique to compare DNA sequences requires large amount of tissue (DNA) for comparison. The other methods use selective agent. In both these methods, prolonged growth and subcultures are involved before selection can be made.

The use of reporter gene eliminates these drawbacks and transformed plants can be recognised easily. A reporter gene is a coding unit whose product is easily assayed (such as GUS whose product can easily be detected by histochemical assay). This gene may be connected to any promoter of interest so that expression of the gene can be used to assay promoter function. In fact, the reporter gene describes the transfer and expression of other promoter. Two genes that are now widely used for this purpose are those coding for β-glucoronidase (GUS) and luciferase (LUC). The coding regions of these non-plant genes have been fused to plant promoters and polyadenylation sequences such that they give high level of expression in plant cells (Table 16.2). The anthocyanin regulatory genes can serve as a unique reporter system in maize and some other cereals. Transformed cells accumulate reddish purple anthocyanin pigments. The anthocyanin genes are extremely sensitive marker used in *Arabidopsis* and *Nicotiana* transgenic plants.

Without selection agent, transformed plants are selected with the help of scoreable gene or reporter gene.

Table 16.2. Reporter genes, assay and identification method in transferred plants.

Reporter genes used for	Substrate and assay	Identification
Chloramphenicol acetyl transferase (CAT)	^{14}C chloramphenicol + acetyl Co-A, TLC separation	Detection of acetyl chloramphenicol by autoradiography
β-glucoronidase (GUS)	Glucoronides (PNPG, X-GLUC, NAG, REG)	Fluorescence detection colorimetric, fluorimetric and histochemical
β-galactosidase (Lac Z)	β-glactoside (X-gal)	Colour of colony
Luciferase (LUC)	Decanal and $FMNH_2$ ATP + O_2 + luciferin	Bioluminiscence (exposure of X-ray films)
Octapine synthase	Arginine pyruvate+NADH	Electrophoresis
Nopaline synthase	Arginine+ketoglutaric acid+NADH	

Examples of Transgenic Plants- Herbaceous Dicot - Tobacco, *Petunia hybrida*, tomato, potato, egg plant, *Arabidopsis thaliana*, lettuce, *Apium graveolens* (celary), sunflower, flax, rape oil seed, cauliflower, cabbage, cotton, soyabean, pea, chicory, carrot, licorice, sweet potato, kiwi, papaya, grape, rose, *Chrysanthemum*, etc.

Woody Dicot - *Populus, Malus, Pyrus communus, Azadirachta indica*

Monocot - *Asparagus, Secale cereale, Oryza sativa, Triticum aestivum, Zea mays, Avena sativa.*

Gymnosperm - *Picea glauca.*

4. INTEGRATION AND EXPRESSION

For expression heterologous proteins in plants, the unique capacity of a soil bacterium *Agrobacterium tumefaciens* to introduce part of its resident plasmid into the plant genomic DNA has been widely utilized. The above bacteria causes a cancerous growth in the infected plant due to the introduction of this DNA fragment derived from the plasmid. The plasmid is called the Ti plasmid and the introduced fragment of DNA is called T-DNA. The tumorigenic genes, i.e., those which cause the tumour are deleted from the T-DNA and then the plasmid is incapable of causing tumours but is still capable of transferring T-DNA to the plant. Any heterologous DNA can be introduced into the plant as part of the above plasmid. To ensure high expression of the heterologous gene, strong promoters derived from plants viruses, like the cauliflower masaic virus are generally used. Several heterologous genes have been expressing plants in the recent years either to improve the field performance of crops or to impart added qualities to the edible part, like essential vitamins or even antigenic proteins, which can serve as vaccines against common disease like cholera. The low cost of growing such plants expression heterologous proteins, some of which can have high commercial values, has made expression such proteins in plants a very attractive proposition. Several plants viruses can in fact multiply in their plant hosts to a very high level at a very short time. Some of these viruses, like the tobacco mosaic virus have been used to clone heterologous genes by replacing some of the non-essential viral genes. Such engineered viruses when allowed to infect plants, are capable of expressing the heterologous gene to very high levels in a matter of few weeks in the infected plants. Some of the characteristic features of DNA integrations are:

1. Integration occurs at random sites in the genome.
2. Linear DNA is better integrated than circular DNA.
3. Multicopy integration usually occurs in tandem at one site (copies attached in series).
4. Higher concentration of DNA results in high integration frequency with multiple copies of gene incorporated.
5. Use of carrier DNA (plasmid) in direct DNA transfer methods enhance multiple copy integration.

QUESTIONS

1. Describe the various methods of direct gene transfer giving suitable examples.
2. Write short notes on the following:
   (a) Transgenic plants.
   (b) Particle acceleration.
   (c) DNA uptake by protoplasts.
   (d) Transgenics for disease resistance.
   (e) Microinjection.
   (f) Electroporation.
   (g) Lasers and Liposomes for transformations.
3. Write short notes on the following:
   (a) *Agrobacterium.*
   (b) Ti plasmid.
   (c) Vector.
   (d) Binary vector.
   (e) T-DNA.
   (f) Vir gene.
   (g) T-DNA transfer.
   (h) Selectable marker.
   (i) Reporter gene
4. What are plasmids? Describe the plasmid mediated gene transfer process.

CHAPTER 17

Transgenic Plants – Impacts and Importance

Since the bringing of agriculture, approximately 10,000 years ago, humans have profoundly modified the shape and form of mot cultivated species. These changes were largely made in pursuit of better adaptation and increased productivity within the context of a cultivated field in contrast to the natural environment. Though the selection of naturally occurring mutants presenting novel and useful phenotypes, countless generations of farmers practiced the earliest and simplest form of genetic modifications. It is only in the last century that a science based approach to the genetic modifications of most cultivated plants was developed largely through the work of plant breeders and biotechnologists.

The methodology and tools available to achieve the hasten increases in productivity or changes in various quality traits have increased dramatically in number and technical sophistication over the last 50-100 years. Sexual crosses between carefully selected parents still play a major role in all present day breeding programme. However, constant efforts are made to increase useful genetic diversity and its exploitation for increasing productivity. Such increases have been created through the use of wide crosses (inter-specific and inter-genetic) and use of *invitro* techniques such as embryo rescue to make use of already existing variations. Other approach to transfer useful traits from one plant to another (by over coming the species specific barrier) through use of somatic cells (e.g., protoplasts fusion) rather than reproductive cells. Various mutagens (irradiation, chemicals, insertional sequences) have also been used extensively to accelerate the identification of desired mutants for breeding or gene discovery. The development of the ability to transfer a gene or genes from one species to another via transgenic is only the most recent addition to a long list of methods which aim to broader the available genetic diversity in a given plant species.

Maize is a good example of continuous efforts for improvement of a crop species. Its ancestral progenitor, teosinte, is phenotypically quite different from modern maize in many important respects. These differences were so striking to the botanists that until relatively recently these two sub-species of *Zea mays* were not even thought to be close relatives. From an agronomic view-point, teosinte is hardly an attractive crop. Its yield is estimated at 100kg/ha whereas modern maize hybrids can yield more than 10 tons/ha or 100 times more. However, the two plants are compatible and can be crossed easily.

Previous methods of genetic improvement relied mostly on the shuffling (or addition in the case of some translocations) of a relatively large number of genes (hundred to thousand) from the same or related plant species. In contrast, nuclear transgenesis involves the addition of a small number of genes (usually less than five, typically two) into the nucleus. Transgensis is thus results in quantitative changes that are quite small in comparison to what occurs through "conventional" methods of genetic improvement. This is not to minimize the potential consequences of this genetic transformation, however as a single gene can have a large impact (be it beneficial or deleterious).

The introduced genes usually serve two distinct purposes, one of the transferred genes (transgenes) is a selectable marker, meaning that it confers on the transformed cells a distinctive phenotype that can be used to select the few genetically modified (GM) cells among a population of untransformed cells. These marker genes typically confer resistance to selective agent, often an antibiotic. The other transgenes is meant to confer a desirable phenotype (a character) e.g., agronomic (herbicide, pest/stress resistance) or related to food quality (self-life, taste, nutritional value).

A second novel feature in transgenic plants is that the introduced genes can originate from almost any living organism or can even be entirely synthesized *de novo*. Thus, transgenesis offers the possibility of introducing relatively 'foreign' DNA into a plant cell.

No matter what the method used to transform plant, such as biological vector (*Agrobacterium tumefaciens*) or a physical device (particle gum), there is no means of controlling either the number of copies, or the location of the introduced material. Therefore, transgenic lines obtained with the same genetic construct need not be identical, either because the expression of transgene(s) differ due to copy number or due to position effect (not all the chromosomal loci are equally favourable to gene expression).

Among the cultivated transgenic crop plants, herbicide tolerance has consistently been the dominant trait followed by Bt based insect resistance, herbicide tolerant soybean, and *canola*, and Bt maize and cotton constitute the four major genetically modified (GM) crops. The two commercialised transgenic vegetable crop are tomato with delayed fruit ripening and potato with insect and virus resistance.

Transgenic crops for -

1. Insect pest control
2. Herbicide resistance
3. virus resistance
4. Abiotic stress tolerance
5. Quality improvement
6. Pharmaceutical and industrial use
7. Economic impact
8. Environmental impact
9. Transplastomic plants

1. INSECT PEST CONTROL

The most talked insect pest control through use of transgene is Bt gene (from *Bacillus thuringiensis*) producing toxin.

The adoption of insect resistant transgenic crops have been increasing annually since the commercial release of first generation maize and cotton expressing a single modified *B. thuringiensis* toxin (Bt) ten years ago. Studies have shown that these Bt crops can be successfully deployed in agriculture, which has lead to a decrease in pesticide usage, and that they are environmental friendly. However, sustainability and durability of pest resistance continued to be discussed. Now, scientists are developing second and third generation insect resistant transgenic plants and examine the proposed models for longevity of such resistance.

First generation transgenic plants - Transgenic plants containing only marker genes, which are useful in the development of transformation systems.

Second generation transgenic plants - Transgenic plants containing, in addition to the selectable marker, one or two transgenic encoding simple agronomic traits (such as pest and herbicide resistance).

Third generation transgenic plants - Transgenic plants that contain multiple transgenes targeting multiple pests and disease, often in a temporal or spatial manner. These might also express additional value added or agronomic traits.

By using a variety of technique, it has become possible to transform plants with foreign genes. Expression of foreign genes in plants makes it possible to produce a very wide range of new plants varieties. Transgenic plants have been developed to be resistant to a range of environmental stresses, including insects, viruses, herbicides, pathogens and salt stress to have flower with modified colour to a modified nutritional content including modifications in amino acids, lipids, discolouration and sweetness.

Bt gene and toxin (*Bacillus thuringiersis*) - Several species of bacteria produce protein in abundance. When insect larvae ingest these bacteria with their food, protein present in bacteria kills larvae. The most widely studied of these bacteria is *Bacillus thuringiensis* or Bt in short. This species lives all over the world. When these bacteria form spores, they also form a large crystal like structure in the bacterial cytoplasm, that is made out of protein. This bacterium comprises a number of different strains and subspecies, each of which produce's different proteins (toxin) that can kill certain specific insects.

Insecticidal toxins from some strains of Bt is shown in Table 17.1.

Table 17.1. Insecticidal toxins from some strains of Bt.

Strain	Toxin class	Protoxin size (KDa)	Target species
Berliner	Cry I	130-140	Lepidoptera
Kurstaki KTO, HD-I	Cry I	130-140	Lepidoptera
Kurstaki HD- I	Cry II	71	Lepidoptera Diptera
Tenebrionis (sandiego)	Cry III	.66-73	Coleopters
Israelensis	Cry IV	68	Diptera

One of the proteins in the crystal like structure is called the Bt-protoxin. When insect larvae eat the bacterial cells along with leaves, the spores and the crystalline like structure containing the proteins are released in the larval gut, where the digestive enzymes cleave the protoxin producing an active toxin.

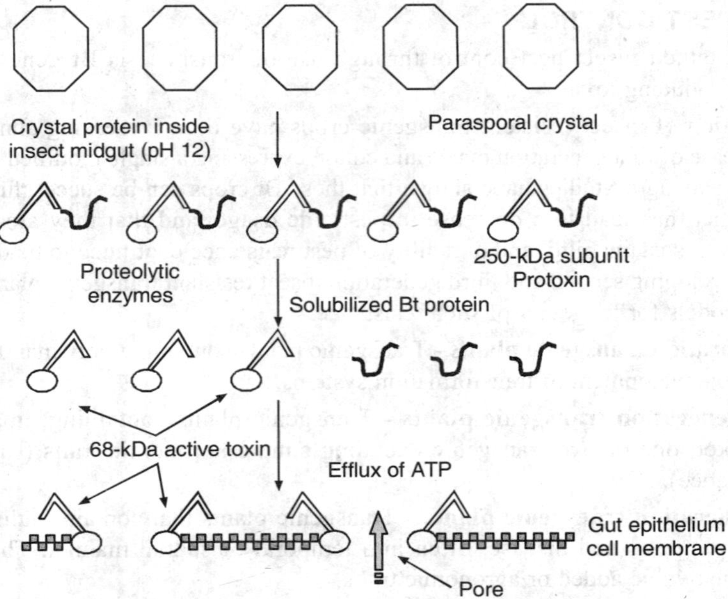

Fig. 17.1. Formation of active toxin from crystal protein and its effect on gut epithelium cell membrane.

Protoxin activated within its gut by the combination of alkaline pH (7.5-8.0) and specific digestive proteases. The toxin binds to the membrane of the epithelium cells of the gut, inserts itself into that membrane and creates an ion channel through which other molecules (e.g., ATP) can freely pass. Punctured by many holes, the gut cell cannot survive long, so the insect larvae starve for lack of nutrition and ultimately die. Because conversion of the protoxin to the active toxin requires both alkaline pH and the presence of specific proteases, such conditions are not present in mammals and hence they are safe from the protoxin.

No significant role for the bacterium has been attributed to the parasporal crystal structure. The parasporal crystal usually consists mainly of protein (~95%) and small amount of carbohydrate (~5%). The crystal protein can generally be dissociated by mild alkali treatment into subunits. The insecticidal toxins of *B. thuringiensis* strains can be grouped into four major classes: Cry-I, Cry II, Cry III and Cry IV. This is based on insecticidal activity against various insect. These toxins are further classified in sub classes and sub groups according to DNA sequence of the toxin gene, e.g. Cry I gene has six sub classes (Cry I A to F) and Cry I has subgroup (Cry IA a to c).

As a result of co-evolution between insects and their pathogens, there is host specificity between Bt toxin and the membranes of the gut cells. The Bt toxin of a particular Bt strain will bind to the gut of Lepidoptera larvae, or only some species of Lepidoptera, but not to other. When toxin does not bind, there is no effect on the cells that line the gut, and the larvae do not die. Thus some Bt toxin will kill lepidoptera (butter flies and moths), other coleopteran (beetles and weevils) and others diptera (mosquitos). For the biological control of insect pests, approximately 1.3×10^8 to 2.6×10^8 spores per sq foot of the target area are applied. Administration of the spores is timed to coincide with the peak of the larval population of the target organism.

Bt. Subspecies kurustaki contains a protoxin gene on one of seven different plasmids that approximately 2.0, 7.4, 7.8, 8.2, 14.4, 45 and 71kb in length. Protoxin is 130 kDa, therefore, not present on small plasmids. This gene has been transferred to other bacteria to kill mosquito larvae as well as gene has been modified to produce toxin during vegetative phase of bacterial growth rather than only during sporulation. Thus it is possible to produce toxin continuously in fermentor by growing bacteria. It has also been attempted to increase the host range.

This Bt toxin has been used in several ways to control the insects. A relatively simple way is to grow the Bt bacteria, dry them out, and prepare the heat killed and dried bacteria in such a way that they can be sprayed or dusted on crops. These preparations are initially highly effective, but the Bt protoxin is not stable after product is sprayed on plants. The Bt protoxin crystals are released from the bacteria and protoxin quickly disappears from the plants.

Scientists at Mycogen, a biotechnology company in San Diego, California (USA), introduced a Bt gene in a different bacterium (*Pseudomonas fluorescens*). These bacteria can readily be grown in large fermentors, killed and than formulated as a spray. With this bacterium the protoxin crystal remain in the bacterial cells, and as a result they are stable even after they have been sprayed on the plants. The spraying Bt toxin works well with insect larvae that live on the surfaces of leaves, but would be less effective with insect larvae that live in soil or larvae living inside the plants. To

Table 17.2. Plants with insecticidal gene derived from *B. thuringiensis*.

Company		Transgenic crop
Monsanto	-	Potato, cotton, tomato, corn
Calgene	-	Cotton, tobacco, potato,
Ciba-geigy	-	Tobacco, corn
Agrigenetics	-	Canola (rape seed)
Campbell Institute	-	Tomato
Rohm & Hass	-	Tobacco

control these insects, scientists have transferred Bt gene using particle gun transfer system in cotton, tomato, tobacco, potatoes, and other crop plants. Transgenic plants produced containing Bt gene are listed in Table 17.2.

Constitutive or tissue specific expression - Although constitutive expression of insecticidal transgene products has provided high levels of resistance in crop plants, tissue-specific or inducible expression might be desirable under some circumstances. Because the epidermal cells are first to be attacked by insects, defence genes expressed under epidermal cell-specific promoters (e.g., CEF6, an enzyme of curricular wax production) might be useful. Phloem -feeding insects can be targeted using the root phloem specific promoter AAP 3, the phloem specific pumpkin promoter PPZ and the rice sucrose synthase RSS promoter. Progress is being made with chemically inducible promoters, including those induced by ethanol, tetracycline, copper, glucocorticosteriod hormones and steroidal and non-steroidal ecdysone agonists.

2. HERBICIDE RESISTANT PLANTS

Certain herbicides can be used as pre-emergence herbicides to kill weeds before the crops are planted. If the crop plants are resistant to these chemicals then they can be used with the crop plant (post emergency). By understanding the mechanism of action of these herbicides and development of resistance by certain bacteria to such chemicals can provide clone for developing herbicide tolerant plants. Some plants or bacteria are resistant because they have an enzyme that detoxifies the herbicides. In other words they possess a gene for this action. Transfer of this gene to a crop plant should protect the crop plant by same action or mechanism. Some plants or bacteria become resistant to herbicide because of mutation in the target enzyme (or gene) and because of this change they are no more sensitive to herbicide or are not damaged by herbicides. The enzyme can work in presence of herbicide. Therefore the detoxifying mechanism or change in affected enzyme can make the organism herbicide tolerant.

Glyphosate (a herbicide) act by inhibiting one of the enzymes that is necessary for the synthesis of amino acids in the chloroplast. Glyphosate, initially produced and marketed by Monsanto under the trade name Roundup®, is widely used as non-selective herbicide. It effectively kills 76 of the world's 78 worst weed species.

Scientist at Monsanto isolated a gene for an enzyme involved in amino acid biosynthesis enzyme EPSP-synthase (5 enol pyruvinyl shikimate 3-phosphate synthase) from resistant *E. coli* bacteria. They modified the gene in such a way that it could be expressed in plants, and then transferred it to plants e.g., tobacco, tomato and soybean. Expression of bacterial gene in plants required a control region that would direct the expression at the gene in the plant (because bacterial control regions do not work in plants). In addition to this the gene had to be modified in such a way that the enzyme, which is synthesized in cytoplasm, would be transported to chloroplast. This is important that when gene of prokaryotic origin is used, the product should be transported to right cellular compartment in the plant. This should not affect the quantity or quality of yield. The gene has been successfully transferred in soybean where the plants showed resistance without change in yield.

Phosphinothricin is a herbicide that acts by inhibiting another enzyme necessary for amino acid biosynthesis (glutamine synthetase) and nitrogen metabolism. This enzyme converts ammonia to glutamate. Inhibiting the activity of this enzyme leads to rapid accumulation of ammonia within the plant cell. Higher concentrations of ammonia are toxic to the cell. Phosphinothricin, produced and marketed by Hoechst AG under the trade name Basta®, is also a very effective non selective herbicide. This product is related to an antibiotic that is also a herbicide, 'produced by the fungus *Streptomyces hygroscopicus*. Scientists at plant Genetic systems, Belgium obtained a gene from this fungus that encodes an enzyme that converts phosphinotricin to a non-herbicidal derivative by combining it with a cell metabolite. This gene, known as bar-gene, has been transferred in tobacco and potato, where it is expressed showing herbicide tolerance in these plants. The yield performance of the plants remained unchanged.

Herbicides are simply chemical compounds that kill or inhibit the growth of plants without deleterious effects on animals. Herbicides usually inhibit processes that are unique to plants, e.g. photosynthesis. Mostly herbicides act as inhibitors of essential enzyme reactions. Any change which can reduce the inhibitory effect of herbicide, will provide increased herbicide tolerance.

Glyphosate acts by inhibiting the enzyme 5 enol pyruvinyl shikimate 3 phosphate synthase (EPSP synthase), an essential enzyme in the biosynthesis of the aromatic amino acid, tysosine, phenylalanine and tryptophan. These are essential components in the diets of higher animals. Therefore higher animals do not contain EPSP synthase, and are not affected by glyphosate.

Glyphosate does inhibit the EPSP synthase of microorganisms as well as those of plants. Selection of organisms is made on inhibitory concentration of herbicides by growing them in presence of herbicide. This way researches isolated glyphosate tolerant mutant of *Salmonella typhimurion*, *Aerobacter acrogens*, and *Escherichia coli*. In bacteria, EPSP synthase is encoded by the aero A gene. When aeroA genes (with plant promoter and adenylation signals) were transferred in plants, the transgenic plants showed increased tolerance to glyphosate. In plants, aromatic amino acids are synthesized in chloroplasts, but gene for EPSP is localized in nucleus. Therefore, a protein is attached to EPSP synthase, which translocate the EPSP synthase into chloroplast, where the protein is removed by cleavage. It has been shown that the petunia transit peptide will target the *E. coli* aeroA gene product into tobacco chloroplasts and will impart glyphosate tolerance.

In another method, glyphosate-tolerant plants have also been produced by using an EPSP synthesis cDNA isolated from a glyphosate tolerant petunia cell culture line. Such lines can be selected by growing cells on medium containing increasing concentration of selection factor, e.g. glyphosate. In the cell line, tolerance resulted from amplification (an increase in copy number) of the EPSP synthesis gene, resulting in over production of EPSP synthase in these cells.

The EPSP synthase cDNA isolated from the cell line was joined to the CaMV 35s promoter and to the Ti nos. polyadenylation signal. The strong CaMV 35s promoter (35s +EPSP synthase + nos) gene was introduced into petunia plants on a Ti vectors, the transgenic developed were tolerant to the four times higher concentration which kills control plants.

Table 17.3. Gene based herbicide resistance in plants

Herbicide	Mode of development of herbicide resistance
Triazines	Resistance is due to an alteration in the psbA gene, which codes for the target of this herbicide, chloroplast protein D1.
Sulphonylureas	Genes encoding resistant version of the enzyme acetolactate synthetase have been introduced into poplar, canola, flax, and rice.
Glyphosate	Resistance is from overproduction of EPSPS, the target of this herbicide.
Bromoxynil	Resistance to this photosystem II inhibitor has been created by transforming tobacco and cotton plants with a bacterial nitrilase gene, which encodes an enzyme that degrades this herbicide.
Phenoxy carboxylic acids (e.g., 2, 4-D and 2, 4, 5-T)	Resistant cotton and tobacco plants have been created by transformation with the rfdA gene from Alcaligenes, which encodes a dioxygenase that degrades this herbicide.
Gluphosinate (Phosphinothricin)	Over 200 different plants have been transformed with either the bar gene from *Streptomyces hygroscopicus* or the pat gene from *S. viridochromogenes*. The phosphinothricin acetyltransferase that these genes encode, detoxifies this herbicide.
Cyanamide	Resistant tobacco plants were produced when cyanamide hydrates gene from the fungus *Myrothecium verrucarla* was introduced. The enzyme encoded by this gene converts cyanamide to urea.

The other examples of herbicides resistance plants are given in the table 26.4. It is evident from these examples that a resistant factor is developed based on mode of action of herbicide, by modifying or over producing the target product.

Canola (*Brassica napus*) cultivars engineered to tolerate the application of broad-spectrum herbicides have been developed both via transgenic and mutagenesis. This type of canola has been adopted rapidly by the Canadian farmers. The proportion of farmers growing transgenic herbicide-resistant canola has increased from 7% in 1995 to 80% in 2000. The use of transgenic herbicide - tolerant canola varieties had increased net return by 32%, had reduced pesticide use by 6000 tones and fuel consumption by 31 million litres. There are undoubtedly very real benefits both for the farmers and for the environment.

3. VIRUS RESISTANT PLANTS

Plants viruses often cause considerable damage and significantly reduce yield. Breeding for disease resistance is the best method to protect plants from viral and other infections. Recently scientists have used the techniques of genetic engineering to develop virus resistant transgenic plants. These methods used immunization with viral coat protein genes, other viral genes, or viral gene antisense sequence to confer resistance.

Potato is one of the most important food crops after cereals and pulses. It is very difficult to improve potato through breeding techniques as it is a tetraploid. Most of the cultivars are susceptible to various diseases caused by fungi, nematodes, and virus. Potatoes are vegetatively propagated. Therefore, seed material (tubers) for planting must be virus free. Potatoes suffer from three important virus diseases called Photo virus X, (PVX, PVY) and potato leaf-roll virus.

The phenomenon of cross protection or immunization of plant is not clearly understood. It is similar to immunization of human being for bacterial disease. When a plant is inoculated with a form of the virus, when virus infects, plant cell start synthesizing coat proteins instead of its own proteins. This cross protection is in some way related to the synthesis of the coat protein by the plant cell.

1986, Roger Beachy and colleagues at Washington University introduced the gene that encodes that coat protein of TMV into tobacco plants, resultantly each and every cell of the transgenic plant start producing coat protein. These plants showed considerable resistance to infection by TMV. The virus was unable to multiply in the cells already containing some coat proteins. Therefore, the number of virus particles per cell remained low in transgenic plants as compared to normal control plants.

Scientists at Mogen International in the Netherlands, used the same approach to make potatoes resistant to PVX. The gene encoding the coat protein of PVX was introduced into two cultivars. The transgenics showed 100 times less virus particles as compared to control plants after two weeks of inoculation. The yield performance of most cultivars was same but potatoes produced were elongated. The viral coast protein gene approach has been used to transfer tolerance to a number of transgenic plants for a number of different crops. Although complete protection is not usually achieved high levels of virus resistance have been reported. Moreover, a coat protein gene from one virus sometimes provides tolerance to a number of unrelated viruses.

In both eukaryotes and prokaryotes, an RNA molecule that is complementary to a normal gene transcript (that is mRNA) is called antisense RNA. The mRNA, being translatable, is considered to be a sense RNA. The presence of antisense RNA can decrease the synthesis of the gene product by forming a duplex molecules with the normal sense mRNA. Thereby, preventing it from being translated. The antisense RNA-mRNA duplex is also rapidly degraded, a response that diminishes the amount of that particular mRNA in the cell. Therefore, in principle it should be possible to prevent plant viruses from replicating and subsequently damaging plant tissues by creating transgenic plants that synthesize antisense RNA that is complementary to virus coat protein mRNA.

The Ti binary vector system was used to transfer both protein producing sense and antisense RNA producing cDNA sequence to separate tobacco cells, from which transgenic plants were regenerated. The transgenic tobacco plants that expressed the cucumber mosaic virus (CuMV) coat protein were produced from viral particle accumulation and did not show symptoms of viral infection, irrespectively of whether the inoculum of the challenge virus was high or low. However, the transgenic tobacco plants expressing the CuMV coat protein antisense RNA were protected only when the concentration of the challenge virus in the inoculums was low. Therefore, this approach is not successful when virus infection is high.

Table 17.3. Virus resistant transgenic plants developed that contain cloned viral coat protein (gene)

Plant species	Virus that provided the coat protein gene
Nicotiana benthamians	Plum pox virus, watermelon mosaic virus 2
Papaya, tobacco	Papaya ring spot virus
Potato	Potato virus (PVX, PVY, PVS)
Rice	Rice stripe virus
Tobacco	Soybean mosaic virus, tobacco streak virus, Tobacco spotted wilt virus, PVX
Tobacco, alfalfa, tomato	Alfalfa mosaic virus
Tobacco, cucumber	Cucumber mosaic virus
Tomato	Tomato mosaic virus

4. ABIOTIC STRESS TOLERANCE

Abiotic stresses such as drought, salinity and extreme temperatures cause significant losses of crop productivity and quality. Development of crops with an inherent capacity to withstand abiotic stress would help stabilize the crop production and significantly contribute to food and nutritional security in developing countries.

Transcriptome engineering or over expression of a master switch gene (such as stress sensors, protein kinases or transcription factors) that regulate several target genes coding for osmolyte biosynthesis enzymes, antioxidant enzymes and stress protein (such as late embryogenesis abundant proteins) is emerging as an important tool to combat abiotic stress. Stress-induced transcription factors such as c-repeat binding protein (CBF) or dehydration responsive element binding proteins regulate the expression of many genes for compatible osmolyte biosynthesis and oxidative stress management. Over expression or stress responsive promoter driven expression of CBF3 gene in transgenic *Arabidopsis* provided protection against multiple environmental stress such as cold, salt and drought. Components of the *Arabidopsis* CBF pathway are conserved in *B. napus*, wheat, rye, and tomato. Transgenic tomato with CBF1 gene and a CAM35s promoter showed significant chilling tolerance.

Transgenic tomato plants expressing *Arabidopsis thaliana* CBF1 gene, showed enhanced tolerance to oxidative stress, as CBF1 over expression induced a high level of expression of a catalase gene in these transgenic tomato plants.

5. QUALITY IMPROVEMENT

The goal of plant biotechnology is not confined to improvements of crop plants for agronomic traits and significant efforts are also being made to improve the nutritional content and organoleptic qualities such as taste and aroma in fruits and vegetables.

Nutritional improvement - Plant produces various compounds such as storage proteins, vitamin, flavonoids, carotenoids that perform vital functions for plants and also have nutritional importance for human beings. Vegetables are sources of minerals, proteins, micronutrients, vitamins, antioxidants, phytosterols and dietary fibre. However, some of the vegetables are deficient in essential amino acids such as methionine and lysine. The amino acid content can be modified or enhanced by expression of synthetic protein, over expression of homologous or heterogonous proteins, modifying the amino acid sequence of the protein or through metabolic engineering.

Potato is an important food crop, the nutritive value of potato protein is diminished due to deficiency in essential amino acids lysine, tyrosine and the sulphur containing amino acids methionine and cysteine. To improve the nutritive value of potato an Amaranthus seed albumin gene AmA1 has been expressed in transgenic potato tubers. This protein is non-allergenic and rich in all essential amino acids corresponding with WHO standards for human diet requirements. Similarly, a 292 bp artificial gene (asp-1) encoding a storage protein composed of essential amino acids was introduced in sweet potato. One of the transgenic lines showed a four fold increase in protein as compared to that of storage roots of control plants.

Carotenoids, such as B-carotene and lycopene, give the fruit its characteristic colour. Carotenoids are good antioxidants and are precursors of vitamin A. These are synthesized through the isoprenoid biosynthetic pathway. Provitamin content of tomato was increased by transferring a bacterial gene encoding for the phytoene - desaturase enzyme that converts phytoene to lycopene into transgenic tomato. These transgenic plants produced three-fold more B-carotene content than that of control plants. Similarly, a six-fold increase in carotenoid content and two to three fold increase in tocopherol content was achieved in transgenic potato plants by antisense technology.

Another group of metabolites exploited for its antioxidant property are the flavonoids. These are a diverse group of polyphenolic secondary metabolites, which impart colour to the fruits. Flavonoids are present only in tomato peel. A transgenic approach has been used to increase the flavonoid content by over-expression of either the enzymes involved in flavonoids biosynthesis or transcription factors that regulate the genes of this pathway. Transgenic tomato plants expressing petunia *CHI-A* gene encoding chalcone isomerase showed significant increase in flavonoids content. Similarly, a 10-fold increase in flavonoid content has been achieved by ectopic expression of the maize transcription factors LC and C1 in transgenic tomato.

Golden rice-with pro-vitamin A - According to the World Health Organization (WHO), vitamin A deficiency (VAD) is the leading causes of preventable blindness in children. For children, a lack of vitamin A causes severe visual impairments and blindness and significantly increases the risk of severe illness and even death from common infections such as diarrhoea and measles.

The genes from daffodil and one from the bacterium *Erwinia uredovora* were inserted in the rice genome. These three genes produce the enzymes necessary to convert GGDP to pro vitamin-A. The inserted genes are controlled by specific promoters such that the enzymes and the provitamin-A are only produced in the rice endosperm.

Provitamin-A is not produced by traditional rice varieties. However, geranylgeranyl diphosphate (GGDP), a compound naturally present in immature rice endosperm, with the help of several enzymes not normally found in rice can be used to produce provitamin-A.

Through the work of two European scientist, Dr. Ingo Potrykus of the Swiss Federal Institute of Technology in Zurich and Dr. Peter Beyer of the University of Freiburg in Germany, rice plants were developed containing two daffodil genes and one bacterial gene that carry out the four steps required for the production of beta-carotene in rice endosperm. Endosperm is the nutritive tissue surrounding the embryo of a seed and makes up the majority of the rice grain that we eat. The resulting plants appear normal expect the after milling (to remove the brown bran), their grain is golden yellow in colour due to the presence of provitamin-A.

When golden rice is ingested, the human body splits the provitamin-A to make vitamin A. Detailed information can only be obtained once the golden rice trait is transferred to local varieties and produced in quantities sufficient to support necessary field experiments. According to Swiss scientist Potrykus, "The intent of golden rice is to supplement to diet with vitamin A, not provide 100% of the Recommended Daily Allowance (RDA)". Potrykus maintains the goal of golden rice having a beneficial effect on vitamin A-deficient people is realistic with experimental golden rice lines ready in the 20-40% RDA range.

Golden rice is the result of an effort to develop rice verities that produce provitamin-A (beta-carotene) as a means of alleviating vitamin A (retinol) deficiencies in the diets of poor and disadvantaged people in developing countries. Because traditional rice varities do not produce vitamin-A, transgenic technologies were required.

Improvement of aroma - The aroma of fruits, vegetables and flowers are mixtures of volatile metabolites such as alcohols, phenols, ethers, adehydes, ketones etc. Some of the short-chain adehydes and alcohols are derived form lipid components by the action of lipases, hydroperoxide lipases and alcohol dehydrogenases. When yeast Δ-9 desaturase gene was transferred in tomato plants, changes in certain flavour compounds such as *cis*-3-hexenol, 1-hexanol, hexanal and cis-3-hexenal was recorded.

Linalool, an acyclic monoterpene alcohol, markedly influences the flavour of tomatoes. Linalool imparts a sweet, floral alcoholic note to fresh tomatoes. Hence linalool levels were altered by engineering the S-linalool synthase (LIS) gene from *Clarkia breweri* in tomato plants. The expression of S-linalool synthase enzyme, which catalyses the formation of linalool, resulted in elevated levels of linalool in the transgenic fruit.

Seedless vegetables - Browning and loss of flavour are two problems associated with potato. Transgenic potato have been generated in which browning is overcome by antisense inhibition of polyphenol oxidase. Cystathionine gamma synthase (CGS) is a key enzyme regulating methionine biosynthesis in plants. To increase the level of soluble methionine in potato, *Arabidopsis thaliana* CGS cDNA was introduced under transcriptional control of the cauliflower mosiac virus 35s promoter into potato. Increase in 2.4 - to 4.4 fold increase in methional level in transgenic potato tubers was recorded.

The seedless nature of parthenocarpic (development of fruit without fertilization) fruits increases consumer acceptance, makes processing of vegetables easier, and also improves the quality of vegetables, e.g., brinjal (where seeds are associated with bitter substances). Parthenocarpy has been shown to be regulated by auxins. Hence, efforts have been made to increase the auxin production or the sensitivity of ovary to auxins, towards inducing parthenocarpy. Expression of *iaaM* gene driven by the ovule specific promoter *defH9* has been shown to confer parthenocarpy to transgenic tomato and eggplant.

In another approach, the *Agrobacterium rhizogenes* derived gene *rol B* has been used for the induction of parthenocarpy in tomato. Transgenic tomato plants transformed with the *rol B* under the control of ovary and young fruit specific promoter *TPRP-F1* developed parthenocarpic fruits.

6. PHARMACEUTICAL AND INDUSTRIAL USE

The ability to transfer gene across different plant species and kingdoms through genetic engineering is being exploited in term of biofarming. Biofarming refers to production of proteins and biomolecules in transgenic plants at agricultural scale. The proteins mainly include antigens, antibodies, enzymes that are of immense importance in therapeutics, pharmaceutical and industrial applications. Though many of these proteins are being made in bacterial, fungal or animal systems, plants are now being preferred for manufacturing these proteins. The use of plants as biofactories is attributed to many factors; (*i*) plants offer cost effective and environmentally safe production of proteins as they use low cost inputs such as light, water and minerals, (*ii*) plants allow mass production, (*iii*) suitable for production of eukaryotic proteins which many require post-translational modifications, oligomerization etc, and (*iv*) plants are not pathogenic to human beings. The feasibility of vegetables as plant factories is very well illustrated in the form of edible vaccines, plantibodies (plant derived antibodies) and plant derived recombinant enzymes.

Terminator gene - One potential use of transgenic technology is to allow seed producers to realize profits from their investments in new product development. Seed companies have preferred to invest heavily in developing new varieties of crops such as corn for which the farmers typically purchases new seeds each years. Two biotech protection methods, dubbed 'Terminator' and 'Traitor' by opponents, may allow companies to increase profits on their cultivars.

Terminator, officially named as "Technology protection system" (TPS), incorporates a trait that kills developing plant embryos, so seed cannot be saved and replanted in subsequent years.

Traitor, officially known as "Trait-specific genetic use restriction technology" or T-Gurt, incorporates a control mechanism that requires yearly application of a preparatory chemical to activate desirable traits in the crop. The farmer can save and replant seeds, but cannot gain the benefits of the controlled traits unless he pays for the activating chemical each year. Both methods avoid the difficulties associated with enforcing 'no replanting' agreements. Because TPS and T-Gurt plants would be transgenic, their commercial use will require approval by the government. Scientist from agricultural research service (USDA) and Delta and Pine Land Company jointly developed this technology in 1998.

The technology protection system (TPS) inserts half a dozen sequences into the DNA of the parent plant that is slated for protection. These DNA sequences are arranged into a system that kills seeds at a prearranged time in their development. The system can be left inactive while the seed company grows several generations of seeds for sale. The system is switched on by soaking the seeds in a special chemical before the seeds are delivered to the farmer for planting.

The special chemical triggers a slow cascade of events that lead eventually to the death of progeny seeds developed on the protected plant. For the purpose of preventing replanting, the progeny seeds should be killed only after they have completed production of all commercially valuable products such as oil. Therefore, the system is designed to take effect only after the crop has grown to maturity in the field and the progeny seeds are nearly ripe.

7. ENVIRONMENTAL IMPACT

The use of Bt gene containing crops has been the most hotly debated issues regarding GM crops. Two different concerns have been broadly raised regarding such engineered insecticide resistance. The first concerns the broader impact of the presence of such insecticidal proteins on other organisms coming in contact with the transgenic crop. The second centres on the possibility of the target insets developing resistance to the insecticidal protein.

The possibility of detrimental environmental impacts of Bt corn become headline news in 1999. A paper was published suggesting that the presence of this insecticidal protein in the pollen of transgenic corn was detrimental to the larvae of the Monarch butterfly (*Danaus plexippus*). This was a laboratory based study and not a field study, even though sparked a controversy about use of transgenic crops and its impact on ecosystem. Later on, based on field studies by American Universities, the issue was settled in 2001. This episode illustrates the fact that the first generation of transgenic crops is largely lacking mechanisms to target gene expression to precise cell organ or cell types. Rather, the introduced transgenes are typically expressed constitutively in all the cells of the plant. In case of Bt corn, the presence of insecticidal cry protein in the pollen, where it serves no useful purpose as the insect attacks the stem of the plant, raised environmental concerns without any reason. If the expression is controlled, particularly in open-pollinated crop like corn, the incidental damages to the environment can be minimize.

The second concern expressed regarding Bt gene was that insect would develop resistance to the insecticidal protein. This would not only make the transgenic crop worthless but might also the usefulness of Bt spray. One of the few tools available to organic farmers in the fight against insect pests. The seed industry is encouraging farmers to keep some area for non-GM crops, where insect can multiply, and also cross with resistant insect, if any, generating susceptible progeny. This will delay the true breeding resistant strain of insect.

8. EDIBLE VACCINES

Edible fruits and vegetables are good choice to develop transgenic oral vaccines. Potato, tomato, banana, grapes are the examples of plant species grown all over the world and particularly in developing countries, where cheap vaccines are required the most. This reduces the cost of purification and down stream processing and transportation. Transgenic plants have been produced in the following other edible plants and species can be suitably modified to desired vaccine production: - apple, asparagus, cabbage, carrot, cauliflower, cucumber, egg plant, papaya, pea etc.

There are currently two methods of protein production from plants:

1. Stable integration of foreign DNA into plant genome introduced either by genetic transformation: *Agrobacterium* mediated or directly by using micro projectile bombardment and,
2. Transient expression of candidate DNA using viral vectors. The stable integration is advantageous because it passes in subsequent generations of large number of transgenic plants, either by vegetative or sexual means and also the possibility to introduce more than one gene for possible multi-component vaccine production. To produce sufficient amount of vaccines by recombinant cell culture technology, fermentation and purification are required which are very expensive. If the antigens are expressed in edible tissues of transgenic plants, it will become a cost effective production and delivery system. This is referred to as 'edible plant vaccine technology'. In addition tissue or organ specific expression of foreign antigens is possible by using tissue specific promoters.

Vaccines are the most important and cost effective sources for fighting infection diseases. Every year, an estimated 17 million people die of infectious diseases which include 7 million children. Although there have been opportunities for the production of cell cultures and recombinant vaccines, there is an increasing demand and the current production facilities are inadequate to supply the vaccines on a large-scale at an affordable price for the people living in developing countries. In recent years, considerable progress has been made in producing functionally active proteins, peptide of medical importance in transgenic plants. The expression of subunit antigens of infectious microorganisms in transgenic plants and their subsequent immunogenic properties led to the production of edible vaccines.

The modern biotechnological tools demonstrated the feasibility of using a genetically engineered food as an inexpensive oral vaccine production and delivery system for diarrhoea disease. Recently clinical trials have been conducted for heat labile enterotoxin (LT-B vaccine against cholera) from *E. coli* in the form of edible vaccine.

In the developing countries, diarrhoea disease is a leading cause of death, especially among children and travellers. Travellers who visit these tropical areas are victim of diarrhoeal diseases because they are frequently exposed to bacterial contaminations in food, water and common places. Enterotoxigenic *E. coli*, which produce a heat Labile (LT) and heat stable (ST) enterotoxin, are the most common causes of travellers diarrhoea throughout the world. LT is comprised of six sub-units. LT-A is an enzymatically active protein which enters the epithelial cells of the gut and initiates cellular metabolic changes that lead to loss of water from cells.

LT-B has five identical enzymatically inactive proteins, which form a pentamer that binds to GM1 gangliosides in the membranes epithelial cells. Binding of LT-B initiates transport of the active subunit inside the cells resulting in diarrhoea and any interference with binding will block the action of toxin. LT-B elicits oral immune response when given orally without any symptoms of disease.

Tobacco plants containing LT-B bacterial gene accumulated LT-B toxin and this LT-B was similar to that produced by bacteria. When mice were orally inoculated were with tobacco derived LT-B, both serum and mucosal antibodies were induced. In another experiment; potato plants were transformed to produce LT-B and upon feeding the transgenic tubers directly to mice, serum antibodies were induced. Production of serum and mucosal antibodies was confirmed on feeding transformed potatoes.

Norwalk Virus

Norwalk virus is the causative agent of acute epidemic gastroenteritis in humans. Recent advances in cloning the Norwalk virus genome and expression of the capsid protein in insect cell cultures have facilitated the study of the virus and the development of candidate vaccines for oral immunization. Norwalk virus capsid protein (NVCP) gene has been transferred and expressed in tobacco leaves and potato tubers. Partially purified antigen from tobacco or potato tubers is used for vaccination.

Hepatitis-B Surface Antigen (HbsAg)

Hepatitis is the single most important cause of viremia in humans and currently there are about 300 million carriers all over the world. The worldwide problem of infection and its association with chronic liver disease has necessitated the development of an effective vaccine. In many parts of the developing world, the expense of immunization programme limits the usage of the currently available serum or yeast could offer as a relatively low-cost method. The transfer of hepatitis-B surface antigen gene in tobacco, expression of recombinant gene in tobacco followed by partial purification of protein from the plant. When this protein was injected into mice, it elicited antibody response similar to that obtained with yeast derived commercially available vaccine. This is clear that gene product obtained from two different organisms has same property and transgenic plants can be used as source of antibodies.

QUESTIONS

1. What are transgenic crops? Give an account of recent developments about transgenic crops for insect pests resistance and nutritional quality.
2. Write short notes on: (a) Golden rice, (b) Terminator gene, (c) Transplastomic plants, (d) industrial applications of transgenic crops, (e) Bt gene, (f) cry protein, (g) herbicide resistance, (h) virus resistance.
3. Discuss the importance of transgenic crops for developing countries.

CHAPTER 18

Transgenic Animals

1. INTRODUCTION

The term *transgenic animal* refers to an animal in which there has been a deliberate modification of the genome (the material responsible for inherited characteristics) in contrast to spontaneous mutation. The foreign gene is constructed using recombinant DNA methodology. In addition to a structural gene, the DNA usually includes other sequences to enable it (*i*) to be incorporated into the DNA of the host and (*ii*) to be expressed correctly by the cells of the host. Foreign DNA introduced into the animal, using recombinant DNA technology must be transmitted through the germ line so that every cell, including germ cells, of the animal contains the same modified genetic material. Much of the research in this area has been centred on developing transgenic mice. The commonly used protocols includes introducing a cloned gene into a fertilized egg by microinjection, implanting the treated fertilized egg into a receptive female, and then testing the offspring to determine whether any of them have the input genes in their cells. By using these experimental strategies, transgenic versions of mice, cattle, sheep, goats, pigs, birds and fish has been generated. It is hoped that transgenic can be used to enhance existing genetic properties of livestock and provide the genetic basis for novel features. In addition, it is anticipated that the mammary glands-especially those of cows, sheep and goats will be used as a biological factory for cloned gene products that can be readily purified in large quantities for milk.

2. METHODS OF CREATION OF TRANSGENIC ANIMALS

Due to practical reasons, i.e., their small size and low cost of housing in comparison to that for larger vertebrates, their short generation time, and their fairly well defined genetics. Mice have become the main species used in the field of transgenics.

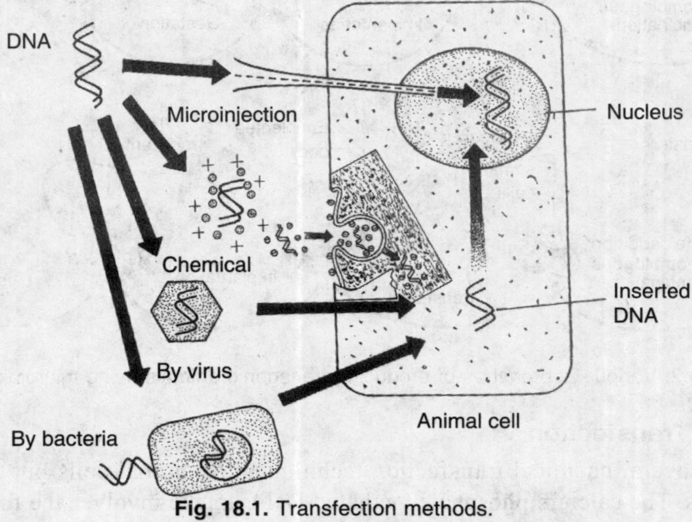

Fig. 18.1. Transfection methods.

The four principal methods used for the creation of transgenic animals (Fig. 18.1) are- (1) physical transfection, (2) chemical transfection (3) virus (retrovirus) mediated gene transfer (transduction) and (4) DNA packaged inside a bacterium (bactofection).

2.1. Physical Transfection

This method involves the direct microinjection of a chosen gene construct (a single gene or a combination of genes) from another member of the same species or from a different species, into the pronucleus of a fertilized ovum. It is one of the first methods that proved to be effective in mammals. The introduced DNA may lead to the over- or under-expression of certain genes or to the expression of genes entirely new to the animal species (Fig. 18.2). The insertion of DNA is, however, a random process, and there is a high probability that the introduced gene will not insert itself into a site on the host DNA that will permit its expression. The manipulated fertilized ovum is transferred into the oviduct of a recipient female or foster mother that has been induced to act as a recipient by mating with a vasectomised male. A major advantage of this method is its applicability to a wide variety of species. Other transfection methods includes particle bombardment, ultrasound and electroporation. These methods are described in chapter on transgenic plant development.

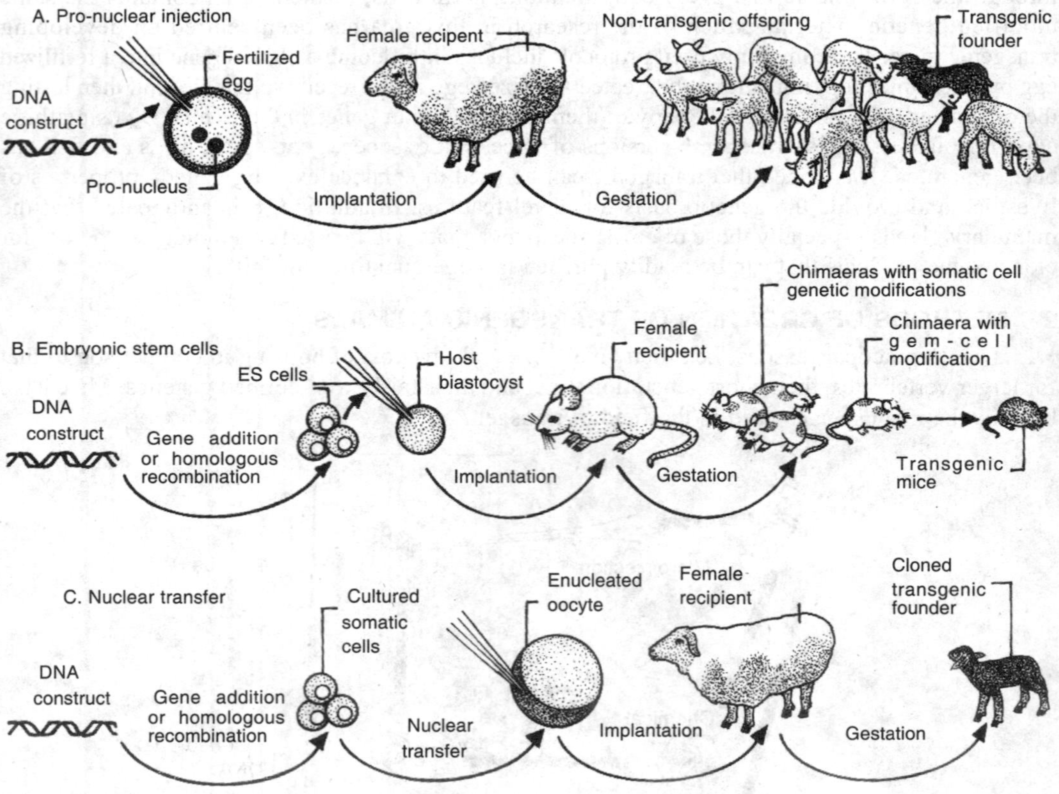

Fig. 18.2. Various approaches of producing transgenic animals using microinjection.

2.2. Chemical Transfection

There are several chemical transfection techniques for animal cells but all are based on similar principles. The calcium phosphate mediated DNA uptake involves the formation of a co-precipitate which is taken up by endocytosis. The ability of mammalian cells to take up exogenous DNA from the cultured medium was first reported in 1962. Formation of a fine DNA and calcium phosphate co-precipitate (should be prepared fresh) facilitates DNA uptake by endocytosis. Some of the DNA fragments which enter the cell may reach the nucleus and integrated. Expression of such genes confers the transfection. The transformation frequency of calcium precipitate method is generally low (1-2%), therefore use of soluble complexes (polyplexes) or liposomes and lipoplexes (fusogenic) phospholipid are used to package DNA inside these vehicles. The desired gene is transferred in plasmid and it may be used directly for chemical transfection or inserted in a bacterium for delivery into a mammalian cell. Yeast cells with the cell wall removed (spheroplast) therefore have been used to introduce YAC DNA into mouse, using liposomes and embryonic stem cells for the production of YAC transgenic mice.

2.3. Retrovirus-mediated Gene Transfer

To increase the probability of expression, gene transfer is mediated by means of a carrier or *vector*, generally a virus or a plasmid. Retroviruses are commonly used as vectors to transfer genetic material into the cell, taking advantage of their ability to infect host cells in this way. Offspring derived from this method are chimeric, i.e., not all cells carry the retrovirus. Transmission of the transgene is possible only if the retrovirus integrates into some of the germ cells and forms rPIC complex (Fig 18.3).

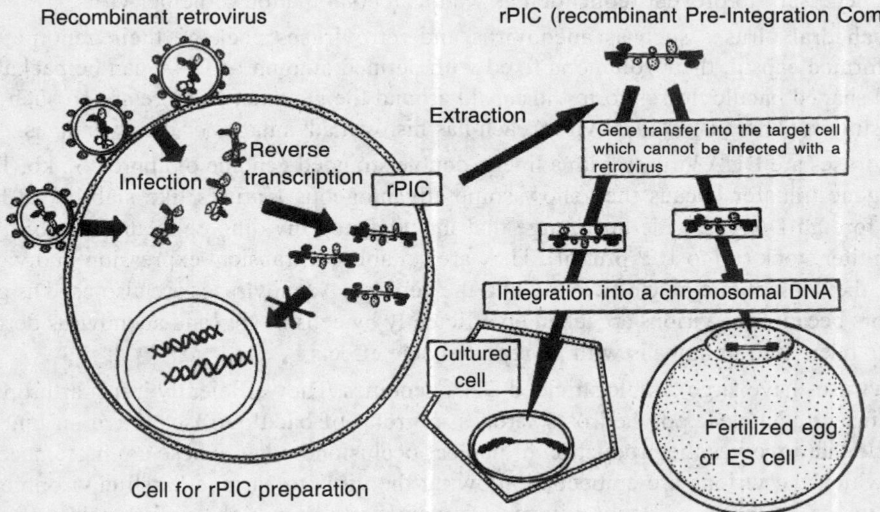

Fig. 18.3. Principles of gene integration by rPIC method.

For any of these techniques the success rate in terms of live birth of animals containing the transgene is extremely low. Providing that the genetic manipulation does not lead to abortion, the result is a first generation (F1) of animals that need to be tested for the expression of the transgene. Depending on the technique used, the F1 generation may result in chimeras. When the transgene has integrated into the germ cells, the so-called germ line chimeras are then inbred for 10 to 20 generations until homozygous transgenic animals are obtained and the transgene is present in every cell. At this stage embryos carrying the transgene can be frozen and stored for subsequent implantation.

There have been numerous reports and applications of transgenic plants in agriculture, mainly to benefit the producer. However, the realization of genetically engineered livestock has been much slower. The production of transgenic animals has focused mainly on producing models (e.g., the mouse) for basic and medical research. In terms of commercially important livestock species, work has revolved around specialized non-agricultural purposes such as pharmaceutical production and xenotransplantation and, to a lesser extent, applied agricultural purposes to improve animal production traits and animal-food products. In this case, one of the most important production animals, the dairy cow, was given enhanced resistance to a common and often devastating infection of the mammary gland, a potential benefit to both the producer and the animal's well-being.

2.4. Virus vector

Viruses have a natural ability to adsorb to the surface of host cells and infect. This property can be exploited to deliver rDNA into animal cells. The viral system is efficient in transfer, expressed well and replicate rapidly in the host cells. Several classes of viral vectors have been developed for use in human gene therapy and at least eight have been used in clinical trials. The general properties of viral vector are-

1. Transgene may be incorporated into viral vectors as additional gene or as replacement to certain genes of viral genome by ligation or homology recombination. If virus can propagate independently it is called helper independent. If essential viral genes are replaced by transgene then virus need a replication gene in trans position (another virus similar to binary vectors) and virus is called 'helper-dependent'.
2. It is necessary to prevent replication as well as recombination of helper virus.

   Icosahedral viruses such as adenovirus and retroviruses package their genome into preformed capsid, their volume is fixed with defined amount of DNA can be packaged. Rod shaped baculoviruses form the capsid around the genome, so there are no such size constraints. There is no ideal virus, each has his own advantages or disadvantages.

Adenoviruses are DNA viruses with a linear, double stranded genome of approx 36 kb. They are used in gene transfer becaus they show some advantageous features like stability, a high capacity for foreign DNA, a wide host range that includes non dividing cells and the ability to produce high titer stock (up to 10^{11} pfu/ml). They are suitable for transient expression in dividing cells because they do not integrate efficiently into the genome. Adenoviruses are also used as gene therapy vectors because the virions are taken up efficiently by cells *in vivo* and adenovirus derived vaccines have been used in humans with no reported side effects.

Baculoviruses have large double stranded DNA genomes. They efficiently infect arthropods, particularly insects. Nuclear polyhedrosis viruses, a group of baculoviruses, have an unusual infection cycle that involves the production of nuclear occlusion bodies. These are proteinaceous particles in which the virions are embedded allowing the virus to survive harsh environmental conditions such as desiccation. Baculoviruses are mainly used for high level transient protein expression in insects and insect cells. Two baculoviruses have been extensively developed as vectors, namely the *Autographa calofornica* multiple nuclear *polyhedrosis virus* and the *Bombyx mori* nuclear polyhedrosis virus.

2.5. DNA Packaged inside a Bacterium (Bactofection)

Generally, *Agrobacterium tumefaciens* mediated transfer of DNA is a common practice in plant system. It has been shown by Kunik and co -workers (2001) that this bacterium can transfer DNA in cultured human cells. It was established in mid-1990's that several bacteria infect human cells and undergo lysis releasing plasmid in host cells e.g., *Salmonella* spp., *Listeria* spp. and *Shigella* spp. The plasmid DNA then finds its way to the host cell nucleus, where it is integrated in the genome and expressed. Another method of DNA transfer is by conjugation [the transfer of DNA through a pilus (plural pilli)]. This pilus is formed by bacterial cell. When live bacteria are used, it is necessary that the bacteria are attenuated. This is because the gene transfer system uses the natural ability of bacteria to infect eukaryotic cells. The bacteria may multiply and destroy host cells.

Application Possibilities for Gene Transfer

In recent years, several application possibilities for gene transfer in domestic animals have been discussed. Until now it has been possible to influence only traits that are based on a single gene or on a limited number of genes. There is only a very limited number of traits of interest to breeders, which are based on a single gene. Various markers used to identify the transgenic animals are presented in Table 18.1.

In these authors' well-known experiments, growth, one of the classical quantitative traits in animal breeding, was changed to become a quasi-qualitative trait through the transfer of a single growth hormone gene which was related to a feedback independent regulation mechanism. Similar consequences can be expected by application of somatotropins in dairy cattle.

As far as gene transfer in cattle is concerned, there are realistic prospects that it will be possible to influence positively different production traits. Traditional selection programmes using conventional breeding techniques have achieved important results and will continue to do so. However, it seems preferable to concentrate the very expensive and complex technique of gene transfer to fields which until now could only be improved with limited success through conventional breeding programmes, such as breeding for increased disease resistance.

Table 18.1. Various markers used in transgenic animals.

Marker	Product	Selection
Ada	Adenosine deaminase	Xyl-A (9-β-D-xylofurenosyl adenosine) and 2′-deoxycoformycin.
Tk	Thymidine kinase	Thynidine and aminopterin to block de novo dTTP synthesis selected on HAT medium.
Ble	Glycopeptide binding progein	Confers resistance to glycopeptides, antibiotics bleomycin, pleomycin, pheomycin, Zeocin.
His D	Histidinol dehydrogenase	Confers resistance to histidinol.
Hpt	Hygromycin	Confers resistance to hygromycin.
Npt II	Neomycin phosphotransferase	Confers resistance to kanamycin, neomycin.
Pac	Puromycin N-acetyl transferase	Confers resistance to puromycin.
trpB	Tryptophan synthase	Confers resistance to indole.

3. TRANSGENIC LIVESTOCK

3.1. Mammary Gland-specific Transgenic Livestock

The modification of milk composition by genetic engineering has been frequently discussed. Milk proteins are made exclusively in the mammary gland and only during lactation. By using the promoter and regulatory components of a gene encoding a milk protein, it is possible to target the expression of a transgene to the mammary gland. Many groups have been working to take advantage of this natural system to produce important human pharmaceuticals in the milk of dairy animals.

Mastitis is a very important issue for the dairy industry. If left untreated, the presence of bacteria in the udder can lead to changes in the milk and overall health of the animal. Such changes include decreased milk production and increased somatic cell count resulting in subsequent economic losses due to: the dumping of milk; culling and animal replacement costs and increased veterinary and labour charges to treat the infection. Mastitis is the most costly disease in the dairy industry, with over 1.7 billion dollars a year in losses in the US alone. It is also the most common reason for antibiotic use in dairy cattle and the most frequent cause of antibiotic residues in milk. In addition, current antibiotic treatments are not totally effective and have been implicated in causing antibiotic resistance.

Growth- Attempts are being made to increase growth and body composition of animals through the transfer of these genes which are responsible for growth-hormone regulation. Experiments with the respective proteohormones in growing animals have shown that such effects can be achieved. These gene transfer experiments have been undertaken particularly in pigs and in sheep.

Disease resistance- Only a very limited number of genes is known that are able to influence the resistance of domestic animals to diseases. A model for such work is the influenza resistance caused by the Mx-gene.

Quality of animal products- The improvement of the quality or composition of animal products through the transfer of respective gene constructs could provide new prospects for animal production.

Gene farming- A suitable combination of tissue-specific promoters and the transfer of these genes into domestic animals may lead to efficient and biologically reliable production of proteins. In particular, efforts have been made to use animals as bioconversion systems.

3.2. Transgenic Animal in Pharmaceuticals

A transgenic animal for pharmaceutical production should (*i*) produce the desired drug at high levels without endangering its own health and (*ii*) pass its ability to produce the drug at high levels to its offspring. The current strategy to achieve these objectives is to couple the DNA gene for the protein drug with a DNA signal directing production in the mammary gland. The new gene, while present in every cell of the animal, functions only in the mammary gland so the protein drug is made only in the milk. Since the mammary gland and milk are essentially "outside" the main life support systems of the animal, there is virtually no danger of disease or harm to the animal in making the "foreign" protein drug. Because of the long time periods involved and low success rates, developing transgenic animals is currently very expensive, as the dollar amounts in Table 18.2 indicate.

Although most protein drugs are made in milk (Table 18.3), a notable exception is human haemoglobin that is made in Swine blood to provide a blood substitute for human transfusion. Because haemoglobin is naturally a blood protein, it is likely to be one of few exceptions to the usual method of production in milk. Furthermore, the economics of blood production are less favorable, because to recover human haemoglobin, the animal producing it must be slaughtered.

Table 18.2. Current market price of the drug and supply produced by one animal.

Drug	Animal	Value/Animal/Yr* ($)
AAT	Sheep	15,000
Factor VIII	Sheep	37,000
Factor IX	Sheep	20,000
Haemoglobin	Pig	3,000
lactoferin	Cow	20,000
CFTR	Sheep, mouse	75,000
Human protein C	Pig	1,000,000

4. TRANSGENIC ANIMAL MODEL DEVELOPMENT CORE

4.1. The Transgenic Mouse

It is designed to support investigators doing biology of aging research by creating mice that have been genetically altered by either inserting a new gene or removing a normal gene. This method has become one of the most exciting approaches of discovering the functions and interactions of genes in mammals. At the University Of Washington, Nathan Shock Center, this transgenic technology is used to develop new animal models for studying genetic mechanisms of the aging process (Fig. 18.4).

Table 18.3. Examples of proteins with therapeutic and industrial value that have been produced (but not commercialized) in the milk of transgenic animals.

Protein	Animal	Uses
Antithrombin III	Goat	Reduce the amount of blood needed in some surgeries
Factor VIII, Factor IX	Goat, Pig, Sheep	Treatment of hemophilia
CFTR	Sheep	Treatment of cystic fibrosis
Lactoferrin	Cow	Natural antibiotic and used in coronary surgery
Alpha-1-antitrypsin	Sheep	Treatment of cystic fibrosis and emphysema
Lysostaphin	Cow	An anti-bacterial compound that prevents mastitis in cows
Spider silk protein	Goat	Production of ultra-strong, lightweight medical and industrial materials

During the previous year, transgenic mouse production has focused on constructs with enhanced defense against free radical injury in aging (e.g., catalase, superoxide dismutase,

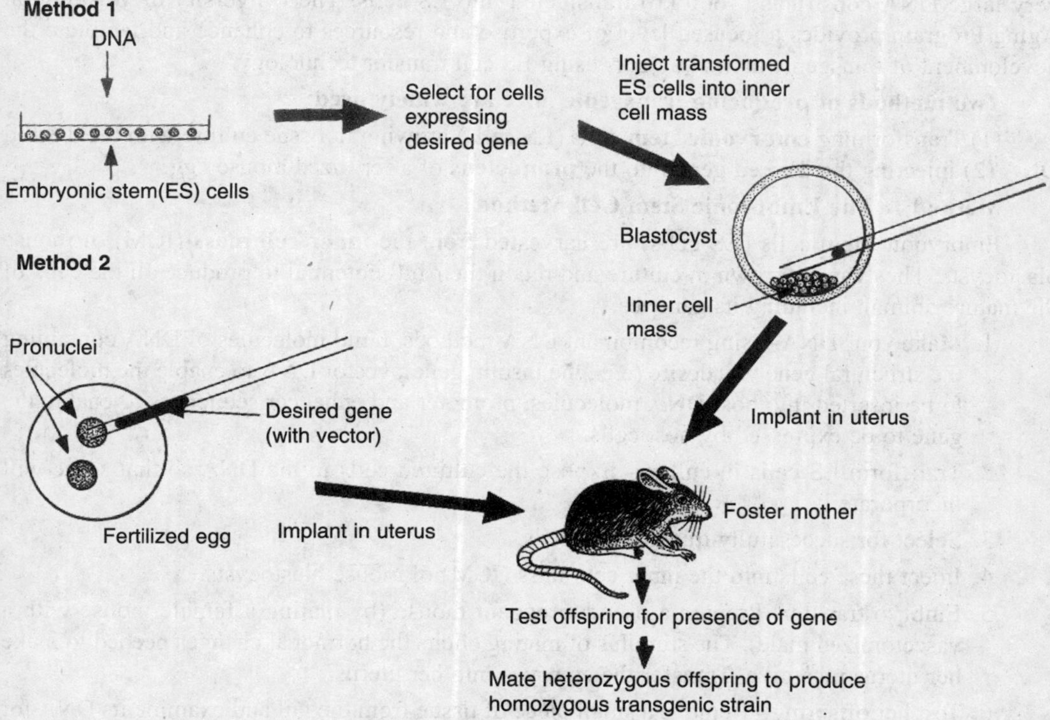

Fig. 18.4. Method of producing transgenic mice.

glutathione S-transferase), Werner Syndrome, adult onset diabetes, Alzheimer's Disease, thrombospondin, and rheumatoid arthritis in aging. Almost 4000 embryos, mainly of the C57BL/6 ingred strain, have been transferred, 498 pups analyzed and at least 40 contained the integrated construct. In addition, this core concentrated an appreciable portion of effort into embryonal stem line (ES) methodologies for generation of knockouts and targeted ES transgenics. This included work to generate mouse models of Werner's Syndrome, models for study of presenillin genes related to Alzheimer's Disease and study of models of thrombospondin in aging. In the past year, a total of 396 embryos were transferred, and 79 pups were born, of which 37 were chimeric. The isolation of mammalian genes is of utmost importance to the biology and medicine of aging because of the contributions these studies can make to the understanding of physiology and development. Techniques for introducing foreign genes into the mouse germ line provide novel approaches for modeling human genetic and chronic degenerative diseases.

Since the initial report in 1980 describing transgenic mice, methods for the direct microinjection of DNA into the pronuclei of fertilized embryos have become established. Foreign genes can be incorporated into somatic germline tissues, with expression of these elements in the progeny of founder mice. The creation of "transgenic" animals that make a specified gene product presents a spectrum of opportunities for basic studies in molecular pathogenesis and pre-clinical investigations applicable to a wide variety of medical problems of aging.

An additional gene transfer technology developed in the 1980's involved the use of stem cells from the early embryo, so-called embryonic stem (ES) cells. The capacity of ES cells to undergo differentiation makes them useful for investigating the effects of genetic modifications of either the gain of function or loss of function. These pluripotent genetically modified ES cells can then be used to make mice with deleted genes (gene knockout) or targeted mutagenesis of genes thought to be involved in the aging process. It is also possible to develop lines of transgenic mice carrying very large DNA constructs (>600 kb) transfected into ES cells. The University of Washington Aging Program provides a focused level of expertise and resources to enhance and facilitate the development of transgenic animal models using ES cell transfer technology.

Two methods of producing transgenic mice are widely used:

(1) Transforming **embryonic stem cells** (**ES** cells) growing in tissue culture with the desired DNA (2) injecting the desired gene into the **pronucleus** of a fertilized mouse egg.

Method 1- The Embryonic Stem Cell Method

Embryonic stem cells (**ES** cells) are harvested from the **inner cell mass** (ICM) of mouse blastocysts. They can be grown in culture and retain their full potential to produce all the cells of the mature animal, including its gametes.

1. Make your DNA-Using recombinant DNA methods, build molecules of DNA containing the structural gene you desire (e.g., the insulin gene), vector DNA to enable the molecules to be inserted into host DNA molecules, promoter and enhancer sequences to enable the gene to be expressed by host cells.
2. Transform ES cells in culture- Expose the cultured cells to the DNA so that some will incorporate it.
3. Select for successfully transformed cells.
4. Inject these cells into the inner cell mass (ICM) of mouse blastocysts.
5. Embryo transfer- Prepare a pseudopregnant mouse (by mating a female mouse with a vasectomized male). The stimulus of mating elicits the hormonal changes needed to make her uterus receptive. Transfer the embryos into her uterus.
6. Test her offspring - Remove a small piece of tissue from the tail and examine its DNA for the desired gene. No more than 10-20% will have it, and they will be heterozygous for the gene.

7. Establish a transgenic strain - Mate two heterozygous mice and screen their offspring for the 1:4 that will be homozygous for the transgene. Mating these will found the transgenic strain.

Method 2 -The Pronucleus Method

1. DNA is prepared as in Method 1.
2. Transform fertilized eggs - Freshly fertilized eggs are harvested before the sperm head has become a pronucleus. The male pronucleus is injected with DNA. When the pronuclei have fused to form the diploid zygote nucleus, the zygote is allowed to divide by mitosis to form a 2-cell embryo. The embryos is implanted in a pseudopregnant foster mother and proceeded as in Method 1.

The Figure 18.5 shows a transgenic mouse (right) with a normal littermate (left). The giant mouse developed from a fertilized egg transformed with a recombinant DNA molecule containing the structural gene for human growth hormone and a strong mouse gene promoter.

The levels of growth hormone in the serum of some of the transgenic mice were several hundred times higher than in control mice.

Fig 18.5. Human growth hormone gene containing transgenic mice on the right.

Random vs. Targeted Gene Insertion

The early vectors used for gene insertion could, and did, place the gene (from one to 200 copies of it) anywhere in the genome. However, if you know some of the DNA sequence flanking a particular gene, it is possible to design vectors that replace that gene. The replacement gene can be one that (a) restores function in a mutant animal or (b) knocks out the function of a particular locus.

In either case, targeted gene insertion requires (a) the desired gene (2) *neo^r^*, a gene that encodes an enzyme that inactivates the antibiotic neomycin and its relatives, like the drug G418, which is lethal to mammalian cells (3) *tk*, a gene that encodes thymidine kinase, an enzyme that phosphorylates the nucleoside analog gancyclovir. DNA polymerase fails to discriminate against the resulting nucleotide and inserts this nonfunctional nucleotide into freshly-replicating DNA. So ganciclovir kills cells that contain the *tk* gene.

Knockout mice are valuable tools for discovering the function(s) of genes for which mutant strains were not previously available. Two generalizations have emerged from examining knockout mice: Knockout mice are often surprisingly unaffected by their deficiency. Many genes turn out not to be indispensable. The mouse genome appears to have sufficient redundancy to compensate for a single missing pair of alleles. Most genes are **pleiotropic**. They are expressed in different tissues in different ways and at different times in development.

The Cre/loxP System

One of the bacteriophages that infects *E. coli*, called P1, produces an enzyme — designated Cre — that cuts its DNA into lengths suitable for packaging into fresh virus particles. Cre cuts the viral DNA wherever it encounters a pair of sequences designated *loxP*. All the DNA between the two *loxP* sites is removed and the remaining DNA ligated together again (so the enzyme is a recombinase).

Using "Method 1" (above), mice can be made transgenic for the gene encoding Cre attached to a promoter that will be activated only when it is bound by the same transcription factors that turn on the other genes required for the unique function(s) of that type of cell; a "target" gene, the one whose function is to be studied, flanked by *loxP* sequences (Fig. 18.6).

In the adult animal, those cells that receive signals (e.g., the arrival of a hormone or cytokine) to turn on production of the transcription factors needed to activate the promoters of the genes whose products are needed by that particular kind of cell will also turn on transcription of the Cre gene. Its protein will then remove the "target" gene under study. All other cells will lack the transcription factors needed to bind to the Cre promoter (and/or any enhancers) so the target gene remains intact.

The result: a mouse with a particular gene knocked out in only certain cells. The Cre/*loxP* system can also be used to remove DNA sequences that block gene transcription. In such a "knockin" mouse, the "target" gene is turned on in only certain cells.

4.2. Gene Transfer in Various other Species

A review of the literature shows that to date the only successful method to produce transgenic rabbits, pigs, sheep, goats and cattle is DNA-microinjection technique into the pronucleus.

Rabbits- Rabbits are used as experimental models in gene transfer experiments. In 1985, the successful production of transgenic rabbits was reported, for the first time, and included the growth hormone construct MT-hGH. The rate of degeneration of rabbit zygotes caused by injection was below 10%. The pre-implantation development capacity of injected zygotes is significantly lower compared with control embryos.

Pigs- Pig zygotes must be centrifuged to show the pronucleus. Fifty percent of the centrifuged non-injected zygotes develop *in vivo* up to the morula or blastocyst stage. After microinjection, 10-20% development to various stages of embryonic development occurs. Of the injected zygotes, 5.6% to 11% developed and led to the birth of piglets. The integration rate in pigs is approximately 10%. Growth-hormone constructs used in initial experiments led to an expression rate of 50%.

The production of transgenic F1 offspring is possible. In the authors' own experiments, inheritance of the transgene could be proved in two out of five animals.

It is not necessary to centrifuge sheep embryos to make the pronucleus visible. According to "Nomarski optics", 80% of the pronuclei can be located if a microscope with interference contrast is available. The capacity for *in vivo* development of sheep zygotes with injection (26 per cent) and without (10 per cent) is half that of pig embryos after similar treatment. Seven days after *in vivo* culture of non-treated and non-*in vitro* cultured sheep zygotes, Rexroad and Wall in 1987 observed a development rate of 86%. An *in vitro* culture of five hours' duration reduced this development rate to 65%, and after the injection of a buffer solution a reduction to 42 per cent was observed. 19% developed to the 32-cell stage after injection of DNA solution.

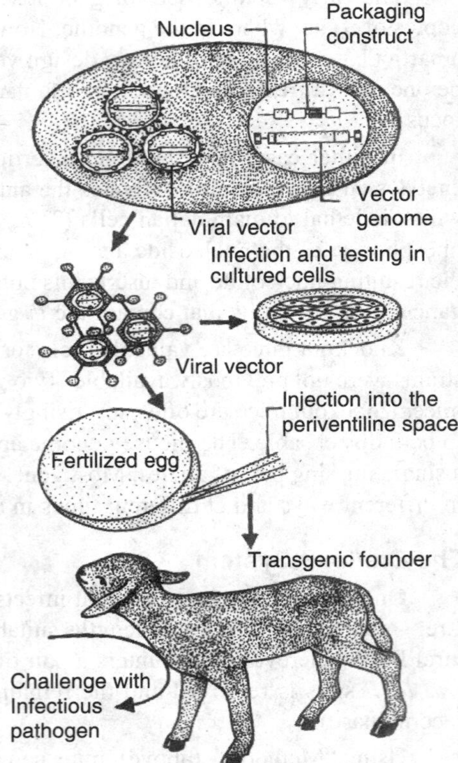

Fig. 18.6. Production of transgenic animals using virus as a vector.

Sheep and Goats- Until recently, the transgenes introduced into sheep inserted randomly in the genome and often worked poorly. However, in July 2000, success at inserting a transgene into a specific gene locus was reported. The gene was the human gene for alpha1-antitrypsin, and two of the animals expressed large quantities of the human protein in their milk.

(1) It was done as sheep fibroblasts (connective tissue cells) growing in tissue culture were treated with a vector that contained these segments of DNA (2) regions homologous to the sheep COL1A1 gene. This gene encodes Type 1 collagen (Its absence in humans causes the inherited disease osteogenesis imperfecta). This locus was chosen because fibroblasts secrete large amounts of collagen and thus one would expect the gene to be easily accessible in the chromatin. A neomycin-resistance gene to aid in isolating those cells that successfully incorporated the vector. The human gene encoding alpha1-antitrypsin. Some people inherit two non- or poorly-functioning genes for this protein. Its resulting low level or absence produces the disease **Alpha1-Antitrypsin Deficiency** (**A1AD** or **Alpha1**). The main symptoms are damage to the lungs (and sometimes to the liver). (1) Promoter sites from the **beta-lactoglobulin** gene. These promote hormone-driven gene expression in milk-producing cells. (2) Binding sites for ribosomes for efficient translation of the mRNAs.

Successfully-transformed cells were then (a) fused with enucleated sheep eggs and (b) implanted in the uterus of a ewe (female sheep). (c) Several embryos survived until their birth, and two young lambs have now lived over a year. (d) When treated with hormones, these two lambs secreted milk containing large amounts of alpha1-antitrypsin (650 μg/ml; 50 times higher than previous results using random insertion of the transgene).

The work on transgenic milk production is expensive requiring large facilities for purifying the protein from sheep's milk. Purification is important because even when 99.9% pure, human patients can develop antibodies against the tiny amounts of sheep proteins that remain.

GTC Biotherapeutics, won preliminary approval to market a human protein, antithrombin, in Europe. Their protein, the first made in a transgenic animal to receive regulatory approval for human therapy and was secreted in the milk of transgenic goats.

Chickens- Chickens have several advantages over other farm animals. (*i*) grow faster than sheep and goats and large numbers can be grown in close quarters (*ii*) synthesize several grams of protein in the "white" of their eggs.

Two methods have succeeded in producing chickens carrying and expressing foreign genes.

1. Infecting embryos with a viral vector carrying the human gene for a therapeutic protein and promoter sequences that will respond to the signals for making proteins (e.g., lysozyme) in egg white.
2. Transforming rooster sperm with a human gene and the appropriate promoters and checking for any transgenic offspring.

Preliminary results from both methods indicate that it may be possible for chickens to produce as much as 0.1 g of human protein in each egg that they lay. Not only should this cost less than producing therapeutic proteins in culture vessels, but chickens will probably add the correct sugars to glycosylated proteins something that *E. coli* cannot do.

Transgenic Fish- Aquatic animals are being engineered to increase aquaculture production, for medical and industrial research, and for ornamental reasons (Fig. 18.7). Genes inserted to promote disease resistance may allow transgenic fish to absorb higher levels of toxic substances, including heavy metals. In turn, consumers of these fish may be ingesting higher amounts of substances such as mercury and selenium. Transgenic fish that have genes from species such as peanuts or shellfish that are common causes of allergic reactions in humans may prompt allergic reactions in an unsuspecting consumer. Transgenic species may behave much like invasive species when interacting with the natural environment. They may compete with native species for resources and pose a threat to the genetic diversity of native populations, especially when genetic modifications such as a rapid growth rate offer advantages over slower-developing native species. Despite industry assurances that transgenic fish would be unable to naturally reproduce or significantly threaten the environment, some scientists are far more doubtful.

Fig. 18.7. Transgenic fish.

The sample bill included in this package addresses these concerns by banning the importation, transportation, possession, spawning, incubation, cultivation, or release of aquatic transgenic animals except under a permit.

Fluorescent cat- Recently South Korean scientist produced transgenic white Turkish angora cats to glow red under ultraviolet light these cats contain a fluorescent gene for flu protein and expressed under skin. Subsequently they produced a number of cloned cats from the skin cells of transformed mother cat. They proposed that such cat could be beneficial in diagnosis of genetic diseases and also showed a way to produce endangered animal by cloning.

QUESTIONS

1. What do you understand by transgenic animals? How they are produced? What are the utilities of such transgenic animals?
2. Discuss different methods to produce transgenic animals?
3. What is the utility of transgenic animals?
4. Write short notes on:
   (a) Retrovirus
   (b) Bactofection
   (c) Transgenic for mastitis
   (d) Lactoferin
   (e) Transgenic mouse
   (f) Transgenic fish

CHAPTER 19

Applications of rDNA Technology in Medicine

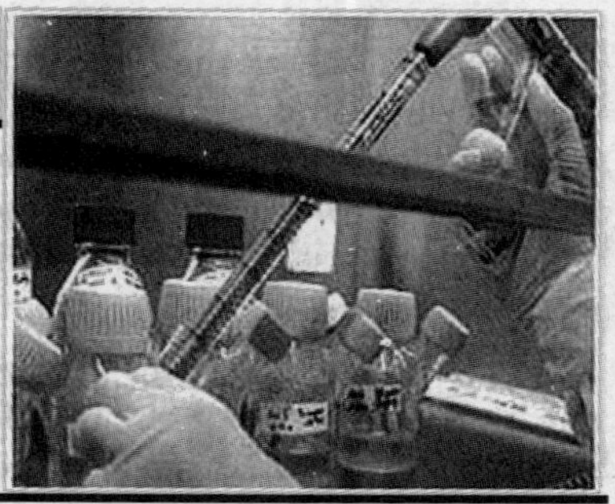

Genes, who are carried on chromosomes, are the basic physical and functional units of heredity. Genes are specific sequences of bases that encode instructions on how to make proteins. Although genes get a lot of attention, it's the proteins that perform most life functions and even make up the majority of cellular structures. When genes are altered so that the encoded proteins are unable to carry out their normal functions, genetic disorders can result. Gene therapy is a technique for correcting defective genes responsible for disease development. Researchers may use one of several approaches for correcting faulty genes:

1. A normal gene may be inserted into a nonspecific location within the genome to replace a nonfunctional gene. This approach is most common.
2. An abnormal gene could be exchanged for a normal gene through homologous recombination.
3. The abnormal gene could be repaired through selective reverse mutation, which returns the gene to its normal function.
4. The regulation (the degree to which a gene is turned on or off) of a particular gene could be altered.

1. DIAGNOSIS OF GENETIC DISEASES

Most of us do not suffer any harmful effects from our defective genes because we carry two copies of nearly all genes, one derived from our mother and the other from our father. The only exceptions to this rule are the genes found on the male sex chromosomes. Males have one X and one Y chromosome, the former from the mother and the latter from the father, so each cell has only one copy of the genes on these chromosomes. In the majority of cases, one normal gene is sufficient to avoid all the symptoms of disease. If the potentially harmful gene is recessive, then its normal counterpart will carry out all the tasks assigned to both. Only if we inherit both copies of the same recessive gene from our parents, we will develop a disease.

On the other hand, if the gene is dominant, it alone can produce the disease, even if its counterpart is normal. Clearly only the children of a parent with the disease can be affected, and then on average only half the children will be affected. Huntington's chorea, a severe disease of the nervous system, which becomes apparent only in adulthood, is an example of a dominant genetic disease.

Finally, there are the X chromosome-linked genetic diseases. As males have only one copy of the genes from this chromosome, there are no others available to fulfill the defective gene's function. Examples of such diseases are Duchenne muscular dystrophy and, perhaps most well known of all, hemophilia.

Queen Victoria was a carrier of the defective gene responsible for hemophilia, and through her it was transmitted to the royal families of Russia, Spain, and Prussia. Minor cuts and bruises, which would do little harm to most people, can prove fatal to hemophiliacs, who lack the proteins

(Factors VIII and IX) involved in the clotting of blood, which are coded for by the defective genes. Sadly, before these proteins were made available through genetic engineering, hemophiliacs were treated with proteins isolated from human blood. Some of this blood was contaminated with the AIDS virus, and has resulted in tragic consequences for many hemophiliacs. Use of genetically engineered proteins in therapeutic applications, rather than blood products, will avoid these problems in the future.

Not all defective genes necessarily produce detrimental effects, since the environment in which the gene operates is also of importance. A classic example of a genetic disease having a beneficial effect on survival is illustrated by the relationship between sickle-cell anemia and malaria. Only individuals having two copies of the sickle-cell gene, which produces a defective blood protein, suffer from the disease. Those with one sickle-cell gene and one normal gene are unaffected and, more importantly, are able to resist infection by malarial parasites. The clear advantage, in this case, of having one defective gene explains why this gene is common in populations in those areas of the world where malaria is endemic.

The diagnosis of inherited disease at the genetic level makes possible for the individual to determine whether they are at risk. DNA analysis can be used for the identification of carriers of hereditary disorders, for prenatal diagnosis of serious genetic conditions, and for early diagnosis before the onset of symptoms. Now-a-days these tests are at the DNA level and are definitive for determining the existence of specific genetic mutations, unlike previous genetic testing which were based on biochemical assay that used to determine either the presence or absence of a gene product.

1.1. Sickle Cell Anemia

Sickle cell anemia is a genetic disease that is the result of a single nucleotide change in the codon for the sixth amino acids of the β-chain of the hemoglobin molecule (from valin to glutamic acid). In this disease the shape of the RBC becomes irregular (sickle shaped). The biological effects of this genetic alteration are severe anemia and progressive damage to the heart, lungs, brain, joints and major organ systems. The anemia is caused by the inability of the mutated hemoglobin to carry sufficient oxygen.

The life expectancy of S/S homozygotes is quite short. Heterozygous individuals A/S (genetic carriers) have normal shaped RBC and no symptoms unless subjected to extreme conditions, where there is low oxygen supply as on high altitude, or extremes of temperature. If both parents are heterozygous (A/S) then 25% of children are (S/S) homozygous. One of the test systems to identify the individuals suffering from sickle cell anemia is as follows-

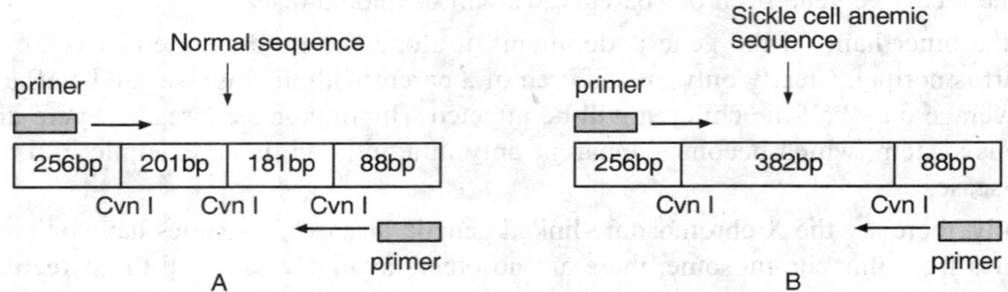

Fig. 19.1. Detection of Sickle cell anemia gene at the DNA level. (A) Normal genotype (B) Mutation at the Cvn I site resulting in sickle cell anemia and loss of restriction site.

It is known that the single nucleotide change in the β-globin gene causes sickle cell anemia. In the disease Cvn I restriction endonuclease site present in the gene gets diminished (Fig. 19.1). This restriction enzyme recognizes the sequence CCTGAGG and cleaves the DNA between the C and T. In the normal gene the DNA sequence is CCTGAGG, whereas, in sickle cell anemic gene the sequence is CCTGTGG. This difference forms the basis for DNA diagnostic assay.

After two primer sequences that flank the CvnI site are added, a small amount of sample DNA can be amplified by PCR. The amplified DNA is digested with CvnI, and the cleavage products are separated by gel electrophoresis and visualized by ethidium bromide staining of the DNA in the gel. If the intact CvnI site is present, then a specific set of DNA fragments is observed and if the CvnI site is absent, different profile of DNA fragment occurs. By this procedure, the genetic makeup of a tested person can be determined (Fig. 19.2).

	Genotype		
Size (bp)	AA	AS	SS
38		▭	▭
25	▭	▭	▭
20	▭	▭	
18	▭	▭	
8	▭	▭	▭

Fig. 19.2. Electrophoretic pattern of sickle cell diseased and healthy restriction digested PCR amplified β-globin DNA. AA, homozygous normal; AS, heterozygous; SS, homozygous sickle cell.

1.2. The PCR/OLA Procedure

All genetic diseases do not produce defect in the restriction site, so they could not be determined easily as in the case of sickle cell anemia. For such diseases, other strategies are required to detect single nucleotide change. One of the procedures used is PCR combined with oligonucleotide ligation assay (OLA).

To understand PCR/OLA procedure, assume that in a normal gene at a specific site (say 127th nucleotide) the nucleotide pair is A-T and in the mutant form the nucleotide pair at this site is G-C. Now, two DNA oligonucleotides are synthesized which are 20 bp long and their nucleotide sequence is complementary to the DNA. First probe (oligonucleotide) X is 20 bp long and is synthesized in such a fashion that its nucleotides sequences are complementary to the adjacent side of 127th nucleotide, including 127th nucleotide (i.e., its last base is at the 3′ end, at position 127). The other probe Y starts at 5′ end is again 20 bp long and its nucleotides are complementary to the other immediately adjacent side of 127th nucleotide, excluding it.

When these two probes are hybridized with target DNA (amplified by PCR) having normal sequence, the nucleotide at the 3′ end of probe X base pairs with 127th nucleotide of the target DNA and probe Y is aligned in such a way that its 5′ end lies next to the 3′ end of probe X. The addition of DNA ligase to the reaction covalently joins probe X and Y. When these two probes are hybridized with mutant gene in which the nucleotide at 127th position is altered as G-C. The nucleotide at the 3′ end of probe X which is 'A' is mismatched and is not able to pair with 127th nucleotide in the target DNA sequence. Probe Y however, is perfectly aligned. In this case, DNA ligase cannot join probe X and probe Y because of the single nucleotide misalignment.

After this procedure, the ligation product are determined by the use of two indicator probes, probe X is labeled at its 5′ end with biotin and probe Y is labeled at its 3′ end with digoxigenin (antibody binding indicator). These molecules help in indentifying the ligated product. The sample in which the ligated product are seen, it means it carries the normal genes sequence and the sample in which the ligation product are not seen, they have the defective gene sequence.

Overall, PCR/OLA system is rapid, sensitive and highly specific. This procedure has now also been automated. The ligase chain reaction is a less sensitive variant of PCR/OLA system. Sample DNA is mixed with an excess amount of a pair of OLA indicator probes in the presence of heat resistance DNA ligase. After an initial ligation reaction at 65 °C, the temperature is raised to 94 °C to denature the probe-target DNA hybrid and then lowered to 65 °C to allow hybridization of the free non-ligated OLA indicator probes to the target DNA. The cycle is repeated 20 times. If the OLA indicator probes match the target DNA perfectly, then ligation will occur at 65 °C during each cycle and after 20 cycles enough ligation product (i.e., probe X joined to Y) will accumulate to be observed by either gel electrophoresis or an ELISA detection system. If no ligation occurs because of a mismatch, then no joined probe product will be produced or detected.

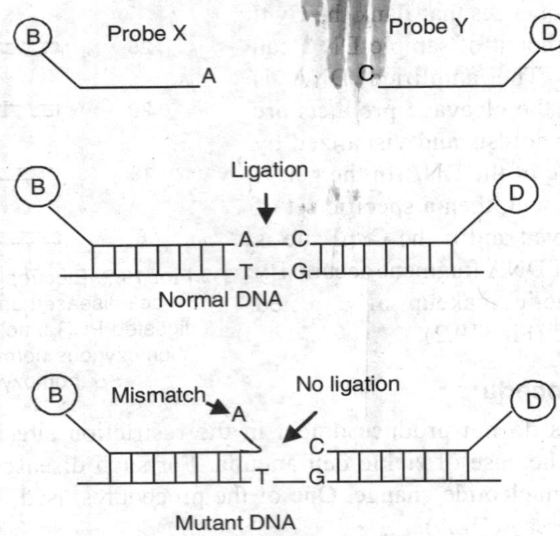

Fig. 19.3. The PCR/OLA Procedure.

1.3. Genotyping with Fluorescence-Labelled PCR Primers

PCR primers labelled with different fluorescent dyes can be used in the development of non-radioactive color-based detection systems. To distinguish between mutant and wild type DNA, PCR is performed with two different primers. One is exactly complementary to the wild type DNA and is labeled at its 5′ end with rhodamine (red). The other is complementary to the mutant DNA and its is labeled at its 5′ end with fluorescein (green). In both cases, amplification is programmed a third unlabeled primer that is complementary to the opposite strand. Since PCR amplification can take place only when the primer is exactly complementary to the target DNA, the presence of these three primers in the same reaction mixture will result in the amplification of either the wild type or the mutant DNA or both, depending on which target DNAs are initially present to act as PCR template. If an individual is homozygous for the wild type DNA, after PCR and removal of unincorporated primer, the reaction mixture will fluoresce red; if a person is homozygous for the mutant DNA, the reaction mixture will fluoresce green; and if he has both mutant and wild type DNA (heterozygous). The reaction mixture will fluoresce yellow. This assay can be automated and can be adapted for any single nucleotide target site of any gene that has been sequenced.

1.4. Mutation at Different Sites within One Gene

Not all genetic diseases are due to a single specific nucleotide change within a gene. In most cases, a variety of different intragenic sites can mutate, and each mutation can cause the same form of the genetic disease. For example, β-thalassemia is a genetic disease that is due to a loss in the activity of β-globin. Heterozygotes (carriers) tend to have only a mild form of anemia. In contrast, people who are homozygous for one of eight or more possible mutations must receive regular blood transfusions and other treatment to survive. Because a mutation at any one of eight or more different sites within the β-globin gene can cause β-thalassemia, testing for a change at only one particular nucleotide site is insufficient; at least eight separate tests are necessary. Although this is feasible, it is costly.

Therefore, a PCR/hybridization strategy that uses one reaction assay system has been devised to screen for mutant nucleotide sites at different locations within a single gene. A set of sequence-specific oligonucleotide probes (each 20 nucleotides in length) is synthesized. Each oligonucleotide can form a perfect match with a segment of the target gene that corresponds to the site of a known mutation. A thymidine homopolymer [poly (dT)] tail approximately 400 mucleotides long is added to the 3′ end of each of the probes. The tail enables the probe DNA to be physically bound to the rest of the ignited discrete spot on a nylon filter while ensuring that the rest of the probe is accessible for hybridization. Segments of the sample (test) DNA that span each of the possible mutant sites within the gene are amplified simultaneously by PCR. In each case, one of each of the pairs of primers is labelled at the 5′ end with biotin. The amplified target DNA is then hybridized to the filter-bound probes under conditions that allow only perfect matches to hybridize. Streptavidin with attached alkaline phosphatase is added during the hybridization reaction. After hybridization, the filter is washed and a colorless substrate is added. A colored spot on the filter appears wherever there is a perfect nucleotide matched between an amplified target DNA segment and one of the specific oligonucleotide probes. One filter can be spotted with a number of sequence specific oligonucleotide probes. Thus, with a single filter assay, one of a number of different mutant sites can be identified.

2. DNA TYPING (DNA FINGERPRINTING)

1. DNA typing is a technique in which biological samples help in solving forensic problems. This technique is used to establish that whether the suspected person has committed a crime or not.
2. For DNA typing, biological samples like blood, skin, semen or hair are collected from the site of crime. A portion of a sample is analyzed to confirm that there is sufficient amount of intact DNA (undegraded) for the further analysis.
3. If there is sufficient amount of intact DNA, it is digested with a restriction enzyme, and the fragments are separated on an agarose gel and transferred by blotting onto a nylon membrane.
4. This nylon membrane is then hybridized sequentially with four or five separate radiolabelled probes that each recognizes a distinct DNA sequence. After each hybridization reaction, the bands where the probe has bound to the digested. DNA sample are visualized by autoradiography and the banding pattern for each sample is noted.
5. Before the next probe is used, the first probe is completely removed from the membrane. Since each hybridization and autoradiography step can take up to 10 to 14 days, the entire process may take many weeks and even several months.
6. A commonly used set of probes for this type of analysis consists of human minisatellite DNAs. These sequences occur throughout the human genome and consist of tandemly repeated sequences. The length of the repeats range from 9 to 40bp, and the number of repeats in the minisatellites ranges from about 10 to 30. Unrelated individuals generally have different length of minisatellites. However, a child will inherit one minisatellite DNA sequence from mother and one from father. Even the same minisatellites DNA sequences can have a different length in different individuals. This variability is due to either a gain or a loss of tandem repeats, probably during DNA replication. These changes do not have any biological effect because the minisatellite DNA do not code for any protein.
7. The human minisatellites DNA sequences are highly variable, and the chance of finding two individuals in the population with the same DNA fingerprint is about one in 10^5 to 10^8. An individual's DNA banding pattern based on minisatellites DNA sequences is almost as unique as his or her fingerprints.

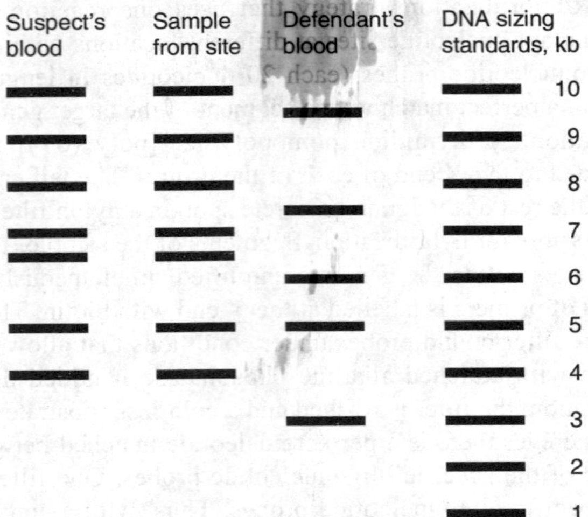

Fig. 19.4. Southern blot of a forensic DNA sample. Here the banding pattern of the suspected person's blood matches with the blood sample collected from the site of crime, thus confirming the victim.

8. In addition to forensic applications of DNA typing, this technique may be used to determine the paternity of a child. The DNA fingerprint banding pattern of the child should be a composite of the patterns of its mother and father.
9. In case when the amount of forensic sample collected is quite less but the DNA is undegraded then in that case, it is possible to amplify small portions of DNA of the minisatellite DNA by PCR (Fig. 19.4).

3. GENE THERAPY

Much attention has been focused on the so-called genetic metabolic diseases in which a defective gene causes an enzyme to be either absent or ineffective in catalyzing a particular metabolic reaction effectively. A potential approach to the treatment of genetic disorders in man is gene therapy. This is a technique whereby a working gene replaces the absent or faulty gene, so that the body can make the correct enzyme or protein and consequently eliminate the root cause of the disease.

The most likely candidates for future gene therapy trials will be rare diseases such as Lesch-Nyhan syndrome, a distressing disease in which the patients are unable to manufacture a particular enzyme. This leads to a bizarre impulse for self-mutilation, including very severe biting of the lips and fingers. The normal version of the defective gene in this disease has now been cloned.

If gene therapy does become practicable, the biggest impact would be on the treatment of diseases where the normal gene needs to be introduced into only one organ. One such disease is phenylketonuria (PKU). PKU affects about one in 12,000 white children, and if not treated early can result in severe mental retardation. The disease is caused by a defect in a gene producing a liver enzyme. If detected early enough, the child can be placed on a special diet for their first few years, but this is very unpleasant and can lead to many problems within the family.

Before treatment for a genetic disease can begin, an accurate diagnosis of the genetic defect needs to be made. It is here that biotechnology is also likely to have a great impact in the near future. Genetic engineering research has produced a powerful tool for pinpointing specific diseases rapidly and accurately. Short pieces of DNA called DNA probes can be designed to stick very specifically to certain other pieces of DNA. The technique relies upon the fact that complementary pieces of DNA stick together. DNA probes are more specific and have the potential to be more sensitive than conventional diagnostic methods, and it should be possible in the near future to distinguish between defective genes and their normal counterparts, an important development.

The transfer of DNA is usually achieved either by mixing the DNA with a substance that enhances its uptake or by packaging it inside a disabled virus (a virus that can carry DNA into a cell but cannot cause a disease). First removing cells from the patient and then re-introducing those that have been appropriately modified may carry out Gene therapy. Alternatively, the direct introduction of DNA into the patient's body may be the only effective strategy.

There are several types of gene therapy:

1. **Gene augmentation therapy:** This is appropriate for the treatment of inherited disorders caused by the loss of a functional gene product. The aim is to add a functional copy of the lost gene back into the genome and express it at sufficient levels to replace the missing protein. It is only suitable if the pathogenic effects of the disease are reversible.
2. **Gene inhibition therapy:** This is suitable for the treatment of infectious diseases, cancer and inherited disorders caused by inappropriate gene activity. The aim is to introduce a gene whose product inhibits the expression of the pathogenic gene or interferes with the activity of its product.
3. **Killing of specific cells:** This is suitable for diseases such as cancer that can be cured by eliminating certain populations of cells. The aim is to express within such cells a suicide gene, whose product is toxic. One approach is the expression of an enzyme that converts a harmless prodrug into a highly toxic molecule. Another is the expression of a protein that makes the cells vulnerable to attack by the immune system. It is very important to ensure that suicide genes are appropriately targeted; otherwise the therapy would result in widespread cell death.
4. **Somatic and germ line gene therapy:** There is an important distinction between somatic gene therapy (DNA transfer to our normal body tissue) and germ line gene therapy (DNA transfer to cells that produce eggs or sperm). The distinction is that the results of any somatic gene therapy are restricted to the actual patient and are not passed on to his or her children.

In most gene therapy studies, a "normal" gene is inserted into the genome to replace an "abnormal," disease-causing gene. A carrier molecule called a vector must be used to deliver the therapeutic gene to the patient's target cells. Currently, the most common vector is a virus that has been genetically altered to carry normal human DNA. Viruses have evolved a way of encapsulating and delivering their genes to human cells in a pathogenic manner. Scientists have tried to take advantage of this capability and manipulate the virus genome to remove disease-causing genes and insert therapeutic genes.

Target cells such as the patient's liver or lung cells are infected with the viral vector. The vector then unloads its genetic material containing the therapeutic human gene into the target cell. The generation of a functional protein product from the therapeutic gene restores the target cell to a normal state.

Some of the different types of viruses used as gene therapy vectors:

3.1. Adenovirus Vector System

Adenoviruses infect a wide range of non dividing human cells and have been used extensively as live vaccines against respiratory infections and gastroenteritis without side effects. These features make adenovirus a likely prospect for delivering genes to target cells. For use as a vector for gene therapy, a cell line that produces the adenoviral E1 gene products, which are essential for adenoviral replication, is cotransfected with two portions of the adenovirus genome.

One of these segments, which is maintained in *Escherichia coli* as a plasmid, carries a therapeutic gene in place of the E1 region and is flanked by adenovirus DNA sequences. The E1 region is located near the 5′ end of the adenovirus genome. The second transfected DNA segment is an adenovirus DNA molecule that lacks part of its 5′ end, including the E1 region, but shares a region of adenoviral DNA with the plasmid that carries the therapeutic gene. A recombination event in the region of overlap between the two transfected pieces of DNA reconstitutes a full length adenovirus genome that contains the therapeutic gene and lacks the E1 region. The E1 gene products that are supplied by the host cell initiate virus production; virus particles are assembled and then released after cell lysis. The DNA cloning capacity of this adenovirus vector system is about 7.5 kb. If recombination does not occur, both transfected DNA molecules are too small to be packaged properly. Moreover, recombination between the E1 DNA region in the genome of the host cell and the recombinant adenovirus DNA to form a replication -competent virus is extremely unlikely.

After infection of a target cell with a recombinant adenovirus, the DNA is passed into the cell nucleus, where the therapeutic gene is expressed. The recombinant DNA construct does not integrate into a chromosome, and consequently, it does not persist for long periods. Therefore, adenovirus based gene therapy requires periodic administration with additional recombinant viruses.

Adenovirus gene delivery systems have been used in gene therapy trials for treating cystic fibrosis.

3.2. ADENO-ASSOCIATED VIRUS VECTOR SYSTEM

Adeno- associated virus is a small, non pathogenic, single-stranded human DNA virus (4.7kb) that can integrate into a specific site on chromosome 19. Productive infection by adeno-associated virus depends on proteins from another virus (helper virus) such as adenovirus, hence the name adeno-associated virus. After the adeno-associated virus enters the nucleus, the polymerase of a host cell converts the adeno-associated virus genome to a double-stranded DNA that is then transcribed.

Its absence of pathogenicity makes adeno-associated virus a good candidate as a vector for the delivery of therapeutic genes. Recombinant adeno associated virus is generated by cotransfection of two plasmids into a host cell that has been infected with adenovirus (helper virus). One of the plasmids carries a therapeutic gene flanked by the inverse terminal repeat sequences (~125bp each) from the adeno-associated virus. The second plasmid contains the two genes (rep and cap) of the adeno-associated virus genome that are responsible for replication of the adeno-associated virus and its capsid, respectively. After lysis of infected host cells, recombinant adeno-associated virus particles are purified from adenovirus by centrifugation and dialysis. Any residual adenovirus (helper virus) in the sample is killed by heat treatment. In the absence of the *rep* gene, the recombinant adeno-associated virus DNA is unlikely to integrate into chromosome 19. On the other hand, without any adeno-associated virus genes, the recombinant vector will not evoke an immune response.

Besides virus-mediated gene-delivery systems, there are several nonviral options for gene delivery. The simplest method is the direct introduction of therapeutic DNA into target cells. This approach is limited in its application because it can be used only with certain tissues and requires large amounts of DNA.

Another non-viral approach involves the creation of an artificial lipid sphere with an aqueous core. This liposome, which carries the therapeutic DNA, is capable of passing the DNA through the target cell's membrane.

Therapeutic DNA also can get inside target cells by chemically linking the DNA to a molecule that will bind to special cell receptors. Once bound to these receptors, the therapeutic

DNA constructs are engulfed by the cell membrane and passed into the interior of the target cell. This delivery system tends to be less effective than other options.

Researchers also are experimenting with introducing a 47th (artificial human) chromosome into target cells. This chromosome would exist autonomously alongside the standard 46 not affecting their workings or causing any mutations. It would be a large vector capable of carrying substantial amounts of genetic code, and scientists anticipate that, because of its construction and autonomy, the body's immune systems would not attack it. A problem with this potential method is the difficulty in delivering such a large molecule to the nucleus of a target cell.

3.3. Current Status of Gene Therapy Research

Gene therapy is "the use of genes as medicine". It involves the transfer of a therapeutic or correct gene into specific cells of an individual in order to repair a faulty gene. Thus it may be used to replace a faulty gene, or to introduce a new gene whose function is to cure or to favourably modify the clinical course of a disease. The scope of this new approach to the treatment of disease is broad, with potential in the treatment of many genetic conditions, some forms of cancer and certain viral infections such as Acquired Immune Deficiency Syndrome (AIDS). At the present time however, gene therapy remains an experimental discipline and much research remains to be done before this approach to the treatment of disease will realize its full potential.

There are over 600 clinical gene therapy trials initiated or approved worldwide. Of these, the majorities are being conducted in the United States and Europe, with only a modest number initiated in other countries, including Australia. Perhaps surprisingly, the majority of trials focus on treating acquired diseases, such as cancer and AIDS, although an increasing number of inherited conditions are being targeted. A form of immune deficiency called adenosine deaminase (ADA) deficiency was the first disease to be treated with a gene therapy approach in humans in the early 1990's, and was also the first condition for which therapeutic gene transfer into stem cells has been attempted in the clinical arena.

Another form of immune deficiency is due to a mutation in a gene located on the X chromosome and is called Severe Combined Immune Deficiency (SCID). This "X-linked condition" only affects boys. The use of gene therapy in the treatment of this condition was started by a French research group Cavazzana-Calvo and co-workers (2000). It was hailed as the first example of a genetic condition being successfully treated by gene therapy and is a milestone in medical history.

The US Food and Drug Administration (FDA) have not yet approved any human gene therapy product for sale. Current gene therapy is experimental and has not proven very successful in clinical trials. Little progress has been made since the first gene therapy clinical trial began in 1990. In 1999, gene therapy suffered a major setback with the death of 18-year old Jesse Gelsinger. Jesse was participating in a gene therapy trial for ornithine transcarboxylase deficiency (OTCD). He died from multiple organ failures 4 days after starting the treatment. His death is believed to have been triggered by a severe immune response to the adenovirus carrier.

The most active research being done in gene therapy for children has been for genetic disorders such as cystic fibrosis. Other gene therapy trials involve children with severe immunodeficiencies, such as adenosine deaminase (ADA) deficiency (a rare genetic disease that makes children prone to serious infection), and those with familial hypercholesterolemia (extremely high levels of serum cholesterol).

Advances in understanding and manipulating genes have set the stage for scientists to alter a person's genetic material to fight or prevent disease. Gene therapy is an experimental treatment that involves introducing genetic material (DNA or RNA) into a person's cells to fight disease. Gene therapy is being studied in clinical trials (research studies with people) for many different types of cancer and for other diseases. It is not currently available outside a clinical trial.

Researchers are studying several ways to treat cancer using gene therapy. Some approaches target healthy cells to enhance their ability to fight cancer. Other approaches target cancer cells, to destroy them or prevent their growth. Some gene therapy techniques under study are described below:

1. In one approach, researchers replace missing or altered genes with healthy genes. Because some missing or altered genes (e.g., *p53*) may cause cancer, substituting "working" copies of these genes may be used to treat cancer.
2. Researchers are also studying ways to improve a patient's immune response to cancer. In this approach, gene therapy is used to stimulate the body's natural ability to attack cancer cells. In one method under investigation, researchers take a small blood sample from a patient and insert genes that will cause each cell to produce a protein called a T-cell receptor (TCR). The genes are transferred into the patient's white blood cells (called T lymphocytes) and are then given back to the patient. In the body, the white blood cells produce TCRs, which attach to the outer surface of the white blood cells. The TCRs then recognize and attach to certain molecules found on the surface of the tumor cells. Finally, the TCRs activate the white blood cells to attack and kill the tumor cells.
3. Scientists are investigating the insertion of genes into cancer cells to make them more sensitive to chemotherapy, radiation therapy, or other treatments. In other studies, researchers remove healthy blood-forming stem cells from the body, insert a gene that makes these cells more resistant to the side effects of high doses of anticancer drugs, and then inject the cells back into the patient.
4. In another approach, researchers introduce "suicide genes" into a patient's cancer cells. A pro-drug (an inactive form of a toxic drug) is then given to the patient. The pro-drug is activated in cancer cells containing these "suicide genes," which leads to the destruction of those cancer cells.
5. Other research is focused on the use of gene therapy to prevent cancer cells from developing new blood vessels (angiogenesis).

Some of the recent developments in gene therapy research include:

1. University of California, Los Angeles, research team gets genes into the brain using liposomes coated in a polymer call polyethylene glycol (PEG). The transfer of genes into the brain is a significant achievement because viral vectors are too big to get across the "blood-brain barrier." This method has potential for treating Parkinson's disease.
2. RNA interference or gene silencing may be a new way to treat Huntington's. Short pieces of double-stranded RNA (short, interfering RNAs or siRNAs) are used by cells to degrade RNA of a particular sequence. If a siRNA is designed to match the RNA copied from a faulty gene, then the abnormal protein product of that gene will not be produced.
3. New gene therapy approach repairs errors in messenger RNA derived from defective genes. Technique has potential to treat the blood disorder thalassaemia, cystic fibrosis, and some cancers.
4. Researchers at Case Western Reserve University and Copernicus Therapeutics are able to create tiny liposomes 25 nanometers across that can carry therapeutic DNA through pores in the nuclear membrane.
5. Sickle cell is successfully treated in mice.

3.4. Limitations of Gene Therapy

1. **Short-lived nature of gene therapy:** Before gene therapy can become a permanent cure for any condition, the therapeutic DNA introduced into target cells must remain

functional and the cells containing the therapeutic DNA must be long-lived and stable. Problems with integrating therapeutic DNA into the genome and the rapidly dividing nature of many cells prevent gene therapy from achieving any long-term benefits. Patients will have to undergo multiple rounds of gene therapy.

2. **Immune response:** Anytime a foreign object is introduced into human tissues, the immune system is designed to attack the invader. The risk of stimulating the immune system in a way that reduces gene therapy effectiveness is always a potential risk. Furthermore, the immune system's enhanced response to invaders makes it difficult for gene therapy to be repeated in patients.
3. **Problems with viral vectors:** Viruses, while the carrier of choice in most gene therapy studies, present a variety of potential problems to the patient --toxicity, immune and inflammatory responses, and gene control and targeting issues. In addition, there is always the fear that the viral vector, once inside the patient, may recover its ability to cause disease.
4. **Multigene disorders:** Conditions or disorders that arise from mutations in a single gene are the best candidates for gene therapy. Unfortunately, some the most commonly occurring disorders, such as heart disease, high blood pressure, Alzheimer's disease, arthritis, and diabetes, are caused by the combined effects of variations in many genes. Multigene or multifactorial disorders such as these would be especially difficult to treat effectively using gene therapy.

3.5. Ethical Considerations for Gene Therapy

While the body has many billions of cells, only a very small proportion of these cells are involved in reproduction, the process by which our genes are handed on to future generations. In males these cells are located in the testes and in females, in the ovaries. These special reproductive cells are called "germ cells". All other cells in the body, irrespective of whether they are brain, lung, skin or bone cells, are known as "somatic cells". In gene therapy, only somatic cells are targeted for treatment. So any changes to the genes of a person by gene therapy will only impact on the cells of their body and cannot be passed on to their children. Changes to the somatic cells cannot be passed on to future generations (inherited). Gene therapy treats the individual and has no impact on future generations. The possible genetic manipulation of the egg or sperm cells (germ cells) remains the subject of intense ethical and philosophical discussion. The strong consensus view at present is that the risks of germ-line manipulation far exceed any potential benefit and should not be attempted.

Gene therapy does have risks and limitations. The viruses and other agents used to deliver the "good" genes can affect more than the cells for which they are intended. If a gene is added to DNA, it could be put in the wrong place, which could potentially cause cancer or other damage. Genes can also be "over-expressed," meaning they can drive the production of so much of a protein that they can be harmful. Another risk is that a virus introduced into one person could be transmitted to other people or into the environment.

Gene therapy trials with some children present an ethical question. We don't test children to see if they are carriers of genetic diseases, because at this point, there's nothing we can do about it. If a child knows he has a problem gene, there's no guarantee he will be affected by it, but he has to live with the knowledge for the rest of his life. For example, the gene for some types of breast cancer has been identified. But if we test a young girl to see if she's a carrier for it, what will she do with that information?

The principle objections are summarized as:

- The technology is imperfect. The effects of gene transfer are unpredictable and, even if the target disease was cured, further defects could be introduced into the embryo.
- Denial of human rights. Individuals resulting from germ line gene therapy would have no say in whether their genetic material should have been modified.
- Potential abuse. Germ line gene therapy could by used not only to eliminate disease, but also to enhance favourable characteristics and suppress unfavourable ones. On a small scale, this would result in a generation of 'designer children', with traits chosen by their parents. On a large scale, gene therapy could result in eugenics - manipulation of the genetic properties of a population.

Some questions to be considered by the society

- What is normal and what is a disability or disorder, and who decides?
- Are disabilities diseases? Do they need to be cured or prevented?
- Does searching for a cure demean the lives of individuals presently affected by disabilities?
- Is somatic gene therapy (which is done in the adult cells of persons known to have the disease) more or less ethical than germline gene therapy (which is done in egg and sperm cells and prevents the trait from being passed on to further generations)? In cases of somatic gene therapy, the procedure may have to be repeated in future generations.
- Preliminary attempts at gene therapy are exorbitantly expensive. Who will have access to these therapies? Who will pay for their use?

3.6. Future of Gene Therapy

To cure genetic diseases, scientists must first determine which gene or set of genes causes each disease. The Human Genome Project and other international efforts have recently completed the initial work of sequencing and mapping virtually all of the 25,000 to 35,000 genes in the human cell. This research will provide new strategies to diagnose, treat, cure, and possibly prevent human diseases. Gene therapy's potential to revolutionize medicine in the future is exciting, and its expectations for curing and preventing childhood diseases are encouraging. One day it may be possible to treat an unborn child for a genetic disease even before symptoms appear.

4. INDUSTRIAL PRODUCTION OF BIOMOLECULES FROM RDNA TECHNOLOGY

Biotechnology essentially involves the industrial application of living organism to produce product or process for the betterment of humanity. Biotechnology based on the principle of recombinant DNA technology (rDNA) started in early 1970 with Paul Berg of Stanford University producing the first recombinant DNA. This was followed by the generation of transformed *Escherichia coli* in 1973 by Herbert Boyer of University of California (San Francisco), which resulted in the production of recombinant human insulin by Eli Lilly in 1982. Since then efforts have been made to genetically engineer most of the living systems such as bacteria, yeast, fungi, plant and animals to have novel gene product (characteristics). Thousands of genes have been cloned and expressed using rDNA technology. In fact, the application of genetic engineering and rDNA technology has led to the generation of new classes of organism called genetically modified organisms (GMO) or live modified organism (LMO). More so, the ability of genetic manipulation of almost all living organism has led to the genomics evolution with far reaching applications of the modern biology system. The most notable applications of the recombinant technology having direct impact on humanity are:

1. Large scale production of therapeutic protein such as insulin, hormones, vaccine and interleukins using recombinant microorganisms.
2. Production of humanized monoclonal antibodies for therapeutic application.

3. Production of insect resistant cotton plant by incorporation of insecticidal toxin of *Bacillus thuringiensis* (plants with Bt gene).
4. Production of golden rice (rice having vitamin A) by incorporating three genes required for its synthesis in rice plant.
5. Bioremediation by the use of recombinant organisms and,
6. Use of genetic engineering techniques in forensic medicine.

One of the greatest benefits of the rDNA technology has been the production of human therapeutics. Therapeutic proteins are preferred over conventional drugs because of their higher specificity and absence of side effects. Therapeutic proteins are less toxic than chemical drugs and are neither carcinogenic (cancer producing) nor teratogenic (tumour producing). Further, once the biologically active form of a protein is identified for medical application, its further development into a medicinal product involves fewer risks than chemical drugs. Notable diseases for which recombinant therapeutics have been produced include diabetes, hemophilia, hepatitis, myocardial infarction and various cancers.

4.1 Industrial Production of Insulin

Insulin is the most important molecule both from human health care and rDNA technology point of view. It has a key role in regulation of glucose metabolism and thus is important for diabetes mellitus. Particularly for type 1 diabetes, which involves the loss of insulin producing â cells in the pancreas, injection of insulin is the primary therapy (type 2, insufficient insulin production). The hormone, produced and secreted by the beta cells of the pancreas' islets of Langerhans, regulates the use and storage of food, particularly carbohydrates. Ever since its discovery and introduction as drug almost 85 years ago (Banting FG and Best CH *et al.* isolated the insulin around 1921), the lives of millions of people with diabetes have been saved, prolonged and improved. Initially, pig insulin and human insulin extracted from whole body was used as the source of insulin, but the scenario changed after the successful expression of human insulin in 1980 in *E .coli*. Approximately, 194 million people worldwide have diabetes and it is estimated that the number of people with diabetes will be more than double by 2030. The current healthcare costs could be as much as US $ 286 billion worldwide, with the majority of these costs linked to treating diabetes-related complications. The current market for injectable insulin is worth more than US$ 7.2 billion globally (about 15% growth per year). Four products represent 50% of the world's injectable insulin market; long-acting insulin glargine (Lantus; Sanofi-Aventis); ultra-short-acting insulin Lispro (Humalog; Lilly); biphasic insulin base / insulin isophane (Actraphane HM; Novo Nordisk). Lantus and Novorapid showed phenomenal annual growth rates in 2005; 47 % and 52 % respectively. In addition, two other fast growers are predicted to increase their market share; biphasic insulin aspart /insulin aspart potamine (Novomix; Novo Nordisk, 72% growth) and long – acting detemir (Levemir; Novo nordisk, 510% growth).

Chemically, insulin is a small, simple protein. It consists of 51 amino acid, 30 of which constitute one polypeptide chain, and 21 of which comprise a second chain (Fig 19.5). The two chains are linked by a disulfide bond. The genetic code for insulin is found in the DNA at the top of the short arm of the eleventh chromosome. It contains 153 nitrogen bases (63 in the A chain and 90 in the B chain). A weakened strain of the common bacterium, *E. coli* an inhabitant of the human digestive tract, is the 'factory' used in the genetic engineering of insulin.

Although bovine (cattle) and porcine (pig) insulin are similar to human insulin, their composition is slightly different. Concerns about purity led to the production of highly purified, mono component insulin in the 1970s. In the 1980s, recombinant-DNA technology and the development of protein-engineering techniques led to the production of human insulin and modified insulin analogues with improved pharmacokinetic properties. Insulin is expressed in recombinant *E. coli* and then subsequently purified and refolded in to bioactive form. In *E. coli*, b-galactosidase is the enzyme that controls the transcription of the genes. To make the bacteria produce insulin, the insulin gene needs to be tied to this enzyme (insulin gene should be attached closely with this gene).

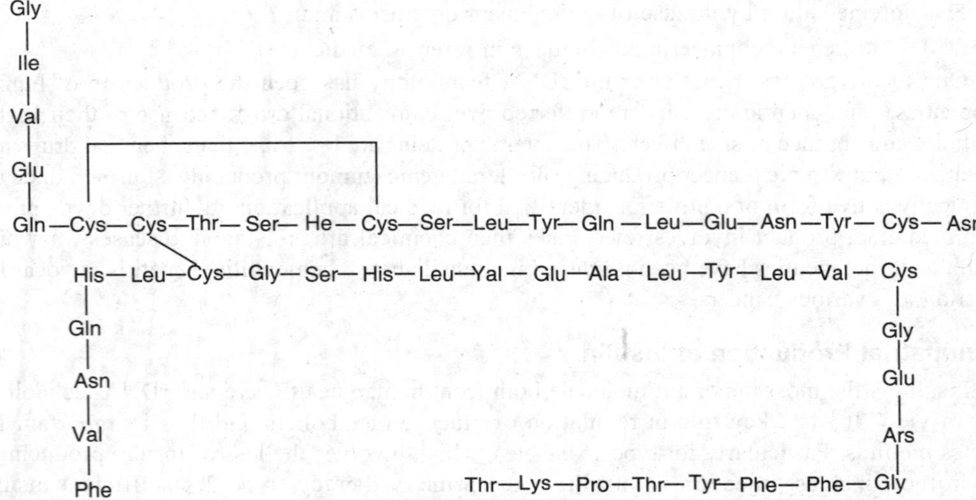

Fig. 19.5 Human insulin gene

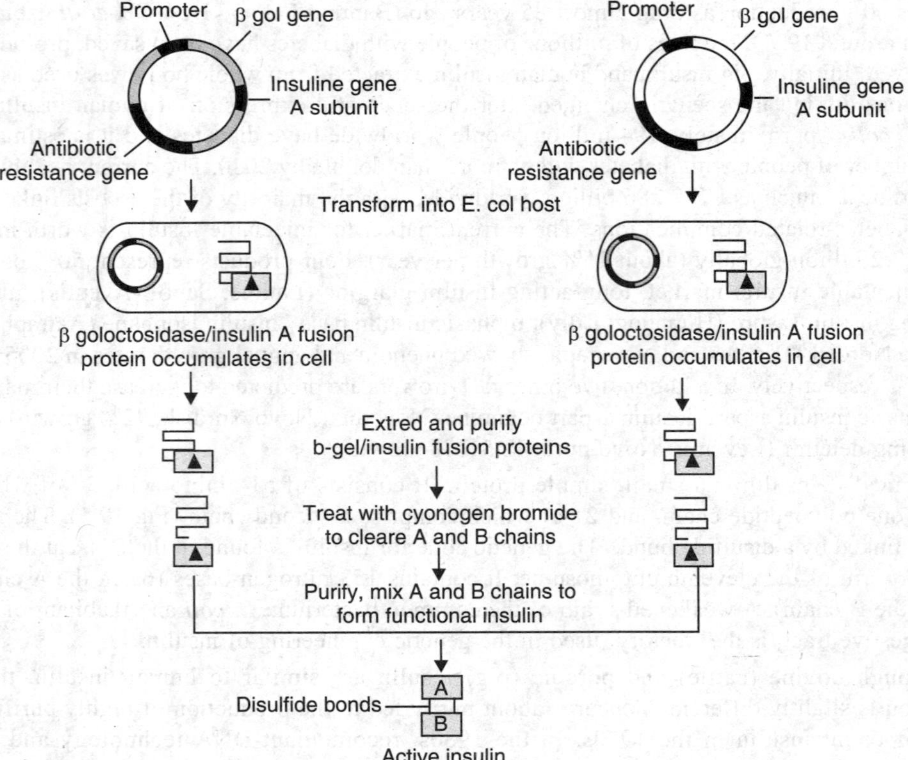

Fig. 19.6 Insulin production process through rDNA technology using E.coli.

The recombinant plasmids are produced and then introduced into *E. coli* cells. Practical use of r-DNA technology in the synthesis of human insulin requires millions of copies of the bacteria whose plasmid has been combined with the insulin gene in order to yield insulin. The insulin gene is expressed as it replicates with the B-galactosidase in the cell undergoing mitosis. (Fig 19.6,7,8). The protein which is formed, consists partly of B-galactosidase, joined to either the A or B chain of insulin. The A and B chains are then extracted from the B-galactosidase fragment and purified. The two chains are mixed and reconnected in a reaction that forms the disulfide cross bridges, resulting in pure*Humulin* - synthetic human insulin. Human insulin is the only animal protein to have been made in bacteria in such a way that its structure is absolutely identical to that of the natural molecule. This reduces the possibility of complications resulting from antibody production. In chemical and pharmacological studies, commercially available Recombinant DNA human insulin has proven indistinguishable from pancreatic human insulin. Initially the major difficulty encountered was the contamination of the final product by the host cells, increasing the risk of contamination in the fermentation broth. This danger was eradicated by the introduction of purification processes. When the final insulin product is subjected to a battery of tests, including the finest radio-immuno assay techniques, no impurities can be detected. The entire procedure is now performed using yeast cells as a growth medium, as they secrete an almost complete human insulin molecule with perfect three dimensional structure. This minimises the need for complex and costly purification procedures.

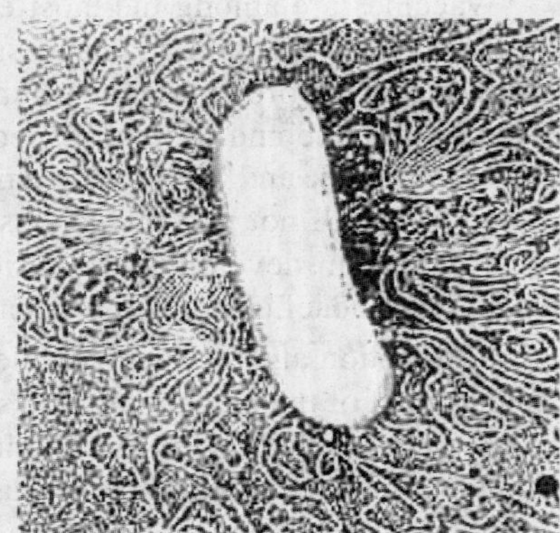

Fig. 19.7 E. coli bacterium cell

Cloning of insulin, expression in different host, fermentation process optimization, refolding and purification of active protein from inclusion bodies and finally formulation of insulin having better pharmacokinetics has provided model system for production of most of the therapeutics using r-DNA technology. New methods of producing insulin were accompanied by advances in formulation, which led to the development of rapid-acting insulins for use at meal times and long-acting insulins for basal insulin requirements. In addition, a series of pre-mixed insulins that combine the various forms are in use. At present, about 180 individual insulin preparations are available worldwide.

Fig. 19.8 Bioreactors used for culture of recombinant bacteria for the production of human insulin.

Insulin in many ways remains as the lead biopharmaceutical molecule. It was first commercialized in1982 using r-DNA technology. It was the first r-DNA molecule which was reengineered to have improved characteristics (fast acting and long acting insulin anlalogs are in industry). It has become the first recombinant DNA derived molecules which has been approved to be delivered by pulmonary route (Exubera, : nasal spray recombinant insulin by Fizer inc.).The world health organization (WHO) estimated around 170 million diabetic globally which will be almost doubled in next 25 years. Considering that around 10 % of these patients are type I diabetic and need insulin treatment, it is estimated that more than 5 tons of insulin will be needed globally with a sales target of $8- 10 billion. This necessitates the easy production and purification of these molecules.

Trials are going on in several countries to produce genetically modified safflower and lettuce plants for a cheaper way to produce insulin. However, purification of insulin protein from other seed/plant proteins is a herculean task and current technology makes them costly.

4.2 BCG VACCINE

Vaccines are among the most efficacious and cost-effective human health tools available. They provide protection against a surprisingly broad spectrum of infectious diseases. Notable recent successes protect against human papillomavirus (Cervarix and Gardasil vaccines from GlaxoSmithKline and Merck, respectively) and rotavirus (Rotarix and RotaTeq vaccines from GlaxoSmithKline and Merck, respectively). However, generating reliable sterilizing or therapeutic immunity is still not possible against a number of latent and chronic pathogens that especially affect people in developing countries. Among those agents are Mycobacterium tuberculosis, human immunodeficiency virus, hepatitis C virus, and the Plasmodium protists that cause malaria.

Protection against such complex pathogens may require activation of cellular immunity, a compartment of the human immune system that most conventional vaccines usually fail to target. An exception to that rule is live attenuated vaccines, but the approach is unavailable or unsuitable for many global communicable diseases. In addition, for certain attenuated viruses, there is a risk of reversion to pathogenic strains and potential residual virulence for some vaccine recipients or contact persons of them. Such risks are unacceptable in the modern era of vaccines, for which increases in both international mobility and numbers of immunocompromised individuals demand a greater degree of safety in the vectors.

Modern vectored vaccines use carriers unrelated to their target pathogens to elicit a protective immune response against recombinant antigens. These combine the advantages of live vaccines with the strong safety profile of highly attenuated vectors and thus may provide novel therapeutic or protective approaches. Promising viral carriers include host-restricted pox viruses such as modified vaccinia Ankara (MVA) that trigger a strong immune response without an ability to replicate in humans. That is an important biological characteristic for safe application in immune compromised recipients and heterogeneous populations. But this type of attenuated vector cannot amplify at the site of inoculation, so relatively high numbers of infectious units per dose must be given to maximize efficacy (although much lower numbers than are required with adenovirus vectors). To maintain large stockpiles for emergency applications, highly efficient production processes are required. Such processes must also be scalable, robust, and preferably autonomous. That is, production should be independent of complex logistics for supply of the cell substrate (as with egg-based technologies) and culture media.

Although some viral vectors have been used in clinical trials, currently no viral-vectored vaccine is licensed for use in humans, although many are licensed for veterinary use.

Bacillus Calmette–Guérin (historically Vaccin Bilié de Calmette et Guérin commonly referred to as Bacille de Calmette et Guérin or BCG) is a vaccine against tuberculosis that is prepared from a strain of the attenuated (weakened) live bovine tuberculosis bacillus, Mycobacterium bovis, that has lost its virulence in humans by being specially subcultured in a culture medium, usually Middlebrook 7H9. Because the living bacilli evolve to make the best use of the available nutrients, they become less well adapted to their traditional environment, human blood, and can no longer induce the disease when introduced into a human host. Still, they are similar enough to their wild ancestors to provide some degree of immunity against human tuberculosis. The BCG vaccine can be anywhere from 0 to 80% effective in preventing tuberculosis for a duration of 15 years; however, its protective effect appears to vary according to geography and the lab in which the vaccine strain was grown.

There are problems. Over the years the original live vaccine has evolved into an array of different strains with unknown properties and immunogenicity. Very little is known about the correlation between dose of BCG and protection against tuberculosis in humans. A lot more is known about the dose of BCG required to give a certain size of response to tuberculin testing but this is not the same thing.

BCG certainly induces tuberculin sensitivity. Sensitivity to a tuberculin skin test (Mantoux or Heaf test) is the commonest and cheapest way of diagnosing and contract tracing cases of tuberculosis. This is how tuberculosis has been managed in the United States.

In India, the BCG Vaccine Laboratory was started in Guindy, Madras in 1948 for the production of BCG vaccine for use in India and also for supply to some of the neighbouring countries. Since 1966, Danish strain 1331 is being used here for the preparation of both the liquid and the freeze-dried BCG vaccines, based on the seed-lot-system. For preparing the liquid and freeze-dried vaccines, the BCG Laboratory, Madras, uses the method followed at the State Serum Institute, Copenhagen, but using Sauton potato medium for maintaining the BCG strain. The prepared vaccine is tested for purity by Ziehl Neelsen smear for acid fast bacilli, and by culture on nutrient broth, thioglycolate medium and Sabouraud's Agar.

A number of different companies make BCG, sometimes using different genetic strains of the bacterium. This may result in different product characteristics. Pacis® BCG, made from the Montréal (Institute Armand-Frappier) strain was first marketed by Urocor in about 2002. Urocor was since acquired by Dianon Systems. Evans Vaccines (a subsidiary of PowderJect Pharmaceuticals Plc, London: PJP). Statens Serum Institute in Denmark markets BCG vaccine prepared using Danish strain 1331. Japan BCG Laboratory markets its vaccine, based on the Tokyo 172 substrain of Pasteur BCG, in 50 countries worldwide. Sanofi Pasteur's BCG vaccine products, made with the Glaxo 1077 strain were recalled in July 2012 due to noncompliance in the manufacturing process.

4.3 INTERFERON

Interferon beta-1b is a protein in the interferon family used to treat the relapsing-remitting form of multiple sclerosis (MS). Interferon beta-1b is a natural human protein (molecular weight ~18,500 Da) that is produced in the body in response to viral infection and has antiviral activity. It has been shown to slow the advance of MS and reduce the frequency of attacks. It is believed that interferon-beta achieves this effect on MS progress via its anti-inflammatory properties. interferon beta-1b drugs have been approved for over 20 years to treat the symptoms of MS. interferon beta-1b is produced as a recombinant protein in the bacterium E. coli by fermentation and subsequent purification to the active drug. The E. coli bacteria produce the interferon beta-1b molecule at low yield and as an insoluble and inactive product. As part of the purification process, the molecule must be restored to its active state, a process known as refolding. Refolding a protein molecule is a difficult, inefficient, and costly process. If interferon beta-1b could be produced by a bacterium at high yield in a fully active form without the necessity of refolding, it would be much less expensive to produce and could be offered to patients at a more affordable price.

The *Pseudomonas fluorescens*-based Pfçnex Expression Technology platform is a proven high-throughput, parallel screening, protein expression strain development platform that has been designed specifically to deliver bacterial production strains expressing high yields of soluble and active protein. It is based on an extensive toolbox of expression strategies that can be flawlessly combined to deliver robust expression strains for the production of recombinant proteins. Scientists at Pfçnex Inc.,˜ $%,a San Diego, California, $%, based protein discovery, development, and production company ,˜$%, have cloned and expressed many difficult-to-express proteins in their soluble and active forms and successfully completed scale-up for manufacturing. Pfçnex scientists, using the platform, cloned the interferon beta-1b ,˜$%,˜ coding gene into 20 unique expression plasmids and transformed them into 30 phenotypically distinct *P. fluorescens* host strains. The resulting 600 expression strains were grown in 96-well plates, and the expression of interferon beta-1b was determined for each strain, all in about one month.

Procedure for obtaining interferon:

(1) **Obtaining an interferon beta gene.** The first stage in the manufacture of Interferon beta-1b is to extract the gene coding for interferon beta from human fibroblast cells.

(2) **Modifying the gene.** The gene is then modified biochemically to produce interferon beta-1b, which is a more stable molecule.

(3) **Making a recombinant DNA molecule.** The modified gene, now coding for interferon beta-1b, is inserted into a plasmid (a circular DNA vector) to produce a recombinant DNA molecule.

(4) **Adding the recombinant DNA molecule to a bacterium.** Recombinant DNA molecules, consisting of the modified gene in a plasmid, are added to an *E. coli* bacterial culture. The bacteria retain their own DNA, and the recombinant molecules exist alongside as additional pieces of genetic material.

(5) **Producing interferon beta-1b by bacterial fermentation.** Large numbers of interferon beta-1b-producing *E. coli* can be produced by bacterial fermentation. As the bacteria multiply, the recombinant DNA and the DNA of the bacterial host are replicated in cycle. The bacterial transcription and translation mechanisms read the genetic information contained in the plasmid as well as their own, and produce interferon beta-1b. Interferon beta-1b is then secreted into the culture medium.

(6) **Extracting interferon beta-1b from the ferment.**
The interferon beta-1b is harvested from the ferment, and then purified.

(7) **Adding inactive ingredients.**
Inactive ingredients, which act as stabilisers, are added to the purified interferon beta-1b. The stabilisers used are mannitol and human albumin. Interferon beta-1b is supplied as a highly purified, sterile, lyophilised (freeze dried) powder.

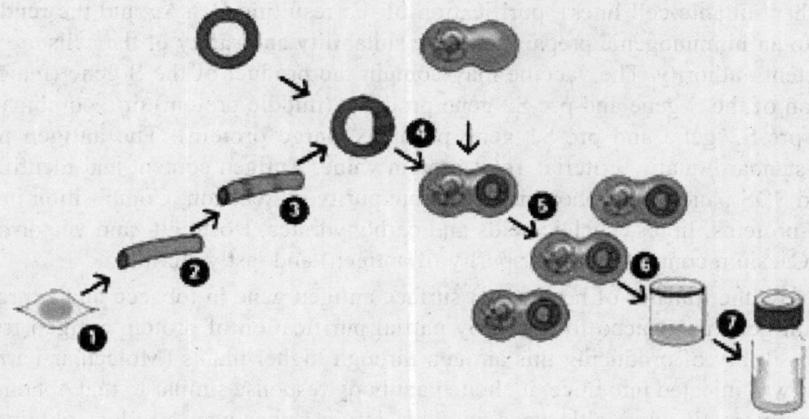

Fig. 19.9 Manufacture of betaferone

Description. Hepatitis - B vaccine (r-DNA) is a non-infectious recombinant DNA Hepatitis B vaccine. It contains purified surface antigen of the virus obtained by culturing genetically-engineered Hansenula polymorpha yeast cells having the surface antigen gene of the Hepatitis B virus. The Hepatitis-B surface antigen (HBsAg) expressed in the cells of Hansenula polymorpha is purified through several chemical steps and formulated as a suspension of the antigen adsorbed on aluminium hydroxide and thiomersal is added as preservative. The vaccine does not contain any material of human or animal origin. The vaccine meets the requirements of WHO when tested by the methods outlined in WHO, TRS 786(1989).

4.4. HEPATITIS B VACCINE

Hepatitis B (HepB) is a major public health problem worldwide. Approximately 30% of the world's population, or about 2 billion persons, have serologic evidence of hepatitis B virus (HBV) infection. Of these, an estimated 350 million have chronic HBV infection and at least one million chronically infected persons die each year from liver cancer and cirrhosis. HBV is second only to tobacco as a known human carcinogen. HepB vaccine is effective in preventing HBV infections when it is given either before exposure or shortly after exposure. At least 85-90% of HBV-associated deaths are vaccine-preventable.

Hepatitis B is a killer, taking the lives of 900,000 people each year. This disease is especially dangerous for infants, since those who are infected when young, may carry the infection for the rest of their lives, often without knowing it. Fortunately, hepatitis B vaccine, if provided to infants, helps protect them against these problems. In effect, it is the world's first anticancer vaccine. Due to the seriousness of hepatitis B disease, and because of the high effectiveness and safety of the vaccine, the World Health Organization (WHO) recommends that it be given to all children worldwide. The hepatitis B vaccine has been available for decades, but introduction into the developing world only began in the late 1980s. Currently, more than 100 countries routinely provide the vaccine, but many still cannot afford to do so.

Definition- Hepatitis B vaccine (rDNA) is a preparation of hepatitis B surface antigen (HBsAg), a component protein of hepatitis B virus; the antigen may be adsorbed on a mineral carrier such as aluminium hydroxide or hydrated aluminium phosphate. The antigen is obtained by recombinant DNA technology.

Hepatitis B vaccine (rDNA) is produced by the expression of the viral gene coding for HBsAg in yeast (*Saccharomyces cerevisiae*) or mammalian cells (Chinese hamster ovary (CHO) cells or other suitable cell lines), purification of the resulting HBsAg and the rendering of this antigen into an immunogenic preparation. The suitability and safety of the cells are approved by the competent authority. The vaccine may contain the product of the S gene (major protein), a combination of the S gene and pre-S2 gene products (middle protein) or a combination of the S gene, the pre-S2 gene and pre-S1 gene products (large protein). The antigen must comply following standard quality criteria: Total protein value, Antigen content and identification using ELISA and SDS page electrophoresis, Antigenic purity by reaction, Composition in terms of the content of proteins, lipids, nucleic acids and carbohydrates, Host-cell- and vector-derived DNA presence, Caesium content (used for purity of antigen) and lastly sterility.

Recently, the transfer of hepatitis-B surface antigen gene in tobacco and expression of this recombinant gene in tobacco followed by partial purification of protein antigen from the plant showed possibility of producing this antigen through higher plants (Molecular Farming). When this protein was injected into mice, it elicited antibody response similar to that obtained with yeast derived commercially available vaccine. This is clear that gene product obtained from two different organisms has same property and transgenic plants can be used as source of antibodies. Attempts are being made to produce many more molecules by cultivation of transgenic higher plants.

QUESTIONS

1. What is gene therapy? Discuss its benefits and disadvantages for the treatment of genetic disease.
2. Discuss the role of rDNA technology in medicine.
3. Write shot notes:
   (a) Insulin production by rDNA
   (b) DNA finger printing
   (c) Hepatitis B vaccine production by rDNA
   (d) Gene therapy
   (e) Diseases of genetic origin.

CHAPTER 20

Intellectual Property Rights and Biotechnology

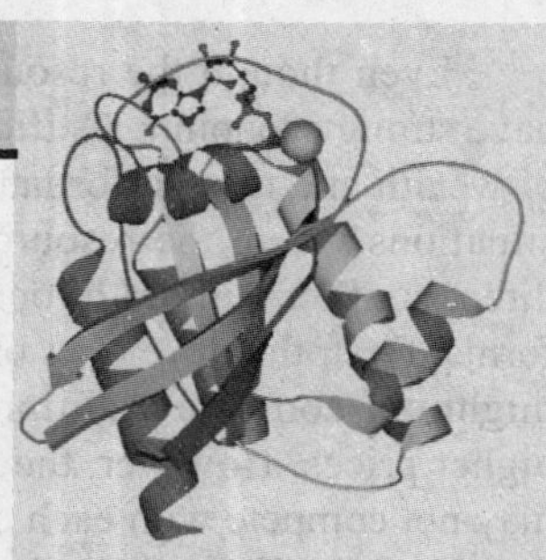

1. INTRODUCTION

Intellectual property refers to creations of the mind: inventions, literary and artistic works, and symbols, names, images, and designs used in commerce. Intellectual property is divided into two categories: *Industrial property*, which includes inventions (patents), trademarks, industrial designs, and geographic indications of source; and *Copyright*, which includes literary and artistic works such as novels, poems and plays, films, musical works, artistic works such as drawings, paintings, photographs and sculptures, and architectural designs. Rights related to copyright include those of performing artists in their performances, producers of phonograms in their recordings, and those of broadcasters in their radio and television programs.

In recent years, instruments enforcing intellectual property rights (IPRs), such as patents and trade secrets, have received attention as mechanisms by which biodiversity resources may be maintained while promoting sustainable development and a more equitable distribution of the resulting benefits among nations. Most of the world's biodiversity-rich countries are underdeveloped and lack the necessary technologies to transform biological resources into products yielding significant measurable benefits. With little or insignificant *in situ* market value, biodiversity-rich wild lands may be expected to succumb to pressure from development activities (e.g., conversion to cropland, inundation of forest lands due to hydroelectric and flood-control projects, etc.).

One way to prevent the destruction of wild lands (and, in turn, biodiversity loss) is to promote biodiversity prospecting which creates new markets for biological resources and generates incentives for their conservation. However, biodiversity prospectors generally are multinational corporations from developed countries. These corporations are reluctant to invest in biotechnologies discovered in developing countries due to poorly defined and enforced intellectual property laws. Several scientists currently are addressing this deficiency in IPR protection. Nations which have become signatories of two major international agreements in recent years: the 1992 Convention on Biological Diversity (CBD) (UNEP, 1992) and the 1993 Trade-Related Intellectual Property Rights (TRIPS) (UN, 1993). These agreements call for establishing a set of suitable intellectual property laws in each nation, depending on the type of intellectual material in question and the economic and technological background of the nation itself.

CBD establishes a formal framework for the reciprocal transfer of biological resources and knowledge (technology) between nations. The convention promotes the idea of biodiversity as a global common heritage, which, therefore, requires biodiversity-rich countries to allow access to biological resources to other countries on 'mutually agreed terms' (UN, 1993). CBD requires technology-rich nations-generally, developed nations to encourage transfer of technology to biodiversity-rich, underdeveloped countries. Thus, the Convention promotes the exchange of biological resources for technology to facilitate biodiversity prospecting, which benefits all nations in the world.

Even though the resource-technology reciprocity and IPR provisions of these agreements have strong economic justification, their use at the global level raises issues concerning the transfer of wealth and respect for national and cultural sovereignty. These issues may become pressing in situations where the resource consumption associated with a patented biotechnology comes in direct conflict with traditional uses of a region's biological resources. Conflict may take different forms. First, the developer of a new biotechnology—who normally has greater financial strength-might out-compete traditional users for raw biological material in the input market by paying higher prices. However, the products produced by the new commercial user and traditional users may not compete with each other in the output market. Second, the commercial product developed may have intellectual similarity with traditional products. Then, conferring IPR to a new product/ technology may limit or prevent traditional consumers from continuing their use of biological resources. Third, the new commercial product may become a substitute for a traditional product and available at a cheaper price. The lower output market price may drive the traditional producers out of business. Such conflicts can alter the underlying market incentives of competing resource users, with likely adverse implications for biodiversity.

2. INTELLECTUAL PROPERTY PROTECTION OF PLANT BIOTECHNOLOGY INVENTIONS

It is becoming more and more difficult to obtain broad claims in patents and the strength of broad claims in issued patents is weakening. A recent example of this is a case in the USA, in which claims to a diagnostic assay - to differentiate one genus of bacteria from other were invalidated.

2.1. Trips

The standard for intellectual property rights is outlined in the global intellectual property treaty agreement of Trade-Related Aspects of Intellectual Property Rights (TRIPS) (http:// www.wto.org/English/docs - e/legal- e/27-trips. pelf). Member countries that have signed this agreement must ensure that the requirements stated in TRIPS are met in their own legislation. TRIPS states that 'patents shall be available for any inventions, whether products or processes, in all fields of technology, provided that they are new, involve an inventive step and are capable of industrial application'. Although seemingly simple and consistent with the requirements for obtaining a patent in most developed countries, this statement is at the heart of most of the controversy relating to biotechnology patenting. An additional paragraph in TRIPS permits several grounds for exclusion in granting patent protection, including exclusion on moral grounds; or diagnostic, therapeutic and surgical methods for the treatment of humans or animals, life forms other than microorganisms and processes for the production of plants or animals. However, these exclusions are optional and vary from country to country. For example, the European Patent Convention (EPC) provides limited moral grounds for exclusion (http://www.european-patent-office.org/ legal/epc/ e/ar53.html), yet no such grounds are defined in the Patent Acts in Canada, Australia, the USA or Japan. TRIPS also states that if protection of plants is not available by patent, then member countries need to provide protection in some other way. A standard method for such alternative protection is plant variety protection, as set out under Union International pour la Protection des Obtentions Vegetables, known more simply as UPOV.

2.2. Plant Variety Protection

UPOV is a global agreement setting out a minimum standard for the protection of plant varieties, similar to that of TRIPS. Member states that have signed the UPOV Convention must ensure that these standards are met within their own legislation (Table 20.1). The two versions of UPOV that are currently in force are set out in the UPOV Conventions of 1978 and 1991 and are

Table 20.1. Summary of four international conventions with implications for IPR linked to living species and materials of relevance to propagation of living species.

Convention of Bio-Diversity (CBD)	Proposed in 1992, it came into force by the end of 1993 ratified by more than 169 countries. The conservation of biological diversity, the sustainable use of its components and fair and equitable sharing of benefits arising out of the utilization of genetic resources are the key features. Bio-diversity and its economic benefits were redefined from the "common heritage of mankind" to "national goods" that nations could protect and trade as commodities. It gives a great deal of emphasis to the rights of developing countries to exploit, and reap benefits from, discoveries made in their own territory.
Union of Protection of Plant Varieties (UPOV) and Plant Breeders Rights	UPOV Convention signed in 1961, came into force in 1968 with revisions in 1972, 1978 and 1991. The 13 essential features of UPOV address the conditions for entitlements to protection to a breeder of a new plant variety provided the variety is distinct from existing, commonly known varieties, is sufficiently homogeneous, stable and new. The plant breeders rights include the requirement of prior authorization of the breeder for the production for the purposes of commercial marketing of the propagating material of the new plant variety, offering for sale of the propagating material, marketing of such material and repeated use of the new plant variety for commercial production of another new variety. For reasons of public interest, the Member States are free to impose restrictions subject to ensuring that breeders receive "equitable remuneration".
Farmer's Rights (UN Food and Agriculture Organization - FAO)	Starting as a concept for debate in 1979 in F AO, it found its way into three FAO conference resolutions 4/89, 5/89 and 3/91. These were negotiated by the Commission on Plant Genetic Resources (PGR) and unanimously adopted by over 160 countries in 1989 and 1991. This defines Farmer's Rights as "rights arising from the past, present, future contribution of farmers in conserving, improving and making available plant genetic resources particularly those in centres of origin/diversity, These rights are vested in the International Community, as trustees for present and future generations of farmers, for the purpose of ensuring full benefits to farmers for supporting the continuation of their contributions". Intentionally acceptable mechanisms for implementation of Farmer's Rights are yet to be arrived at.
Trade Related Intellectual Property Rights (TRIPS)	Article 27.1, 27.2 and 27.3 are most relevant to a discussion on patents in the field of biology and biotechnology. Section 27.1 requires members to grant patents for inventions in all fields of technology provided they are new, involve an inventive step and are capable of industrial application. Article 27.2 states "members may exclude from patentability inventions, prevention within their territory of commercial exploitation of which is necessary to protect ordre public or morality, including to protect human, animal or plant life or health or to avoid serious prejudice to the environment, provided that such exclusion is made not merely because the exploitation is prohibited by domestic law", Article 27.3 allows members to exclude from patentability inventions related to diagnostic, therapeutic and surgical methods for treatment of humans or animals. It also allows exclusion of plants and animals other than microorganisms, and essentially biological processes for production of plants or animals other than non-biological and microbiological processes. However members are required to provide for the protection of plant varieties, either by patents or an effective slligeneris system or by a combination thereof.

similar in that they provide protection to a plant variety that is distinct from existing known varieties and that is uniform, stable and novel. However, there are several significant changes in the 1991 Act. For example, the definition of propagating material has been tightened and provisions relating to farmers' rights (or privilege) - the rights that permit a farmer to replant seed for personal use - have been defined. Although plant variety protection meeting the standard set out under UPOV is accepted in 50 countries, such protection has not been uniformly accepted and many countries with strong histories of farmers' privilege have yet to accede to the convention (http://allafi:-ica.com/stories/ 200205090709.html). A rigorous debate also continues over the effect of plant variety protection and associate Material Transfer Agreements, on sharing and developing new germplasm.

UPOV 1978 provides an exclusive right to produce and offer for sale propagating materials of a plant variety but not the harvested end product, for example a fruit. Furthermore, the right pertains only to a commercial end product and not to non-commercial uses. As a result, replanting seed is implicitly allowed under UPOV 1978, leading to farmers' privilege. Also provided in the 1978 convention was a breeders' exemption permitting the use of protected varieties as a germplasm source to develop new plant varieties. The 1991 Act introduced several significant amendments. The number of plant genera and species that could be protected under UPOV was increased from the select list of plants in UPOV 1978 to all plants. Furthermore, the 1991 Act provides the option to protect all aspects of the production and reproduction of a plant variety, thereby removing farmers' privilege. However, the application of this provision is discretionary for each member state of UPOV and a country can provide an exemption in its laws to permit farmers' privilege, if desired. Another important change pertains to providing protection to plant varieties that are 'essentially derived' from a protected variety. An 'essentially derived' plant is one that comprises the properties of the protected variety along with only a minor change. The introduction of a gene using recombinant techniques into a protected plant variety might not be sufficient to exceed the 'essentially derived' criteria unless the gene alters the variety in a significant manner (http://www.upov.int/eng/convntns/ 19911pd£'act1991.pdt).

On 3 November 2001, the International Treaty on Plant Genetic Resources (see http://www.fao.org/ag/ cgrfa/news.htm) was adopted by 116 countries; there were two abstentions (the USA and Japan). Before the Treaty comes into effect 40 countries must ratify it. This Treaty pertains to ensuring that the raw materials used to develop new crop varieties remain publicly available. In so doing, the Treaty promotes the conservation of plant genetic resources for food and agriculture. The aim of the Treaty is to ensure farmers' privilege and to develop a multilateral system comprising an aggregate of genetic material from the member countries, so that, after paying a fee, members can have access to the genetic material. The preface to the Treaty indicates that 'nothing in this Treaty shall be interpreted as implying in any way a change in the rights and obligations of the Contracting Parties under other international agreements'. However, there is an active debate as to whether the Treaty will remain subordinate to TRIPS and UPOV. The USA abstained from signing this treaty partly because of the lack of clarity in the intellectual property provisions.

2.3. GATT (General Agreement of Tariffs and Trade)

During Urugway conference, WTO (World Trade Organization) was created. General Agreement of Tariffs and Trade (GATT) was framed by WTO in 1948 and was meant to be a temporary arranged to settle amicably, among countries, disruptes regarding who gets what share of world trade. This is achieved by determining both tariff rates and quantitative restrictions on imports and exports globally. In 1994, about 100 countries signed this agreement including the then president of USA, Mr. Bill Clinton. This was to be effective from 1-1-95 in phases. Its new quarter is in Geneva, Switzerland. Although GATT has made the world a better place to do business by allowing more free and fruitful flow of goods and services, this benefit has unfortunately gone mainly to developed countries to the disadvantage of the countries in the third world.

2.4. US Plant Patents

Another way to protect plant-related subject matter includes a 'plant patent', a unique form of protection offered in the USA. A US plant patent is available for a plant that reproduces through asexual reproduction but it does not include a tuber-propagated plant (Section 161 http:// www.uspto.gov /web / offices /pac/mpep/consolidated_laws.pdf). Although not a common form of plant protection, it is used to protect ornamental and fruit-producing trees, roses, poinsettias, strawberries and other plants that reproduce asexually. A plant patent is different from a regular utility patent.

In the autumn of 2000, the USPTO (United States Patent and Trademark Office) began rejecting plant patents with a UPOV-based certificate that had been issued before filing for the corresponding plant patent application if the UPOV-based application had been >1 year before the plant patent application had been filed. This interpretation of a UPOV-based disclosure had not been made previously because it was not considered 'enabling', that is, the disclosure of a plant variety within a Plant Breeder Right's certificate did not provide enough information to enable someone 'one of skill in the art' - to produce the plant variety. The position taken by the USPTO is in direct opposition to that decided in re LeGrice but the USPTO argued that rejection on these grounds is consistent with *Ex parte Thomson*. However, it should be noted that the only public disclosure made in *re LeGrice* was a notice in a publication, which is arguably a non-enabling disclosure, whereas in *ex parte Thomson*, seeds were made publicly available for >1 year before application for a plant patent, clearly placing one of skill in the art in possession of the invention.

The applicant of a US plant patent is provided with a 1 year period of grace. Strong pressure from the industry resulted in a review of USPTO's position and the issue of a preliminary statement recanting its position (http://www.uspto.gov/inappright.html) and, in May 2002, an amendment to the US Patent Act was proposed (it is still under discussion), providing a 10-year period of grace. Even so, the question as to whether a public, non-enabling disclosure of a plant is sufficient to permit one of skill in the art to be in possession of the invention, as is the case in *re LeGrice*, was not addressed.

2.5. Utility Patents

Plants can also be protected using a regular (utility) patent in countries that permit patenting of plant or higher life forms (HLFs). This is a more common method for protecting whole novel plants, plant genes, methods for creating novel plants and novel applications for an existing plant. However, the costs are greater and the process more involved than plant variety protection. Many major jurisdictions permit the patenting of non-human HLFs, including Europe, the USA, Japan and Australia.

In the USA, patents have been granted to HLFs since the 1980 landmark decision in Diamond v/s Chakrabarty (http:// people.bu.edu/ ebortman/ index/chakrabarty.html). The recent Supreme Court decision in the case of Pioneer Hi-Bred International, Inc., v/s J.E.M. AG Supply. Inc. et.al. further established that such protection is valid for plants, even if protection of a plant is available through either plant variety protection or plant patent protection. This case also confirmed that plants are a composition of matter, as ruled earlier by the US Patent Board of Appeal.

The scope of protection offered by a utility patent is broader than that available under plant variety protection. As noted above, a farmer saving and replanting seed, and a breeder producing a new variety, can do so without infringing a plant variety certificate. However, if a utility patent, the patent owner, protects the plant or licensee has the right to exclude the making, using or selling of the plant or seed, making a user buy seed every year.

2.6. Gene Patenting

Although patents have been granted on nucleotide sequences for >30 years, there has been much recent controversy surrounding the patenting of genes. Genome sequencing initiatives, coupled with improved techniques for identifying and sequencing genes, has resulted in an exponential increase in the number of gene patents in the last decade. As a result, the obscure world of gene patenting is now being scrutinized closely in many different sectors, not least because the effect of these patents is felt in everyday life, especially healthcare. For example, in Europe, a European Parliament resolution regarding the patenting of BRCA 1 and BRCA 2 (breast cancer associated) genes was passed calling on the EPO (European Patent Office) to ensure that all patent applications in Europe do not violate the principle of non-patentability of humans, their genes or cells in their natural environment. The resolution identified two European patents related to BRCA 1 and BRCA 2 and asked that an official objection be filed against these patents. The importance of intellectual property in India is well established at all levels- statutory, administrative and judicial. India ratified the agreement establishing the World Trade Organisation (WTO). This Agreement, inter-alia, contains an Agreement on Trade Related Aspects of Intellectual Property Rights (TRIPS) which came into force from 1st January 1995.

2.7. Patenting of Life Forms and GMO

Life forms such as microorganisms, plants and animals, are not patentable in India under the provisions Indian patent Act (1970). However, patent can be obtained for various biotechnological processes and product applications within the scope of International conventions. In America, Europe and other developed countries, microorganisms isolated from nature or are obtained by simple manipulations are not patentable. But microorganism obtained by novel techniques like genetic engineering are patentable. The first patent of GMO (Genetically Modified Organisms) was allowed by US supreme court in 1980 as described in utility patent. A maize plant over producing tryptophan amino acid was patented in USA in 1985. This was beginning of patenting of high organisms for patenting. For animals, a patent was granted in 1988 for 'oncomouse', genetically modified mouse in USA. In USA, non-naturally occurring non-human multicellular organisms are now considered patentable by US patent and trademark office. This clearly excludes humans and human parts. There is long debate about patenting of life forms including GMO and several organizations and religious groups are opposing the patenting of these life forms.

2.8. Copyrights

India's copyright law, laid down in the Indian Copyright Act, 1957 as amended by Copyright (Amendment) Act, 1999, fully reflects the Berne Convention on Copyrights, to which India is a party. Additionally, India is party to the Geneva Convention for the Protection of rights of Producers of Phonograms and to the Universal Copyright Convention. India is also an active member of the World Intellectual Property Organisation (WIPO), Geneva and UNESCO.

The copyright law has been amended periodically to keep pace with changing requirements. The recent amendment to the copyright law, which came into force in May 1995, has ushered in comprehensive changes and brought the copyright law in line with the developments in satellite broadcasting, computer software and digital technology. The amended law has made provisions for the first time, to protect performer's rights as envisaged in the Rome Convention.

2.9. Trade Secrets

Trade secrets, often include private proprietary information that allows a definite advantage to the owner. This can be illustrated by the popular example of Coca-Cola brand syrup formula which is not known publically under trade-secret. Trade secrets in the area of biotechnology may include material like (*i*) hybridization conditions (*ii*) cell lines (*iii*) corporate merchandising plan or (*iv*) customer lists.

Unlike patents, trade secrets have an unlimited duration and therefore may not be required to satisfy the more difficult conditions laid down for patent applications. Disclosure of a trade secret and its unauthorized use can be punished by the court and the owner may be allowed compensation. However if a trade secret becomes public knowledge by independent discovering or other means, it is no longer protectable.

QUESTIONS

1. What is IPR? Give an account of various international conventions for the protection of IPR.
2. Discuss the IPR in comparison to developed and developing countries.
3. Write short notes- a. IPR, b. TRIPS, c. GATT, d. UPNOV, e. Patent, f. patenting of GMO, g. trade secrets.

CHAPTER 21

Physical and Genetic Mapping

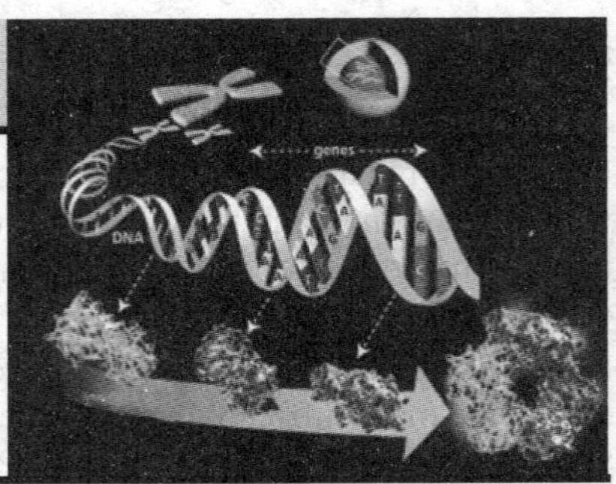

1. INTRODUCTION

Genetic mapping is based on the use of genetic techniques to construct maps showing the positions of genes and other sequences features on a genome. These genetic techniques include cross-breeding experiments or, in the case of humans, the examination of family histories. Genetic mapping is based on the principles of inheritance as first described by Gregor Mendel in 1865 and genetic linkages. Genetic maps are created to locate the genes or characters on the chromosome for their utilization in genetic studies. Physical maps are created to identify certain markers to detect or diagnose the specific character.

1.1. Genetic Linkage

Genetic linkage occurs when particular genetic loci or alleles for genes are inherited jointly. Genetic linkage was first discovered by the British geneticists William Bateson and Reginald Punnett shortly after Mendel's laws were rediscovered. Genetic loci on the same chromosome are physically connected and tend to stay together during meiosis, and are thus genetically linked. For example, in fruit flies the genes affecting eye color and wing length are inherited together because they appear on the same chromosome. Alleles for genes on different chromosomes are usually not linked, due to independent assortment of chromosomes during meiosis. Because there is some crossing over of DNA when the chromosomes segregate, alleles on the same chromosome can be separated and go to different daughter cells. There is a greater probability of this happening if the alleles are far apart on the chromosome, as it is more likely that a cross-over will occur between them. The relative distance between two genes can be calculated using the offspring of an organism showing two linked genetic traits, and finding the percentage of the offspring where the two traits do not run together. The higher the percentage of descendants that does not show both traits, the further apart on the chromosome they are.

Among individuals of an experimental population or species, some phenotypes or traits occur randomly with respect to one another in a manner known as independent assortment. Today scientists understand that independent assortment occurs when the genes affecting the phenotypes are found on different chromosomes or separated by a great enough distance on the same chromosome that recombination occurs at least half of the time. But in many cases, even genes on the same chromosome that are inherited together produce offspring with unexpected allele combinations. This results from a process called crossing over. At the beginning of normal meiosis, a chromosome pair (made up of a chromosome from the mother and a chromosome from the father) interwine and exchange sections or fragments of chromosome. The pair then breaks apart to form two chromosomes with a new combination of genes that differs from the combination supplied by the parents. Through this process of recombining genes, organisms can produce offspring with new combinations of maternal and paternal traits that may contribute to or enhance survival.

1.2. Genetic Map

A genetic map is a linkage map of a species or experimental population that shows the position of its known genes and/or genetic markers relative to each other in terms of recombination frequency during crossover of homologous chromosomes. The greater the frequency of recombination (segregation) between two genetic markers, the farther apart they are assumed to be. Conversely, the lower the frequency of recombination between the markers, the smaller the physical distance between them. Historically, the markers originally used were detectable phenotypes (enzyme production, color, shapes etc.) derived from coding DNA sequences. Now, noncoding DNA sequences such as microsatellites or those generating restriction fragment length polymorphisms (RFLPs) have been used. Genetic maps help researchers to locate other markers, such as other genes by testing for genetic linkage of the already known markers. A genetic map is not a physical map or gene map.

To be useful in genetic analysis, a gene must exist in at least two forms, or alleles; each specifying a different phenotype. Earlier only those genes could be studied whose specifying phenotypes were distinguishable by visual observation. This approach soon became outdated as in many cases a single phenotypic character could be affected by more than one gene. For example, in 1922, 50 genes had been mapped onto the four fruit fly chromosomes, but nine of these genes were for eye color.

The observations by Thomas Hunt Morgan that the amount of crossing over between linked genes differs (partial linkage) led to the idea that crossover frequency might indicate the distance separating genes on the chromosome. Morgan's student Alfred Sturtevant developed the first genetic map, also called a linkage map.

1.3. Recombination Frequency

Sturtevant assumed that crossing over was a random event, there being an equal chance of it occurring at any position along a pair of lined-up chromatids. He proposed that the greater the distance between linked genes, the greater the chance that non-sister chromatids would cross over in the region between the genes. By working out the number of recombinants it is possible to obtain a measure for the distance between the genes. This distance is called a genetic map unit (m.u.), or a centimorgan and is defined as the distance between genes for which one product of meiosis in 100 is recombinant. A recombinant frequency (RF) of 1% is equivalent to 1 m.u. A linkage map is created by finding the map distances between a number of traits that are present on the same chromosome, ideally avoiding having significant gaps between traits to avoid the inaccuracies that will occur due to the possibility of multiple recombination events.

Recombination frequency is the frequency that a chromosomal crossover will take place between two loci (or genes) during meiosis. Recombination frequency is a measure of genetic linkage and is used in the creation of a genetic linkage map. During meiosis, chromosomes assort randomly into gametes, such that the segregation of alleles of one gene is independent of alleles of another gene. This is stated in Mendel's second law and is known as the law of independent assortment. The law of independent assortment always holds true for genes that are located on different chromosomes, but for genes that are on the same chromosome, it does not always hold true.

As an example of independent assortment, consider the crossing of the pure-bred homozygote parental strain with genotype *AABB* with a different pure-bred strain with genotype *aabb*. A and a and B and b represent the alleles of genes A and B. Crossing these homozygous parental strains will result in F1 generation offspring with genotype AaBb. The F1 offspring AaBb produces gametes that are *AB*, *Ab*, *aB*, and *ab* with equal frequencies (25%) because the alleles of gene A assort independently of the alleles for gene B during meiosis. Note that 2 of the 4 gametes (50 %)-

Ab and aB-were not present in the parental generation. These gametes represent **recombinant gametes**. Recombinant gametes are those gametes that differ from both of the haploid gametes that made up the diploid cell. In this example, the recombination frequency is 50% since 2 of the 4 gametes were recombinant gametes.

The recombination frequency will be 50% when two genes are located on different chromosomes or when they are widely separated on the same chromosome. This is a consequence of independent assortment. When two genes are close together on the same chromosome, they do not assort independently and are said to be **linked**. Linked genes have a recombination frequency that is less than 50%.

As an example of linkage, consider the classic experiment by William Bateson and Reginald Punnett. They were interested in trait inheritance in the sweet pea and were studying two genes-the gene for flower color (*P*- purple and *p*- red) and the gene affecting the shape of pollen grains (*L*- long and *l*- round). They crossed the pure lines *PPLL* and *ppll* and then self-crossed the resulting *PpLl* lines. According to Mendelian genetics, the expected phenotypes would occur in a 9:3:3:1 ratio of PL:Pl:pL:pl. To their surprise, they observed an increased frequency of PL and pl and a decreased frequency of Pl and pL (Table 21.1).

Table 21.1. Bateson and Punnett experiment.

Phenotype and genotype	Observed	Expected from 9:3:3:1 ratio
Purple, long (PPLL)	284	216
Purple, round (PPll)	21	72
Red, long (ppLL)	21	72
Red, round (ppll)	55	24

Their experiment revealed **linkage** between the *P* and *L* alleles and the *p* and *l* alleles. The frequency of *P* occurring together with *L* and with *p* occurring together with *l* is greater than that of the recombinant *Pl* and *pL*. The recombination frequency cannot be computed directly from this experiment, but it is less than 50%. The progeny in this case received two dominant alleles linked on one chromosome (referred to as **coupling** or cis **arrangement**). However, after crossover, some progeny could have received one parental chromosome with a dominant allele for one trait (e.g., Purple) linked to a recessive allele for a second trait (eg round) with the opposite being true for the other parental chromosome (e.g., red and long). This is referred to as **repulsion** or a **trans arrangement**. The phenotype here would still be purple and long but a test cross of this individual with the recessive parent would produce progeny with much greater proportion of the two crossover phenotypes. While such a problem may not seem likely from this example, unfavorable repulsion linkages do appear while breeding for disease resistance in some crops.

When two genes are located on the same chromosome, the chance of a crossover producing recombination between the genes is directly related to the distance between the two genes. Thus, the use of recombination frequencies has been used to develop **linkage maps** or **genetic maps**.

1.4. Genetic Mapping in Bacteria

Bacteria are haploid organisms and do not undergo meiosis. So for creating their genetic maps geneticists made use of other methods to induce crossovers between homologous segments of bacterial DNA. They used the three methods of recombination that occurs in bacteria. (a) Conjugation-Two bacteria come into physical contact and one bacterium (the donor) transfers DNA to the second bacterium (the recipient). The transferred DNA can be a copy of some or possibly the donor cell's entire chromosome, or it could be a segment of chromosome DNA up to 1 mb in length integrated in a plasmid. The latter is called episome transfer. (b) Transduction-It involves transfer of a small segment of DNA up to 50 kb or so, from donor to recipient via a bacteriophage. (c) Transformation- The recipient cell takes up from its environment a fragment of DNA, rarely longer than 50 kb, released from a donor cell.

In bacteria, the phenotype studied are the biochemical characteristics like ability to synthesize tryptophan in the dominant or wild type strain and inability to synthesize tryptophan in other strain, which is the recessive allele. The gene transfer is usually set up between a donor strain that possesses dominant gene to the recipient strain that possesses recessive gene. The transfer into the recipient is monitored by looking for attainment of the biochemical function specified by the gene being studied. This can be understood by (Fig. 21.1). Here, the functional gene for tryptophan synthesis from a wild strain is being transferred to recipient that lacks the functional copy of that gene (trp^-). This recipient is called as auxotroph (bacteria which can survive only if provided with tryptophan). The wild strain (trp^+) do not require tryptophan for its survival. After the transfer, two crossovers are needed to integrate the transferred gene into the recipient cell's chromosome, converting the recipient from trp^- to trp^+.

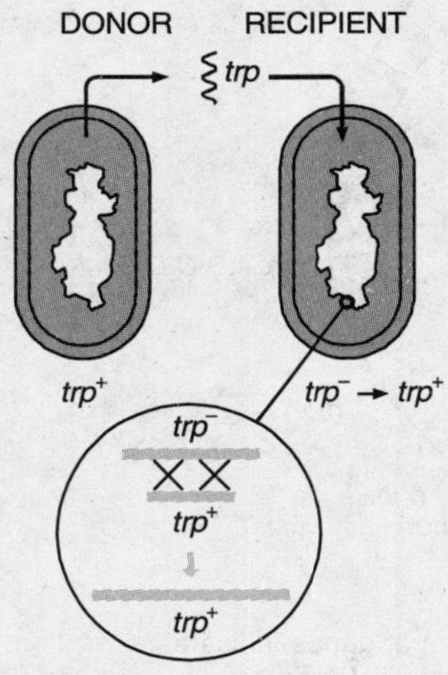

Fig. 21.1. Transfer of DNA between donor and recipient bacteria.

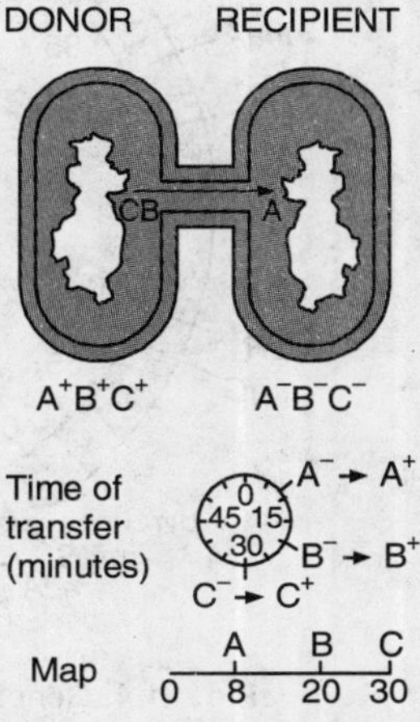

Fig. 21.2. Sequential transfer of markers during conjugation.

The precise detail of the map depends on the method of gene transfer being used. During conjugation, DNA is transferred from donor to recipient in the same way that a string is pulled through a tube. The relative positions of markers on the DNA molecule can therefore be mapped by determining the times at which the markers appear in the recipient cell. For example in Fig. 21.2, markers A, B and C are transferred after 8, 20 and 30 minutes of beginning of conjugation. The entire *E. coli* DNA takes approx. 100 minutes to transfer.

In case of transformation and transformation mapping enable genes that are relatively close together to be mapped, because the transferred DNA segment is short (<50kb), so the probability of two genes being transferred together depends on how close together they are on the bacterial DNA (Fig. 21.3).

Elie Wollman and Francois Jacob (1950s) conducted first genetic mapping experiments in bacteria. They studied linear transfer of genes in conjugation experiments between *Hfr* (Hfr- High frequency of recombination) and F^- (F^- fertility factor)

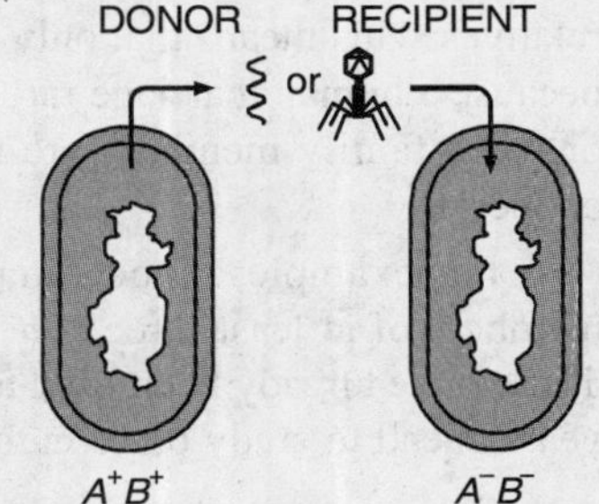

Frequency with which $A^-B^- \rightarrow A^+B^+$ depends on how close together *A* and *B* are on the chromosome

Fig. 21.3. Co-transfer of closely linked markers during transduction or transformation.

strains of *E. coli*. During the experiment they interrupted conjugation between bacteria at specific times termed as "Interrupted mating". They noticed that the time it takes for genes to enter a recipient cell is directly related to their order along the chromosome. This experiment derived them to give the hypothesis that (a) The chromosome of the *Hfr* donor is transferred in a linear manner to the *F–* recipient cell (b) The order of genes along the chromosome can be deduced by determining the time required for various genes to enter the recipient.

Conjugation studies have been used to map over 1,000 genes along the circular *E. coli* chromosome. The genetic maps are scaled in minutes e.g., *E. coli* chromosome is 100 minutes long, conjugative transfer of the complete chromosome takes approximately 100 minutes (Fig. 21.4).

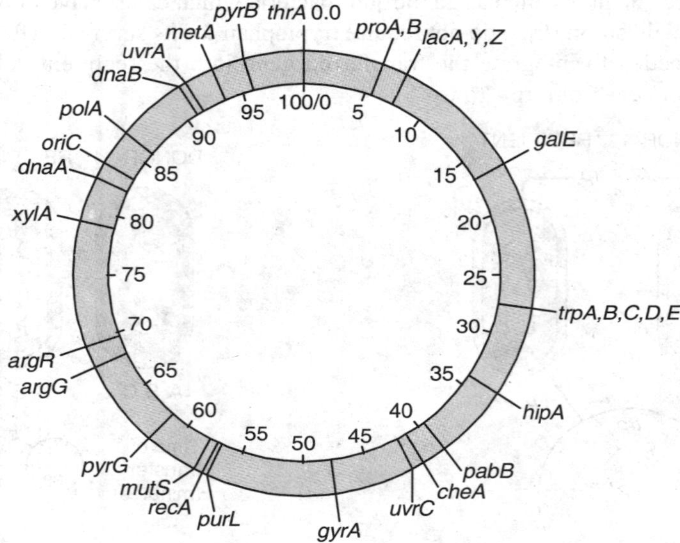

Fig. 21.4. Genetic map of *E. coli*; Positions of genes indicate the relative times at which they are transferred.

1.5. Gene Mapping in Humans by Pedigree Analysis

To map human chromosomes, obviously one cannot perform controlled mating experiments. However, it is possible to estimate map positions by examining linkage in several generations of relatives. This means that only limited data are available, and their interpretations is often difficult because a human marriage rarely results in a convenient test cross, and often the genotypes of one or more family members are unobtainable because those individuals are dead or unwilling to cooperate.

For example, blood samples from several large Mormon families in Utah, where all the members of at least three generations were alive to be sampled, have been collected and stored. These have already been used to establish genetic linkage relationships and will be available in the years ahead to study other human genes as they are identified.

2. MOLECULAR MARKERS IN PHYSICAL MAPPING

Different types of molecular markers are used to understand and ascertain relationship in different organisms/individuals as well as to detect or diagnose character. These markers are to locate certain characteristics on the gel (banding pattern) which can be used to detect a specific character/defect in the genome. Unlike genetic mapping, physical mapping is not to locate the genes/characters on a genome, but to create a unique pattern by processing the genomic DNA. There are several molecular markers available which are used depending upon the objective of the work and facilities available at the centre. Use of these markers to create maps (e.g., electrophoretic patterns) of an organism is known as '**physical mapping**'. Molecular markers used in physical mapping are described below. New technologies are also developed simultaneously to resolve biological problems and help legal proceedings.

2.1. Restriction Fragment Length Polymorphism (RFLP)

RFLP is a method used by molecular biologists to follow a particular sequence of DNA as it is passed on to other cells. RFLPs can be used in many different settings to accomplish different objectives. RFLPs can be used in paternity cases or criminal cases to determine the source of a DNA sample. RFLPs can be used to determine the disease status of an individual. RFLPs can be used to measure recombination rates which can lead to a genetic map with the distance between RFLP loci measured in centiMorgans.

RFLP, as a molecular marker, is specific to a single clone/restriction enzyme combination. It is a difference in homologous DNA sequences that can be detected by the presence of fragments of different lengths after digestion of the DNA samples in question with specific restriction endonucleases. Most RFLP markers are co-dominant (both alleles in heterozygous sample will be detected) and highly locus-specific. An RFLP probe is a labelled DNA sequence that hybridizes with one or more fragments of the digested DNA sample after they were separated by gel electrophoresis, thus revealing a unique blotting pattern characteristic to a specific genotype at a specific locus. Short, single- or low-copy genomic DNA or cDNA clones are typically used as RFLP probes. The RFLP probes are frequently used in genome mapping and in variation analysis (genotyping, forensics, paternity tests, hereditary disease diagnostics, etc.) (Fig. 21.5).

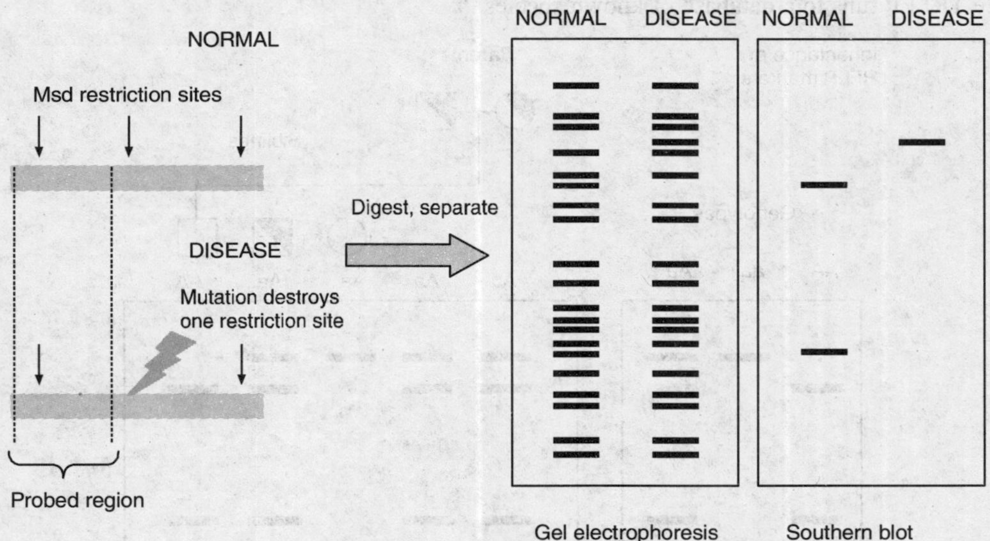

Fig. 21.5. Disease diagnosis by RFLP.

i. Procedure

Usually, DNA from an individual specimen is first extracted and purified. Purified DNA may be amplified by polymerase chain reaction (PCR). The DNA is then cut into *restriction fragments* using suitable endonucleases, which only cut the DNA molecule where there are specific DNA sequences, termed recognition sequence or restriction sites that are recognized by the enzymes.

These sequences are specific to each enzyme, and may be either four, six, eight, ten or twelve base pairs in length. The more base pairs there are in the restriction site, the more specific it is and the lower the probability that it will find a place to be cut. The restriction fragments are then separated according to length by agarose gel electrophoresis. The resulting gel may be enhanced by Southern blotting. Alternatively, fragments may be visualised by pre-treatment or post-treatment of the agarose gel, using methods such as ethidium bromide staining or silver staining respectively.

RFLPs have provided valuable information in many areas of biology, including: screening human DNA for the presence of potentially deleterious genes (Fig. 21.6). Providing evidence to establish the innocence of or a probability of the guilt of, a crime suspect by DNA "fingerprinting". The distance between the locations cut by restriction enzymes (the *restriction sites*) varies between individuals, due to insertions, deletions or transversions. This causes the length of the fragments to vary, and the position of certain amplicons differs between individuals (thus *polymorphism*). This can be used to genetically tell individuals apart. It can also show the genetic relationship between individuals, because children inherit genetic elements from their parents. Mitochondrial DNA RFLP analyses can lead to the determination of maternal relationships. Fragments may also be used to determine relationships among and between species by comparison of the resulting haplotypes (abridged for 'haploid genotype'). RFLP is a technique used in marker assisted selection. Terminal Restriction Fragment Length Polymorphism (TRFLP or sometimes T-RFLP) is a molecular biology technique initially developed for characterizing bacterial communities in mixed-species samples. The technique has also been applied to other groups including soil fungi.

The technique works by PCR amplification of DNA using primer pairs that have been labelled with fluorescent tags. The PCR products are then digested using RFLP enzymes and the resulting patterns visualized using a DNA sequencer. The results are analyzed either by simply counting and comparing bands or peaks in the TRFLP profile, or by matching bands from one or more TRFLP runs to a database of known species.

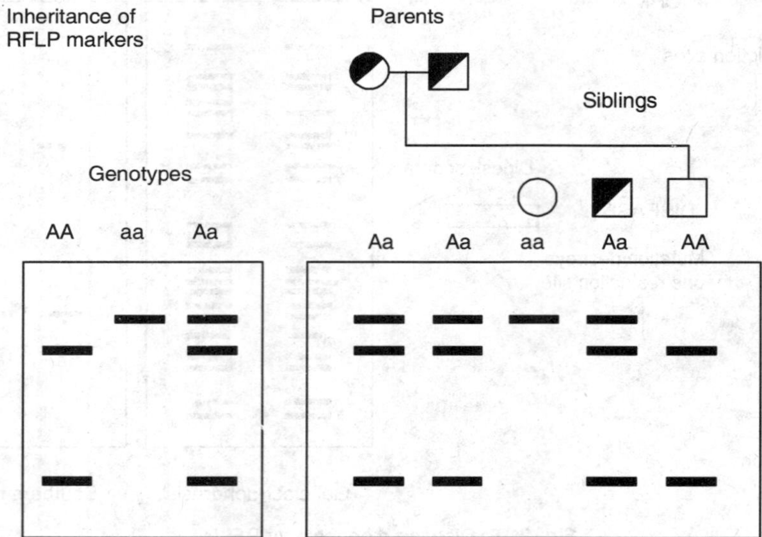

Fig. 21.6. Inheritance of RFLP markers.

ii. Measurement of distance between two RFLP loci

To calculate the genetic distance between two loci, you need to be able to observe recombination. Traditionally, this was performed by observing phenotypes but with RFLP analysis, it is possible to measure the genetic distance between two RFLP loci whether they are a part of genes or not.

Let's look at a simple example in fruit flies. Two RFLP loci with two RFLP bands possible at each locus (Fig. 21.7).

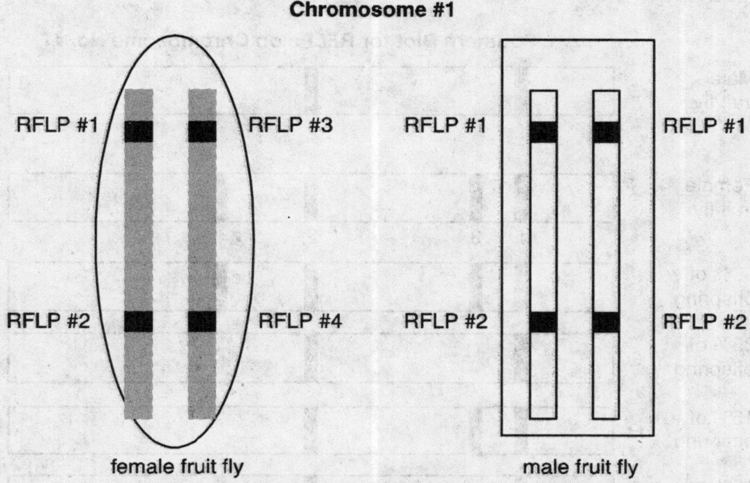

Fig. 21.7. RFLP loci on male and female chromosome.

These loci are located on the same chromosome for the female (left) and the male (right). The upper locus can produce two different bands called 1 and 3. The lower locus can produce bands called 2 or 4. The male is homozygous for band 1 at the upper locus and 2 for the lower locus. The female is heterozygous at both loci. Their RFLP banding patterns can be seen on the Southern blot below (Fig. 21.8).

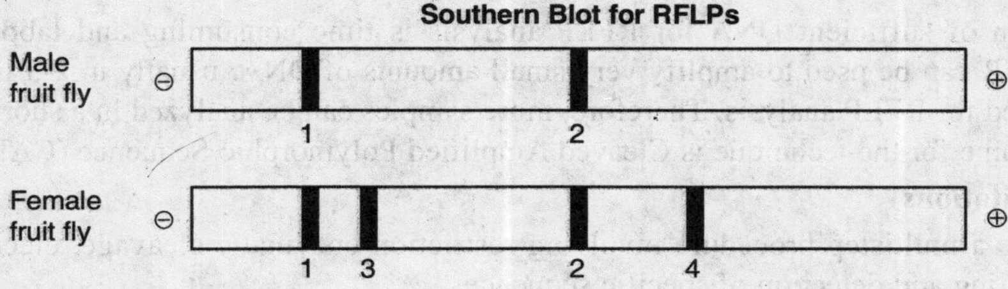

Fig. 21.8. RFLP loci on gel of male and female chromosome.

The male can only produce one type of gamete (1 and 2) but the female can produce four different gametes. Two of the possible four are called parental because they carry both RFLP bands from the same chromosome; 1 and 2 from the left chromosome or 3 and 4 from the right chromosome. The other two chromosomes are recombinant because recombination has occurred between the two loci and thus the RFLP bands are mixed so that 1 is now linked to 4 and 3 is linked to 2.

Type of chromatid	Alleles
Parental	RFLP 1 and 2
Parental	RFLP 3 and 4
Recombinant	RFLP 1 and 4
Recombinant	RFLP 3 and 2

When these two flies mate, the frequency of the four possible progeny can be measured and from this information, the genetic distance between the two RFLP loci (upper and lower) can be determined (Fig. 21.9).

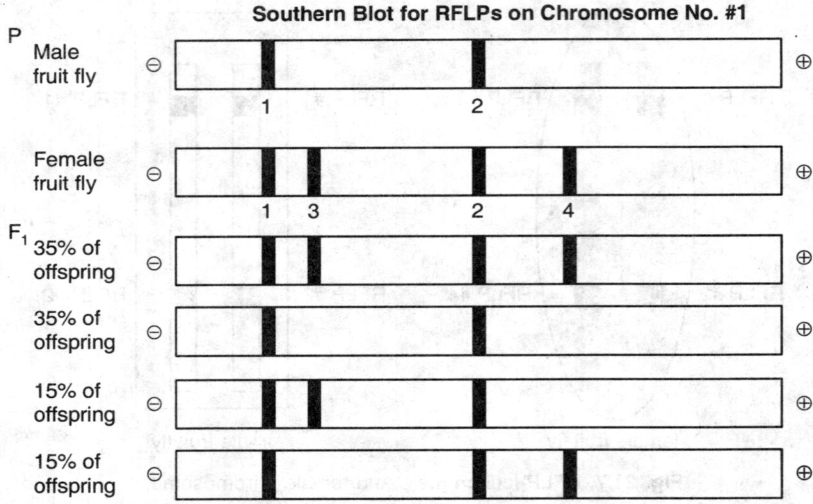

Fig. 21.9. Southern blot for RFLP's on chromosome no. 1.

In this example, 70% of the progeny were produce from parental genotype eggs and 30% were produced by recombinant genotype eggs. Therefore, these two RFLP loci are 30 centiMorgans apart from each other.

iii. PCR-RFLP

Isolation of sufficient DNA for RFLP analysis is time consuming and labor intensive. However, PCR can be used to amplify very small amounts of DNA, usually in 2-3 hours, to the levels required for RFLP analysis. Therefore, more samples can be analyzed in a shorter time. An alternative name for the technique is Cleaved Amplified Polymorphic Sequence (CAPS) assay.

iv. Limitations

RFLP is a multistep procedure involving restriction enzymatic cleavage, electrophoresis, southern blotting and detection of specific sequences.

It is a time consuming process.

2.2. Random Amplified Polymorphic DNA (RAPD)

This technique can be used to determine taxonomic identity, assess kinship relationships, detect interspecific gene flow, analyse hybrid speciation, and create specific probes. Advantages of RAPDs include suitability for work on anonymous genomes, applicability to work where limited DNA is available, efficiency and low expense. It is also useful in distinguishing individuals, cultivars or accessions. RAPDs also have applications in the identification of asexually reproduced plant varieties for forensic or agricultural purposes, as well as ecological ones.

In RAPD by using different primers, molecular characters can be generated that are diagnostic at different taxonomic levels. This is really a stripped-down version of PCR but uses a single sequence in the design of the primer (i.e., two primers are still needed for PCR: the same primer is used at either end).

The primer may be designed specifically, but could be chosen randomly and is used to amplify a series of samples which will include both the material of interest as well as other control samples with which the experimental material needs to be compared. Choice of primer length will be critical to the determination of band complexity in the resulting amplification pattern. Eventually a particular probe will be found that is able to distinguish between the sample of interest and those that are different.

i. Procedure

Unlike traditional PCR analysis, RAPD (pronounced 'rapid') does not require any specific knowledge of the DNA sequence of the target organism: the identical 10-mer primers will or will not amplify a segment of DNA, depending on positions that are complementary to the primers' sequence. For example, no fragment is produced if primers annealed too far apart or 3′ ends of the primers are not facing each other. Therefore, if a mutation has occurred in the template DNA at the site that was previously complementary to the primer, a PCR product will not be produced, resulting in a different pattern of amplified DNA segments on the gel (Fig. 21.10).

RAPD is an inexpensive yet powerful typing method for many bacterial species

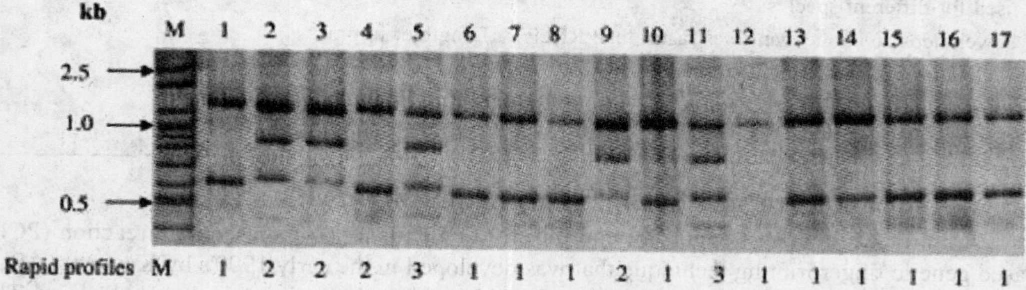

Fig. 21.10. Silver-stained polyacrylamide gel showing three distinct RAPD profiles generated by primer OPE15 for *Haemophilus ducreyi* isolates from different countries. Selecting the right sequence for the primer is very important because different sequences will produce different band patterns and possibly allow for a more specific recognition of individual strains.

RAPD amplification products can be either variable (polymorphic) or constant (non-polymorphic). In a RAPD analysis of several individuals within a species, and species within a genus, constant fragments diagnostic for a genus may be identified, as well as fragments which are polymorphic between species of the genus.

RAPDs can be applied to analyse fusion of genotypes at different taxonomic levels. At the level of the individual, RAPD markers can be applied to parentage analysis, while at the population level, RAPD can detect hybrid populations, species or subspecies. The detection of genotype hybrids relies on the identification of diagnostic RAPD markers for the parental genotypes under investigation. However RAPD markers tend to underestimate genetic distances between distantly related individuals, for example in inter-specific comparisons. It is wise to be cautious when using RAPD for taxonomic studies above the species level. Conventional RFLP techniques are ill-suited for the analysis of paternity and estimation of reproductive success in species with large offspring clutches, because of the need to determine paternity for each individual offspring. RAPD fingerprinting provides a ready alternative for such cases. Synthetic offspring may be produced by mixing equal amounts of the DNA of the mother and the potential father. The amplification products from the synthetic offspring should ideally contain the full complement of bands that appear in any single offspring of these parents (Table 21.2).

ii. Limitations of RAPD

1. Nearly all RAPD markers are dominant, i.e., it is not possible to distinguish whether a DNA segment is amplified from a locus that is heterozygous (1 copy) or homozygous (2 copies). Co-dominant RAPD markers, observed as different-sized DNA segments amplified from the same locus, are detected only rarely.

2. PCR is an enzymatic reaction, therefore the quality and concentration of template DNA, concentrations of PCR components, and the PCR cycling conditions may greatly influence the outcome. Thus, the RAPD technique is notoriously laboratory dependent and needs carefully developed laboratory protocols to be reproducible.
3. Mismatches between the primer and the template may result in the total absence of PCR product as well as in a merely decreased amount of the product. Thus, the RAPD results can be difficult to interpret.

Table 21.2. Comparison between RAPD and RFLP.

RAPD	RFLP
Quantity of DNA required for analysis is small, i.e. 10-50 ng.	Large quantity of purified DNA required, i.e. 2-10 ng.
Same primers with arbitary sequences can be used for different species.	Different species specific probes are required.
Fewer steps so about five times faster than RFLP	Lengthier in process
Technique comparatively less reliable	More reliable
Cannot detect allelic variants	Can detect allelic variants
1-10 loci detected	1-3 loci detected

2.3. Amplification Fragment Length Polymorphism (AFLP)

Amplified Fragment Length Polymorphism (**AFLP**) is a polymerase chain reaction (PCR) based genetic fingerprinting technique that was developed in the early 1990's by **Keygene**. AFLP can be used in the fingerprinting of genomic DNA of varying origins and complexities. The amplification reaction is rigorous, versatile and robust, and appears to be quantitative. While AFLP is capable of producing very complex fingerprints (100 bands where RAPD produces 20), it is a technique that requires DNA of reasonable quality and is more experimentally demanding. AFLP uses restriction enzymes to cut genomic DNA, followed by ligation of complementary double stranded adaptors to the ends of the restriction fragments. A subset of the restriction fragments are then amplified using 2 primers complementary to the adaptor and restriction site fragments. The fragments are visualized on denaturing polyacrylamide gels either through autoradiographic or fluorescence methodologies.

i. Procedure

AFLP-PCR is a highly sensitive method for detecting polymorphisms in DNA. The technique was originally described by Vos and Zabeau in 1993. The procedure of this technique is divided into three steps (Fig. 21.11):

1. Digestion of total cellular DNA with one or more restriction enzymes that cuts frequently (*Mse*I, 4 bp recognition sequence) and one that cuts less frequently (*Eco*RI, 6 bp recognition sequence). The resulting fragments are ligated to end-specific adaptor molecules.
2. Selective amplification of some of these fragments with two PCR primers that have corresponding adaptor and restriction site specific sequences.
3. Electrophoretic separation of amplicons on a gel matrix, followed by visualisation of the band pattern.

In a second, "selective", PCR, using the products of the first as template, primers containing two further additional bases, chosen by the user, are used. The *Eco*RI-adaptor specific primer used bears a label (fluorescent or radioactive). Gel electrophoretic analysis reveals a pattern (fingerprint) of fragments representing about 1/4000th of the *Eco*RI-*Mse*I fragments.

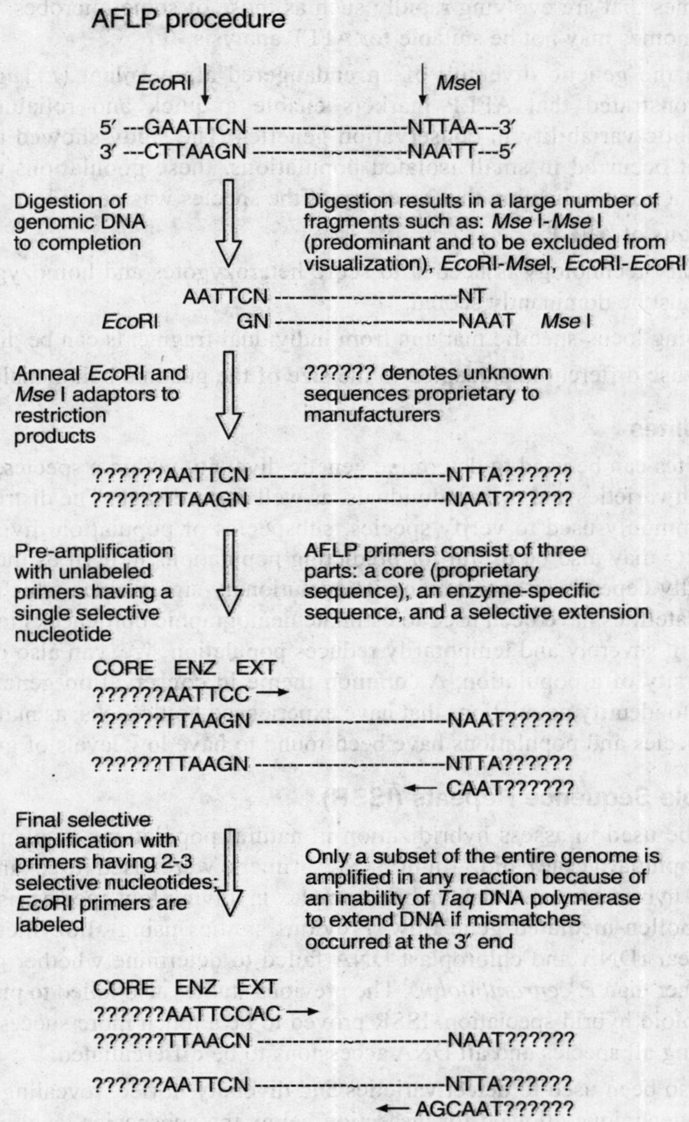

Fig. 21.11. AFLP procedure.

AFLP's, can be co-dominant markers, like RFLP's. Codominance results when the polymorphism is due to sequences within the amplified region. Yet, because of the number of bands seen at one time, additional evidence is needed to establish that a set of bands result from different alleles at the same locus. If, however, the polymorphism is due to presence/absence of a priming site, the relationship is dominance. The non-priming allele will not be detected as a band. Compared to RAPD, fewer primers should be needed to screen all possible sites.

AFLP can be used for mapping, fingerprinting and genetic distance calculation between genotypes. The advantage of AFLP is its high multiplexity and therefore the possibility of generating high marker densities. One limitation of the AFLP technique is that fingerprints may share few common fragments when genome sequence homology is less than 90%. Therefore, AFLP cannot be used in comparative genomic analysis with hybridization-based probes or when comparing genomes that are evolving rapidly such as those of some microbes. Conversely, very homogeneous genomes may not be suitable for AFLP analysis.

A study on the genetic diversity of an endangered alpine plant (*Eryngium alpinum* L. (Apiaceae) demonstrated that AFLP markers enable a quick and reliable assessment of intraspecific genetic variability in conservation genetics. The study showed that although the endangered plant occurred in small isolated populations, these populations contained a high genetic diversity, a good indication that recovery of the species was possible.

ii. Limitations of AFLP

1. Proprietary technology is needed to score heterozygotes and homozygotes. Otherwise, AFLP must be dominantly scored.
2. Developing locus-specific markers from individual fragments can be difficult.
3. Need to use different kits adapted to the size of the genome being analyzed.

2.4. Microsatellites

Microsatellites can be used to determine genetic diversity within a species, as well as being able to distinguish varieties and even individuals, as well as parentage. The distribution of genetic variability is commonly used to verify species, subspecies or population division. Monitoring change in diversity may also be useful for predicting populations in peril as the persistence of a population partially depends on maintaining its evolutionary significance which requires genetic variation. Microsatellites have been used to estimate demographic bottlenecks in some species. A bottleneck, when it severely and temporarily reduces population size, can also drastically reduce the genetic diversity of a population. A common theme in conservation genetics is the use of genetic variation to identify populations that have experienced bottlenecks, as numerous threatened or endangered species and populations have been found to have low levels of genetic variation.

2.5. Inter-Simple Sequence Repeats (ISSR)

ISSRs can be used to assess hybridization in natural populations of plants, as a study on *Penstemon* (Scrophulariaceae) did. Eight ISSR primers were used to examine patterns of hybridization and hybrid speciation in a hybrid complex involving four species, as well as allowing examination of pollen-mediated gene flow. Previous studies using allozymes, restriction-site variation of nuclear rDNA and chloroplast DNA failed to determine whether gene flow occurs among species other than *P. cenranthifolius*. The previous studies also failed to provide support for hypotheses of diploid hybrid speciation. ISSR proved to be a much more successful technique in this study, allowing all species and all DNA accessions to be differentiated.

ISSR has also been used to detect varieties and diversity in rice, revealing much more data than RFLPs. The technique allowed for dissection below the subspecies level and this gives it a good level of applicability in the study of rare or endangered plants.

ISSRs have been used in conjunction with RAPD data to determine the colonization history of *Olea europapea* in Macronesia, along with lineages in the species complex. The two techniques have also been utilized in examining the historical biogeography of Sea rocket (*Cakile maritima*) and Sea Holly (*Eryngium maritimum*), comparing different and only distantly related taxa of broadly similar extant distribution. The trees generated by the different methods were largely similar topologically. Using the result, dispersal routes of the species along a linear coast line could be construed. Joint use of RAPD and ISSR has also been used to examine clonal diversity in *Calamagrostis porteri* ssp. *insperata* (Poaceae), a rare grass that has little or no sexual reproduction, and spreads by vegetative reproduction.

The relative advantages and disadvantages of various molecular markers in physical mapping are summarized in Table 21.3. This information suggests that RFLP, SSR and AFLP markers are most effective in detecting polymorphism. However, given the large amount of DNA required for RFLP detection and the difficulties in automating RFLP analysis, AFLP and SSR are currently most popular markers.

Table 21.3. Comparison of most commonly used marker systems.

Features	RFLP	RAPD	AFLP	SSR	SNP
DNA required (μg)	10	0.02	0.5-1.0	0.05	0.05
DNA quality	High	High	Moderate	Moderate	High
PCR based	No	Yes	Yes	Yes	Yes
No. of polymorph loci analysed	1.0-3.0	1.5-5.0	20-100	1.0-3.0	1.0
Ease of use	Not easy	Easy	Easy	Easy	Easy
Amenable to automation	Low	Moderate	Moderate	High	High
Reproducibility	High	unreliable	High	High	High
Development cost	Low	Low	Moderate	High	High
Cost per analysis	High	Low	Moderate	Low	Low

The main uses of these markers include-

1. Assessment of genetic variability and characterization of germplasm.
2. Identification and fingerprinting of genotypes.
3. Estimation of genetic distances between population, inbreeds and breeding material.
4. Detection of monogenic and qualitative trait loci.
5. Marker assisted selection.
6. Identification of sequences of useful candidate genes.

3. RESTRICTION MAPPING OF DNA FRAGMENTS

Restriction mapping

Genetic mapping using RFLPs as DNA markers can locate the positions of polymorphic restriction sites within a genome, but very few of the restriction sites in a genome are polymorphic, so many sites are not mapped by this technique. We increase the marker density on a genome map by using an alternative method to locate the positions of some of the non polymorphic restriction sites. This is what restriction mapping achieves, although in practice the technique has limitations that means it is applicable only to relatively small DNA molecules.

Methodology for restriction mapping

The simplest way to construct a restriction map is to compare the fragment sizes produced when a DNA molecule is digested with two different restriction enzymes that recognize different target sequences. An example using the restriction enzymes EcoRI and BamHI is shown in figure. 21.12. First, the DNA molecule is digested with just one of the enzymes and the sizes of the resulting fragments measured by agarose gel electrophoresis. Next, the molecule is digested with the second enzyme and the resulting fragments again sized in an agarose gel. The results of subsequent use of two enzymes give clear picture about restriction sites creating a large number of fragments but this method do not allow their relative positions to be determined. Additional information is therefore obtained by cutting the DNA molecule with both enzymes together. In the example shown in Figure 21.12, the **double restriction** enables three of the sites to be mapped. However, a problem arises with the larger EcoRI fragment because this contains two BamHI sites and there are two alternative possibilities for the map location of the outer one of these. The problem dissolved by going back to the original DNA molecule and treating it again with BamHI on its own, but this time preventing the digestion from going to completion by, for example, incubating the reaction for only a short time rousing a suboptimal incubation temperature. This is called a partial restriction and leads to a more complex set of products. The complete restriction products now being supplemented with partially restricted fragments that still contain one or more uncut BamHI sites. In the example shown in Figure 21.12, the size of one of the partial restriction fragments is diagnostic and the correct map can be identified.

A partial restriction usually gives the information needed to complete a map, but if there are many restriction sites then this type of analysis becomes bulky, simply because there are many different fragments to consider. An alternative strategy is simpler because it enables the majority of the fragments to be ignored. This is achieved by attaching a radioactive or other type of marker to each end of the starting DNA molecule before carrying out the partial digestion. The result is that many of the partial restriction products become "invisible" because they do not contain an end-fragment and so do not show up when the agarose gel is screened for labeled products (Fig. 21.12). The sizes of the partial restriction products that are visible enable unmapped sites to be positioned relative to the ends of the starting molecule.

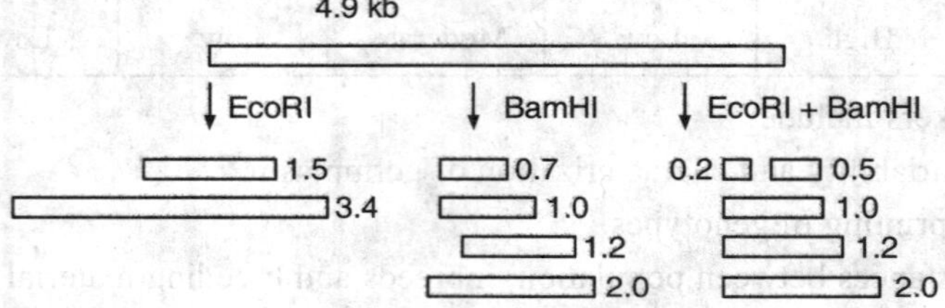

Interpretation of double restriction by EcoRI and BamHI

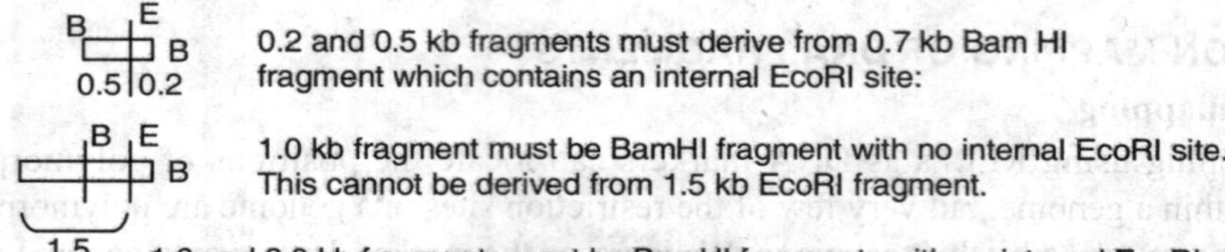

0.2 and 0.5 kb fragments must derive from 0.7 kb Bam HI fragment which contains an internal EcoRI site:

1.0 kb fragment must be BamHI fragment with no internal EcoRI site. This cannot be derived from 1.5 kb EcoRI fragment.

1.2 and 2.0 kb fragments must be BamHI fragments with no internal EcoRI site. They must lie within the 3.4 kb EcoRI fragment. There are two possibilities as shown below:

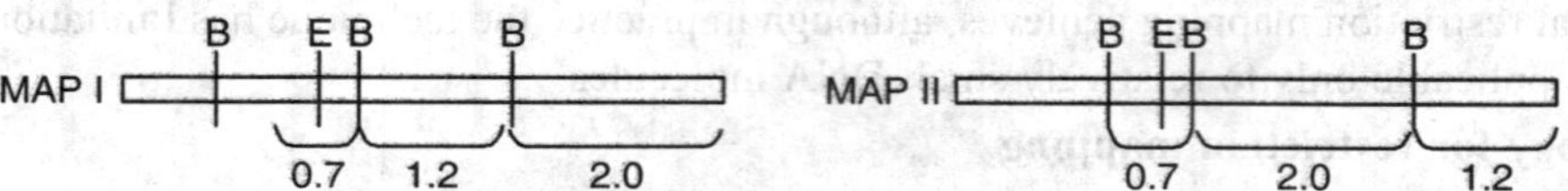

Predicted result of apartial and BamHI restriction
If map one is correct then partial restrict products will include a fragment of 1.2 + 0.7 = 1.9 kb
If map second is correct then the partial restriction products will include a fragment of 2.0 + 0.7 = 2.7 kb
Conclusion: Map second is correct

4.9 kb

BamHI,
Suboptimal digestion

1.0 1.7 3.7
0.7 2.7 3.9
2.0 3.2 4.9
1.2

Fig. 21.12. Preparation of restriction maps by restriction endonuclease.

The scale of restriction mapping is limited by the sizes of the restriction fragments.

Restriction maps are easy to generate if there are relatively few cut sites for the enzymes being used. However, as the number of cut sites increases, so also do the numbers of single, double and partial-restriction products whose sizes must be determined and compared in order for the map to be constructed. Computer analysis can be brought into play but problems still eventually arise. A stage will be reached when a digest contains so many fragments that individual bands merge on the agarose gel, increasing the chances of one or more fragments being measured incorrectly or missed out entirely. If several fragments have similar sizes then even if they can all be identified, it may not be possible to assemble them into a clear map.

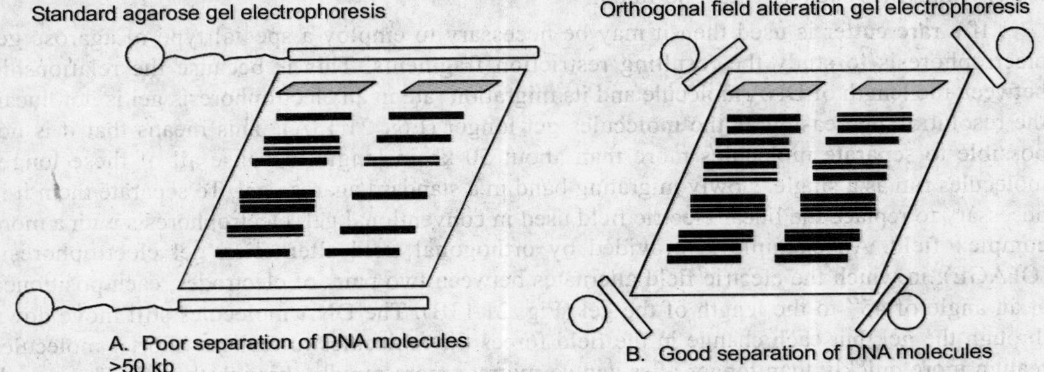

Fig. 21.13. A,B Conventional and non-conventional agarose gel electrophoresis.

Restriction mapping is therefore more applicable to small rather than large molecules, with the upper limit for the technique depending on the frequency of the restriction sites in the molecule being mapped. In practice, if a DNA molecule is less than 50 kb in length it is usually possible to construct a clear restriction map for a selection of enzymes with six nucleotide recognition sequences. Restriction maps are equally useful after bacterial or eukaryotic genomic DNA has been cloned, if the cloned fragments are less than 50kb in length, because a detailed restriction map can then be built up as a preliminary to sequencing the cloned region. This is an important application of restriction mapping in projects sequencing large genomes. Restriction mapping can be used for mapping of entire genomes larger than 50kb by slightly eliminating the limitations of restriction mapping by choosing enzymes expected to have infrequent cut sites in the target DNA molecule. These "rare cutters" fall into two categories:

1. A few restriction enzymes cut at seven- or eight-nucleotide recognition sequences. Examples are Sapl (5′-GCTCTTC-3′) and SgfI (5′-GCGATCGC-3′). The enzymes with seven-nucleotide recognition sequences would be expected, on average, to cut a DNA molecule with GC content of 50% once every 47 = 16,384 bp. The enzymes with eight nucleotide recognition sequences should cut once every 48 = 65,536 bp. These figures compare with 46 = 4096 bp for enzymes with six-nucleotide recognition sequences, such as BamHI and EcoRI. Cutters with seven-or eight-nucleotide recognition sequences are often used in restriction mapping of large molecules, but the approach is not as useful as it might by simply because not many of these enzymes are known.
2. Enzymes can be used whose recognition sequences contain motifs that are rare in the target DNA. Genomic DNA molecules do not have random sequences and some molecules are significantly deficient in certain motifs. For example, the sequence 5′-CG-3′ is rare in the genomes of vertebrates because vertebrate cells possess an enzyme that adds a methyl group to carbon 5 of the C nucleotide in this sequence. Domination of the

resulting 5-methylcytosine gives thymine. The consequence is that during vertebrate evolution, many of the 5′-CG′3 sequences that were originally in these genomes have become converted to 5′-TG-3′ . Restriction enzymes that recognize a site containing 5′-CG-3′ therefore cut vertebrate DNA relatively infrequently. Examples are SmaI (5′-CCCGGG-3′), which cuts human DNA once every 78 kb on average, and BssHII (5′-GCGCGC′3′), which cuts once every 390 kb.

The potential of restriction mapping is therefore increased by using rare cutters. It is still not possible to construct restriction maps of the genomes of animals and plants, but it is feasible to use the technique with large cloned fragments, and with the smaller DNA molecules of prokaryotes and lower eukaryotes such as yeast and fungi.

If a rare cutter is used then it may be necessary to employ a special type of agarose gel electrophoresis to study the resulting restriction fragments. This is because the relationship between the length of DNA molecule and its migration rate in an electrophoresis gel is not linear, the resolution decreasing as the molecules get longer (Fig. 21.13A). This means that it is not possible to separate molecules more than about 50 kb in length, because all of these longer molecules run as a single, slowly migrating band in a standard agarose gel. To separate them it is necessary to replace the linear electric field used in conventional gel electrophoresis with a more complex field. An example is provided by orthogonal field alternation gel electrophoresis (OFAGE), in which the electric field alternates between two pairs of electrodes, each positioned at an angle of 45″ to the length of the gel (Fig. 21.13B). The DNA molecules still move down through the gel, but each change in the field forces the molecules to realign. Shorter molecules realign more quickly than longer ones and so migrate more rapidly through the gel. The overall result is that molecules much longer than those separated by conventional gel electrophoresis can be resolved. Related techniques include CHEF (contour clamped homogeneous electric fields) and FIGE (field inversion gel electrophoresis).

QUESTIONS

1. What are molecular markers? Describe applications of molecular markers in solving biological problems.
2. What is genetic mapping? How it different from physical mapping?
3. What is physical mapping of a genome? Describe in brief techniques used in physical mapping.
4. What is RFLP? Compare this technique with RAPD.
5. Write short notes on:
   (a) RAPD
   (b) RFLP
   (c) ISSR
   (d) Microsatellite
   (e) AFLP
   (f) Polymorphism

CHAPTER 22

Biotechnology: Scope and Applications

1. INTRODUCTION

Biotechnology is defined as the 'application of scientific and engineering principles to the processing of material by biological agents to provide goods and services'. The Spinks Report (1980) defined biotechnology as 'the application of biological organisms, systems or processes to the manufacturing and service industries'. United States Congress's Office of Technology Assessment defined biotechnology as 'any technique that used living organisms to make or modify a product, to improve plants or animals or to develop microorganisms for specific uses'. The document focuses on the development and application of modern biotechnology based on new enabling techniques of recombinant-DNA technology, often referred to as genetic engineering.

The history of biotechnology begins with zymotechnology, which commenced with a focus on brewing techniques for beer. By World War I, however, zymotechnology would expand to tackle larger industrial issues, and the potential of industrial fermentation gave rise to biotechnology. The oldest biotechnological processes are found in microbial fermentations, as born out by a Babylonian tablet circa 6000 B.C. unearthed in 1881 and explaining the preparation of beer. In about 4000 B.C. leavened bread was produced with the aid of yeast. The Sumerians were able to brew as many as twenty types of beer in the third millennium B.C. In the 14th century, first vinegar manufacturing industry was established in France near Orleans. In 1680 Antony Van Leeuwenhook first observed yeast cells with his newly designed microscope. In 1857, Louis Pasteur highlighted the lactic acid fermentation by microbe. By the end of 19th century large number of industries and group of scientists were involved in the field of biotechnology and developed large scale sewage purification system employing microbes were established is Germany and France. In 1914 to 1916, Delbruck, Heyduck and Hennerberg discovered the large-scale use of yeast in food industry. In the same period, acetone, butanol and glycerin were obtained from bacteria. In 1920, Alexander Fleming discovered penicillin and large scale manufacturing of penicillin started in 1944. Table 22.1 presents chronological history of biotechnology.

1.1. Fermentation to Produce Foods

Fermentation is perhaps the most ancient biotechnological discovery. Over 10,000 years ago mankind was producing wine, beer, vinegar and bread using microorganisms, primarily yeast. Yogurt was produced by lactic acid bacteria in milk and molds were used to produce cheese. These processes are still in use today for the production of modern foods. However, the cultures that are used have been purified and often genetically refined to maintain the most desirable traits and highest quality of products.

1.2. Industrial Fermentation

In 1897 the discovery that enzymes from yeast can convert sugar to alcohol lead to industrial processes for chemicals such as butanol, acetone and glycerol. Fermentation processes are still in use today in many modern biotech organizations, often for the production of enzymes to be used in pharmaceutical processes, environmental remediation and other industrial processes.

1.3. Food Preservation

Drying, salting and freezing foods to prevent spoilage by microorganisms were practiced long before anyone really understood why they worked or even fully knew what caused the food to spoil in the first place.

1.4. Quarantines

The practice of quarantining to prevent the spread of disease was in place long before the origins of disease were known. However, it demonstrates early acceptance that illness could be passed from an infected individual to another healthy individual, who would then begin to have symptoms of the disease.

1.5. Selective Plant Breeding

Crop improvement, by selecting seeds from the most successful or healthiest plants, to obtain a new crop having the most desirable traits, is a form of early crop technology. Farmers learned that using only the seeds from the best plants would eventually enhance and strengthen the desired traits in subsequent crops. In the mid-1860's, Gregor Mendel's studies on inheritable traits of peas improved our understanding of genetic inheritance and lead to practices of cross-breeding (now known as hybridization).

Table 22.1. Historical development of biotechnology.

Year	Development
Before 6000 B.C.	Yeast employed to make wine and beer
4000 B.C.	Leavened bread produced with the aid of yeast
1521	Aztecs harvested algae from lakes as a source of food.
1670-1680	Copper mined with aid of microbes, Rio Tinto, Spain.
	Antoine van Leeuwenhoek first observed microbes with newly designed microscope.
1869	Johann Meischer isolated DNA from the nuclei of white blood cells.
1876	Louis Pasteur identifies extraneous microbes as a cause of failed beer fermentations
1890	Alcohol first used to fuel motors
1893	Fermentation process patented by Koch, Pasteur.
1897	Edurad Buchner discovered that enzymes extracted from yeast can convert sugar into alcohol.
1910	Thomas H. Morgan proved that genes are carried on chromosomes "Biotechnology" term coined.
	Large-scale sewage purification systems employing microbes are established.
1912-1914	Three important industrial chemicals- acetone, butanol and glycerol were obtained from bacteria.
1918	Germans use acetone produced by plants to make bombs.
1920	Alexander Fleming discovered penicillin.
	Plant hybridization.
1928	Yeast grown in large quantities for animal and glycerol.
1938	Proteins and DNA studied by X-ray crystallography.
1941	George Beadle proposed "one gene, one enzyme" hypothesis.
1943-1953	Linus Pauling described sickle cell anemia calling it a molecular disease. Cortisone made in large amounts.
	DNA identified as the genetic material.

1944	DNA as genetic material and transformation factor in bacteria (Avery, Macleod and McCarty).
1946	Bacterial recombination discovered (Lederberg and Tatum)
1951	Lambda bacteriophage discovered (Lederberg)
1953	Double helix structure of DNA revealed (Watson and Crick).
1956	Isolation of DNA polymerase I, enzymatic synthesis of DNA (Kornberg).
1958	Semiconservative replication of DNA (Messelson and Stahl).
1960	Isolation of mRNA.
1961	Nucleic acid hybridization (Marmur and Doty), Operon model (Jacob and Monod) and *in vitro* protein synthesis allows codon assignment (Nirenberg and Mathaei).
1962	Mining of uranium with the aid of microbes begin in Canada.
1973	Beginning of genetic engineering. Stanley Cohen produced first recombinant DNA organism.
	Brazilian government initiates major fuel programme to replace oil with alcohol.
1975	Hybridomas which make monoclonal antibodies were first created
1976	US National Institute of Health introduces guidelines on genetic engineering
1977	Human Growth Hormone produced by bacterial cells.
1978	Genetic engineering techniques used to produce human insulin in *E. coli* by Genentech Inc.
1979	Genentech Inc. produce human growth hormone and two kinds of interferon DNA from malignant cells transformed a strain of cultured mouse cells - new tool for analyzing cancer genes.
1980	Rank Hovis McDougall receives permission to market fungal food for human consumption in UK.
	The US Supreme Court rules in Diamond v. Chakrabarty that genetically engineered microorganisms can be patented.
1981	Monoclonal antibodies receive US approval for use in diagnosis.
1982	The Food and Drug Administration approves the first biotechnology therapy, a human insulin drug made by Genentech.
1983	Syntex Corporation received FDA approval for a monoclonal antibody-based diagnostic test for *Chlamydia trachomatis*.
1984	Chiron Corp. announced the first cloning and sequencing of the entire human immunodeficiency virus (HIV) genome.
	Stanford University received a product patent for prokaryote DNA.
	Charles Cantor and David Schwartz developed pulsed-field gel electrophoresis.
	Animal interferon's approved for protection against cattle disease.
1985	Genetically engineered plants resistant to insects, viruses, and bacteria were field tested for the first time.
	Axel Ullrich reported the sequencing of the human insulin receptor
	Plants can be patented.
1986	Orthoclone OKT3® (Muromonab-CD3) approved for reversal of acute kidney transplant rejection.
	The EPA approved the release of the first genetically engineered crop, gene-altered tobacco plants
1987	Genentech received FDA approval to market rt-PA (genetically engineered tissue plasminogen activator) to treat heart attacks.
	Recombivax-HB® (recombinant hepatitis B vaccine) approved.

1989	Epogen® (Epoetin alfa) a genetically engineered protein introduced, providing a means to help patients with kidney failure.
1990	GenPharm International, Inc. created the first transgenic dairy cow. The cow was used to produce human milk proteins for infant formula.
	The Human Genome Project, the international effort to map all of the genes in the human body, was launched.
	The first successful field trial of genetically engineered cotton plants was conducted by Calgene Inc. The plants had been engineered to withstand use of the herbicide Bromoxynil.
1993	Kary Mullis won the Nobel Prize in Chemistry for inventing the technology of polymerase chain reaction
1994	The first genetically engineered food product, the Flavr Savr tomato, gained FDA approval.
1996	Dolly the sheep is cloned.
1997	Researchers at Scotland's Roslin Institute report that they have cloned a sheep--named Dolly--from the cell of an adult ewe. Polly the first sheep cloned by nuclear transfer technology bearing a human gene appears later.
	A group of Oregon researchers claim to have cloned two Rhesus monkeys.
	Rituxan, the first antibody-based therapy for cancer (for patients with non-Hodgkin's lymphoma) was approved.
	A new DNA technique combines PCR, DNA chips, and computer programming providing a new tool in the search for disease-causing genes.
1998	Human embryonic stem cell lines are established.
	University of Hawaii scientists, clone three generations of mice from nuclei of adult ovarian cumulus cells.
2000	"Working draft" of the human genome's 3.15 billion letters is completed after a decade of research. •
2003	Broad Institute is founded in Cambridge to give scientists access to the human genome project, and to understand the molecular basis of disease.
2007	Craig C. Mello, a University of Massachusetts researcher, shares the Nobel Prize with Andrew Fire of Stanford University for discovering a special kind of RNA that can shut down individual genes.

In last fifteen years progress have been made by microbiologists and genetic engineers, and we are hopeful to solve many fold problems of the present day, specially energy and food crisis to cater the need of growing population of the world. Mineral ore deposits are also becoming more scarce and expensive to recover from earth's crust. Microorganisms can be used to enhance to recovery of metals from low-grade ores and from effluents containing undesirable quantities of heavy metals or other toxins. When these technologies are applied at industrial level, they constitute bio-industry (Table 22.2).

2. SCOPE OF BIOTECHNOLOGY

Genetic engineering in biotechnology stimulated hopes for both therapeutic proteins, drugs and biological organisms themselves, such as seeds, pesticides, engineered yeasts, and modified human cells for treating genetic diseases. The field of genetic engineering remains a heated topic of discussion in today's society with the advent of gene therapy, stem cell research, cloning, and genetically-modified food. Biotechnology is the applied science and has made advances in two major areas, viz., molecular biology and production of industrially important biochemical's. The

Table 22.2. Principal products of bio-industry.

Technology	Examples of Products
Fermentation technology	Antibiotics, vitamins, enzymes, amino acids, citric acid, lactic acid, malic acid, ethanol, acetone, butanol, biogas, biopesticides nucleotides, steroids, alkaloids, diagnostic reagents etc.
Enzymatic engineering	Isoglucose, glucose syrup etc.
Genetic engineering and Cell cultures	Single cell protein, clones, Interferon's, vaccines, blood products, monoclonal antibodies.

scientists are now diverting themselves toward biotechnological companies; this has caused the development of many biotechnological industries. In USA alone more than 225 companies have been established and successfully working, like Biogen, Cetus, Geneatech, Hybritech, etc. In world, USA, Japan, and many countries of Europe are leaders in biotechnological researchers encouraged by industrialists.

These companies are working for human welfare and opted following areas for research and development:

- Automated bioscreening for therapeutic agents.
- Bioprocessing alkenes to valuable oxides and glycols.
- Developing immobilized cell and enzyme systems for chemical process industries.
- Engineering of a series of organisms for specific industrial use.
- Genetical improvement of microorganisms for production of pharmaceutical products.
- Human gene therapy.
- Improved production of Vitamin B12.
- Large-scale production of fructose from inexpensive forms of glucose.
- Manufacturing ethanol by continuous fermentation.
- Microbiological based production of human insulin and interferon's.
- Microbiologically upgradation of hydrocarbons.
- Production and development of vaccine to prevent calibacillosis.
- Production of biopesticide and biofertilizers.
- Production of diagnostic kits for toxoplasmosis identification.
- Production of monoclonal antibodies for organ transplant tissue typing.
- Production of photosynthetically efficient plants.
- Production of transgenic plants and animals.
- Production of xanthan gum in oil fields for recovery of crude mineral oils.

The advances in recombinant DNA technology have occurred in parallel with the development of genetic processes and biological variations. The development of new technologies have resulted into production of large amount of biochemically-defined proteins of medical significance and created an enormous potential for pharmaceutical industries.

Biotechnology in itself is a vast subject and its scope is extended to various braches of biology. The details about these applications are given in various chapters in this book. This includes plant tissue culture, production of transgenics in animal and plants, applications in medicine as tools and therapeutics, creation of new enzymes and their immobilization for industrial use, development of monoclonal antibodies and control of pollutions, etc.

2.1. Industrial Applications of Biotechnology

The industrial application of molecular biotechnology is often subdivided, so that we speak of red, green, gray or white biotechnology. This distinction relates to the use of the technology in the medical field (in human and animal medicine), agriculture, the environment and industry. Some companies also apply knowledge deriving from molecular biotechnology in areas that cut across these distinctions (e.g., in red and green biotechnology, sequencing services). According to an investigation by Ernst and Young relating to the German biotech industry, 92% of companies are currently (2004) working in the field of red biotechnology, 13% in green, and 13% in gray or white biotechnology.

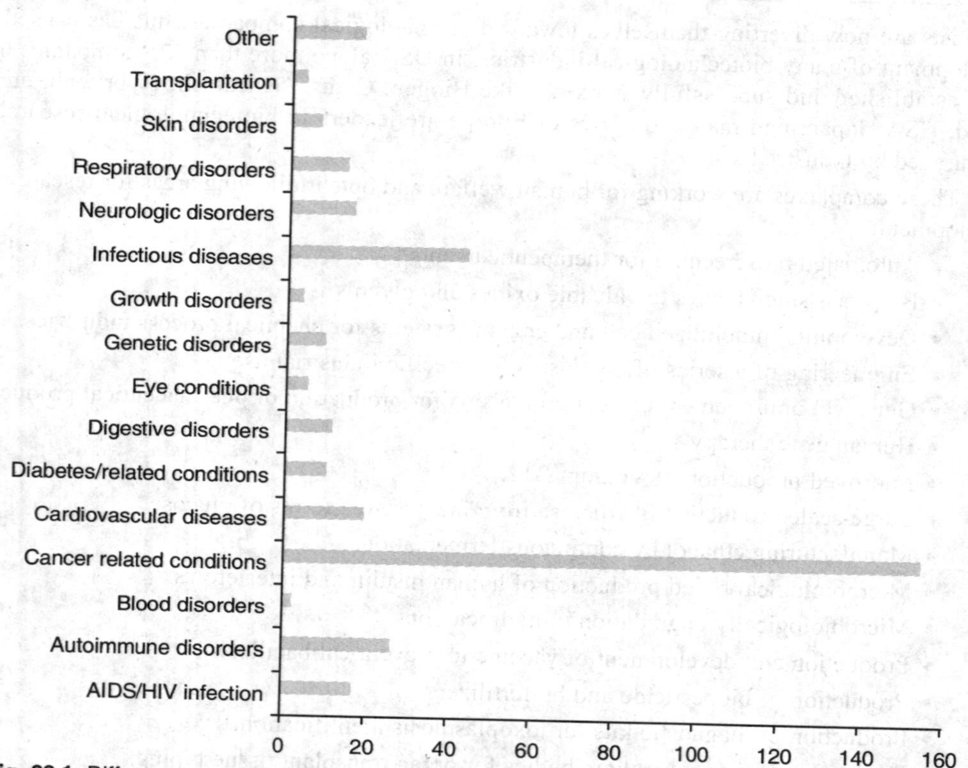

Fig. 22.1. Different areas of medicine in which efforts are being made with the help of biotechnology.

2.2. Biotechnology in Medicine

Biotechnology products for therapeutic use include a very diverse range of products, as outlined in Tables 22.4, 22.5. Some products are intended to mimic the human counterpart, whereas others are intended to differ from the human counterpart and may be analogues, chemically modified (e.g., pegylated) or novel products (e.g., single chain or fragment antibody products, gene transfer vectors, tissue-engineered products). Most of these products are regulated as medicinal products; however, the regulatory status of others such as some cell therapies and tissue: organ-based products differs globally and falls within the borderline between the practice of medicine, medical devices and medicinal products. Different areas of medicine in which biotechnology is used to develop diagnostic kits and cure are presented in the Figure 22.1.

Biotechnology-derived pharmaceuticals may be derived from a variety of expression systems such as *Escherichia coli*, yeast, mammalian, insect or plant cells, transgenic animals or other organisms. The expressed protein or gene may have the identical amino acid or nucleotide sequence as the human endogenous form, or may be intentionally different in sequence to confer some technical advantage such as an optimised pharmacokinetic or pharmacodynamic profile. The glycosylation pattern of protein products is likely to differ from the endogenous human form due to the different glycosylation preferences of the expression system used. Furthermore, intentional post-translation modifications or alterations may be made such as pegylation. It is important for the toxicologist to be aware of the nature of the product to be tested in terms of primary, secondary and tertiary structure, and any post-translational modifications such as glycosylation status, particularly as these may be altered if the manufacturing system is modified.

Table 22.4. Biotechnological products in medicine.

Class	Products
Hormones	Follicle stimulating hormone, growth hormone, insulin, insulin analogues
Growth factors	Platelet-derived growth factor, nerve growth factor, insulin growth factor-1
Cytokines	Interferones, interleukins, colony stimulating factor, erythropoietin
Vaccines	Conventional, recombinant protein antigen, modified bacteria or viruses
Nucleic acid based products	Gene therapy, DNA vaccines, ribozymes
Cell, tissue & organs	Autologous, xenoxenix
Others	Clotting factors, enzymes

2.3. Red Biotechnology

Within the field of red biotechnology, which deals with applications in human and animal medicine, there are various further distinctions that can be made: biopharmaceutical drug development, drug delivery cell and gene therapies, tissue engineering/regenerative medicine, pharmacogenomics (personalized medicine), system biology, and diagnosis using molecular medicine.

2.4. Biopharmaceutical Drug Development

In the field of biopharmaceutical drug development, it is the development of therapeutic human proteins by recombinant methods. (Table 22.5) for use as medicines that has the longest tradition. As mentioned above, recombinant human insulin was the first recombinant medicine in the world, produced by Genentech and brought to market in 1982. Today, recombinant human insulin has almost completely driven the other preparation of insulin (isolated from human or animal tissues) from the market.

Table 22.5. Selected examples of recombinant proteins with indication and manufacturer.

Drug	Product name	Indication	Manufacturer
Human insulin	Humulin	Diabetes mellitus type I	Eli Lilly
Somatotropin	Humatrope	Inadequate growth	Eli Lilly
Erythropoietin alpha	Erypo/Epogen	Anemia	Jansen-Cilag/Amgen
Factor VIII	Bioclate/Kogenate	Hemophilia	Centeon/Bayer
Interferon alpha 2 a	Roferon A	Cancer	Roche
Interferon beta 1 b	Betaferon	Mulitples sclerosis	Schering
Tissue plasminogen activator tPA (alteplase)	Actilyse	Thrombolytic agent	Boehringer Ingelheim

The first therapeutic antibodies, especially monoclonal antibodies, have been on the market since the late 1990s. In 2002, antibodies were (along with vaccines) the most important therapeutic class of drugs under development and there are also more recent market studies more than 100 antibodies or antibody fragments were at the clinical development stage in 2002 and research and development is being carried out on around 470 more in about 200 companies around the world (Table 22.6,7). Since the introduction of therapeutic antibodies onto the market, they have achieved significant turnovers, which are growing continually. The market for 2008 is estimated at a volume of US $16.7 billion (from Data-monitor, November 2003). Today, in addition to proteins, which currently play the most significant role in the biopharmaceutical field, new types of drugs based on RNA (antisense drugs, ribozymes, aptamers, Spiegelmers and RNA interference) are also being developed on the basis of advances in knowledge on molecular biotechnology.

Table 22.6. Selected examples of approved monoclonal antibodies.

Drug	Product name	Indication	Manufacturer
Abciximab	Reopro	Anticoagulant	Eli Lilly
Centocor Europe			
Trastuzumab (anti-HERA2-a)	Herceptin	Breast cancer	Roche
Adalimumab (anti-TNF-alpha)	Humira	Rheumatoid arthritis	Abbott
Infliximab (anti-TNF-alpha)	Remicade	Crohn's disease	Centocor
Alematuzumab (anti CD52)	Campth	Leukemia	Millennium &Ilex

Table 22.7. Selected examples of therapeutic RNAs on the market or under development.

Principle of action	Product name/production stage	Indication	Company
Antisense	Vitravene/market	CMV retinitis	ISIS pharmaceuticals
Antisense	Affinitak/phase II	Cancer	ISIS pharmaceuticals
Antisense	Alicaforsen/phase III	Crohn's disease	ISIS pharmaceuticals
Antisense	AP 12009/phase II	Brain tumor	Antisense pharma
Ribozyme	ANGIOZYME/phase II	Intestinal cancer	Sirna therapeutics

2.5. Drug Delivery

Closed linked to the development of therapeutic agents are the means of achieving their targeted delivery to their site of action. These drug delivery systems are mainly used for drugs whose physical and chemical characteristics make them insufficiently stable in reaching their site of action intact. They can also be used to transport drugs in a targeted way to particular sites of action (tissue specific targeting), or to overcome biological barriers such as the intestinal wall or the bloodbrain barrier.

2.6. Green Biotechnology

Green biotechnology is the application of biotechnology processes in agriculture and food production. The main dominant forces in green biotechnology today are agro giants with a world-wide area of operation such as BASF, Bayer CropScience, Monsanto and Syngenta. They are concentrating considerable attention on molecular plant biotechnology, which is seen as a future growth factor in agroindustry. The traditional pesticide market, on the other hand has been stagnating for years.

2.7. Transgenic Plants

The main emphasis in modern plant biotechnology is the production of transgenic plants. The first use of gene technology to bring about changes in plants became possible at the beginning of the 1980s, around ten years after the first experiment with bacteria. The market value of transgenic plants is estimated to be in excess of 2 billion euros, according to the calculation of the German Federal Office for the Environment. These figures relate to transgenic crop plants, which were being grown on an area totalling about 40 million hectares worldwide in 1999 and 2000.

2.8. Novel and Functional food

New types of foodstuffs with novel properties are often called functional food. Another category that is often mentioned in this context is nutraceuticals. These are foods that have a medicinal effect.

2.9. Livestock Breeding

Modern biotechnology is being employed commercially to introduce novel performance features in productive livestock. The transgenic specimens then display for example different wool characteristics for sheep, or improved milk characteristics in cattle.

2.10. Grey/White Biotechnology

The terms Grey and White Biotechnology have been coined for the application of biotechnological processes in environmental and industrial production contexts. The latter is primarily focused on the production of fine chemicals, in particular technical enzymes.

2.11. Technical Enzymes

Modern biotechnology already dominates the technical enzymes market. They can be found as proteases, lipases, cellulases and amylases for example in modern detergents, where the serve, amongst other purposes as protein and fat solubilizers.

2.12. Safety Concerns

There are a number of safety issues relating to biotechnology products that differ from those raised by low molecular weight products and need to be taken into account when designing the safety evaluation programme for a biotechnology derived pharmaceutical product. The quality and consistency of the product requires careful control in terms of product identity, potency and purity because of concerns about microbiological safety, impurities arising from the manufacturing process (e.g., host-cell contaminants, endotoxin, residual DNA levels and process chemicals), and the fidelity of the protein sequence and post-translational modifications during process improvements and scale-up. The immunogenic nature of heterologous proteins, vectors, cells, tissues and process contaminants must also be considered in the design of the safety evaluation programme and appropriate monitoring for anti-product antibodies, particularly neutralising antibodies included in toxicity studies to aid interpretation of the findings. For gene transfer products, there are concerns about the distribution and persistence of vector sequences, the potential for expression of vector sequences in non-target cells:tissues and, in particular, the potential for inadvertent gonadal distribution and germ-line integration. In 1997, the Food and Drug Administration (FDA) became aware that preclinical studies from multiple clinical trial applications indicated evidence of vector DNA in animal gonadal tissues following extragonadal administration. These positive polymerase chain reaction (PCR) signals were for DNA extracts from whole gonads subsequent to vector administration. The observations involved multiple classes of vectors, formulations and routes of administration.

QUESTIONS

1. Define biotechnology. Give landmarks which helped in the development of biotechnology.
2. Discuss in brief, application of biotechnology and safety issues involved.
3. Write short notes on:
   (a) Discovery of double helix.
   (b) DNA sequence method.
   (c) Use of computers in biotechnology.
   (d) Cloning of animals.
   (e) Human genome project.
   (f) Discovery of PCR.

CHAPTER 23

Tools of Molecular Biology

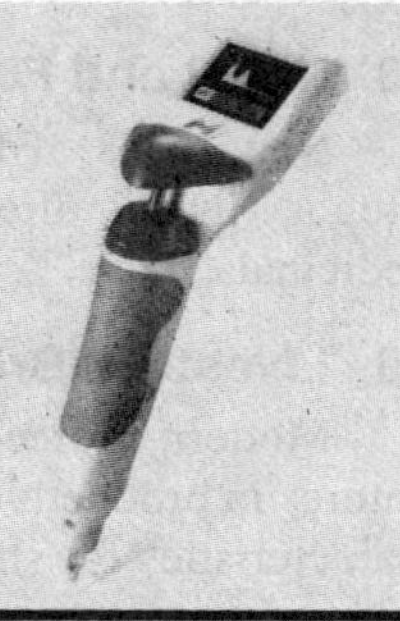

1. CELL FRACTIONATION

Organelles are membrane-enclosed vesicles inside all eukaryotic cells that function in a variety of important cellular processes. A method is described here for isolation of various subcellular fractions from rat liver by a technique termed differential centrifugation. These fractions can be used for assay for the presence of specific organelles using enzyme assays.

The arrangement of macromolecules within a cell is as important to cellular function as their catalytic activities. Cellular compartmentalization provides efficiency by bringing together related compounds that can interfere with each other (i.e., lysosomal hydrolytic enzymes). Cellular compartmentalization is accomplished in part by various subcellular organelles. The method utilizes differential centrifugation to isolate components of different densities. With this technique, the heaviest or most dense organelles, nuclei pellet in less time and with less force than is required to pellet lighter organelles such as mitochondria. First, a cell homogenate is made by rupturing the cell membranes in the tissue. The homogenate is then centrifuged for a short period of time to remove cell debris and nuclei. The supernatant is then transferred to another tube and centrifuged longer to pellet the lighter mitochondria.

The method of fractionation depends upon tissue type and organelle to be isolated. Homogeneous cell populations from cell culture are well suited for cell fractionation. Some tissues, like those in the liver also have one cell type that predominates, so are also well suited. Most chlorophyll-free plant tissues are acceptable for preparing mitochondria, but recently-harvested plant tissues are usually required. Once a cell type is chosen, it is important to obtain the organelles in a biochemically active, morphologically whole state (Table 23.1). Homogenizers are used to break open the cells without damaging the organelles. Homogenizers have a precise clearance between the glass tube and pestle, which breaks the cell membrane leaving the smaller organelle membranes intact. The homogenization buffer is a solution which often includes sucrose to partially dehydrate the organelles, keeping them intact.

No technique used to isolate organelles is perfect. It is very difficult to get pure unbroken preparations of any organelle. Techniques providing optimal isolation of one organelle may completely rupture another organelle. Thus methods are often used to measure the contamination of one organelle fraction by another. This can be done by analyzing each organelle fraction for organelle-specific marker enzymes.

Table 23.1. Relative density of different fractions.

Subcellular fraction	Relative density
Nuclei	100
Mitochondria	50
Lysosomes	30
Microsomes	25
Cytosol	20

1.1. Method of Tissue Homogenization

1. Transfer about one-half (~2 gm) of the tissue to a glass homogenizer on ice. Do not use a tight-fitting ground-glass homogenizer in this step since the organelle membranes may disrupt partially.
2. Add 18 ml of ice-cold homogenization buffer to the homogenizer. Note: the Calcium in the buffer is used here to stabilize the organelle membranes.
3. Homogenize 3-5 strokes until a homogeneous homogenate is made. Keep the homogenizer on ice. Caution: Excessive grinding or heating can damage and inactivate subcellular fractions.
4. Transfer the homogenate into a plastic 50 ml tube on ice.
5. Repeat steps 6-9 (combining all homogenates) until all the minced liver tissue has been homogenized (Fig. 23.1).

1.2. Isolation of Nuclei

6. Centrifuge the tube at 700 x g for 10 min at 0 ºC. The pellet contains the cell debris and nuclei. The supernatant contains all the lighter cellular fractions.
7. Without disturbing the crude nuclear pellet, carefully decant the supernatant into another tube on ice. Since the mitochondria are present in supernatant, this is used for the Isolation of the Mitochondria, while pellet can be used for the isolation of the nuclei.
8. To continue with the nuclear isolation: Nuclear Wash:
   a. Add 10 ml of ice-cold homogenization buffer to your crude nuclear pellet.
   b. Use a teflon homogenizer to resuspend the pellet in the buffer.
   c. Place two layers of cheesecloth in a funnel over a tube. Filter the 10 ml suspension through the cheesecloth.
   d. Rinse the debris in the filter with another 10 ml of homogenization buffer collecting the filtrate in the tube. Discard the filter with its cellular debris.
   e. Centrifuge the combined filtrates (20 ml) at 1500 x g for 10 min at 0 ºC. The pellet contains the washed nuclear fraction. Without disturbing the nuclear pellet, discard the supernatant.
9. Add 5.0 ml of ice-cold homogenization buffer to the nuclear pellet.
10. Use a teflon homogenizer to resuspend the pellet in the buffer.
11. Using a pipette, aliquot the nuclear suspension into two glass test tubes.
12. Label the two glass tubes "Nuclei", and store them at –20 °C. Caution: Glass tubes should not be greater than half full, or they may crack upon freezing.

1.3. Isolation of Mitochondria

13. Obtain the supernatant from step 12 in a tube on ice.
14. Centrifuge at 5,000 x g for 15 min at 0 ºC. The pellet contains the crude mitochondrial fraction. The supernatant contains microsomes, membrane fragments, ribosomes, cytoplasmic enzymes, etc.
15. Without disturbing the mitochondrial pellet, carefully decant the post-mitochondrial supernatant (and the pink partially sedimenting layer) into a graduated cylinder.
16. Pour the supernatant into a clean tube on ice.
17. The supernatant can be used for the "Isolation of the Microsomes".
18. To continue with mitochondrial isolation: Mitochondrial Wash:
    a. Add 20 ml of ice-cold homogenization buffer to the crude mitochondrial pellet.
    b. Using a teflon homogenizer, resuspend the pellet in the buffer.

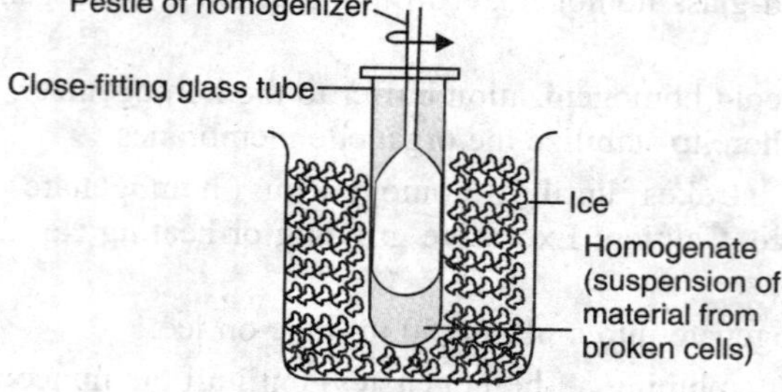

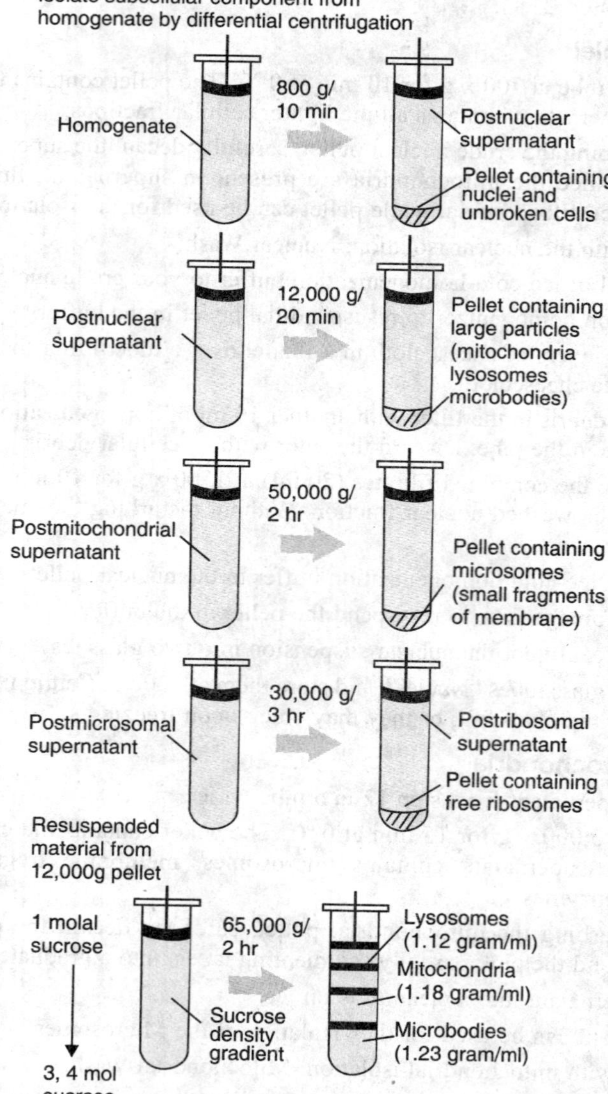

Fig. 23.1. Cell fractionation by centrifugation.

c. Pour the suspension back into the tube, add another 20 ml of buffer, mix, and centrifuge at 234,000 x g for 10 min at 0 °C. The pellet contains the washed mitochondria.

d. Without disturbing the mitochondrial pellet, decant and discard the supernatant.

19. Add 5.0 ml of ice-cold homogenization buffer to the mitochondrial pellet.

20. Using a teflon homogenizer, resuspend the pellet in the buffer. This is how using differential centrifugation, various organelles can be isolated and studied.

2. FLUORESCENT *IN SITU* HYBRIDIZATION (FISH)

FISH (Fluorescent *in situ* hybridization) is a cytogenetic technique which can be used to detect and localize the presence or absence of specific DNA sequences on chromosomes. It uses fluorescent probes which bind only to those parts of the chromosome with which they show a high degree of sequence similarity. FISH has evolved into a core technique spanning many areas of diagnosis and research, including cancer cytogenetics, prenatal screening of chromosomal aberrations, molecular pathology and developmental molecular biology. The general principle of FISH, hybridizing fluorescently-labelled loci-specific, telomeric or centromeric DNA probes to chromosomes, can be applied to a range of specimens such as metaphase chromosome spreads or to interphase nuclei and cells. Fluorescence microscopy can be used to find out where the fluorescent probe bound to the chromosome. FISH is often used for finding specific features in DNA. These features can be used in genetic counselling, medicine, and species identification.

First, a probe is constructed. The probe has to be long enough to hybridize specifically to its target (and not to similar sequences in the genome), but not too large to impede the hybridization process, and it should be tagged directly with fluorophores, with targets for antibodies or with biotin. This can be done in various ways, for example nick translation and PCR using tagged nucleotides.

Then, an interphase or metaphase chromosome preparation is produced. The chromosomes are firmly attached to a substrate, usually glass slides. Repetitive DNA sequences must be blocked by adding short fragments of DNA to the sample. The probe is then applied to the chromosome DNA and incubated for ~12 hours while hybridizing. Several wash steps remove all unhybridized or partially hybridized probes. The results are then visualized and quantified using a microscope that is capable of exciting the dye and recording images.

If the fluorescent signal is weak, amplification of the signal may be necessary in order to exceed the detection threshold of the microscope. The signal strength depends on many factors; probe labelling efficiency, the type of probe and the type of dye affect the fluorescent signal. Fluorescently-tagged antibodies or streptavidin are bound to the dye molecule. These secondary components are selected so that they have a strong signal.

Fibre FISH-An alternative to interphase or metaphase preparations, fibre FISH, interphase chromosomes are attached to a slide in such a way that they are stretched out in a straight line, rather than being tightly coiled, as in conventional FISH, or adopting a random conformation, as in interphase FISH. This is accomplished by applying mechanical shear along the length of the slide; either to cells which have been fixed to the slide and then lysed, or to a solution of purified DNA. A technique known as chromosome combing is increasingly used for this purpose. The extended conformation of the chromosomes allows dramatically higher resolution - even down to a few kilobases. The preparation of fibre FISH samples, although conceptually simple, is a rather skilled art, and only specialized laboratories use the technique routinely.

The incentive for multiple labelling of chromosomes and subsequent detection in multiple colors has given rise to Multi-color FISH (M-FISH), fuelled by the desire to maximize the efficiency of the FISH process when applied to situations such as the study of the complex translocations often found in cancer cytogenetics. It is possible to detect multiple targets with fewer fluorophores (e.g., 3 fluors for 7 targets) based on labelling the probes combinatorially, with more than one fluor in various ratios. To simultaneously discriminate all 24 human chromosomes, 5 fluorochromes are required, each fluorescing in a distinct spectral region.

Complex "acquired" chromosome abnormailities are common in cancer cytogenetics, ranging from amplification of chromosome number to various defined translocations. One technique which is particularly suited to study of tumour DNA is Comparative Genomic Hybridisation (CGH), a development of chromosome painting, which detects a deviation of the ratio of two fluorochromes to indicate amplification or deletion along the chromosome strand. FISH can also be instrumental for gene mapping. Genes and DNA segments are continually being assigned to chromosome bands, but it is often necessary to determine the precise arrangement of each sequence relative to another at a more detailed level.

Genomic *in situ* hybridization (GISH) is becoming an important technique to characterize parental genomes in wide hybrids and to detect introgressed segments and chromosome rearrangements. Total genomic DNA as a probe in *in situ* hybridization has been used to detect parental chromosomes in wheat-rye hybrids and in discriminating closely related Triliceae species. *In situ* **hybridization** using genomic probes has been demonstrated to be an effective technique for the detection of meiotic pairing between wheat-rye. However, GISH has not been used to characterize meiotic pairing in *Oryza*.

3. NUCLEIC ACID PURIFICATION, YIELD ANALYSIS OF PURIFIED DNA

3.1. Purification Genomic and Plasmid DNA

Genomic and plasmid DNA can be isolated from cells by disrupting cells and then exploiting the differences in properties between DNA, protein and other constituents.

There is no universal method for purification of genomic DNA. The method for purification of DNA depends on the nature of the source material. DNA from plants is more difficult to obtain in clonable form than from animals due to presence of secondary metabolites and polysaccharides and the extent of interfering substances present in that source. The first reported purification of chromosomal DNA was in 1944 by Avery and co-workers from the "S" cells of *Diplococcus pneumoniae* Type II. These cells contained a very large amount of capsular polysaccharide and the method was developed to purify the DNA and remove the contaminant. From about fifteen years Avery's method with some modification continued to be used for DNA solution. It was gradually realized that there was need to improve upon the yield and also the size of DNA. The deoxynucleases that were released on cell lysis degraded a large part of DNA, Marmur introduced the use of alkaline buffer (0.1 M Tris, pH 9.0) and sodium dodecyl sulphate. Both the high pH and sodium dodecyl sulphate were found to inactivate the nucleases. In addition, lysis operation to be performed at low temperature (0-2 °C) to slow down the enzymes and the subsequent extraction with chilled phenol saturated with the pH 9.0 buffer, which denatured the proteins bound to DNA and consequently increased the yield of DNA. The DNA was precipitated from the aqueous layer with ethanol as was done by Avery. No protein could be detected by the Lowry test in the DNA sample, but some RNA was present. All methods subsequently developed for isolation of chromosomal DNA from bacteria, viruses and animal cells in culture have been largely derived from the method of Marmur. EDTA has been introduced in the lysis medium to inactivate nucleases which are known to be dependent on Mg^{++}. In some methods phenol has been replaced by a mixture of equal volumes of phenol and chloroform. Since phenol treatment is thought to shear large DNA molecules, methods have been developed avoiding phenol and such methods probably yield better quality DNA at the expense of quanity.

When DNA is desired to be isolated from animal or plant tissue, the above mentioned steps have to be preceded by procedure to obtain cell powder from the ground tissue. The tissue is minced finely and then disintegrated thoroughly in a Warring blender with liquid nitrogen. This step is sometimes performed in a mortar and pestle. The powder left behind after evaporation of the liquid nitrogen is extracted with the alkaline buffer (pH 8.0 to 9.0) containing EDTA and the detergent. Subsequently the mixture is treated with protease and ribonuclease followed by phenolisation. A problem that is quite often encountered with plant tissue is the presence of a large amount of polysaccharides. Plant DNA free from polysaccharides is obtained by cesium chloride ethidium bromide equilibrium centrifugation as described under purification of plasmid DNA.

Different types of isolation procedures

1. SDS-phenol extraction

This is the simplest and fastest method for isolation of DNA and is widely used with filamentous fungi and plant material. The technique uses SDS with phenol to denature and dissolve macerated hyphae leaving the DNA intact. DNA is then precipitated from solution with a precise volume of isopropanol.

2. SDS-proteinase K treatment

This technique is applicable to most materials and relies on proteinase K and SDS to dissolve the sample and digest the protein component without affecting the DNA. The sample is then extracted with protein denaturants (phenol, phenol-choloroform, chloroform) and the aqueous liquid containing the DNA is dialyzed and precipitated with ethanol or isopropanol in sodium or ammonium acetate solution. RNAse may or may not be used depending on different protocols used.

3. CTAB treatment

Isolation of DNA from plants is preferably achieved by treatment with CTAB. This detergent carries a positive charge which interacts with the negative charge of DNA in high salt concentration and forms a soluble complex. In subsequent steps a decrease in the salt concentration causes precipitation of DNA, leaving other compounds, especially polysaccharides, in solution. The preparation of DNA from animal tissue which is in mucopolysaccharides could also be done by this technique which yields good quality material.

3.2. Plasmid DNA Isolation

Plasmids are extra chromosomal double stranded circular DNA molecules present in bacteria and lower eukaryotes which exist in super helical state *in vivo*, the size of plasmid being in the range between 1 kb and more than 300 kb. Chromosomal DNA is much larger and though could also be circular is made linear by shear pressure during cell lysis and the DNA isolation procedure. These long linear molecules invariably get entangled in the debris of lysed cells.

The early procedures of plasmid isolation were developed in the late 1960s in the laboratory of Donald R Heiliski in the University of California at San Diego. Cell lysis was achieved by using EDTA, lysozyme and mild non ionic detergent like Triton X-100 at pH 8.0 and avoiding vigorous shaking. A centrifugal step was then incorporated to preferentially sediment the chromosomal DNA molecules entangled in the cellular debris. The supernatant solution contained the plasmids, which was treated successively with ribonuclease and proteinase. Treatment with phenol, extraction with chloroform - isoamyl alcohol and precipitation of the plasmid DNA with chilled ethanol followed.

In another procedure the yield was much better increased by quickly bringing the lysozyme treated cell suspension to boil, holding at 100 °C for 40 second and then immediately immersing in ice cold water. The chromosomal DNA was removed as usual by centrifugation. It may be noted here that though the hydrogen bonds between the two strands of duplex DNA are disrupted by alkali or bonds are reformed instantly on restoration of favorable pH or temperature, because of there closed circular nature. Preparations by both these methods, however, contained higher proportion of contaminating linear fragments of chromosomal DNA.

3.3. Purification of Plasmid DNA

3.3.1. Isopycnic Ultracentrifugation

Plasmid DNA can be purified by equilibrium centrifugation in a Cesium chloride (CsCl) ethidium bromide gradient. Dyes like ethidium bromide and propidium bromide binds to duplex DNA by intercalating between the two strands. This intercalation is much more in case of linear duplex DNA, apparently because of easy access through the two open ends and least with covalent, closed circular supercoiled plasmid DNA. The bound dye caused a greater reduction in the apparent buoyant density of the linear DNA compared to that of the circular DNA and hence the plasmids could be separated from the linear chromosomal DNA by equilibrium centrifugation the CsCl density gradient, (Fig. 23.2) shows the relative positions of proteins, linear DNA or nicked plasmid DNA, supercoiled plasmid DNA and single stranded RNA. The plasmid DNA was recovered by puncturing the centrifuge tube with a hypodermic needle (not too fine) just under the plasmid band which could be visualized by low wave length ultraviolet light. The principle behind centrifugal purification has been described.

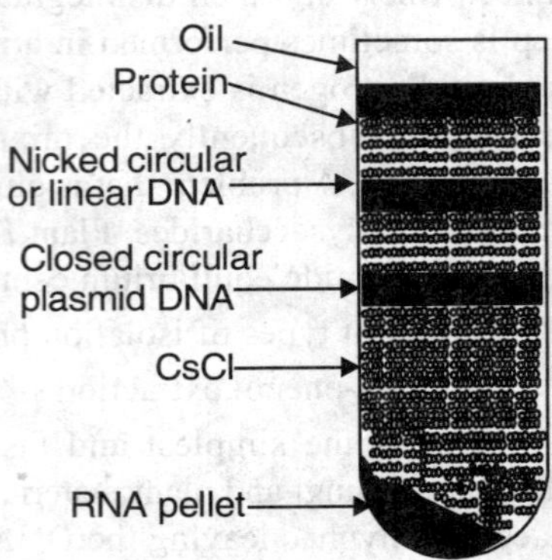

Fig. 23.2. CsCl- Ethidium bromide gradient centrifugation showing plasmid DNA in the cell lysate. DNA molecules are separated on the basis of their relative size and density.

There are many genetic engineering operations in which very pure plasmid DNA is not required. Various small columns are commercially available which can be effectively used to obtain a considerably pure plasmid DNA preparation from a crude preparation even from a cell lysate. These columns contain among other materials, silica based resin or membrane anion exchange systems, paramagnetic particles or just plain glass powder.

3.3.2. Adsorption chromatography

In chromatography, Hydroxyapatite is used which binds to the proteins and nucleic acids at low phosphate concentration. If the phosphate concentration is stepwise raised, proteins and RNA are eluted first followed by small RNA molecules i.e. plasmid DNA and finally high molecular weight i.e., chromosomal DNA at the highest phosphate concentration. When clear lysate is loaded onto hydroxyapatite column, separation occurs based on above-mentioned principle.

3.3.3. Preparative electrophoresis

Plasmid molecule especially in their complete superhelical CCC form can be easily separated from RNA and large chromosomal DNA by electrophoresis in agrose gel. It can yield DNA in mg quantity and can also give an idea of the molecular weight of the DNA by comparing it with the molecular weight marker.

3.4. Isolation of RNA

Cells contain ribonucleases and hence inactivating this enzyme during cell lysis is crucial. Agents that effectively inactivate ribonucleases are therefore, added during cell lysis. The ionic detergents sodium dodecyl sulphate and sodium lauryl sarkosinate are known to inhibit ribonucleases and are frequently used. Reducing agents like β-merceptoethanol interferes with the tertiary structure of ribonuclease and could be effectively used for inactivating the enzyme.

Once the cell lysate is obtained, RNA is isolated through steps similar to that of DNA isolation. DNA in the cell lysate is sheared by repeated ejection through a fine hypodermic needle, proteolytic enzymes added and then the RNA is extracted with phenol or a phenol chloroform mixture. The next critical step here is to bring the pH down to pH 4.7-5.2 and also to saturate the phenol to be used for extraction with acetate buffer pH 4.7-5.2 instead of Tris buffer pH 8.0. This drives the DNA to the interphase of phenol and water and leaves the RNA in the water phase. The RNA is precipitated as usual from the aqueous phase by ethanol or isopropanol.

Ribosomal RNA constitutes 80-85% of total cellular RNA, transfer RNA and small nuclear RNA (in case of eukaryotes) constitute about 10-15% while messenger RNA constitutes are about 1-5%. For the most genetic engineering experiments it is the messenger RNA that is important.

3.5. Purification of Messenger RNA

Bacterial messenger RNA can be separated from ribosomal and transfer RNAs by velocity sucrose density gradient centrifugation or gel electrophoresis. Method for purification of eukaryotic mRNA was developed by Philip Leder in 1972 by taking advantage of the fact that mRNA most often has a polytail. The polyA RNA could be separated from other species of RNA by affinity chromatography though a column of oligo (dT) cellulose. A polyU sepharose column has also been used by some.

3.6. Analysis of Yield of Nuclei Acid

3.6.1. Analysis of Yield DNA

The analysis of yield of purified DNA is best performed by the diphenylamine method. The intensity of the blue color formed at acidic pH is determined by monitoring absorption at 595 nm. This assay is specific and can be used to assay DNA even in a crude DNA preparation. A quick but less sensitive and much less specific method that is quite often used is to determine absorption of the nucleic acid itself at 260 nm. This assay is based on the fact that the purine and pyrimidine base present in nucleic acids absorb maximally around 260 nm. This assay would obviously be positive for both DNA and RNA, free nucleotides and nucleosides and even purine and pyrimidine based, but in a purified DNA or RNA sample the assay is reliable. Another quick but very approximate assay of double stranded DNA is by visual estimation of the ultraviolet fluorescence of DNA in a droplet elicited by ethidium bromide.

3.6.2. Analysis of Yield RNA

It is obvious from the preceding subsection that yield of RNA can also be determined by monitoring absorption at 260 nm as long as it is pure sample of RNA. A more specific and sensitive method is based on the monitoring of the color developed by reaction with orcinol. The RNA sample is boiled with freshly recrystallized orcinol in the presence of FeCl and HCl. The green color, developed is monitored at 655 nm. Ribose linked only to purines contribute to the color. Specificity of this reaction is not as good as the diphenyl amine reaction for DNA 2-Deoxyribose DNA methyl pentose, hexuronic acid and aldoheptoses also produce some green colour giving absorbance in that region.

4. BLOTTING TECHNIQUES

The term "blotting" refers to the transfer of biological samples from a gel to a membrane and their subsequent detection on the surface of the membrane.

4.1. Southern Blotting

Southern blotting uses DNA molecule as marker (developed by E.M. Southern). Restriction enzyme digestion of genomic DNA results in so many fragments that a stained electrophoretic gel shows a smear of DNA. A probe can identify one fragment in this mixture, with the use of the technique called Southern Blotting. After DNA segments are fractionated on the gel, one absorbent membrane is laid on the gel and DNA bands are transferred (blotted) on to the membrane by capillary action. These DNA bands remain in the same relative position on the membrane. The membrane is emerged in a labeled probe (a defined radiolabeled nucleic acid segment that is used to identify a specific DNA molecules bearing the complimentary sequence, produced by PCR or traditional cloning method). An autoradiogram is used to demonstrate the presence of any bands on the gel that are homologous to the probe (Fig. 23.3). The cloned DNA finds widespread application as a probe for detecting either a specific clone or a specific DNA/RNA fragment. In all these cases the ability of nucleic acids to find and bind to complementary sequences in solution is being exploited.

4.2. Northern Blotting

Northern blotting uses mRNA as marker. The southern blotting can be extended to detect a specific RNA molecule from a mixture of RNAs fractioned on a gel. This technique is called Northern Blotting to contrast it with the Southern's technique for DNA analysis. The fractioned RNA is blotted on to a membrane and probed in the same way as for Southern blotting. One application of Northern analysis is to determine whether specific gene is transcribed in certain tissue or under certain environmental conditions. RNAse is extracted from appropriate cell sample, then electrophoresed, blotted and probed with the cloned gene in question. A positive signal shows presence of the transcript (Fig. 23.3).

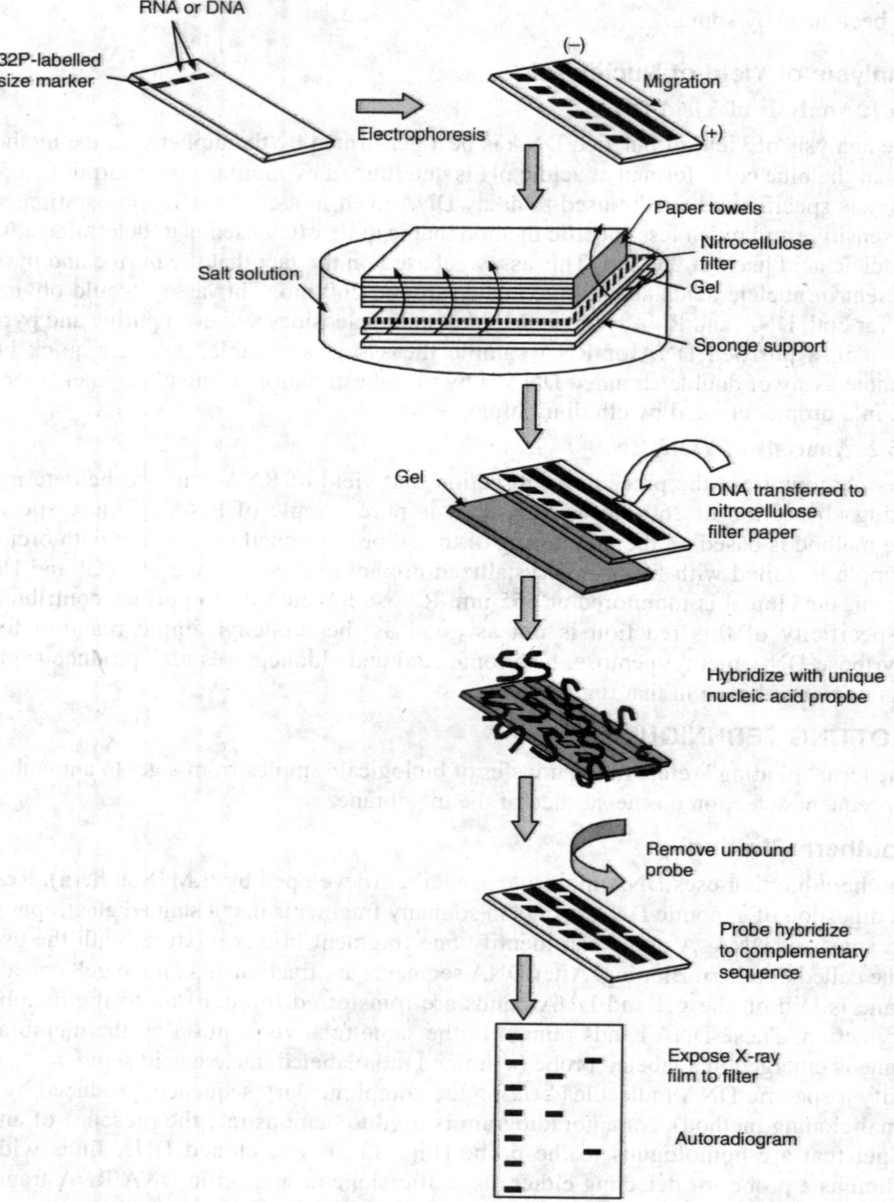

Fig. 23.3. Blotting technique (DNA/RNA is used termed as Southern/Northern technique).

4.3. Western Blotting

Western blotting uses protein as marker compound. Western blotting (also called immunoblotting because an antibody is used to specifically detect its antigen) was introduced by Towbin et al. in 1979 and is now a routine technique for protein analysis. The specificity of the antibody-antigen interaction enables a single protein to be identified in the midst of a complex protein mixture. Western blotting is commonly used to positively identify a specific protein in a complex mixture and to obtain qualitative and semiquantitative data about that protein.

The first step in a Western blotting procedure is to separate the macromolecules using gel electrophoresis. Following electrophoresis, the separated molecules are transferred or blotted onto a second matrix, generally a nitrocellulose or polyvinylidene fluoride (PVDF) membrane. Next, the membrane is blocked to prevent any nonspecific binding of antibodies to the surface of the membrane. The transferred protein is complexed with an enzyme-labelled antibody as a probe. An appropriate substrate is then added to the enzyme and together they produce a detectable product such as a chromogenic or fluorogenic precipitate on the membrane for colorimetric or fluorimetric detection, respectively. The most sensitive detection methods use a chemiluminescent substrate that, when combined with the enzyme, produces light as a byproduct. The light output can be captured using film, a CCD camera or a phosphorimager that is designed for chemiluminescent detection. Whatever substrate is used, the intensity of the signal should correlate with the abundance of the antigen on the blotting membrane.

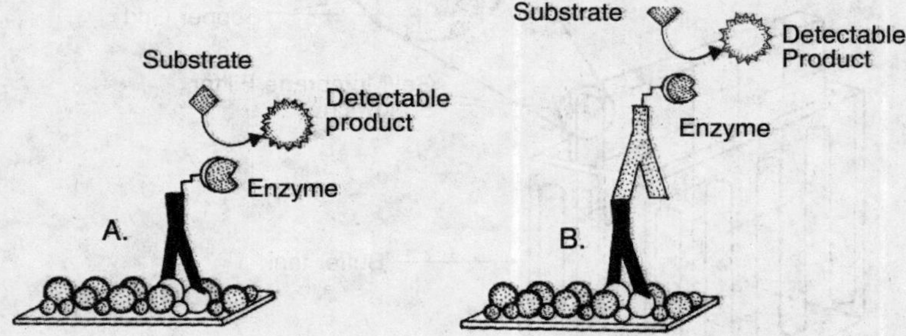

Fig. 23.4. A. In the direct detection method, labeled primary antibody binds to antigen on the membrane and reacts with substrate, creating a detectable signal. B. In the indirect detection method, unlabeled primary antibody binds to the antigen. Then, a labeled secondary antibody binds to the primary antibody and reacts with the substrate.

Detailed procedures for detection of a Western blot vary widely. One common variation involves direct vs. indirect detection as shown in Figure 23.4. With the direct detection method, the primary antibody that is used to detect an antigen on the blot is also labelled with an enzyme or fluorescent dye.

In the indirect detection method, a primary antibody is added first to bind to the antigen. This is followed by a labeled secondary antibody that is directed against the primary antibody. Labels include biotin, fluorescent probes such as fluorescein or rhodamine, and enzyme conjugates such as horseradish peroxidase or alkaline phosphatase. The indirect method offers many advantages over the direct method.

4.4. Protein Transfer to a Membrane

Following electrophoresis, the protein must be transferred from the electrophoresis gel to a membrane. There are a variety of methods that have been used for this process, including diffusion transfer, capillary transfer, heat-accelerated convectional transfer, vacuum blotting transfer and electroelution. The transfer method that is used most commonly used for proteins is electroelution or electrophoretic transfer, because of its speed and transfer efficiency. This method uses the electrophoretic mobility of proteins to transfer them from the gel to the matrix. Electrophoretic transfer of proteins involves placing a protein-containing polyacrylamide gel in direct contact with a piece of nitrocellulose or other suitable, protein-binding support and "sandwiching" this between two electrodes submerged in a conducting solution (Fig. 23.5). When an electric field is applied, the proteins move out of the polyacrylamide gel and onto the surface of the membrane, where the proteins become tightly attached. The resulting membrane is a copy of the protein pattern that was found in the polyacrylamide gel itself.

Transfer efficiency can vary dramatically among proteins, based upon the ability of a protein to migrate out of the gel and its propensity to bind to the membrane under a particular set of conditions. The efficiency of transfer depends on factors such as the composition of the gel, complete contact of the gel with the membrane, the position of the electrodes, the transfer time, size and composition of proteins, field strength and the presence of detergents. Optimal transfer of proteins is generally obtained in low ionic strength buffers and with low electrical current.

The most commonly used membranes for Western blotting include nitrocellulose, polyvinylidene difluoride (PVDF) and nylon membranes.

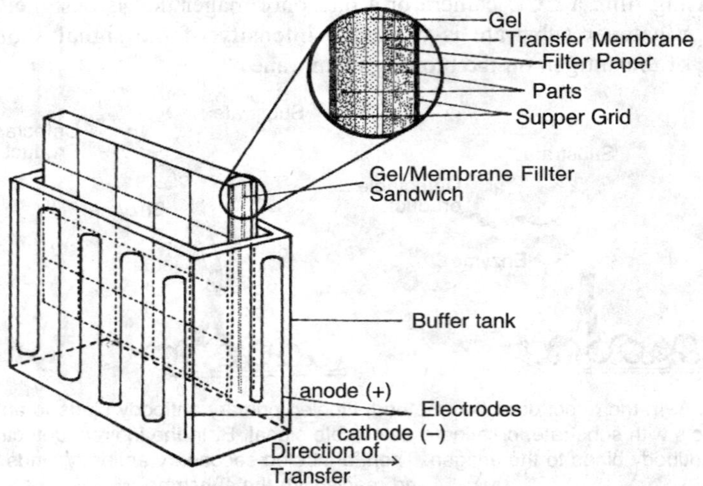

Fig. 23.5. Electrophoretic transfer.

After transfer and before proceeding with the Western blot, it is often desirable to stain all proteins on the membrane with a reversible stain to check the transfer efficiency. Although the gel may be stained to determine that protein left the gel, this does not ensure efficient binding of protein on the membrane. Ponceau S stain is the most widely used reagent for staining proteins on a membrane. However, it has limited sensitivity, does not photograph well, and fades with time. MemCode Reversible Stain is a superior alternative for staining protein on nitrocellulose or PVDF membranes. MemCode Stain detects low nanogram levels of protein, is easily photographed, does not fade with time and takes less than 30 minutes to stain, photograph and erase.

4.5. Blocking Nonspecific Binding Sites

In a Western blot, it is important to block the unreacted sites on the membrane to reduce the amount of nonspecific binding of proteins during subsequent steps in the assay. A variety of blocking buffers ranging from milk or normal serum to highly purified proteins have been used to block unreacted sites on a membrane. The blocking buffer should improve the sensitivity of the assay by reducing background interference. Individual blocking buffers are not compatible with every system. For this reason, a variety of blockers in both Tris buffered saline (TBS) and phosphate buffered saline (PBS) are available. The proper choice of blocker for a given blot depends on the antigen itself and on the type of enzyme conjugate to be used. For example, with applications using an alkaline phosphatase conjugate, a blocking buffer in TBS should be selected, because PBS interferes with alkaline phosphatase. The ideal blocking buffer will bind to all potential sites of nonspecific interaction, eliminating background altogether without altering or obscuring the epitope for antibody binding.

A complete line of blocking buffers for Western blotting includes BLOTTO, Casein, BSA, SEA BLOCK and the exclusive SuperBlock Blocking Buffers. A comparison of three blotting techniques is given in the Figure 23.6.

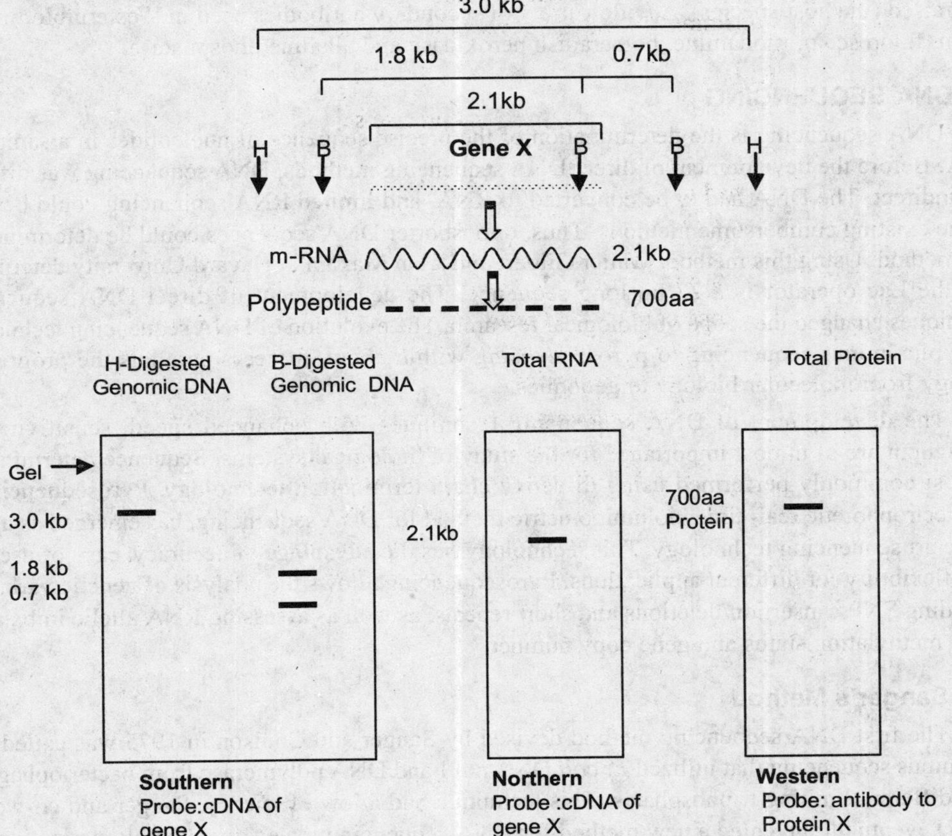

Fig. 23.6. Comparison of Southern, Northern and Western analyses of gene X. B and H are target sites for restriction enzyme B and H. Positions of bands on Southern blot cannot be compared with positions of bands on Northern and Western blots.

4.6. Washing

Like other immunoassay procedures, Western blotting consists of a series of incubations with different immunochemical reagents separated by wash steps. Washing steps are necessary to remove unbound reagents and reduce background, thereby increasing the signal:noise ratio. Insufficient washing will allow high background, while excessive washing may result in decreased sensitivity caused by elution of the antibody and/or antigen from the blot. As with other steps in performing a Western blot, a variety of buffers may be used. Occasionally, washing is performed in a physiological buffer such as Tris buffered saline (TBS) or phosphate buffered saline (PBS) without any additives. More commonly, a detergent such as 0.05% Tween 20 is added to the buffer to help remove nonspecifically bound material.

4.7. Primary and Secondary Antibodies

The choice of a primary antibody for a Western blot will depend on the antigen to be detected and what antibodies are available to that antigen. A huge number of primary antibodies are available commercially and can be identified quickly by searching sites. Both polyclonal and monoclonal antibodies work well for Western blotting. Polyclonal antibodies are less expensive and less time-consuming to produce and they often have a high affinity for the antigen. Monoclonal antibodies are valued for their specificity, purity and consistency that result in lower background. A wide variety of labelled, secondary antibodies can be used for Western blot detection. The choice of secondary antibody depends upon the species of animal in which the primary antibody was raised (the host species). A wide variety of secondary antibodies used in Western blotting are biotin, fluorescein, rhodamine, horseradish peroxidase and alkaline phosphatase.

5. DNA SEQUENCING

DNA sequencing is the determination of the precise sequence of nucleotides in a sample of DNA. Before the development of direct DNA sequencing methods, DNA sequencing was difficult and indirect. The DNA had to be converted to RNA, and limited RNA sequencing could be done by the existing cumbersome methods. Thus, only shorter DNA sequences could be determined by this method. Using this method, Walter Gilbert and Alan Maxam at Havard University determined that the Lac operator is a 27 bp long sequence. The development of direct DNA sequencing techniques changed the scope of biological research. The evolution of DNA sequencing technology from plus-minus sequencing to pyrosequencing within about 20 years parallels the progress in biology from molecular biology to genomics.

The development of DNA sequencing techniques with enhanced speed, sensitivity and throughput are of utmost importance for the study of biological systems. Sequence determination is most commonly performed using di-deoxy chain termination technology. Pyrosequencing, a nonelectrophoretic real- time bioluminometric method for DNA sequencing, has emerged as a state of the art sequencing technology. This technology has the advantage of accuracy, easy of use, and high flexibility for different applications. Pyrosequencing allows the analysis of genetic variations including SNPs, insertion/deletions and short repeats, as well as assessing RNA allelic imbalance, DNA methylation status and gene copy number.

5.1. Sanger's Method

The first DNA sequencing method devised by Sanger and Coulson in 1975 was called plus and minus sequencing that utilized *E. coli* DNA pol I and DNA polymerase from bacteriophage T4 with different limiting triphosphates. This technique had a low efficiency. Sanger and co-worker (1977) eventually invented a new method for DNA sequencing via enzymatic polymerization that basically revolutionized DNA sequencing technology. The most popular method for doing this is called the **dideoxy method** or Sanger method (named after its inventor, Frederick Sanger, who was awarded the 1980 Nobel prize in chemistry [his second] for this achievement). Finding a single gene amid the vast stretches of DNA that make up the human genome - three billion base-pairs' worth - requires a set of powerful tools. These tools include genetic maps, physical maps and DNA sequence - which is a detailed description of the order of the chemical building blocks, or bases, in a given stretch of DNA.

Scientists need to know the sequence of bases because it tells them the kind of genetic information that is carried in a particular segment of DNA. For example, they can use sequence information to determine which stretches of DNA contain genes, as well as to analyze those genes for changes in sequence, called mutations, that may cause disease.

The first methods for sequencing DNA were developed in the mid-1970s. At that time, scientists could sequence only a few base pairs per year, not nearly enough to sequence a single gene, much less the entire human genome. By the time the HGP began in 1990, only a few laboratories had managed to sequence a mere 100,000 bases, and the cost of sequencing remained very high. Since then, technological improvements and automation have increased speed and lowered cost to the point where individual genes can be sequenced routinely, and some labs can sequence well over 100 million bases per year.

DNA is synthesized from four deoxynucleotide triphosphates. The top formula shows one of them: deoxythymidine triphosphate (dTTP) (Fig. 23.7). Each new nucleotide is added to the 3′ - OH group of the last nucleotide added.

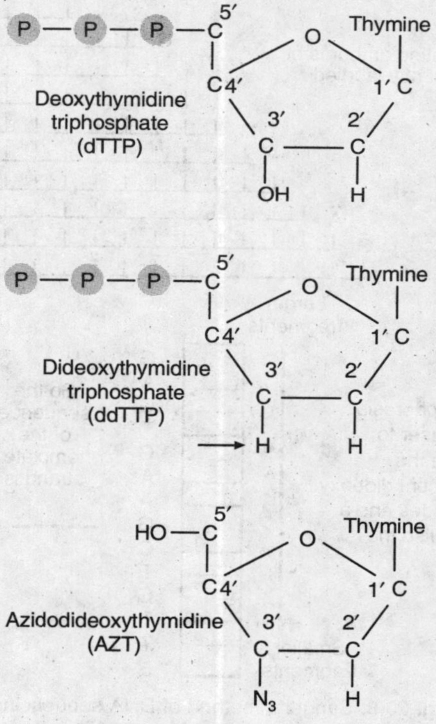

Fig. 23.7. Structure of dideoxynucleotides.

The dideoxy method gets its name from the critical role played by **synthetic** nucleotides that **lack** the –OH at the 3′ carbon atom. A dideoxynucleotide (dideoxythymidine triphosphate - ddTTP as shown here) can be **added** to the growing DNA strand. When it is added it **stops chain elongation** because there is no 3′ -OH for the next nucleotide to be attached. For this reason, the dideoxy method is also called the **chain termination method**.

The bottom formula shows the structure of azidothymidine (AZT), a drug used to treat AIDS. AZT (which is also called **zidovudine**) is taken up by cells where it is converted into the triphosphate. The reverse transcriptase of the human immunodeficiency virus (HIV) prefers AZT triphosphate to the normal nucleotide (dTTP). Because AZT has no 3′ -OH group, DNA synthesis by reverse transcriptase halts when AZT triphosphate is incorporated in the growing DNA strand. Fortunately, the DNA polymerases of the host cell prefer dTTP, so side effects from the drug are not so severe as might have been predicted.

5.1. The Procedure

The DNA to be sequenced is prepared as a single strand (Fig. 23.8). This template DNA is mixed with the following-

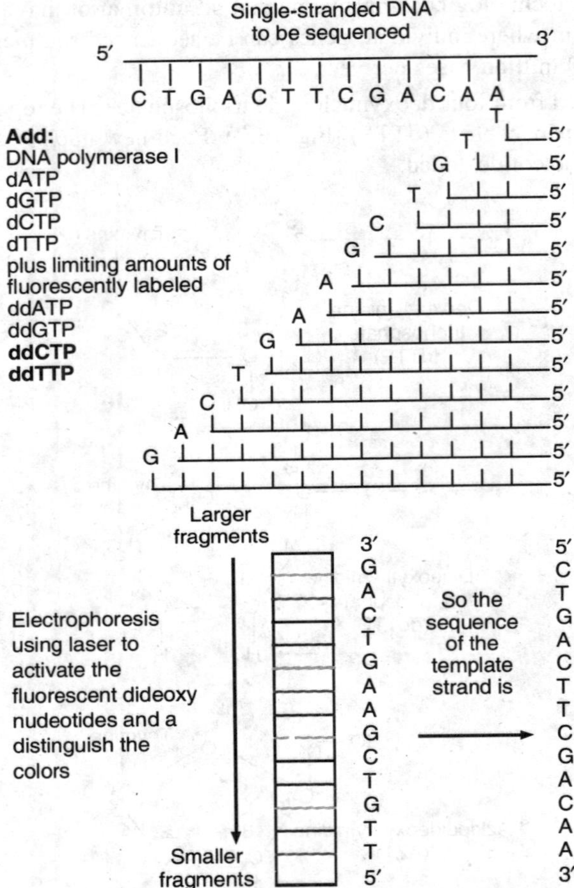

Fig. 23.8. Sanger's method of DNA sequencing.

- a mixture of all four **normal** (deoxy) nucleotides in ample quantities
  - dATP
  - dGTP
  - dCTP
  - dTTP
- a mixture of all four **dideoxy**nucleotides, each present in limiting quantities and each labeled with a "tag" that fluoresces a different color:
  - **ddATP**
  - **ddGTP**
  - **ddCTP**
  - **ddTTP**
- DNA polymerase I

Because all four normal nucleotides are present, chain elongation proceeds normally until, by chance, DNA polymerase inserts a dideoxy nucleotide instead of the normal deoxynucleotide. If the ratio of normal nucleotide to the dideoxy versions is high enough, some DNA strands will succeed in adding several hundred nucleotides before insertion of the dideoxy version halts the process.

At the end of the incubation period, the fragments are separated by length from longest to shortest. The resolution is so good that a difference of one nucleotide is enough to separate that strand from the next shorter and next longer strand. Each of the four dideoxynucleotides fluoresces a different color when illuminated by a laser beam and an automatic scanner provides a printout of the sequence.

5.2. Maxam and Gilbert Method

In 1977, Maxam and Gilbert described a sequencing method based on chemical degradation at specific locations of the DNA molecule. The end labeled DNA fragments are subjected to random cleavage at adenine, cytosine, guanine or thymine positions using specific chemical agents and the products of these fours reactions are separated using polyacrylamide gel electrophoresis (PAGE). As in Sanger method, the sequence can be easily read from four parallel lanes in the sequencing gel. Double stranded or single stranded DNA from chromosomal DNA can be used as template. Originally, end labeling was done with ^{32}P phosphate or with a nucleotide linked to ^{32}P and enzymatically incorporated into the end fragment. The read length is upto 500bp. The chemical reactions in the technique are slow and involved hazardous chemicals that require special handling in the DNA cleavage reaction. As in Sanger's method, additional cautions in Maxam and Gilbert method include purification and separation of DNA fragments and higher analysis time. Therefore, this technology is not suitable for high throughput large–scale investigation.

5.3. Hybridization Method

Ed Southern's (1990) sequencing by hybridization technique relies on detection of specific DNA sequences using hybridization of complementary probes. It utilizes a large number of short nested oligonulceotides immobilized on a solid support to which the labeled sequencing template is hybridized. The target sequence is deduced by computer analysis of the hybridization pattern of the sample DNA. DNA sequence can also be analyzed by sequencing by synthesis. Sequencing by hybridization makes use of a universal DNA microarray, which harbors all nucleotides of length *k* (called "*k*-words", or simply words when *k* is clear). These oligonucleotides are hybridize to an unknown DNA fragment, whose sequence one would like to determine. Under ideal conditions, this target molecule will hybridize to all words whose Watson-Crick complements occur somewhere along its sequence. Thus, in principle, one would determine in a single microarray reaction the set of all *k*-long substrings of the target and try to infer the sequence from those data. The average length of a uniquely resconstructible sequence using an 8-mer array is <200 bases, far below a single read length on commercial gel-lane machine. The main weakness of sequencing by hybridization is ambiguous solutions-when several sequences have the same spectrum; there is no way to determine the true sequence.

5.4. Pal Nyren's Method

In 1996, Pal Nyren's group reported that natural nucleotide can be used to obtain efficient incorporation during a sequencing-by-synthesis protocol. The detection was based on the pyrophosphate (inorganic biphosphate) released during the DNA polymerase reaction, the quantitative conversion of pyrophosphate to ATP by sulfurylase and the subsequent production of visible light by firefly luciferase. The first major improvement was inclusion of dATPαS in place of dATP in the polymerization reaction, which enabled the pyrosequencing reaction to be performed in homogeneous phase in real time. The non-specific signals were attributed to the fact that dATP is a substrate for luciferase. Conversely, dATPαS was found to be inert for luciferase, yet could be incorporated efficiently by all DNA polymerases tested. The second improvement was the introduction apyrase to the reaction to make a four-enzyme system. Apyrase allows nucleotides to be added sequentially without any intermediate washing step.

Pyrosequencing nonelectrophoretic real-time DNA sequencing method is based on sequencing by synthesis based on the pyrophosphate (inorganic biphosphate) released during the DNA polymerase reaction. In a cascade of enzymatic reaction, visible light is generated that is proportional to the number of incorporated nucleotides. The cascade starts with a nucleic acid polymerization reaction in which inorganic biphosphate (PPi) is released as a result of nucleotide incorporation by polymerase. The released PPi is subsequently converted to ATP by ATP sulfurylase, which provides the energy to luciferase to oxidize luciferin and generate light. The light so generated is captured by a CCD camera and recorded in the form of peaks known as pyrogram (compared with electropherograms in Sanger's method). Because the added nucleotide is known the sequence of template can be determined. Standard pyrosequencing uses the Klenow fragment of *E. coli* DNA pol I, which is relatively slow polymerase. The ATP sulfurylase used in pyrosequencing is a recombinant version from the yeast and the luciferase is from the American firefly. The overall reaction form polymerization to light detection take place within three to four seconds at real time. One pmol of DNA in a pyrosequencing reaction yields 6×10^{11} ATP molecules which in turn, generate more than 6×10^{9} photons at a wavelength of 560 nm. This amount of light is easily detected by a photodiode, photomultiplier tube or a CCD camera. Pyrosequencing technology has been further improved into array-based massively parallel microfluidic sequencing platform.

5.5. Automatic DNA Sequencer

A variant of the above dideoxy-method was developed, which allowed the production of automatic sequencers. In this new approach, different fluorescent dyes are tagged either to the oligonucleotide primer (dye primers) in each of the four reaction tubes (blue for A, red for C, etc), or to each of the four ddNTPs (dye terminators) used in a single reaction tube: when four tubes are used, they are pooled. After the PCR reaction is over, the reaction mixture is subjected to separation of synthesized fragments through electrophoresis (Fig. 23.9). Depending upon the electrophoretic system used, whether slab gel electrophoresis or capillary electrophoresis, following two types of automatic sequencing systems have been designed.

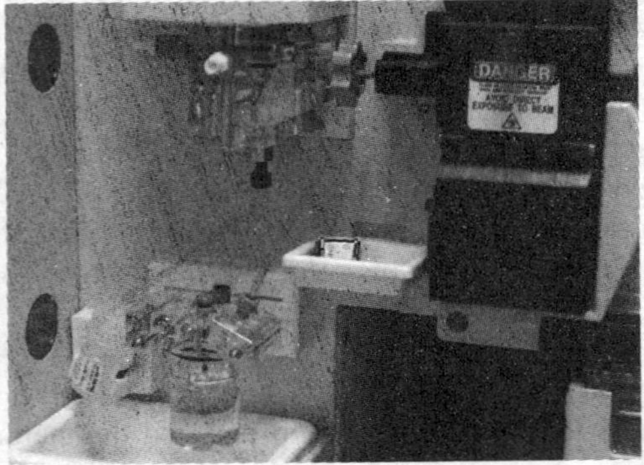

Fig. 23.9. Automated DNA sequencer (left) and details of sample loop (right).

5.6. Slab Gel Sequencing Systems

These systems make use of ultrathin (75 μm) slab gels and involve running of atleast 96 lanes per gel. In these systems, automation in sample loading of sequencing gels has also been achieved, by using a plexiglass block having wells that are same distance apart as the comb teeth cut in a porous membrane that is used as a comb for drawing samples by capillary action. Each well in plexiglass block is filed with a sample (PCR dideoxy-reaction mixture), so that when the porous membrane comb is lowered onto the sample wells in the pexiglass the samples are drawn up automatically into the comb teeth by capillary action. Using this approach of employing porous combs, automated loading of up to 192, 384 or 480 samples per gel has been achieved. The porous comb with the samples is placed between the glass plates of the gel apparatus above the flat surface of the polymerized gel and the samples are driven from the comb into the gel by electrophoresis.

5.7. Capillary Gel Electrophoresis

In these systems, slab gel electrophoresis is replaced by capillary gel electrophoresis to analyse DNA samples. In these systems, instead of scanning DNA as it migrates through 96 lanes each in a series of 96 capillary tubes, DNA fragments pass are scanned. In the original models of the above old slab gel machines, gels must be poured and reagents frequently reloaded, interrupting the sequencing. In capillary gel sequencing systems, on the other hand, the robot moves the DNA samples and reagents through the tubes continuously, requiring attention only once a day. The system produces a steady flow of data, each signal representing one of the four DNA bases (adenine, cytosine, guanine and thymine).

DNA sequencing by PCR- DNA sequencing can also be carried out using asymmetric PCR method.

Examples of sequencing- Beginning in the late 1990s, the scientific community witnessed a remarkable climax of accomplishments related to DNA sequencing. In addition to the historic sequencing of the human genome, sequences have now been generated for the genomes of several key model organisms, including the mouse (*Mus musculus*); the rat (*Rattus norvegicus*); two fruit flies (*Drosophila melanogaster* and *D. pseudoobscura*); two roundworms (*Caenorhabditis elegans* and *C. briggsae*); yeast (*Saccharomyces cerevisiae*) and several other fungi; a malaria-carrying mosquito (*Anopheles gambiae*) along with a malaria-causing parasite (*Plasmodium falciparum*); two sea squirts (*Ciona savignyi* and *C. intestinalis*); a long list of microbes; and a couple of plants, including mustard weed (*Arabidopsis thaliana*) and rice (*Oryza sativa*). Sequencing work is well underway on the honey bee (*Apis mellifera*), and is just getting started or expected to begin soon on the chimpanzee (*Pan troglodytes*), the cow (*Bos taurus*), the dog (*Canis familiaris*) and the chicken (*Gallus gallus*).

6. POLYMERASE CHAIN REACTION

A powerful technique for amplifying DNA was discovered by Kary Mullis of Cetus corporation, USA in 1985. The prestigious journal 'Science' published Polymerase Chain Reaction (PCR) as major scientific development and Taq DNA polymerase enzyme used in the technique was described as 'molecule of the year' in 1989. PCR is so powerful that from single copy of DNA molecule, millions of copies can be obtained with high accuracy, specificity and in very short time. So in few years since its invention, the polymerase chain reaction (PCR) has become a convenient and trustworthy alternative to gene cloning using vectors.

6.1. Mechanism of Basic PCR

PCR can be defined as recurring amplification of a particular region of the genome preselected with the help of primers flanking the region of the genome to be amplified. The DNA replication involves polymerization of nucleotides, using template DNA strand with the help of enzyme DNA polymerase and suitable primers. In living cells small single stranded RNA molecule acts as primer, while in PCR short DNA molecules

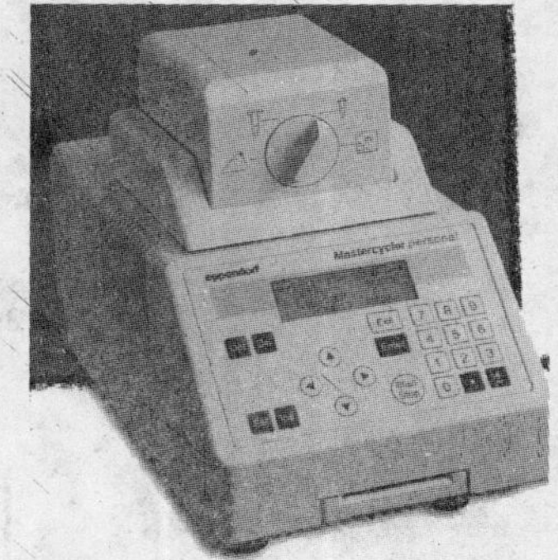

Fig. 23.10. PCR machine.

(which are complementary to two sites- either side of the piece of DNA to be amplified) are added as primers. Primers added are usually synthesized oligonucleotides. In PCR this reaction takes place in an 'eppendorf tube' instead of living cells (Fig. 23.10).

The purpose of a PCR (Polymerase Chain Reaction) is to make a huge number of copies of a gene. This is necessary to have enough starting template for sequencing.

6.2. The Cycling Reactions

There are three major steps in a PCR, which are repeated for 30 or 40 cycles in order to obtain large number of copies of desired DNA template (Fig. 23.11). This is done on an automated cycler, which can heat and cool the eppendorf tubes with the reaction mixture in a very short time.

6.3. PCR or Thermal Cycler Machine

1. Denaturation at 90-98 °C:

During the denaturation, the double strands of DNA separate. All enzymatic reactions stop (for example: the extension from a previous cycle). Higher temperature may be required to denature template and/or target DNA that are rich in G+C (>55%). For routine amplification of linear DNA template whose G+C is 55% or less denaturation is done at 94 °C for 45 seconds.

2. Annealing at 40-60 °C:

Temperature 54°C anneals two complementary primers to ends of separated single-strands of target DNA. The primers are jiggling around, caused by the Brownian motion. Ionic bonds are constantly formed and broken between the single stranded primer and the single stranded template. The more stable bonds last a little bit longer (primers that fit exactly) and on that little piece of double stranded DNA (template and primer), the polymerase can attach and starts copying the template. Once there are a few bases built in, the ionic bond is so strong between the template and the primer, that it does not break anymore.

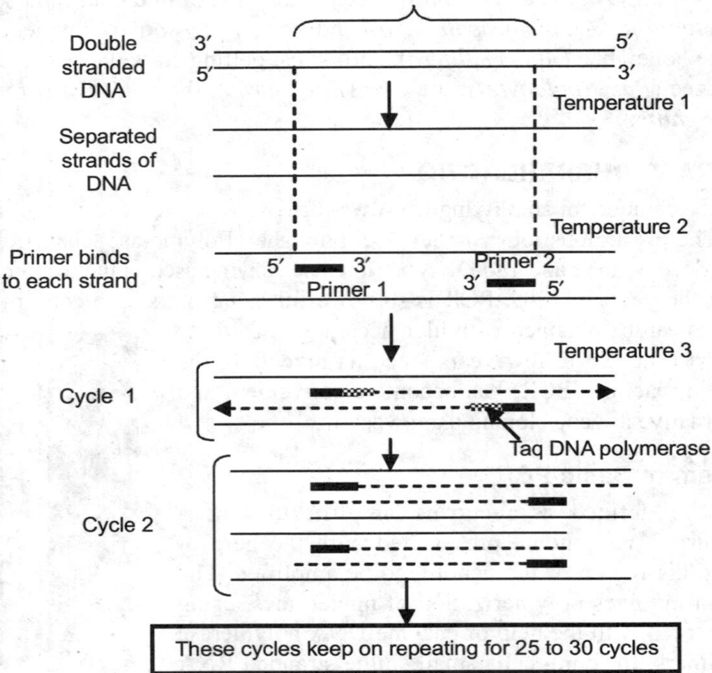

Fig. 23.11. Schematic presentation of PCR reaction and replication of DNA.

3. Extension at 72 °C:

This is the ideal working temperature for the polymerase. At this temperature thermostable Taq DNA polymerase use single stranded DNA strands of target and primers to synthesize new strands. The primers have a strong ionic attraction to the template than the forces breaking these attractions. Primers that are on positions with no exact match, get loose again (because of the higher temperature) and don't give an extension of the fragment. The bases (complementary to the template) are coupled to the primer on the 3′ side (the polymerase adds dNTP's from 5′ to 3′, reading the template from 3′ to 5′ side, bases are added complementary to the template).

This three temperature cycle is repeated 25-30 times and the result is million -fold increase of the copies of target DNA. Reaction is done in DNA thermal cycler using step cycle program set to denature at 94 °C for 45 sec, anneal at 54 °C for 20 sec and extend at 72 °C for 30 sec.

Earlier DNA polymerase enzyme from *E.coli* was used. But as it is sensitive to heat as heat destroyed the enzyme new enzyme had to be added. Use of thermostable DNA polymerase enzyme of *Thermophilus aquaticus* (from hot springs) was reported by Saiki and co-workers in 1988 and that was the major breakthrough in PCR development. It could give more yield, could avoid repeated addition of enzyme, was highly specific and longer fragments upto 10 kb can be amplified instead of less than 400bp earlier. This enzyme act best at 72 °C and the denaturation of 90 °C dose not destroy its enzymatic activity. Other thermostable DNA polymerase enzymes used today in PCR are Pfu DNA polymerase from *Pyrococcus furiosus*, Vent polymerase from *Thermococcus litoralis* and Tbr polymerase from *T. brockianus* etc and they are found to work more efficiently.

6.4. Comparison of PCR with Gene Cloning

PCR will ultimately replace gene cloning to a large extent except some applications of gene cloning like production of a protein in large quantities for human use. The final result of both gene cloning and PCR can be said to be a selective amplification of specific sequences. Some major point of comparison between the both are as shown in Table 23.2.

Table 23.2. Difference between PCR and cloning.

Parameter	PCR	Cloning
Selection of desired DNA template	First step	Last step
Concentration of starting material	Nanogram	Microgram
Biological material required	DNA polymerase	Restriction enzymes, ligase, vector, bacteria
Manipulation	In vitro	In vivo
Automation	Total	None
User skill	Not required	Necessary
Probability of error	less	more
Labour intensive	No	Yes
Cost	Less	More
Time required for a typical experiment	3-5 hours	3-5 days

6.5. Applications

1. PCR for screening of uncharacterized mutations: Small insertions or deletions can be simply detected by designing primers from regions closely flanking the mutation site and distinguishing the normal and mutant alleles by size on polyacrylamide or agarose gels. If the mutation changes a restriction site, mutant and normal alleles can be distinguished by amplifying across the mutant site and digesting the PCR product with relevant restriction endonuclease.

2. Isolation of genomic DNA: PCR allows isolation of DNA fragments from genomic DNA by selective amplification of a specific region of DNA. This use of PCR augments many methods, such as Southern and Northern blotting and DNA cloning that require large amounts of DNA, representing a specific DNA region. PCR supplies these techniques with high amounts of pure DNA, enabling analysis of DNA samples even from very small amounts of starting material.

3. Amplification and quantification of DNA: Because PCR amplifies the regions of DNA that it targets, PCR can be used to analyze extremely small amounts of sample. This is often critical for forensic analysis, when only a trace amount of DNA is available as evidence. PCR may also be used in the analysis of ancient DNA that is thousands of years old. These PCR-based techniques have been successfully used on animals, such as a 40,000 years old mammoth, and also on human DNA. Viral DNA can likewise be detected by PCR.

4. Human Health and the Human Genome: PCR has very quickly become an essential tool for improving human health and human life. Medical research and clinical medicine are profiting from PCR mainly in two areas: detection of infectious disease organisms, and detection of variations and mutations in genes, especially human genes. The primers used need to be specific to the targeted sequences in the DNA of a virus, and the PCR can be used for diagnostic analyses or DNA sequencing of the viral genome.

The method is especially useful for searching out disease organisms that are difficult or impossible to culture, such as many kinds of bacteria, fungi, and viruses, because it can generate analyzable quantities of the organism's genetic material for identification.

In short, if a disorder is caused by an infectious agent, PCR can, in principle, search out the culprit. More than 60 PCR protocols for identifying pathogens have been described to date.

QUESTIONS

1. What is cell fractionation? What are its uses in biological science?
2. What are GISH and FISH techniques? How homologous sequences are identified?
3. How genomic and plasmid DNA is isolated and quantified?
4. What is blotting technique? Compare different blotting techniques.
5. What is PCR reaction? How it has revolutionized genetic engineering?
6. Describe the Sanger's method of DNA sequencing.

CHAPTER 24

Plant Tissue Culture: History

1. INTRODUCTION

The technique of plant cell culture occupies a key role in the second green revolution in which gene modifications and biotechnology are being used to improve crop yield and quality. This technology has been employed to a wide range of crop plants and tree species to solve the biological problems. This state of art technology has become possible by continuous efforts of many scientists, who carried out the basic work. A brief history of this development is described below.

1.1. Historical Developments

The history of plant tissue culture begins with the concept of cell theory given independently by Schleiden 1838 and Schwann 1839, that established the cell as a functional unit. This implies that cells are autonomous. The concept was tested experimentally by Haberlandt after 130 years, who conceived the idea of culturing plant cells. The developments in culturing of plant cells, tissues and organs are associated with the developments in our knowledge about nutritional requirements of plant cell, discovery of growth regulating factors, analytical tools and techniques, and development of microscopy.

Though, earlier attempts to grow plant cells met with failures, success was achieved in growing animal cell culture at that time. Earlier works involving plant tissue were mainly concerned with nutritional requirement of cells to make them divide and sustain growth. The work progressed rapidly after the discovery of auxin and cell division factor. Early studies also frequently used complex nutritional factors like coconut water, yeast extract, malt extract and casein hydrolysate. The developments after World War II were rapid as many inventions developed for the war purposes found their applications in biological sciences.

It was untiring efforts of Robbins, Kotte, White, Gautheret, Heller, Van Over-beek, Steward, Caplin, Miller, Nitsch, Reinert, Street, Morel, Skoog, Hildebrandt, Melcher, Cocking, La Rue and others that plant tissue culture, which was initiated as a tool has become a powerful technology in last three decades.

With the better understanding of the technique of plant tissue culture and nutritional requirements, of plant cell, it was possible to develop newer technologies by culturing plant organs (anther, ovary, ovule, petal, leaf and meristem) leading to establishment of new research lines viz., haploids, virus-free plants, *in vitro* fertilization, embryo rescue and direct regeneration from leaf disc for genetic engineering. Subsequently, it has become possible to grow isolated epidermal cells, gland cells or even protoplasts and to regenerate from such specialized cells or individual protoplasts. Regeneration of plants and production of useful metabolites through plant biotechnology has become an industrial application. Demonstration of variability in cell cultures in relation to secondary metabolites has given the concept of morphological variations and hence 'Somaclonal variation'. The significant contributions of selected scientists are given in the following paragraphs:

Gottlieb Haberlandt

With the establishment of cell theory proposed by Schleiden & Schwann (1839), cell is considered as structural and functional unit of living organisms. In plants, totipotency of excised plant tissues and cells was established subsequently. It was Gottlieb Haberlandt, an Austrian scientist, who conceived the idea of culturing isolated plant cells in the nutrient solutions. Based on cell theory, he assumed that there was no limitation of divisibility, therefore, he started with isolated mesophyll cells using Knop's nutrient solution, sucrose, aspargine and peptone working at Graz, Austria. Haberlandt's vision of the totipotency of plant cells represents the actual beginning of tissue culture. Haberlandt (1902) in his famous paper described the cultivation of mesophyll cells of *Lamium purpureum* and *Eichhornia crassipes*, epidermal cells of *Ornithogalum* and hair cells of *Pulmonaria*. The cells survived 3-4 weeks but without cell division, there was increase in cell size. He conducted several experiments and carefully recorded his observations and interpreted the results. Using pieces of mature potato tubers he observed that cell division occurred in small, thin tissue almost without exception when the explant contained a vascular strand.

Gottlieb Haberlandt

P. Nobecourt

Pierre Roger Gautheret

If Haberlandt had used in place of highly difficult tissues, some other materials like willow or carrot, which are known to proliferate without growth substances, he would have obtained the first tissue culture. The other reason for the failure was the incomplete information at that time about nutrition of plant tissues and lack of knowledge about plant growth regulators. Haberlandt worked for several years at Graz, Viena, Austria and then in 1909, moved to Berlin as professor of botany. In Berlin he suggested hormone like substances are responsible for wound healing. At that time hormones were not known in plants. He was member of Austrian and German botanical societies and Academies. He was a multi-talented, imaginative and impressive scientist. The details of his work, biography and achievements are given in a recently published book "Plant tissue culture:100 years since Gottlieb Haberlandt", by Laimer and Rucker published by Springer, Wien.

P. Nobecourt

P. Nobecourt a French plant pathologist, announced simultaneously in 1939 with White and Gautheret, the possibility of cultivating plant tissues for unlimited period. It was possible with the use of recently discovered indole acetic acid (IAA) by F. W. Went (1932). This success opened new avenues of cultivating plant cells for different studies. This success was announced a few months before the beginning of Second World War. And for six years, American & French scientists worked without knowledge of their mutual results.

Pierre Roger Gautheret

Like many others, Pierre Roger Gautheret was working earlier at the University of Sorbonne, Paris, France. Gautheret also tried to cultivate isolated cells and root tips without getting tissue cultures. Following these failures, he turned towards the tissue participating in wound healing. He first used piece of cambium cut from trees, especially *Acer pseudoplatanus*, *Sambucus nigra* and *Salix capraea*. Preliminary attempts with liquid medium failed completely. Later explants were placed on the surface of medium solidified with agar. He did not expect any good results and kept those cultures in a cupboard. He was surprised to find very healthy callus on the explant after two months. He observed living turgescent dividing cells. Over the next 5 years the recognition of the importance of B vitamins in yeast extract and the auxin (IAA) allowed significant advances to be made. It was because of these factors that Gautheret could achieve the proliferation and division of cambial tissue of *Salix capraea* and *Populus alba* for several months. In 1939, Gautheret reported the propagation of carrot as the first plant tissue culture of unlimited growth. IAA stimulated the growth of undifferentiated tissue on cut surfaces of sterile explants. This tissue, termed callus, was similar in appearance to wound tissue and subsequently it was found that callus could be subcultured indefinitely. His results were immediately verified in Italy by Grelli. Finally, the possibility of cultivating plant tissues for unlimited periods was announced almost simultaneously by White (1939), Nobecourt (1939) and Gautheret (1939). In spite of difficulties encountered during the war, Gautheret was able to receive some collaborators in his laboratory and one of them was Georges Morel. Gautheret was working on fleshy organs, especially Chicory, Jerusalum artichoke and *Brassica napus* and Morel established strains of lianas, herbs and trees.

During the study of normal and tumour culture, the development of auxin non-requiring tissues were observed by French and American scientists. Gautheret discovered the reasons for this difference (Kulescha and Gautheret, 1948), the normal tissues maintained *in vitro* produces an insignificant amount of auxin while the tumour tissues produce rather important amount of this growth substance. Later on, hyper-auxinity has been established in tumour tissues. The same was observed in case of habituated tissues not requiring exogenous supply of auxin. He proposed the term 'habituation' for such cultures.

In 1942, Gautheret observed development of buds in tissue culture of *Ulmus* and of plantlets in Chicory tissue cultures, establishing the totipotency of the cells. The various investigations carried out by Gautheret during 30s and 40s were mainly concerned about nutrition and cell differentiation of cultivated cells. His other major contributions are establishment of hormones autonomous cultures (habituated cultures) from hormone requiring cultures and study of tumour cells. He also worked on root culture of *Helianthus tuberosus* and demonstrated that light, temperature and other factors affect the root growth. His important contributions are cited in following two monumental works.

1. Gautheret R.J. (1955). The nutrition of plant tissue cultures. *Annual Review Plant Physiol*. **6**:433-484.
2. Gautheret R.J. (1983) Plant tissue culture: A History. *Bot. Mag. Tokyo*. 96: 393-410.

Georges Morel

G. Morel was among the first to culture monocotyledonous tissues. He developed the method of meristem culture for the elimination of viruses and the micropropagation of orchids and discovered the two unique opines of crown gall tissues. These opines have become marker for analysis of transformed cultures induced by *Agrobacterium tumefaciens* and *A. rhizogenes*.

Morel, in 1950, obtained the indefinite growth of monocotyledonous tissues such as *Gladiolus*, *Iris* and *lily* on the medium enriched with natural extract (coconut milk, yeast extract). He also cultivated Royal Fern on this medium.

Morel and Martin (1952) developed meristem culture technique and recovered *Dahlia* shoots, free from viruses, by meristem tip culture. In 1955, they recovered virus free potato. This attained wide application of plant tissue culture to raise virus free plants in agriculture. The first industrial application was the work of George Morel who successfully multiplied tropical orchids through division of protocorms, little differentiated structures developing naturally on orchid embryos. In 1960, Morel observed the emergence of such bodies after carrying out a cymbidium shoot-tip culture. He also noted that protocorms could be sectioned into quarters and subcultured, each section regenerating a new protocorm, within a few weeks which, in turn, could be divided. The so-obtained protocorms subsequently evolved into young plantlets. This remarkable property found almost immediate commercial use since orchid producers had to cope with slow multiplications rate through cluster division. Around 1970, first commercial multiplication laboratory was established in USA for orchids.

In 1971, he studied *Agrobacterium tumefaciens*, crown gall inducing bacteria, on the dicots and isolated octopine and nopaline from the *A. tumefaciens* infected plants. These chemicals are produced only when Ti-plasmid of the bacteria integrates with the genome of the host plant cells. They are also known as opines. This investigation opened new avenue to use Ti plasmid as a vector for gene transfer after removing opines genes from the plasmid. Morel joined the laboratory of Professor R. Gautheret in France during the World War II. He established cultures of lianas, herbs and trees. Gautheret described him as able and enthusiastic worker.

Philip R. White

P. R. White, an American scientist who worked at Rockfeller Institute, New Jersey, USA, was one of the pioneer tissue culturist of early period. In 1934 he reported for the first time successful continuous cultures of tomato root tips and obtained indefinite growth of roots. Initially he used salts of Knop's solution, sucrose and yeast extract in his medium. Later, yeast extract was replaced by three vitamins, viz. pyrodoxine, thiamine and nicotinic acid. This synthetic medium has been proved to be one of the basic medium for a variety of tissue cultures. White maintained some of his root cultures till before his death in 1968.

White

With the advice of Stanley, a famous scientist who isolated tobacco mosaic virus (TMV), White started work on the dual culture. He established cultures of tomato roots infected with TMV. Cultures of infected roots gave rise to healthy roots by root tip cultures. The success of White's experiment opened the field of root cultures, which have been explored by many workers to solve morphological, physiological and pathological problems.

In 1939, White had presented his work in tobacco tissues at a meeting of the Botanical Society of America. He reported for the first time, long term callus cultures which were established from tumour tissues of *Nicotiana glauca x N. longsdorfii*. He was successful due to the incorporation of auxin in the medium. Similar results were also observed by Gautheret and Nobecourt in the same year by conducting independent research on carrot. White has contributed in formulating root culture medium, which is known by his name as White's medium (1943). This medium consists of essential elements, glycine, nicotinic acid, thiamine, pyridoxine, sucrose etc.

The work of White can be categorized into three aspects, continuous growth of excised tomato root tips in liquid medium, *in vitro* cultivation of viruses on excised roots and growth of tumour tissues. On March 25, 1968, he died after suffering from acute hepatitis. His book on tissue culture was the only book on the subject.

White P. R. (1963). The cultivation of animal and plant cells. Ronald Press, New York.

F.C. Steward

F.C. Steward was one of the pioneers of plant tissue culture and contributed a lot by way of understanding the requirements of plant tissue cultures and developing techniques for the different applications.

F.C. Steward

Working at Cornell University, Ithaca, USA, he had developed a School of Plant Physiologists involved in plant tissue culture work. Caplin and Steward (1948) used coconut water for the first time in plant tissue culture and obtained vigorous proliferation of carrot explants. It was also evident from this finding that the stimulating substance of coconut milk was not in auxin. Steward and co-workers were the pioneer in establishing cell cultures of single cells and clumps in liquid medium (Steward and Shantz, 1955) and developed various types of vessels for this purpose including the rotating nipple flasks Fig. 24.1.

Fig. 24.1. Nipple flask designed by Steward.

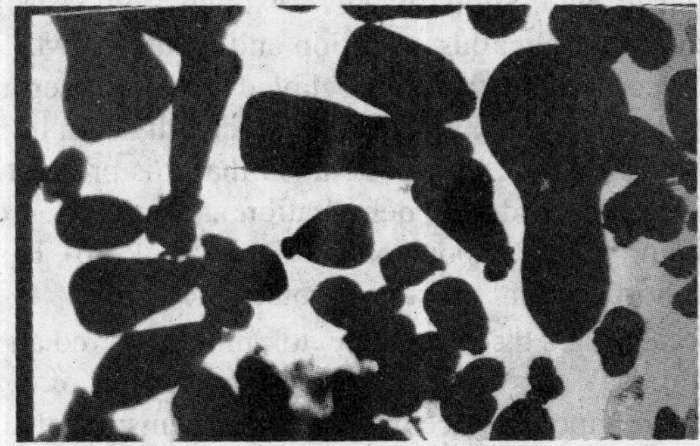

Fig. 24.2. Somatic embryos in carrot.

The another landmark comes in the form of somatic embryogenesis in carrot cell suspension cultures and production of complete plants (Steward and Ammirato, 1969). Steward had taken cell suspension of callus tissues derived from wild carrot embryos and plated them out on a coconut milk medium solidified with agar. From this plated cell population, thousands of embryo developed each being derived from one or a few cells of the callus tissue. This has not only opened new avenues of micropropagation of plant species at an enormous rapid rate but also completely established the totipotency of the cell (Fig. 24.2).

His volumes on 'Plant physiology, a treatise' were one of the most widely used reference books on plant physiology and tissue culture.

Jakob Reinert

Reinert, a German Scientist, was a pioneer in the field of plant culture and contemporary with Steward, Cocking and Street. In 1959, he was culturing undifferentiated parenchyma tissue of carrot root in complex medium containing sugar, salt and coconut milk. On this medium the callus had been sub-cultured repeatedly, without evident morphological change. He transferred the tissue to a synthetic medium containing an elaborate mixture of known substances, including amino acids, vitamins and hormones. The tissue became more granular and showed evidence of differentiated regions. This culture, on transfer to a medium lacking high levels of auxin, produced young plants. Histological analysis showed bipolar embryo formation in the system, which produced plants on proper medium. This had proved the development of a complete plant from mature undifferentiated parenchyma and in other words 'the totipotency' of cells.

Besides the embryogenesis in carrot, Reinert worked several years on developing technology for wide ranging applications of plant tissue culture, which includes cryopreservation of cells, regeneration of plantlets, anther culture, process of androgenesis, development of haploids, and protoplasts culture.

The book 'Plant cell, tissue and organ culture' by Reinert and Bajaj (1977) Springer Verlag, Berlin Heidelberg, was the first compilation of state of art literature on plant tissue culture.

Folke Skoog

Folke Skoog

Professor Folke Skoog was one of the leading plant physiologists of the world. Skoog was already renowned for his pioneering work with auxin, a plant growth hormone, when he joined the University of Wisconsin, Madison, U.S.A. While at Wisconsin, he discovered a major new class of plant hormones, the cytokinins, which stimulate the division of plant cells, and regulate plant growth and development. During the cultivation of tobacco pith cells, Skoog and co-workers observed that the incorporation of purine base- adenine in the culture medium enhances callus growth. Similarly, DNA extracts from yeast and Herring sperm whale also favoured callus initiation and callus growth of tobacco pith cells. This leads to the isolation of a compound from autoclaved Herring sperm whale DNA called kinetin (6-furfuryl amino purine). Though kinetin enhances cell division in pith cells, it does not present in this form in plant cells. Therefore, they concluded that this product was formed during autoclaving of whale DNA. It is produced due to dehydration and migration of a deoxyribore moiety from the 9- position of adenine to the N^6-position (Miller et al., 1956). His work has had a profound impact on agricultural and horticultural practices around the world.

Skoog's discovery of cytokinins triggered an international flood of publications that continues to this day. His work showed that a number of cytokinins occur naturally, and that at least one of these compounds occurs in every organism tested, from bacteria to humans.

With colleague Nelson J. Leonard at the University of Illinois, Skoog synthesized and tested hundreds of compounds for cytokinin activity, and established the principles that govern relationships between plant structure and activity. These discoveries are generally held to be one of the major advances made in the plant sciences during the last 50 years.

Skoog was also a pioneer in investigating how to control the formation of roots, stems and leaves from undifferentiated cells in plant tissue cultures. Tissues could be made to develop as undifferentiated masses of cells, or to become roots, stems or a combination of roots and stems, resulting in complete plants. Skoog and Miller (1957) demonstrated for the first time that a ratio of auxin and cytokinin can control the root, shoot and callus formation. Skoog's theory that plant development is controlled by hormone levels and other factors led to the realization that whole plants can be generated from cultured cells. This laid the groundwork for the production of transgenic plants and other advances in biotechnology.

In 1962 he and graduate student Toshio Murashige published a culture medium for optimal plant tissue growth that remains in use till date. This is a high salt medium, which was modified by Linsmaier and Skoog (1965) and relatively low salt formulation was evolved.

Skoog was born in Halland, Sweden on July 15, 1908. He decided to take up residence in the United States during a visit to California in 1925, and he became a U.S. citizen 10 years later. He received a bachelor of science degree in chemistry from the California Institute of Technology in 1932. He earned his Ph.D. from that institution in 1936. While at the institute, he worked closely with prominent plant physiologists Kenneth Thimann and Fritz Went (known for their work and discovery of auxin).

Between 1937-1941, Skoog was on staff at Harvard University, and from 1941-1944, at Johns Hopkins University. He served as a chemist and technical representative of the U.S. Army in Europe between 1944-1946. Skoog played an active role in a number of biological societies. He served as president of the American Society of Plant Physiologists, the Society of Developmental Biology, the American Society of General Physiologists and the International Plant Growth Substances Association. He was a member of the American Academy of Arts and Sciences, and the National Academy of Sciences. He was awarded the National Medal of Science in 1991 during a ceremony at the White House.

Renowned plant physiologist and National Modal of Science recipient Folke Karl Skoog, professor of botany at the university of Wisconsin, Madison, U.S.A. for 32 years, died Feb. 15, 2001 after a long illness. He was 92.

E.C. Cocking

Professor Edward Cocking, a Fellow of Royal Society of London, works at the Centre for crop nitrogen fixation at Nothingham University, Nottingham, U.K. He has established an International network for research on nitrogen fixation in the worlds major non-legume crops (especially rice, wheat, maize, sorghum, and oilseed rape). This involves basic research on the interaction of crops with azorhizobia for the establishment of endophytic nitrogen fixation.

E.C. Cocking

Professor Cocking and his associates have been instrumental in developing many techniques of plant cell and protoplasts culture. In 1960s, for the first time a method of isolation of protoplasts in large quantities was developed using enzymes obtained from *Myrothecium* fungus. They obtained protoplasts from root tips for *Lycopersicon esculentum* using cellulase produced by the fungus. Protoplasts isolation and their use in somatic hybridization opened new avenues of improving plant species. Subsequently, the technique found wide applications because of availability of cellulase at commercial level. At that time they started attempting fusion of legume cells containing bacteria with non-legume cells towards developing non-legume nitrogen fixing plants. They continue these efforts with the establishment of research centre for nitrogen fixation studies.

They (Bhojwani and Cocking, 1972) were also successful in isolating protoplasts from pollen grains and pollen mother cells. They used protoplasts for understanding tobacco mosaic virus infection and multiplication in plant cells. Development of technique of protoplasts isolation has many applications in cell biology, plant physiology and genetic engineering as mentioned in the chapter on protoplast culture. Besides many articles on protoplasts culture, his some of the recent publications are as follows.

1. Blackhall, NW; Jotham, JP; Azhakanandam, K; Power, JB; Lowe, KC; Cocking, EC; Davey, MR (1999): Callus initiation, maintenance and shoot induction in rice. In: Methods in Molecular Biology. Vol. 111. (Ed: Hall,RD) Humana Press Inc, Totowa NJ, 19-29.
2. Azahakanandam, K; McCabe, MS; Lowe, KC; Cocking, EC; Power, JB; Davey, MR (1999): Gene transfer into and molecular characterisation of T0, T1 and T2 transgenic rice plants mediated by *Agrobacterium tumefaciens*. Journal of Experimental Botany 50, 20.
3. Kennedy IR and Cocking EC (1997) Biological Nitrogen Fixation: The Global Challenge and Future Needs. Position Paper of the Meeting at the Rockefeller Foundation Bellagio Conference Center, 8-12 April, 1997.

Indra K. Vasil and Vimla Vasil

Prof. Indra K. Vasil and Dr. Vimla Vasil are the product of Delhi School of Morphology headed by the then Professor P. Maheshwari. They joined the laboratory of Professor A. C. Hildebrandt at Madison, Wisconsin, as post-doctoral fellow. Vasil, V. and Hildebrandt were the first to prove the early cell theory of Schwann by regenerating a tobacco plant from a single cell derived from a suspension of cultured cells. The landmark work of Vimla Vasil on culture of single isolated cells of tobacco has opened new applications of cell cultures in genetics, morphogenesis and vegetative propagation and proved the totipotency of cell. Single cells of hybrid tobacco callus were grown in micro-chambers in the absence of any other cells, in fresh liquid medium containing coconut milk. These produced a colony of 50-75 cells in 10-25 days. Upon transfer from the micro-chambers to agar medium, cell clumps produced callus in 3 months, which ultimately produced complete plantlets (Vasil V. and Hildebrandt, 1965).

I.K. Vasil

During his stay at Wisconsin, I.K. Vasil studied cell differentiation and morphogenetic behaviour in *Petroselinum hortense* and *Cichorium endivia*. He was successful in producing *C. endivia* plantlets from freely suspended cells grown *in vitro*.

Presently I.K. Vasil and V. Vasil are Professor in the University of Florida at Gainsville, USA. I.K. Vasil has contributed significantly on the cell culture, protoplasts cultures and somatic embryogenesis of cereals (monocotyledons), particularly *Pennisetum americanum*, *Panicum maximum*, *Triticum aestivum*, and *Zea mays*. He has edited a series of books on plant tissue culture and somatic cell genetics of higher plants published by Academic Press.

Monocotyledons, particularly the cereals were considered unsuitable material to regenerate via somatic embryogenesis. I.K Vasil and co-workers demonstrated that the young leaves, inflorescences or embryos are useful in initiating embryogenic cultures. The comprehensive work showed that selection and maintenance of embryogenic cultures at an early stage is important for maintaining long term embryogenic potential.

Establishment of embryogenic and non-embryogenic lines in monocotyledons and concept about these cultures were given by I.K. Vasil. Embryogenic cultures are compact, dry, amorphous and white in colour as compared to non-embryogenic cultures which are watery, dirty white to light brown in colour and soft in nature. It is necessary to isolate and maintain separately embryogenic cultures.

Riker and Hildebrandt

Professor Albert C. Hildebrandt was a professor at the University of Wisconsin, Madison, U.S.A. As a graduate student and through ensuing years as a staff member, Hildebrandt worked closely with Professor A.J. Riker, who had a keen interest in the effect of the crown gall bacterium, *Agrobacterium tumefaciens*, on the host plants and its possible relation to cancer in animals. Professor Riker was a plant pathologist and used plant tissue culture to understand the basic biology of infection of plant pathogens (Hildebrandt and Riker, 1958). As an approach to the study of crown gall disease, Hildebrandt began the study of the growth *in vitro* of excised plant tissue, both healthy and diseased. Prof. Hildebrandt became an internationally recognized authority on tissue culture. He was very active in the international tissue culture association. He contributed chapters to several books and manuals concerning tissue culture techniques and their uses.

The tissue culture work began with the study of the isolation and growth of tissues in culture, including the effects of nutrition and environment. Isolations were made from healthy stems, galls, apical meristems and anthers. The effects of bacteria and viruses on these callus tissues were widely studied. Tissues were induced to differentiate back to entire plants through the use of plant growth substances. This was done with a number of plant species- tobacco, sunflower, potato, geranium, gladiolus, African violets, poplar and cassava. Hildebrandt, with his students, then developed methods for isolating single cells from tissue cultures (micro-chamber, nurse tissue culture etc), growing them in liquid culture and again inducing them to differentiate back to the original plant (Sievert and Hildebrandt, 1965). A great deal of still- and cine-photomicroscopy was done on the growth of isolated cells and their relation to disease organisms.

This work on regeneration of plants from single cells (Vasil and Hildebrandt, 1965) became the basis for the eventual development of the technique for the plant transformation in the biotechnology industry. Prof. Hildebrandt also proposed for the first time the possibility of fusion of somatic cells, the somatic hybridization. Hildebrandt formulated a plant tissue culture medium known as Schenk and Hildebrandt (1971) medium.

The isolation of single cells from apical meristems and anthers was used as a means to produce virus-free plants. Hildebrandt did this for several plants, including, geranium, gladiolus and poplar. Virus-free stock was propagated and made available to commercial growers. Related to this work was the study of several virus diseases of gladiolus and geranium.

Hildebrandt was resource person for information on diseases of floricultural crops for growers of the state.

Hildebrandt was born in State college, Pennsylvania, on April 10, 1916. He obtained his Ph.D. degree in plant pathology in 1945 and joined the faculty in 1949 as assistant professor and rose to the rank of full professor in 1960. He retired in 1978. Emeritus Prof. Hildebrandt, age 84, died on January 4, 2001.

Some important citation of the group are as follows:

1. Hildebrandt AC and Riker AJ 1958 Viruses and single cell clones in plant tissue culture. Fed. Proc. 17:986-993.
2. Tamaoki T, Hildebrandt AC, Burris RH and Riker AJ 1961 Oxidation of reduced diphosphopyridine nucleotide by mitochondria from normal and crown-gall tissue cultures of tomato. Plant Physiology 36:347-351.
3. Hildebrandt AC 1977 Single cell culture, protoplasts and plant viruses. In Plant cell, tissue, and organ culture, J Reinert & YPS Bajaj, Springer-Verlag, Heidelberg.

Toshio Murashige

Professor Toshio Murashige is working as Professor of plant sciences at the University of California, Riverside, U.S.A. As a graduate student he worked with Professor Skoog on nutrition of plant cells using tobacco pith cells as an experimental material. During that period plant tissue culture was in its developmental stage. Only White's medium was known. They cultured tobacco pith cells on White's medium and evaluated, one by one, inorganic and organic nutrients and formulated a medium, known as Murashige and Skoog's medium (Murashige T. and Skoog, F. 1962 A revised medium for rapid growth and bioassay with tobacco tissue cultures. Physiologia Plantarum 15:473-497). In the study, the optimal concentration of each individual element was determined empirically. This is a synthetic medium and is designed to assay the effect of organic supplements. This is the most widely used medium for plant tissue culture work. Later on, other media were developed based on this formulation.

After joining the University of California, he worked on developing micropropagation technique and established three stage micropropagation system. These are : (*i*) establishment of aseptic culture, (*ii*) Multiplication of propagule and, (*iii*). Preparation for reestablishment of plants in soil. According to him each stage has a different nutrient and light requirement. He worked on multiplication of several plant species belonging to diverse taxa e.g., ornamentals like *Dracaena*, *Scindapcus*, *Syngonium*. He also worked on process of somatic embryo formation using carrot and citrus plants and described changes at molecular levels during the process.

The following two contributions are well appreciated among several hundred articles he has published.

1. Murashige T. 1974. Plant propagation through tissue culture. Annual Review of Plant Physiology 25:134-66.
2. Murashige T. 1978 The impact of plant tissue culture on agriculture, TA Thorpe (ed) IPTCA, Calgary, Canada.

Presently he is working as emeritus scientist on problem of rejuvenation in adult tree explants. Rejuvenation of trees is evident by restored organogenesis, leaf morphology and shoot vigor, and diminishing leaf abscission, chlorosis and tissue and culture medium discoloration *in vitro*. Investigation of the underlying mechanism, using *in vitro* growing shoots, disclosed identical agarose gel electrophoretic patterns of mitochondrial DNA extracted from juvenile and rejuvenated shoots and distinct from the mitochondrial DNA of adult shoots.

Huang LC, Hsaio CK, Lius S, Liu SF, Huang BL, and Murashige T. 1992. Restoration of vigor and rooting competence in stem tissues of mature citrus by repeated grafting of shoot apices onto freshly germinated seedlings *in vitro*. *In vitro* Cell. Dev. Biology - Plant 28P:30-32.

P. Maheshwari

P. Maheshwari

The most outstanding Indian embryologist, P. Maheshwari was trained at Allahabad by Dudgeon. Maheshwari remained as expert of Botany for almost three decades. After completing the D.Sc. Thesis (1929) at Allahabad, he moved to the Agra College (1931-1937), and after serving brief tenures at Allahabad (1937-1939), Lucknow (1939) and Dacca (1939-1949), at the invitation of the Vice-Chancellor (Sir Maurice Gwyier), he joined the University of Delhi (1949-1966) and established a very productive school of embryology. The earlier work of Prof. Maheshwari was concerned with embryology, which he started in 1930 at Agra. By 1933, a flourishing school of Angiosperm Embryology was established and in coming years, this centre became internationally known. B.M. Johri, V. Puri, B. Singh and H.R. Bhargava were the prominent students trained at the above centre. Johri's work on the Alismaceae and Butomaceae done between 1933 and 1938 is among the first important piece of research in embryology in the country. Occurrence of pollen grains in the stylar canal of a typical angiosperm – *Butomopsis lancealata* by Johri (1936) was an appreciable discovery.

P. Maheshwari worked at Haward University in 1945 and devoted most of his time to complete the book entitled, "An introduction to the embryology of angiosperms" (1950). He had also edited a few symposium volumes in 1962-63. He also established the international society of plant morphologists and started an international journal "Phytomorphology."

It was due to continuous efforts of P. Maheshwari that India acquired an international status in embryology. In 1965, he was elected fellow of the Royal Society of London.

In India, studies on tissue culture started in mid 1950 at the Department of Botany, University of Delhi under the directions of Professor P. Maheshwari. This was beginning of experimental embryology. Some faculty members who were trained abroad to acquire knowledge were - S. Narayanswami who worked with C.D. La Rue at Michigan, and B.M. Johri who worked with H.E. Street at Swansea.

Over the past four decades no other institution in India has made more significant contributions than the Department of Botany, University of Delhi, in the use of tissue culture methodologies for morphogenetic studies involving ovary, ovule, embryo, endosperm, cotyledons, and other reproductive organs. These works deal with isolation, culture, *in vitro* fertilization, and development of plantlets from vegetative tissues. The successful *in vitro* growth of ovary, ovule, and embryo parts in fruit set and seed development in response to exogenous growth regulators has been achieved. Emphasis was also laid on the nucellar culture for obtaining true clones of Citrus and the endosperm culture for the production of triploid plant in case of *Citrus*, *Oryza* and *Putrangiva roxburghii*.

S. Guha Mukherjee and S.C. Maheshwari

Plant tissue culture studies led to the demonstration for the first time of the development of haploids through anther and pollen culture by Sipra Guha and Maheshwari, S.C. (1964-67). This research finds many applications in crop and tree breeding programmes, being a quicker and simpler method of producing homozygous lines than conventional breeding programme. Professor S.C. Maheshwari was working in the Botany Department of Delhi University. Later on he joined the newly formed department of Molecular biology of Delhi University at South Campus. He contributed significantly in the development of these departments in the University. The University Grants Commission and Department of Science and Technology have recognised the research credential of Prof. Maheshwari and granted him a centre for research on molecular biology. During his long research career he worked on various aspects of plant growth and developments like protoplasts culture, characterization of chloroplast DNA, regeneration in cereals, gene isolation and sequencing, northern analysis, transcript mapping, genomic mapping and gene delivery using electroporation in plants like spinach, mung bean, and rice towards better understanding the mechanism of gene expression during development. Presently he is working in International Centre for Genetic Engineering and Biotechnology at New Delhi.

Prof. Sipra Guha Mukherjee worked on regeneration of plants and mechanism of regeneration involving various enzymes, membrane phospholipids and second messengers at school of life sciences, J.N. University, at New Delhi (She died in 2007).

The other centres which gradually developed as National Institutes in India are: National Chemical laboratory, Pune; Bhabha Atomic Research Centre, Bombay; Indian Agricultural Research Institute, New Delhi; Indian Institute of Science, Bangalore; National Botanical Research Institute, Lucknow; Bose Institute, Calcutta and Central Institute for Medicinal and Aromatic Plants, Lucknow. The universities at Jaipur, Jodhpur, Udaipur, Dharwar, Varanasi, Calcutta, Baroda, Pune, Hissar and Jawaharlal Nehru University, New Delhi have significantly added to our knowledge using plant biotechnology in understanding the process of growth differentiation and metabolism. The newly established department of biotechnology of the Ministry of Science and Technology has identified the key areas of research. This has encouraged the biotechnological research in India. At I.A.R.I., New Delhi, National facility for plant tissue culture repository (cryopreservation unit) has been established with the National Centre of Plant Genetic Resources, New Delhi.

QUESTIONS

1. Write short notes on the followings, describing their significant contributions:
   (*i*) Roger Gautheret (*ii*) P. Maheshwari
   (*iii*) P.R. White (*iv*) F. Skoog
   (*v*) J. Reinert (*vi*) T. Murashige
   (*vii*) G. Morel (*viii*) A.C. Hildebrandt
   (*ix*) F.C. Steward (*x*) E.C. Cocking
   (*xi*) I. K. Vasil
2. Describe in brief landmark discoveries in the field of plant tissue culture.
3. What are significant contributions of Delhi school of plant morphology?
4. Name major centres of plant biotechnological research in India.

CHAPTER 25

Tools and Techniques for Plant Tissue Culture

1. pH AND pH METER

The hydrogen ion concentration of most solutions is extremely low. In 1909, Sorenson introduced the term pH as a convenient way of expressing hydrogen ion concentration.

pH of a solution is strictly defined as the negative logarithm of the hydrogen ion activity. But in practice usually hydrogen ion concentration is taken.

$$pH = -\log_{10}(H^+) = 7$$

The pH of pure water is 7 at 25 °C. Generally glass distilled water is used for the preparation of culture medium. However, some times buffered solutions may be used for the same to keep the pH of the medium constant.

1.1. Measurement of pH

An approximate idea of the pH of a solution can be obtained using indicators. These are organic compounds of natural or synthetic origin whose colour is dependent upon the pH of the solution. Indicators are usually weak acids, which dissociate in solution. A standard pH meter has two electrodes, one glass electrode for measuring pH and the other calomel reference electrode (Fig. 25.1). Reference electrode is filled with saturated KCl solution.

$$\text{Indicator} = \text{Indicator}^- + H^+$$

The pH probe measures pH as the activity of hydrogen ions surrounding a thin walled glass bulb at its tip. The probe produces a small voltage (about 0.06 volt per pH unit) that is measured and displayed as pH units by the meter. The meter circuit is fundamentally no more than a voltmeter that displays measurements in pH units instead of volts. The input impedance of the meter must be very high because of the high resistance — approximately 20 to 1000 MΩ — of the glass electrode probes typically used with pH meters. The circuit of a simple pH meter usually consists of operational amplifiers in an inverting configuration, with a total voltage gain of about – 17. The inverting amplifier converts the small voltage produced by the probe (+ 0.059 volt/pH) into pH units, which are then offset by seven volts to give a reading on the pH scale. For example:

- At neutral pH (pH 7) the voltage at the probe's output is 0 volts.
- At alkaline pH, the voltage at the probe's output ranges from + 0 to + 0.41
- At acid pH, the voltage at the probe's output ranges from – 0.41 volts to – 0.
- Nowadays in more sophisticated pH meters, the both types of the electrodes are combined in one.

1.2. Applications

1. To adjust pH of different solutions, preparation of buffers and culture media.
2. Determination of pH of cells (cell sap) and in analytical techniques.
3. To monitor pH of the medium in a bioreactor.

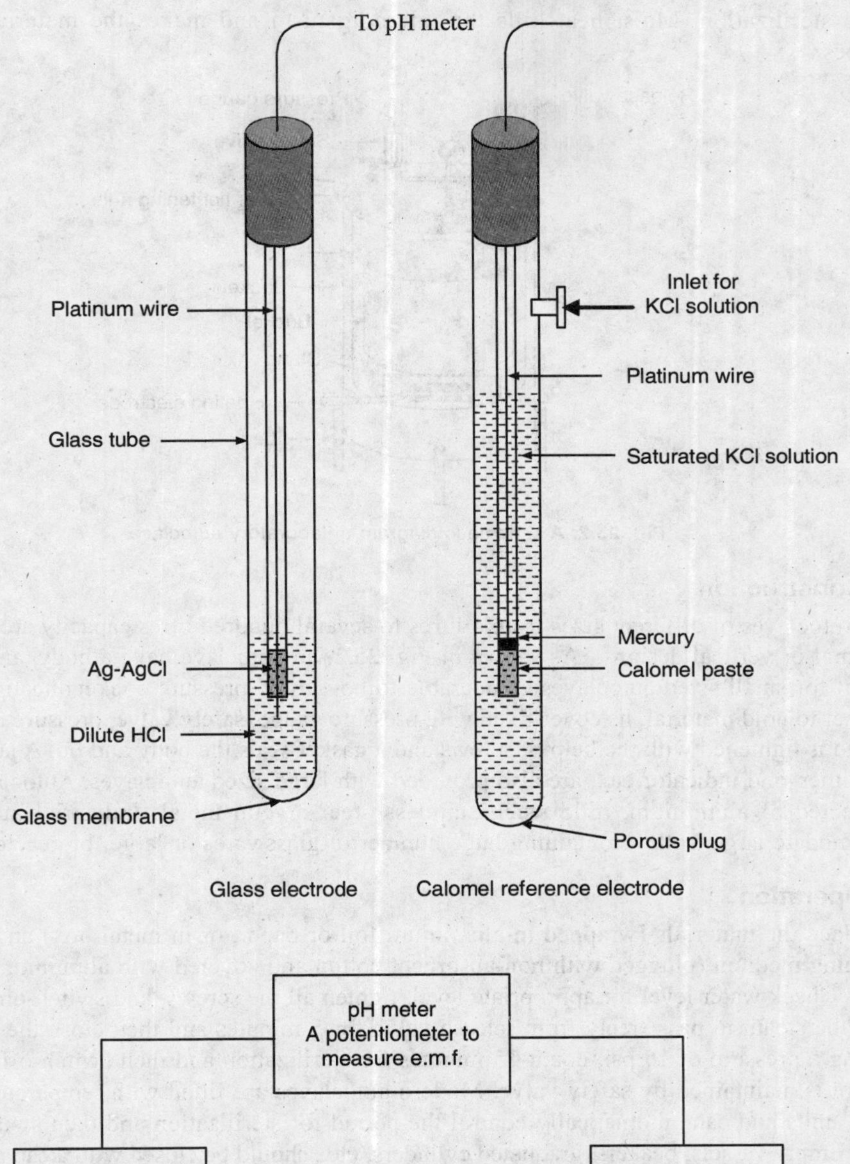

Fig. 25.1. The construction of pH meter electrodes and basic principle.

2. AUTOCLAVE

Autoclave is used to sterilize medium, glassware and tools for the purpose of plant tissue culture. The same equipment is used in hospitals to sterilize gauge, cotton, tools and linen, etc. Sterilization of material is carried out by increasing moist heat (121°C) due to increased pressure inside the vessel (15-22 psi, pounds per square inch or 1.02 to 1.5 kg/cm^2) for 15 minutes for routine sterilization. Moist heat kills the microorganism and makes the material free from microbes.

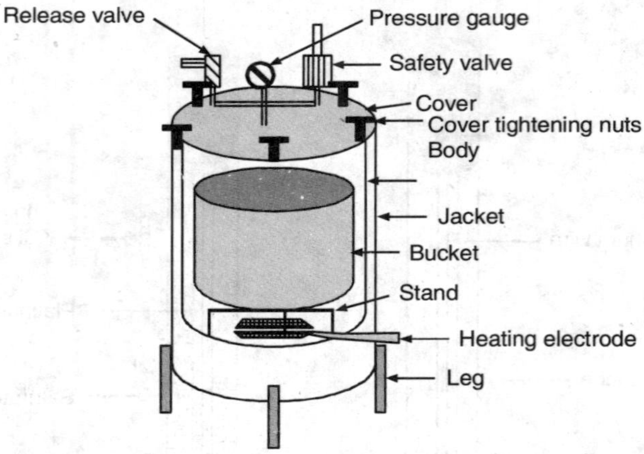

Fig. 25.2. A schematic diagram of laboratory autoclave.

2.1. Construction

Autoclaves of different sizes from 5 litres to several hundred litres capacity are available in horizontal or vertical designs. As shown in Fig. 25.2, an autoclave have a body, an internal (or external for small sized autoclaves comparable to household pressure cooker) heating system, a container to hold material, its cover fixed with pressure gauge, safety valve, pressure release valve etc. Lid is tightened with the help of screws and a gasket seals the body and lid. A jacket, paddle lifter, timer, and indicator etc., are also provided with large sized autoclaves. Autoclaves may be constructed of aluminium, mild steel, stainless steel or gun metal. Industrial autoclave can accommodate large trolley containing huge number of glasswares or large bioreactors.

2.2. Operation

Place the materials (wrapped in aluminium foil or paper, or in metal box) and glasswares containing medium (plugged with non-absorbent cotton and covered with aluminium foil) in the bucket. Check water level for appropriate level, tighten all the screws, and switch-on the current. Allow the steam to pass freely from release valve for 5 minutes and then close the valve. After attaining a pressure of 15 psi, count 15 minutes for sterilization and then switch-off the current. Pressure is maintained by safety valve. Modern autoclaves are fitted with temperature and time control units and can automatically control the period for sterilization and then switch-off them selves. Empty vessels, beakers, graduated cylinders, etc., should be closed with a cap or aluminum foil. Tools should also be wrapped in foil or paper or put in a covered sterilization tray. It is critical that the steam penetrate the items in order for sterilization to be successful.

Table 25.1. Volume of solution and duration for its sterilization.

ml/Container	Duration at 121°C (min)
20-50	15
75	20
250-500	25
1000	30
1500	35
2000	40

For large sized vessels and large volume flasks containing high amount of liquid, duration for sterilization should be increased accordingly. Table 25.1 shows the relationship between volume of the solution and duration for its sterilization at 15 psi. The vessels should not be filled more than 1/3 of its capacity for proper sterilization.

2.3. Precautions

1. Check water level each time, the heating elements should remain immersed in the water.
2. Check spring of safety valve frequently and clean opening whenever necessary.
3. Opposite screws of the lid should be tightened simultaneously.
4. Do not over tighten the screws to avoid damage to the gasket.
5. Use permanent marker to mark your flasks and its medium.
6. All the electrical equipment should be properly earthen to avoid electric shock.

3. PLANT GROWTH CHAMBER

Plant growth chambers can be constructed in a suitable sized room or can be purchased as commercially available equipment. Thermal insulation of walls increases the efficiency of the cooling system. Essentially plant growth chamber has three environmental control systems:

1. Light-intensity and duration cycle control.
2. Temperature control and regulation.
3. Humidity control and regulation.

All the modern instruments are electronically controlled precision instruments with sophisticated sensors and timers to regulate the desired set values.

3.1. Light

Light is fixed in the roof of equipment or in shelves. Light is provided by commercially available light sources like cool white fluorescent tubes and incandescent lamps in a ratio of 3:1 and usually a light intensity of 2000-2500 lux (about 200-250 candles or 30 μ mol m^{-2} s^{1}) is used. The duration of light and dark cycle is adjusted as per requirement, usually 16 hours light cycle is given. Nowadays warm fluorescent tubes are also available which provides wide spectrum as compared to cool white fluorescent tubes and mixing of incandescent light is not required with former tubes. Light intensity can also be regulated by photoperiod simulators. Thus, light quality, intensity and period is controlled and regulated by the instrument as per set valves.

3.2. Temperature

In modern equipment, temperature is precisely regulated by good quality (platinum) temperature sensor. In all cases, air conditioning units provide the cooling. It is always advisable to keep one spare compressor unit, for emergency, to avoid delay in repairs and damage to cultures. Usually temperature of 22-28 °C is used for growing plant tissue culture. Temperatures should be measured in a constructed growth chamber at different levels and places, viz., light racks, central and corners to have a correct temperature setting.

3.3. Humidity

Humidity inside the growth chamber is provided by humidifier (a mist generating system) and controlled by humidistate. Usually 60% RH (relative humidity) is used to maintain healthy growth. Low RH may cause early drying of medium while high humidity may cause fungal growth in the environment and on a various articles.

Thus, in a growth chamber, light, temperature and humidity are precisely controlled and cultures are grown in a controlled environment. All the controls are set on control panel.

4. LAMINAR AIR FLOW BENCH

Laminar air flow (LAF) bench is the main working table for aseptic manipulations related to plant tissue culture. This is an equipment fitted with High Efficiency Particulate Air (HEPA) Filters, which allow air to pass but retain all the particles and micro-organisms. These HEPA Filters have a very small pore size (0.3 μm) with 99.97-99.99% efficiency. Air blown by a blower is passed through these filters, thus, always generally a positive pressure of air is maintained from inside to outside. This positive pressure does not allow any particle to enter in the working area of LAF bench. Air pressure inside the instrument is measured by a manometer and pressure of more than 12 bars shows choking of filters and at this point filters should be replaced. Equipment is fitted with UV light and visible light source. UV is switched-on for 30 minutes before starting the work to make area free from microbes. LAF Bench of steel and wooden cabinets is available with different working table size and for vertical (downward air flow) or horizontal (horizontal flow) model. There are several manufacturers in India for this instrument (body) but HEPA filters are imported (Fig. 25.3). HEPA filters are also used to create 'clean area' for culture rooms and inoculation room etc. If LAF bench is placed in such a clear area, efficiency and life of the equipment are increased.

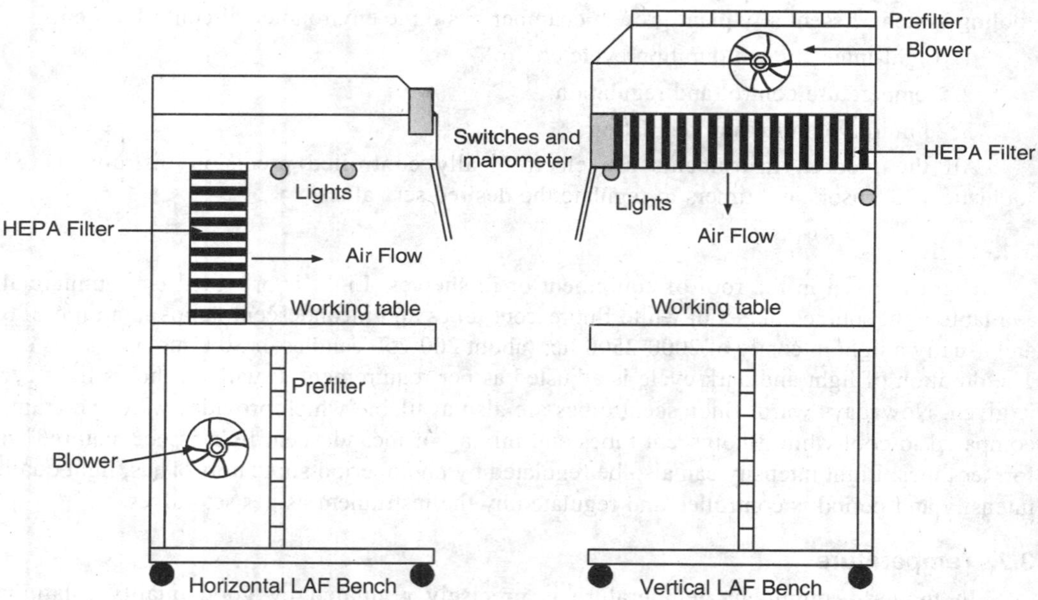

Fig. 25.3. Line diagram of horizontal laminar air flow bench.

5. MICROSCOPY

5.1. Electron Microscopy

Electron microscopy permits a detailed study of sub-cellular organelles as its resolving power is much greater than that of the light microscope. Max Knoll and Ernst Ruska in 1931, at Technical University in Berlin, constructed electron microscope (EM). In the EM, streams of electrons are deflected by an electrostatic or electromagnetic field in the same way that a beam of light is refracted by a lens. Electron beam is generated by heating a filament in vacuum, which are accelerated by a potential and shows properties similar to light (λ = 0.005nm of electrons and 550 nm for light). Though appears to be similar there are great differences in light and electron microscopes, the principal being the image formation (Fig. 25.4.). In electron microscope, image is produced by electron scattering. Electron dispersion is a function of the thickness and molecular packing of the object and depends especially on the atomic number of the atoms in the object. The higher the atomic number, the greater is the dispersion. There are three general types of electromagnetic lenses. The one is placed between the source of illumination and the specimen. This focuses the beam of electrons on specimen and functions in a similar manner as the condenser in light microscope. The other two lens systems are on the opposite side of specimen which magnify the image in similar way as objective and ocular in light microscope. The final image is produced on a screen coated with a phosphorus compound which fluoresces upon irradiation by electrons. These images are recorded on photographic films. Air molecules in the microscope interfere with the movement of electrons. To prevent this, a high vacuum ($10^{-4} - 10^{-6}$ mm Hg) is created inside the microscope. In light microscope, magnification is largely determined by the objective and the maximum magnification of 100-120 can be obtained. This is multiplied by ocular lens by a factor of 5-15, reaching a total magnification of 500-1500x.

The resolving power of Transmission Electron Microscope (TEM) is very high. The image generated by objective can be multiplied several hundred times by projector coil, e.g., 100 objective 200x projector coil = 20,000x. In TEM, this can be reached up to 10,000,000x. Electron microscope has a greater depth of field as compared to light microscope. This property of EM is used to develop three dimensional images of cell organelles as ribosomes and protein structure.

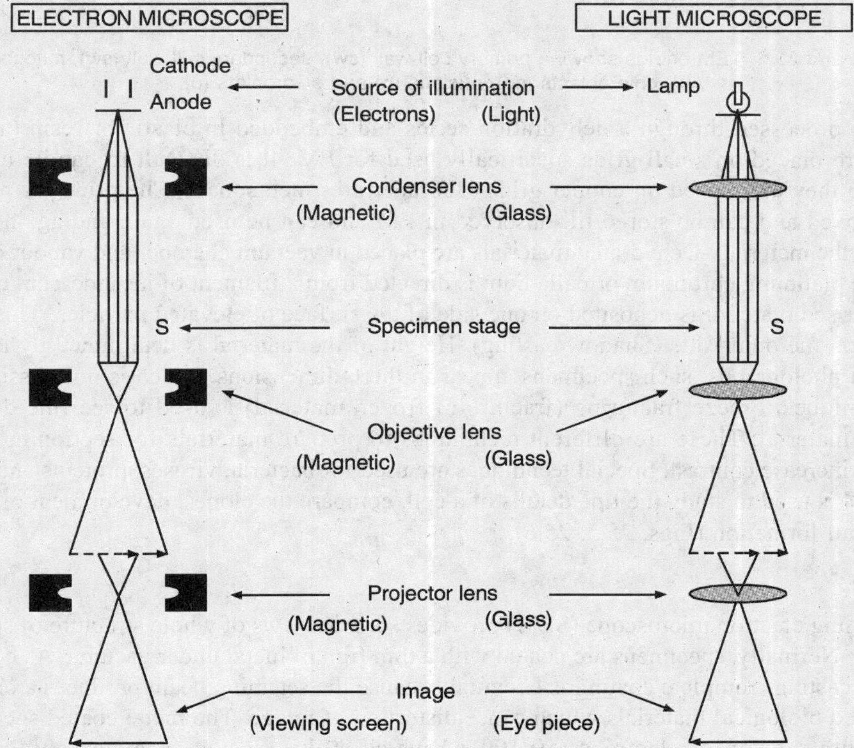

Fig. 25.4. Diagrammatic comparison of image formation in electron microscope and light microscope.

The preparation of specimen is similar to dehydration used for histology for light microscopy. The sections prepared for EM should be thin (less than 0.5 or 500 nm), otherwise, material will appear completely dark. Such thin sections (50-300 nm) are prepared with the help of ultramicrotome, fitted with glass or diamond knife. To cut sections, material is fixed in osmium

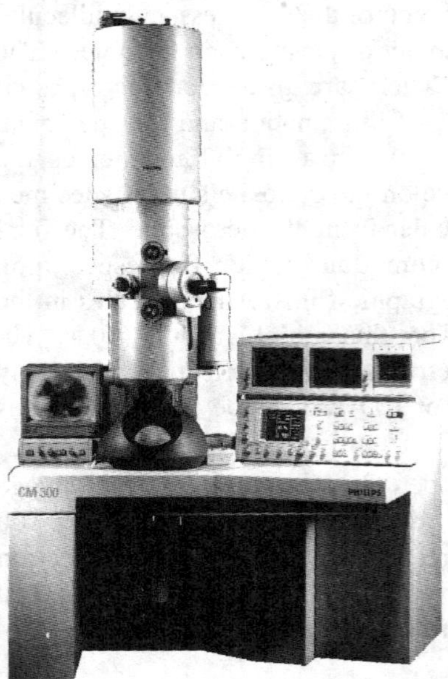

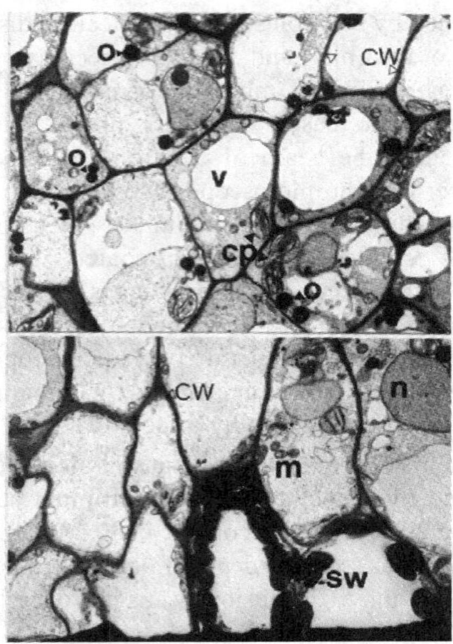

Fig. 25.5 and 25.6. TEM photos showing primary cell wall (cw), secondary cell wall (sw), mitochondria (m), chloroplasts (cp), nucleus (n) and oil droplets (o).

tetraoxide, processed through a dehydration series and embedded in plastic or resin (araldite). Sections are placed on small grids specifically used for EM. It is difficult to handle ultra thin sections so they are placed on copper grids (400 apertures/inch square).The grids are placed in labelled boxed and can be stored till observed in EM and can be used again and again without damaging the material. Dehydrated materials are placed in vacuum chamber and vapour of heavy metals like platinum, chromium or palladium is directed from a filament of incandescent tungsten. The vapourised material is deposited on one side of the surface of elevated particles. This creates a shadow on the other side (Shadow casting). Height of the material is determined by length of shadow. In photographs, such specimens appear in three dimensions, which is not possible with other techniques. Freeze fracturing (fracture of frozen material) is used to see fine details of replicated material. These are different techniques to prepare materials for sectioning and for staining to increase contrast. Special techniques are used for bacteria, viruses, proteins and nucleic acids. TEM is used to study the fine details of a cell, compare the clones, development of plastids and cell wall formation (Figs. 25.5, 25.6).

5.2. SEM

Scanning electron microscope (SEM) provides surface views of whole structure of specimen (Fig. 25.7). Normally, specimens are coated with a thin film of metal under vacuum. As compared to shadow casting, complete coating is essential because the scanning beam produces a charge on the uncoated biological materials which cause distortion of image. The metal coated specimen is scanned with a beam of electrons (50-100 Å) which strikes on the specimen and emits the secondary electrons. These electrons pass through a cathode tube and image is produced on the screen of Cathode Ray Tube.

In plant system, SEM is used to see the nature of appendages, pollen morphology and morphogenesis in plant tissue culture. It is also used to study the structures of metals, crystals and in forensic sciences.

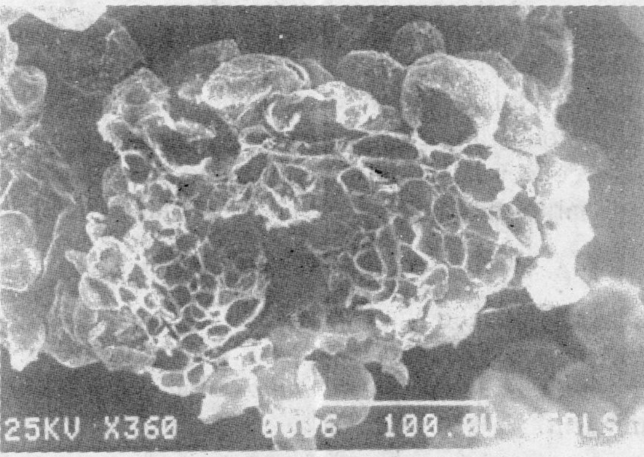

Fig. 25.7. SEM photograph of semiorganized and fractured callus.

5.3. Light Microscopy

Bright field microscopy is absolutely indispensable tool for cell biologists. This is required for routine observations of cells, cellular differentiation and pigmentation. Fluorescence microscopy has become a powerful tool for cell biologists, particularly for the selection of fluorescing secondary metabolite rich cells. Only these two techniques are discussed here in brief.

According to wave theory, light is propagated from one place to another as wave travelling in a hypothetical medium. Light waves are described in terms of their amplitude, frequency, and wavelength. Amplitude is the maximum displacement of light path from the position of equilibrium. Frequency of light is the number of complete cycles occurring in a second. The property of light waves inversely associated with the frequency is the wavelength and is defined as the distance between corresponding points on a wave or distance between two successive peaks or crests. The velocity of light is about 186300 miles or 3×10^{15} kms per second.

A good microscope has not only good magnifying power but also good resolving power to provide finer details of the object. Thus, the basic difficulty in designing a microscope is not the magnification, but the ability of lens system to distinguish two adjacent points as distinct and separate. This ability is known as resolving power of the microscope. The resolving power of a microscope depends upon the wavelength of light and numerical aperture (N.A.). The minimum resolvable distance between two luminous points (v) is given by the following formula –

$$V = \frac{0.16\lambda}{\text{N.A.}} \quad \text{where, } \lambda = \text{wavelength}$$

Thus, shorter the wavelength of light used and lower the N.A., greater is the resolving power. The limit of resolution of a microscope is approximately equal to 0.5 /N.A., which for a light microscope is approximately 200 nanometers (nm) or about the size of many bacterial cells.

The numerical aperture of a lens is dependent on the refractive index (η = the ratio of the speed of light in a given medium to the speed of light in a vacuum) of the medium filling the space between the specimen and the front of the objective lens and on the angle of the most oblique rays of light that can enter the objective lens (θ). It is given by formula [N.A. = $\eta\ x \sin\theta$]. That is why immersion oil is placed between the object and oil immersion lens (100 X objective).

6. COLORIMETER

The most commonly used method for determining the concentration of biochemical compounds is colorimetry. It uses the property of light such that when white light passes through a coloured solution, some wavelengths are absorbed more than others. Hyaline solution can be made coloured by specific reactions with suitable reagents. These reactions are generally very sensitive to determine quantities of material in the region of millimole per litre concentration. The big advantage is that complete isolation of the compound is not necessary and the constituents of a complex mixture such as blood can be determined after little treatment. The depth of colour is directly proportional to the concentration of the compound being measured, while the amount of light absorbed is proportional to the intensity of the colour and therefore, to the concentration.

6.1 Monochromator

A **monochromator** is an optical device that transmits a mechanically selectable narrow band of wavelengths of light or other radiation chosen from a wider range of wavelengths available at the input. The name is from the Greek language *mono-*, single, and *chroma*, colour, and the Latin suffix *-ator*, denoting an agent.

The main goal of a monochromator is to separate and transmit a narrow portion of the optical signal chosen from a wider range of wavelengths available at the input. In the simplest case the monochromator is composed from two slits (entrance and exit) and a dispersion element (prism or diffraction grating). In both these elements, one takes advantage of the dependence of the refraction angle (prism) or the reflection angle (grating) on the radiation wavelength. In the case of prism, the larger the photon energy (shorter wavelength), the smaller is the refraction angle. The main goal of the entrance slit is to define the geometric properties of the investigated irradiation. The dispersion or diffraction is only controllable if the light is collimated, that is if all the rays of light are parallel, or practically so. The goal of the prism is to disperse the light into a rainbow. At the exit slit, the colours of the light are spread out (in the visible this shows the colours of the rainbow). Because each colour arrives at a separate point in the exit slit plane, there are a series of images of the entrance slit focused on the plane. Because the entrance slit is finite in width, parts of nearby images overlap. The light leaving the exit slit contains the entire image of the entrance slit of the selected colour plus parts of the entrance slit images of nearby colours. A rotation of the dispersing element causes the band of colours to move relative to the exit slit, so that the desired entrance slit image is centred on the exit slit. The range of colours leaving the exit slit is a function of the width of the slits. The entrance and exit slit widths are adjusted together. It is obvious, therefore, that the monochromator refracting power increases with the decrease of the slit width and with the increase of the distance between the slit and the prism.

"Optical density" can also refer to index of refraction.

In spectroscopy, the **absorbance** (also called **optical density**) of a material is a logarithmic ratio of the radiation falling upon a material, to the radiation transmitted through a material. Absorbance measurements are often carried out in analytical chemistry.

Absorbance is a quantitative measure expressed as a logarithmic ratio between the radiation falling upon a material and the radiation transmitted through a material.

$$A_\lambda = -\log_{10}\left(\frac{I_1}{I_0}\right)$$

where A_x is the absorbance at a certain wavelength of light (λ), I_1 is the intensity of the radiation (light) that has passed through the material (transmitted radiation), and I_0 is the intensity of the radiation before it passes through the material (incident radiation).

The term *absorption* refers to the physical process of absorbing light, while *absorbance* refers to the mathematical quantity. Also, absorbance *does not always measure absorption*: if a given sample is, for example, a dispersion, part of the incident light will in fact be scattered by the dispersed particles, and not really absorbed. However, in such cases, it is recommended that the term "attenuance" (formerly called "extinction") be used, which accounts for losses due to scattering and luminescence.

Absorbance	Transmittance (I_1/I_0)
3	0.001
2	0.01
1	0.1
0.9	0.13
0.75	0.18
0.5	0.32
0.25	0.56
0.1	0.79
0	1

A **chromophore** is the part of a molecule responsible for its colour. The colour arises when a molecule absorbs certain wavelengths of visible light and transmits or reflects others. The chromophore is a region in the molecule where the energy difference between two different molecular orbits falls within the range of the visible spectrum. Visible light that hits the chromophore can thus be absorbed by exciting an electron from its ground state into an excited state.

In biological molecules that serve to capture or detect light energy, the chromophore is the moiety that causes a conformational change of the molecule when hit by light.

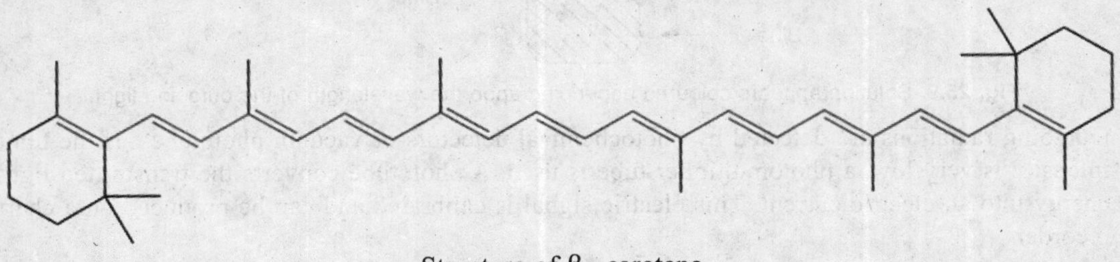

Structure of β –carotene

6.2. Source of Radiation

A lamp is usually used as the source of radiation. The wavelength of light depends upon the quality of source lamp.

Colorimetry is an analytical device to determine the amount of an unknown substance in a solution. The device works on optical properties of the substance. Basically we measure the amount of light of a particular wavelength absorbed by that solution, and then relate solute concentration to the absorbance. A simplified schematic presentation of the instrument is as follows:-

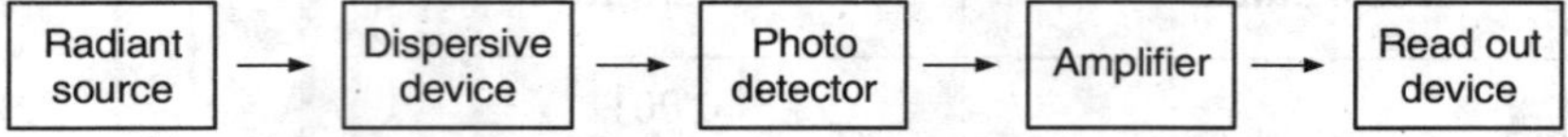

The relationship between concentration of solution and light it absorbs is given by the following laws.

Lambert's Law - When a ray of monochromatic light passes through an absorbing medium its intensity decreases exponentially as the length of the absorbing medium increases.

Beer's Law - When a ray of monochromatic light passes through an absorbing medium its intensity decreases exponentially as the concentration of the absorbing medium increases.

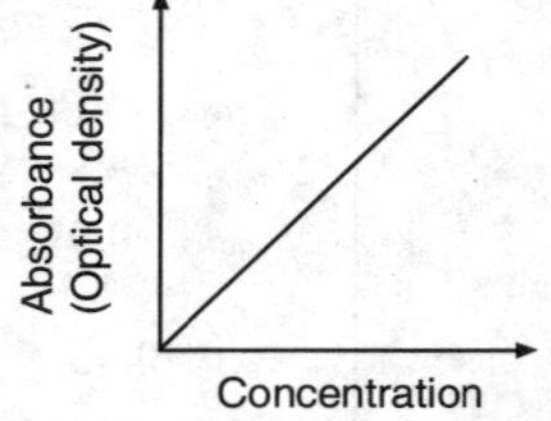

Fig. 25.8. Relationship between concentrations and absorbance.

Transmittance - The ratio of intensities is known as the transmittance and this is usually expressed as percentage. Absorbance or extinction is known as the optical density of the substance. Absorbance increases with increase in concentration of the solution (Fig. 25.8).

6.3. Detection of Radiation

Detection wavelength is selected on the basis of wavelength of light coming out of a solution (Fig. 25.9). Absorption of a particular wavelength of light is characteristic of a compound.

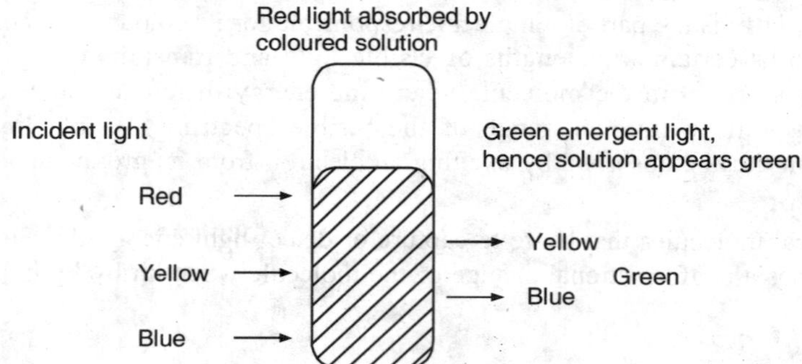

Fig. 25.9. Solution appears coloured depending upon the wavelength of the outgoing light.

Outgoing radiations are detected by photochemical detectors or vacuum phototubes. If the light intensity is very low, a photomultiplier tube is used. A phototube converts the transmitted light energy into an electric current. This electric signal is calibrated and can be produced on a chart recorder.

A better method is to split the light beam, pass one part through the sample and the other through the blank, and balance the two circuits to give zero (a double beam system). The extinction is determined from the potentiometer reading, which balances the circuit.

6.4. Applications

Measurement of concentrations - Spectrophotometers are widely used for quantitative measurements of substance concentration in the solutions by measuring absorbance at the optimal wavelength. This is used to determine concentration in fractions obtained in column chromatography, TLC and in other solutions.

Absorption spectra - Identification and structure evaluation of various isolated pure compounds can be done by UV-Visible spectrophotometry. All the compounds have very specific absorption spectra depending on the molecular structure. Comparison of spectra helps in identification of compounds. The wavelength at which maximum absorption takes place is represented as λ_{max} of that compound in a given solvent.

Enzyme kinetics - In enzymatic reactions, changes in absorbance are recorded to determine the reaction velocity and concentration of the product formed/substrate utilized. Spectrophotometers coupled with microprocessor or computer can store data, figures and are helpful in comparing the effect of different variables.

7. CENTRIFUGATION

A centrifuge is an instrument which produces centrifugal force by rotating the samples around a central axis with the help of an electric motor. Centrifuges can be categorized as the clinical type (5-10,000 rpm), refrigerated high-speed centrifuges (10,000-20,000 rpm) and ultra - centrifuges (20,000 to 80,000 rpm). With increase in rpm, the friction of rotor with air produces so much of heat that they have to be run under refrigeration (so called refrigerated centrifuge) and both refrigeration and vacuum are used in ultracentrifuge, which runs at very high rpm. For these high speeds, even the rotor has to be made of special metal to withstand the great force.

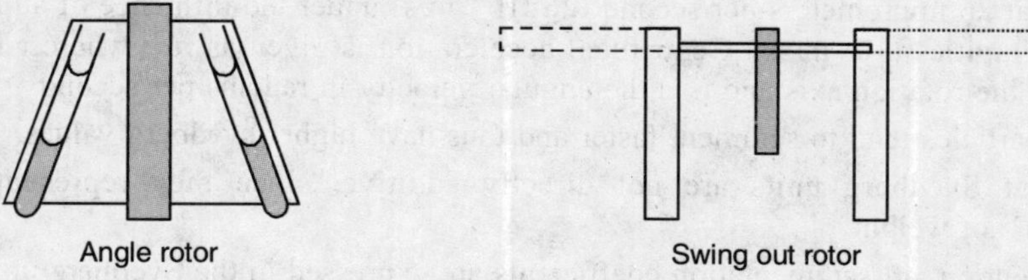

Fig. 25.10. Different types of rotors used in centrifugation.

There are two types of rotors, angle head and swing (Fig. 25.10). In the former, the samples are kept at an angle of about 30° to the vertical axis whereas, in the latter the samples while spinning are horizontal. Simple calculations show that for the same radius, the swinging bucket method produces more gravitational force. Ultracentrifuges are of two types - analytical and preparative model.

7.1. Analytical Model

This consists of rotors and tubes, called cells. The instrument is designed to allow the operator to follow the progress of the substances in the cells, while the process of centrifugation is in progress. By estimating sedimentation velocity during the process, the molecular weight, purity etc. can be determined.

7.2. Preparative Model

This is used for purification of the components of macromolecules or other substances and all determinations are made at the end of centrifugation. The instrument has no monitoring device, while large centrifugal forces are set for a fixed time period.

Centrifugation is the most widely used technique for separation of various metabolites and also used to separate non-miscible liquids during extraction of secondary metabolites, e.g., water (aqueous), chloroform (organic) mixture. A wide variety of centrifuges are available, ranging in capacity and speed. During the process of centrifugation, solid particles experience a centrifugal force, which pulls them outwards, i.e., away from the center. The velocity with which a given solid particle moves through a liquid medium is related to angular velocity.

7.3 WORKING PRINCIPLE

A **svedberg** unit (symbol **S**, sometimes **Sv**) is a non-SI unit for sedimentation rate. The sedimentation rate for a particle of a given size and shape measures how fast the particle 'settles', or **sediments**. It is often used to reflect the rate at which a molecule travels to the bottom of a test tube under the centrifugal force of a centrifuge. The svedberg is technically a measure of time, and is defined as exactly 10"13 **seconds** (100 fs).

The Svedberg unit (S) offers a measure of particle size based on its rate of travel in a tube subjected to high g-force.

The unit is named after the Swedish chemist Theodor Svedberg (1884–1971), winner of the 1926 Nobel Prize in chemistry for his work on colloids and his invention of the ultracentrifuge. 50S is the larger subunit of the 70S ribosome of prokaryotes. It is the site of inhibition for antibiotics such as macrolides, chloramphenicol, clindamycin, and the pleuromutilins. It includes the 5S ribosomal RNA and 23S ribosomal RNA.

The Svedberg coefficient is a nonlinear function. A particle's mass, density, and shape will determine its S value. It depends on the frictional forces retarding the particle's movement, which in turn are related to the average cross-sectional area of the particle.

The sedimentation coefficient is the ratio of the speed of a substance in a centrifuge to its acceleration in comparable units. A substance with a sedimentation coefficient of 26S (26×10"13 s) will travel at 26 micrometers per second (26×10"6 m/s) under the influence of an acceleration of a million gravities (10^7 m/s^2). Centrifugal acceleration is given as rù^2; where r is the radial distance from the rotation axis and ù is the angular velocity in radians per second.

Bigger particles tend to sediment faster and thus have higher svedberg values.

Note that Svedberg units are not directly additive, since they represent a rate of sedimentation, not weight

By convention, all sedimentation coefficients are expressed in the Svedberg units.

Protocols for centrifugation typically specify the amount of acceleration to be applied to the sample, rather than specifying a rotational speed such as revolutions per minute. This distinction is important because two rotors with different diameters running at the same rotational speed will subject samples to different accelerations. During circular motion the acceleration is the product of the radius and the square of the angular velocity ω, and the acceleration relative to "g" is traditionally named "relative centrifugal force" (RCF). The acceleration is measured in multiples of "g" (or × "g"), the standard acceleration due to gravity at the Earth's surface, a dimensionless quantity given by the expression:

$$RCF = \frac{r\omega^2}{g}$$

where

g is earth's gravitational acceleration,

r is the rotational radius,

ω is the angular velocity in radians per unit time

This relationship may be written as

$$RCF = 1.11824396 \times 10^{-5} r_{\text{cm}} N^2_{\text{RPM}}$$

where

r_{cm} is the rotational radius measured in centimetres (cm), and

N_{RPM} is rotational speed measured in revolutions per minute (RPM).

7.4. Applications

Centrifugation is widely used in analytical techniques, preparation of extracts, separation of non-miscible liquid mixtures, purification of enzymes, inhibitors and removing particles (silica, etc.) from the solvents. Centrifugation with heating and vacuum (suction) is used for concentrating extraction (instrument called sample concentrator) and rapidly removing solvents. All the enzymatic work requires refrigerated centrifuges to keep the samples cool during centrifugation and to protect enzymes from inactivation by heat.

8. CHROMATOGRAPHY

Chromatography (meaning 'coloured writing') is a technique to separate molecules on the basis of differences in size, shape, mass, charge and adsorption properties. The term chromatography was used by the Russian botanist Tswett to describe the separation of plant pigments on a column of alumina. There are different types of chromatography but they all involve interactions between these components : the mixture to be separated, a solid phase, and a solvent. The magnitude of these interactions depends upon the particular method used. Usually column chromatography is used to separate large quantities of compounds, whereas, paper chromatography (PC) or thin-layer chromatography (TLC) in one or two dimensions, is used for analytical work.

The mobile phase can be a gas or a liquid, whereas the stationary phase can only be a liquid or solid. In liquid column chromatography (LCC), separation involves a simple partitioning between two immiscible liquid phases, one stationary and the other mobile, the process is called liquid-liquid (or partition) chromatography (LLC). When physical surface forces are mainly involved in the retentive ability of the stationary phase, the process is denoted liquid-solid (or adsorption) chromatography (LSC). In ion exchange chromatography (IEC), ionic components of the sample are separated by selective exchange with counter ions of the stationary phase. Use of exclusion packings as the stationary phase brings about a classification of molecules based largely on molecular geometry and size.

8.1. Paper Chromatography

Principle - Cellulose in the form of paper sheets makes an ideal support medium where water is adsorbed between the cellulose fibres and forms a stationary hydrophilic phase.

The suitably concentrated mixture is spotted onto the paper, dried with a hair-drier and the chromatogram is developed by allowing the solvent to flow along the sheet. The solvent front is marked and after drying the paper, the positions of the compounds present in the mixture are visualized by a suitable staining reaction. The ratio of the distance moved by a compound to that moved by the solvent is known as the R_f value and is more or less constant for a particular compound, solvent system and paper under carefully controlled conditions of solute concentration, temperature and pH.

Sample - Generally, alcoholic extracts of plant material with or without partial purification are used for chromatography. Biological materials should be desalted before chromatography by electrolysis or electrodialysis. Excess salt results in a poor chromatogram with spreading of spots and changes in their R_f values. It can also affect the chemical reactions used to detect the compounds being separated. The sample (10-20 μl) is then applied to the paper with a micropipette or capillary.

Paper - Whatman No. 2 is the paper most frequently used for analytical purposes. Whatman No. 3 MM is a thick paper and is best employed for separating large quantities of material, the resolution is, however, inferior to Whatman No. 1. For rapid separation, Whatman Nos. 4 and 5 are convenient, although the spots are less well defined. In all cases, the flow rate is faster in the 'machine direction', which is normally noted on the box containing the paper. The paper may be impregnated with a buffer solution before use or chemically modified by acetylation. Ion exchange papers are also available commercially. For the separation of lipids and similar hydrophobic molecules, silica-impregnated papers are available commercially.

Solvent - This choice, like that of the paper, is largely empirical and will depend on the mixture investigated. If the compounds move close to the solvent front in solvent 'A' then they are too soluble, while if they are crowded around the origin in solvent 'B' then they are not sufficiently soluble. A suitable solvent for separation would, therefore, be an appropriate mixture of 'A' and 'B', so that the R_f values of the components of the mixture are spread across the length of the paper. R_f value is defined as [R_f = the distance moved by solute / the distance moved by solvent front]. This value is a constant for a particular compound under standard conditions and closely reflects the distribution coefficient for that compound. It is recommended that the developing chamber should be saturated with solvent, by using filter paper lining inside the chamber.

8.2. Two Dimensional Chromatography

The mixture is separated in the first solvent, which should be volatile, then after drying, the paper is turned through 90° and separation is carried out in the second solvent. After location, a map is obtained and compounds can be identified by comparing their position with a map of known compounds developed under the same conditions (Fig. 25.11).

Detection of spots - Most compounds are colourless and are visualized by specific reagents. The location reagent is applied by spraying the paper under a fumigation hood or rapidly dipping it in a solution of the reagent in a volatile solvent. Viewing under ultraviolet light is also useful since some compounds, which absorb strongly show up as dark spots against the fluorescent background of the paper. Other compounds show a characteristic fluorescence under ultraviolet light.

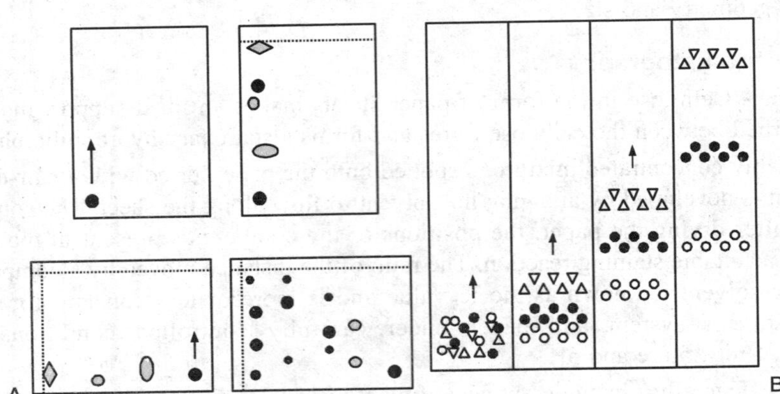

Fig. 25.11. Separation of a mixture by (A) two-dimensional chromatography and, (B) principle behind the chromatographic separation.

8.3. Thin Layer Chromatography

Principle - Separation of compounds on a thin layer of adsorbing material is similar in many ways to paper chromatography, but has the added advantage that a variety of supporting media can be used so that separation can be by adsorption ion exchange, partition chromatography, or gel filtration depending on the nature of the medium employed. The method is very rapid compared to paper chromatography and many separations can be completed within an hour. Compounds can be detected at a lower concentration than on paper as the spots are very compact.

Furthermore, separated compounds can be detected by corrosive sprays and elevated temperatures with some thin layer materials, which of course is not possible with paper.

Production of thin layer - The R_f value is affected by the thickness of the layer below 200 μm and a thickness of 250 μm is suitable for most separations. There are several good spreaders available in the market, which, can produce an even layer of required thickness by adjusting the thickness control screw. Calcium sulphate is sometimes incorporated into the adsorbent to bind the layer to the plate and, because of this, it is advisable to work rapidly once the adsorbent is mixed with water. There are now a number of prepared thin layer plates using different adsorbents on various supporting materials such as glass, plastic, and aluminium that are available commercially and these may be more convenient to use than trying to prepare plates in the laboratory.

Development - It is essential to make sure that the atmosphere of the separation chamber is fully saturated with the solvent mixture, otherwise R_f values will vary widely from tank to tank. However, horizontal chambers of high performance TLC (HPTLC) are very useful and convenient as they require less time, solvent and no stabilization time. Development of the plate is usually by the ascending technique and is very rapid.

9. THERMOMETER

The thermometer is a device that measures temperature or temperature gradient using a variety of different principles; it comes from the Greek roots *thermo*, heat, and *meter*, to measure. A thermometer has two important elements: the temperature sensor (e.g., the bulb on a mercury thermometer) in which some physical change occurs with temperature, plus some means of converting this physical change into a value (e.g., the scale on a mercury thermometer). Industrial thermometers commonly use electronic means to provide a digital display or input to a computer.

The Alcohol thermometer or *Spirit thermometer* is an alternative to the Mercury-in-glass thermometer, and functions in a similar way. An organic liquid is contained in a glass bulb which is connected to a capillary of the same glass and the end is sealed with an expansion bulb. The space above the liquid is a mixture of nitrogen and the vapour of the liquid. For the working temperature range, the meniscus or interface between the liquid is within the capillary. With increasing temperature, the volume of liquid expands and the meniscus moves up the capillary. The position of the meniscus shows the temperature against an inscribed scale.

The liquid used can be pure ethanol or toluene or kerosene or Isoamyl acetate, depending on manufacturer and working temperature range. Since these are transparent, the liquid is made more visible by the addition of a red or blue dye. One half of the glass containing the capillary is usually enamelled white or yellow to give a background for reading the scale. Temperature is measured by maximum-minimum thermometer or a continuous rotary drum chart type thermometer. The U-shaped maximum-minimum thermometer is commonly used for determining diurnal maximum and minimum range of temperature in the culture room. During the dark period, temperature remains slightly (1-2 °C) lower than light period (all bulbs and tube lights of the culture room increase the temperature). Therefore, chokes of the tube lights are fitted outside the culture room. The indicators of the maximum-minimum thermometer are moved by mercury column and they remain at that position until moved by the observer with the help of magnet. Their positions indicate the minimum and maximum temperature in the previous 24 h. After recording temperature indicators are reset to mercury level (Fig. 25.12).

10. HYGROMETER

Hygrometers are instruments used for measuring humidity. A simple form of a hygrometer is specifically known as a "psychrometer" and consists of two thermometers, one of which includes a dry bulb and the other of which includes a bulb that is kept wet to measure *wet-bulb temperature*. Evaporation from the wet bulb lowers the temperature, so that the wet-bulb thermometer usually shows a lower temperature than that of the dry-bulb thermometer, which measures *dry-bulb temperature*. Relative humidity is computed from the ambient temperature as shown by the dry-bulb thermometer and the difference in temperatures as shown by the wet-bulb and dry-bulb thermometers. Relative humidity can also be determined by locating the intersection of the wet- and dry-bulb temperatures on a psychrometric chart.

Dial type hair hygrometer is a convenient tool to measure humidity in the culture room. Humidifier is used to maintain 60% relative humidity (RH) in the culture room. Humidistate is a bimetallic thermocouple device to control and regulate the function of humidifier to maintain the humidity. Distilled water should be filled in the humidifier. The RH present in the culture room is measured by hair hygrometer. As the name suggests a chemically treated hair elongates with increased humidity and shortens with dryness (similar to mercury in thermometer). It is calibrated from 0 to 100% RH. At lower humidity, medium dries rapidly whereas at higher humidity chances of fungal growth over all surfaces and cotton plugs is increased. Therefore, about 60% RH is maintained in the culture room (Fig. 25.12).

11. LUX METER

Light is the form of radiant energy, i.e., electromagnetic radiation of specific wavelength. Visible light as we perceive, is located in narrow wavelength region of spectrum between 380 to 760 nm. Light has dual characters, displaying both wave properties (refraction, diffraction, interference and polarization phenomena) and particle properties (light is radiated in discrete amounts of energy or photons).

$$\text{Light intensity} = \frac{R \times 100}{A}$$

R = Reading of lux meter

A= Reflected light

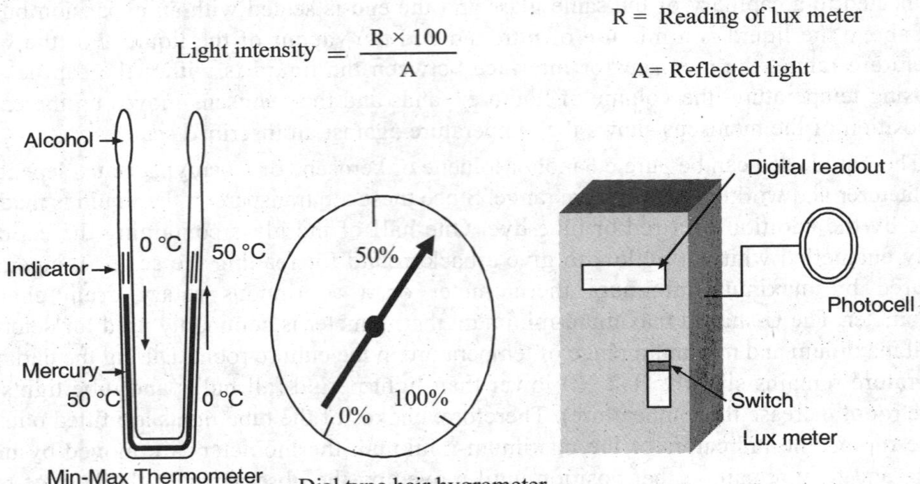

Fig. 25.12. Various accessories required in culture room.

Properties of light should be defined either as irradiance (radiant flux intercepted per unit area; unit = w/m^2) or an illuminance (luminous flux intercepted per unit area, unit lux, 10.76 lux = 1 foot candle). It is to note that an irradiance measurement is not spectrally defined, whereas, an illuminance measurement indicates the level of visible light as the human eye would see it. Artificial light is provided by cool, white, fluorescent tubes and/or incandescent bulbs. Of different classes of artificial light sources, fluorescent sources have been used almost exclusively for plant tissue culture because they are more efficient at producing broad band visible light than incandescent bulbs and are available in low output wattage than lamps.

Light intensity can be measured by photometer or lux meter. A photometer consists of photoelectric cell and a micro-ammeter. Photoelectric cell is sensitive for light and converts light into current. Microammeter shows reading due to this current and its needle moves. Nowadays digital read out is given by appropriately converting the current into digital signal (Fig. 25.12). High intensity is proportional to the current generated in the photoelectric cell by falling light. The unit of illumination is called as 'lux' and the scale of ammeter is calibrated in lux.

Usually 2000 to 2500 lux is provided to the cultures maintained in light in the culture room. The instrument is a sensitive tool and handled with care. It should not be exposed to the sunlight without switching the proper reading switch to high light illumination.

12. METHODS OF STERILIZATION

12.1. Asepsis

Plant tissue culture requires contamination free environment, tools and cultures or strict maintenance of germ free system in all the operations, known as asepsis. Particularly in commercial production units, the contamination of one batch of the cultures may result in heavy financial losses or even loss of a culture strain. Therefore strict control measures are enforced to maintain the entry of the personnel and living materials. The basic rules and practices of asepsis are followed in all the tissue culture laboratories.

Plant tissue culture media are rich in nutrients and very suitable for the growth of microbes also. These microorganisms grow faster, consume nutrients rapidly and suppress the growth of plant tissues (by over growth). This saprophytic unwanted growth of microbes is formed as contamination. There are many ways by which these microbes suppress the growth of plant cells and tissues, e.g., release of enzymes and toxins (which inhibit the growth of plant cells), selective absorption of specific nutrients, and some tissue infection. To achieve success in cultivation of higher plant parts, it is essential to exclude these contaminating microorganisms and hence aseptic techniques must be employed to save cultures. This has led to the development of 'clean area concept'. All the area, tools and working places should be free from microbes to have less and less problems of contamination. Following care should be taken:

1. Minimize the air current in the working area so it is possible to avoid spores of contaminating microorganisms to move in along with the air currents over the sterile areas. At least, fan should not be used in laminar air flow bench room (inoculation room). Preferably, all the places should be air-conditioned.
2. Store properly the prepared media, nutrients and tools in cabinets.
3. Use separate area for cleaning and washing and for the preparation of medium.

UV tubes are also fixed in Laminar air flow bench placed in inoculation chamber. All these UV tube lights should be used frequently before inoculations. UV lights of corridors may be left open during nights.

12.2. Surface Sterilization of Explants

Most contamination is introduced with the explant because of inadequate sterilization or just very dirty material. It can be fungal or bacterial. This kind of contamination can be a very difficult problem when the plant explant material is harvested from the field or greenhouse. Initial contamination is obvious within a few days after cultures are initiated. Bacteria produce "ooze" on solid medium and turbidity in liquid cultures. Fungi look "furry" on solid medium and often accumulate in little balls in liquid medium.

Bacteria are the most frequent contaminants. They are usually introduced with the explant and may survive surface sterilization of the explant because they are in interior tissues. So, bacterial contamination can first become apparent long after a culture has been initiated.

All explants have to be sterilized before transferring them on to the medium. A suitable sized explant can be sterilized by any one of the following solutions -

1. 1 % solution of sodium hypochlorite (commercial bleach).
2. 7 % saturated solution of calcium hypochlorite.
3. 1 % solution of bromine water.
4. 70% ethyl alcohol.
5. 0.2% mercuric chloride.
6. 10% hydrogen peroxide solution.
7. 1% silver nitrate solution.

Suitable sized plant material is sterilized as follows:

1. Clean the working area with ethanol, and start the air flow of the laminar air flow bench.
2. Sterilized petridishes, distilled water, scalpel, filter paper sheets (suitable sized and autoclaved) alcohol, disinfectant (mercuric chloride 0.01-0.1% aqueous solution, or 20% sodium hydrochloride), forceps, bead sterilizer or beaker or coupling jar, sprit lamp or gas burner are collected on Laminar flow bench. Put on the UV light of the bench and put it off after 30 minutes.

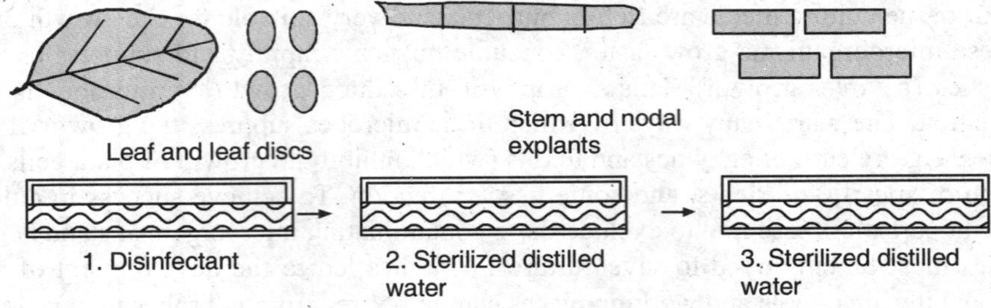

Fig. 25.13. Method of surface sterilization of explants.

3. Bring the plant material to be sterilized on the bench and prepare pieces for sterilization.
4. Clean the working area and hands with alcohol, put on mask and cap, and light the sprit lamp.
5. Keep 3-4 petridishes in a line, add disinfectant in 1st plate and autoclaved distilled water in subsequent plates.
6. Place plant pieces in 1st plate and immerse the material with the help of forceps for 5-10 minutes depending upon the disinfectant used.
7. Transfer material from 1st to 2nd plate, rinse gently and pass to 3rd and 4th plate, one by one with thorough rinsing.
8. Finally, drain the distilled water or place the material in a fresh petridish or filter paper and prepare suitable sized explants (Fig. 25.13).

12.3. Mode of Action of Disinfectants

The various metallic ions can be arranged in a series of decreasing antibacterial activity. Hg^{2+} and Ag^{+}, are effective at less than one part per million (ppm), because of their high affinity for sulphahydryl group. Bacteria are killed by Ag containing 10^5 to 10^7 Ag^+ ions per cell. The concentration required for killing is markedly affected by inoculum size.

Chlorine was the anticeptic introduced (as chlorinated lime) by O.W. Holmes in Boston in 1835 and by Semmelweis in Vienna in 1847, to prevent transmission of puerperal sepsis by the physicians hand. Chlorine combines with water to form hypochlorous acid (HOCl) a strong oxidizing agent.

$$Cl_2 + H_2O = HCl + HOCl$$
$$Cl_2 + 2NaOH = NaCl + NaOCl + H_2O$$

Sodium hypochlorite (NaOCl) solutions are (200 ppm chlorine) are used to sanitize clean surfaces in the food and the dairy industries and in restaurants. Sodium hypochlorite is commercial bleach available in the market as 3-6% solution and can be used directly. It is prepared by passing chlorine in to dilute solution of sodium hydroxide. Powdered bleach is also added into water supply for killing microbes.

12.4. Equipment and Medium Sterilization

All the manipulations of plant tissue culture methods are carried out in aseptic conditions. All the materials used are therefore, free from microbes or sterilized. Various methods of sterilization are used depending upon the material and type of sterilization. Essentially, method for sterilization is different for living materials than tools and glass ware. Now a days many items are available as ready to use, pre-sterilized (by gamma radiation) and disposable. Use of such disposable materials have facilitated the work. In laboratory, in principle, three types of sterilization is used -

1. Dry heat
2. Wet heat
3. Filter sterilization.

1. **Dry heat:** Glassware, metal tools and other articles, which do not get charred by high temperature, are put in containers or wrapped in paper or thick aluminium foil and placed in dry oven and sterilized for a period of not less than three hours at a temperature of 140-160 °C. Media and plasticware can not be sterilized by this method.

2. **Wet heat:** The most popular method of sterilization both equipment and media is autoclaving at 121 °C with a pressure of 15 psi (pounds per square inch) for 15 min (1.02 kg/cm^2). Modern autoclaves are capable of providing saturated steam treatment ranging from 70-132 °C, which is a pressure of up to 25 psi. All the vessels containing medium should be placed vertically and should not be filled more than 40% to their total capacity.

3. **Filter sterilization:** Filter sterilization or cold sterilization is used when a solution or medium can not be sterilized by autoclaving. It is the property of the filter (porosity 0.22 to 0.45 μm) to retain all the microorganism and make the solution free from microbes. This exclusion of microorganisms makes the solution sterilized without heating or autoclaving. Only liquids can be sterilized by this method and not the plant materials or other things. This is commonly used where thermolabile or heat sensitive chemicals like plant growth regulators (IAA, GA_3, Zeatin, Abscisic Acid), some biochemicals, enzymes for protoplasts isolation, antibiotics or nutrient solution for specific experiment are used to avoid decomposition during autoclaving. After sterilization, solution is added to the medium. In case of static medium, compound is added before solidification (cooling) after autoclaving.

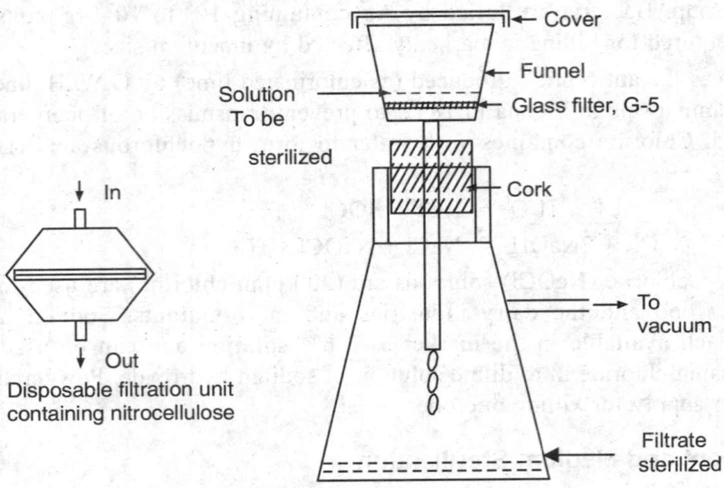

Fig. 25.14. Filtration units of various types.

Glass and membrane filters (nitro-cellulose membranes) are commercially available. Glass (sintered glass filters, Borosil) are available in different grades (G-1 to 5) and the G-5 (1-2 µm) is used for bacteriological purposes (Fig. 25.14). These filters have a porous glass disc for filtration of liquids and gases as a filter media which is non-corrosive and reusable. The grades are classified by maximum pore size which is obtained by measuring the pressure at which the first air bubble breaks away from filter under certain conditions. The pressure differential is then used to calculate the equivalent capillary diameters in microns. The desired pore size is obtained by suitably controlling the grain size, firing time and temperature and the thickness of the disc. These filters are resistant to heat, solvents and detergents.

Membrane type filters (of different diameter) and porosity (0.22 to 0.45 µm) are available as disposable or reusable filters, or as filter holders containing a filter disc (Fig. 25.14). These filters can be connected through tubing or syringe to pass solution or gases.

Filters are wrapped in aluminium foil or paper, autoclaved at 121°C for 15 min and then taken to laminar air flow bench. Solution is placed in funnel (glass type filter) or syringe (membrane type filter) and solution is allowed to pass through filter under suction (glass) or pressure (membrane). A membrane filter or glass tube, filled with cotton may be placed between vacuum flask and suction pump to avoid entry of air from the out side during operation. Sterilized filtrate is collected in flask and dispensed in the liquid or molten medium with the help of sterilized pipette fitted with cotton plug. Volume added to the medium is adjusted to arrive at a final concentration. Membrane type filters are also used at inlets and outlets of various types including air, in a bioreactor system and to draw samples.

There are a variety of wet and dry heat treatments, radiations, filtration and gas and chemical treatments available for direct sterilization of material. Gas treatments are rarely used in the laboratory. Ultraviolet light treatment of working surfaces and sterile rooms are used in the labs while gamma radiation is used in the industry for the preparation of pre-sterilized disposable plastic-wares.

12.5. Sterilization of Tools

Surgical blades and scalpels are not sterilized by dry heat because the high temperature makes the cutting edge dull. Such articles including spatula and forceps are usually immersed in 80% v/v (volume by volume) ethyl alcohol until required, and sterilized during use by frequent immersion in alcohol and flaming. Nowadays, bead sterilizer (works on principle of dry heat) is available for sterilizing such tools. After an initial stabilization time of 30 minutes during which equipment attains a temperature of about 250 °C, units ensure total sterilization by destruction of all microorganisms within seconds. This unit can conveniently placed on Laminar air-flow bench.

Use of fumigation (formaldehyde, sulphur etc.) have been discouraged. It is used some times to clear the laboratory or working area from microorganisms. The place is not used for 3-4 days following fumigation. Culture rooms are fumigated only after removal of cultures. Generally, wiping of working places with 80% alcohol is sufficient for regular use.

13. MEDIUM AND ITS PREPARATION

The media used by earlier workers were based on Knop's solution. Subsequently media developed by White (1943) and Heller (1953) were used. Murashige and Skoog's medium (Murashige and Skoog, 1962) is a land mark in plant tissue culture research and is the most frequently used medium for all types of tissue culture work. Based on its composition, other media were evolved to meet the diverse experimental and species specific requirement. There are - Linsmaier and Skoog (1965), B5 medium of Gamborg *et. al.* (1968), SH medium of Schenk and Hildebrandt (1972), Nitsch and Nitsch (1969) medium, and woody plant medium (WPM) of Llyod and McCown, (1980).

The methodology of plant tissue culture has advanced to the stage, where tissues from virtually any plant species can be cultured successfully. The successful plant tissue culture depends upon the choice of nutrient medium. The cells of most plant species can be grown on completely defined media. All the media consist of mineral salts, a carbon source (generally sucrose), vitamins and growth regulators. The MS medium designed for tobacco is now used widely for various species, in callus and cell cultures.

13.1. Medium Composition

The nutrient medium for most plant tissue cultures is comprised of five groups of ingredients – inorganic nutrients, carbon source, vitamins, growth regulators and organic supplements.

Inorganic Nutrients - Inorganic nutrients consist of macro- and micro-elements as their salts. Usually nutrient media contain 25 mM each of nitrate and potassium. For regular culture and cell cultures, the combined nitrogen level (nitrate and ammonium nitrogen) may reach up to 60 mM. Ammonium is essential for most cultures but in lower concentrations than that of nitrate nitrogen. A concentration of 1-3 mM of calcium, magnesium and sulphate, is always adequate. The required micronutrients include I, B, Mn, Zn, Mo, Cu, Co and Fe. Role of different nutrients is given in the Table 25.2.

Table 25.2. Some of the important elements supplied to culture medium for plant nutrition and their physiological function.

Element	Function
Nitrogen	Component of proteins, nucleic acids and some coenzymes. Element required in greatest amount.
Potassium	Regulates osmotic potential, principal inorganic cation.
Calcium	Cell wall synthesis, member function, cell signalling.
Magnesium	Enzyme cofactor, component of chlorophyll.
Phosphorous	Component of nucleic acids, energy transfer, component of intermediates in respiration and photosynthesis.
Sulphur	Component of some amino acids (methionine, cysteine) and some cofactors.
Chlorine	Required for photosynthesis.

Iron	Electron transfer as a component of cytochromes.
Manganese	Enzyme cofactor.
Cobalt	Component of some vitamins.
Copper	Enzyme cofactor, electron-transfer reactions.
Zinc	Enzyme cofactor, chlorophyll biosynthesis.
Molybdenum	Enzyme cofactor, component of nitrate reductase.

Carbon Source - Glucose, fructose, maltose or sucrose (2-4%) can be used as source of energy or carbon but sucrose is the preferred source for most of the cultures. The sucrose in the medium is rapidly converted into glucose and fructose. The glucose is absorbed first followed by fructose.

Table 25.3. Composition of Murashige & Skoog's medium and B_5 medium.

Salts	MS		B_5	
A. Macronutrient	mg/l	mM	mg/l	mM
NH_4NO_3	1650	20.6	-	-
KNO_3	1900	18.8	2500	25
$CaCl_2.2H_2O$	440	3.0	150	1.0
$MgSO_4.7H_2O$	370	1.5	250	1.0
KH_2PO_4	170	1.25	-	-
$(NH_4)_2SO_4$	-	-	134	1.0
$NaH_2PO_4.H_2O$	-	-	150	1.1
B. Micronutrient	mg/l	µM	mg/l	µM
KI	0.83	5.0	0.75	4.5
H_3BO_3	6.2	100	3.0	50
$MnSO_4.4H_2O$	22.3	100	-	-
$MnSO_4.H_2O$	-	-	10	60
$ZnSO_4.7H_2O$	8.6	30	2.0	7.0
$Na_2MoO_4.2H_2O$	0.25	1.0	0.25	1.0
$CuSO_4.5H_2O$	0.025	0.1	0.025	0.1
$CoCl_2.6H_2O$	0.025	0.1	0.025	0.1
Na_2.EDTA	37.3	100	37.3	100
$FeSO_4.7H_2O$	27.8	100	27.8	100
Sucrose (g)	30,000		20,000	
pH	5.7 - 6.0		5.5 - 5.8	
C. Vitamins				
Inositol	100		100	
Nicotinic acid	0.5		1.0	
Pyridoxin. HCl	0.5		1.0	
Thiamine. HCl	0.1		10.0	
D. Amino acid				
Glycine	2.0		-	
E. Plant Growth Regulators				
NAA/IAA/2,4,-D	0.01-10.0		0.01-10.0	
Kinetin	0.04-1 0.0		0.04- 10.0	

Vitamins and amino acids - Thiamine, pyridoxine and nicotinic acid are commonly used as vitamins in B_5 and MS media. The former is required for most cultures while latter two promote cell growth. Amino acids and organic supplements - Amino acids serve as source of reduced nitrogen. In case of inadequate nitrogen, complex organic nitrogen supplement like casein hydrolysate (0.1-1 g/l) may be supplemented. Glycine is commonly used amino acid. Other organic supplements are coconut milk, yeast extract, peptone and malt extract. However, synthetic media are preferred and organic supplement of unknown chemical nature is used only when it is essential.

Plant Growth Regulators (PGR) - A balanced combination of PGR is required for sustained growth. Two types of combinations are used; one for cell proliferation consisting of (preferably) 2,4- dichlorophenoxy acetic acid (2,4-D) or 1-naphthalene acetic acid (NAA) and a cytokinin (kinetin, benzyl adenosine, 2-isopentyladenosine, zeatin, thidiazuron), another for regeneration essentially containing low auxin [(NAA, IAA, Indole butyric acid (IBA)] and a cytokinin in high amount, but not 2,4-D as an auxin. 2,4-D is known to induce cell proliferation but suppresses differentiation in dicot plants. However, 2,4-D and 2,4,5-T (2,4,5-trichlorophenoxy acetic acid) are effective in inducing somatic embryogenesis in cereal (monocots) and herbaceous dicot cultures.

13.2. Medium Preparation

A convenient approach to prepare a medium is to have stock solutions of all the nutrients in a 10x or 50x concentration. Medium is prepared by suitably diluting the appropriate amount of stock solutions for desired volume of the medium. It may be advantageous to have separate stock solutions of calcium salt and potassium iodide. All the ingredients are mixed, sugar added and pH is adjusted to 5.8-6.0 and medium is poured in the culture vessels. All the vessels are plugged with non-absorbent cotton, covered with aluminium foil and autoclaved at 121 °C for 15 min. Prepared media can be stored for a few weeks before inoculation. Liquid medium for a given material is same as static medium used for callus cultures except gelling agent- agar. All the ingredients should be thoroughly mixed before dispensing in the vessels.

QUESTIONS

1. Write short notes on:
   (a) Autoclave
   (b) Laminar flow bench
   (c) Growth chamber
   (d) Thin layer chromatography
   (e) Lux meter
   (f) Electron microscopy
   (g) pH meter
   (h) Com,lorimeter
   (i) Filter sterilization
   (j) Clean area concept
   (k) Asepsis
   (l) Explant sterilization
2. Compare a compound microscope with an electron microscope.
3. What do you understand by λ_{max} of a compound?
4. What are HEPA filters?
5. Define RCF. How this is important in separation of particles?
6. What is mode of action of mercuric chloride and sodium hypochlorite?
7. Describe and compare glass and membrane filters.
8. Describe MS medium. What precautions are required during the preparation of medium?
9. Name different media for plant tissue culture and differentiate them.
10. What do you understand by macro- and micro-nutrients of a medium?

CHAPTER 26

Basic Plant Tissue Culture

1. CULTURE INITIATION

Plant parts are used to initiate and establish *in vitro* (in glass) growing cultures. In comparison to *in vitro* grown cultures, in case of field or pot grown plants, growth is affected by various soil and environmental factors of unknown nature. Therefore, it is difficult to study the factors affecting the growth, differentiation and metabolism of the plant. Plant tissue culture provides an excellent system to study the plant growth and differentiation in totally controlled conditions. Cultures are grown in a completely defined chemical and physical environment. Thus, by changing the factors, one by one, the influence of a particular factor can be determined very easily. The field grown (*in vivo*) plants are autotrophic, i.e., they synthesize their own food by photosynthesis. In contrast to this, sugar is essential in the medium for the growth of *in vitro* cultures, i.e., they are unable to synthesize their carbohydrate requirement (even though cultures are grown in light, photosynthesis is almost negligible).

Explant type- The present knowledge permits the use of any plant part as a source of material to initiate cultures. The plant part used for this purpose is known as an 'Explant.' Nodal and inter nodal segments of stem, apical and axillary bud, leaf, leaf disc, petiole, anther, pollen, flower bud, petal, inflorescence, ovule, ovary root and even isolated epidermal peel, gland and trichome have been used as an explant.

Preparation of explant- *In vivo* material, when brought to laboratory, it is always preferable to bring material under ice/dry ice to minimize catabolic activity. Material is washed thoroughly with tap water and liquid detergent (e.g., Tween-80), and then washed thoroughly. After proper cleaning and removal of adhering particles, material is sterilized and used. Some hard seeds (like leguminous and hard seed coat seeds) may require pretreatment (2-10 min) with 50% sulphuric acid to break seed coat dormancy. This should be done with care and seeds should be washed thoroughly after acid treatment under running tape water for 1-2 hours. When axillary or apical buds are used, extra leaves are removed, and pre-treated with a wetting agent (quick dip in 70% ethanol or 5-10 min in detergent solution) to facilitate penetration of disinfectant. Explants are prepared in suitable size for the purpose of sterilization, inoculation and suitability to generate culture. Stem segments consist of single node or multiple nodes, or inter nodes are used while leaves can be used as whole leaf or leaf disc can be prepared.

When tuberous materials like Jerusalem artichoke (*Helianthus tuberosus*), sugar beet roots (*Beta vulagaris*) or carrot roots (*Daucus carota*) are used as an explant, large pieces/complete tubers are sterilized and then with the help of scalpel and cork borer, tissue cylinders and discs from cylinders are prepared under aseptic conditions. The explant preparation and sterilization are carried out under laminar flow bench. Sterilization of explants is already described in the previous chapter using sodium hypochlorite or mercuric chloride.

2. CALLUS CULTURE

Cell from any plant species can be cultured aseptically on or in a nutrient medium. The

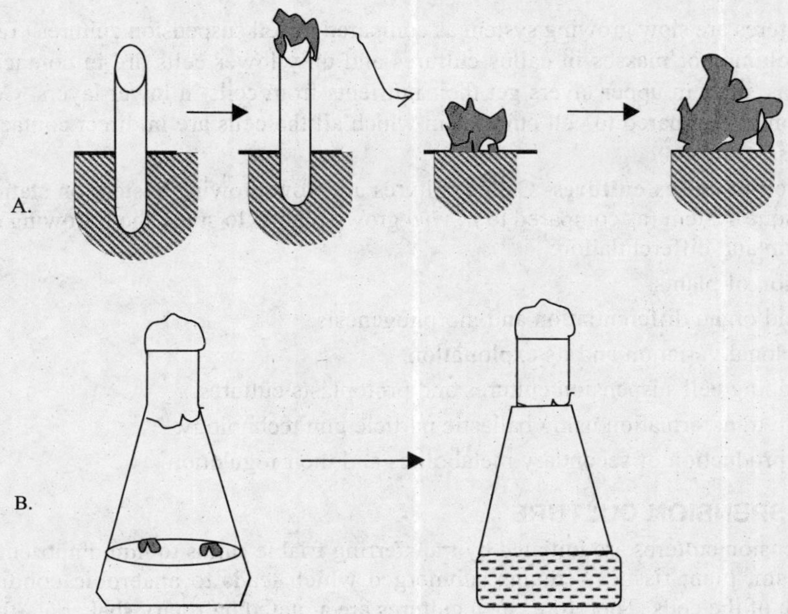

Fig. 26.1. Initiation of callus (A) from an explant and (B) cell suspension cultures from callus.

cultures are initiated by planting a sterilized tissue (an explant) on an agar medium. Within 2-4 weeks, depending upon plant species, a mass of unorganized cells (callus) is produced. Such a callus can be subcultured indefinitely by transferring a small piece on to the fresh agar medium.

When a suitable sized and sterilized explant is transferred aseptically under a laminar flow bench on to an appropriate nutrient medium (for example, MS medium or B5 medium) containing appropriate combination of plant growth regulators, it produces callus (Fig. 26.1). The callus is produced from outer layers of cortical cells in a stem explant by repetitive division of the cells. These dividing cells generate pressure on the epidermis, which ultimately ruptures exposing newly formed callus. By continuous division of cells produces a mass of cells or callus on the explant. This callus is separated from the explants and transferred on to fresh medium. After attaining growth, callus is subcultures at regular interval of 3 to 4 weeks by dividing it in to pieces (250 to 500 mg each) and transferred on to fresh medium. The callus piece used to inoculate a flask is known as 'inoculum'. Inocula (plural of inoculum) of similar shape and size are used in all the subcultures and experiments.

Table 26.1. Comparison of callus and cell suspension cultures.

Parameters	Callus	Cell suspension
Growth	Slow	Fast
Cell to cell contact	Cells in contact	Dissociated
Medium	Only lower layer is in contact with the medium.	All cells are in direct contact with the medium
Precursors	Not available to all cells.	Available to all cells.
Subculture period	Long, 4-8 weeks	Short, 7-21 days.
Accumulation of metabolites	Higher than cell suspension.	Lower that callus culture.
Scale-up in bioreactor.	Not possible	Cell suspensions are grown in bioreactor.

Callus cultures are slow growing system as compared to cell suspension cultures (Table 26.1). Cells grow as clumps or masses in callus cultures and only lower cells are in contact with the medium whereas, cells in upper layers get their nutrients from cells in lower layers. Cells are in close association as compared to cell cultures in which all the cells are in direct contact with the medium and dissociated.

Application of callus cultures- Callus cultures are slow growing system on static medium and offers a unique system (as compared to *in vivo* grown plants) to study the following aspects of plant metabolism and differentiation:

1. Nutrition of plants.
2. Cell and organ differentiation and morphogenesis.
3. Somaclonal variation and its exploitation.
4. Developing cell suspension cultures and protoplasts cultures.
5. Genetic transformation using ballastic particle gun technology.
6. In the production of secondary metabolites and their regulation.

3. CELL SUSPENSION CULTURE

Cell suspension cultures are initiated by transferring friable callus to liquid nutrient medium. In liquid medium, plant tissues remains submerged which leads to anaerobic conditions and ultimately death of the cells. Therefore, such cultures are agitated on rotary shaker at 80-150 rpm, with an orbital diameter of 2.5-5.0 cms. Agitation serves both to aerate the cultures and to disperse the cells. Cells from the inoculum are separated during this process and a suspension of cells is produced. The division rates of suspension culture cells at the expotential phase are typically higher than callus cells, but doubling times are slow than bacterial cells and usually vary from 24-72 hrs. It is a common observation that if relatively small number cells are transferred (low inoculum density) to a new medium (either static or liquid), they may fail to divide, whereas a larger quantity of tissue transferred from the same culture may proliferate rapidly on the same medium. This observation has led to the concept of 'critical initial cell density. This is defined as the smallest inoculum per volume of medium, from which a new culture can be reproducibly grown.

There are a few conditions, which determine the critical initial density of cells. These are:

(1) The cultures physiological characteristic.

(2) The length of time and conditions under which the culture was previously maintained.

(3) The composition of the fresh medium.

The third point is of interest. As the isolated cells are failed to grow on fresh medium, ‘conditioned medium’ or ‘nurse tissue’ conditions are used to grow isolated cells or protoplasts. A ‘conditioned medium’ is the medium on which some tissues were previously grown. This makes the minor adjustment in the nutrients and chemical substances released in the medium by the callus, promotes the growth of isolated cells of protoplasts. In cell suspension cultures, cells grow as isolated single cell and cell aggregates of a few cells to a few hundred cells. Cells aggregation vary from species to species and some times it is difficult to maintain fine cell suspension culture, e.g., in members of the family Rutaceae.

Cell cultures are subcultured by dilution of the stock culture, 5 to 10 times v/v (volume by volume), depending upon the growth of cells. As mentioned above, growth of cell suspension cultures is always higher than callus cultures and, therefore, required rapid subculture (7-21 days) as compared to callus cultures (4-8 weeks).

Cell cultures in the liquid medium are grown as:

(a) Batch cultures in shake flasks and bioreactor.

(b) Continuous culture system.

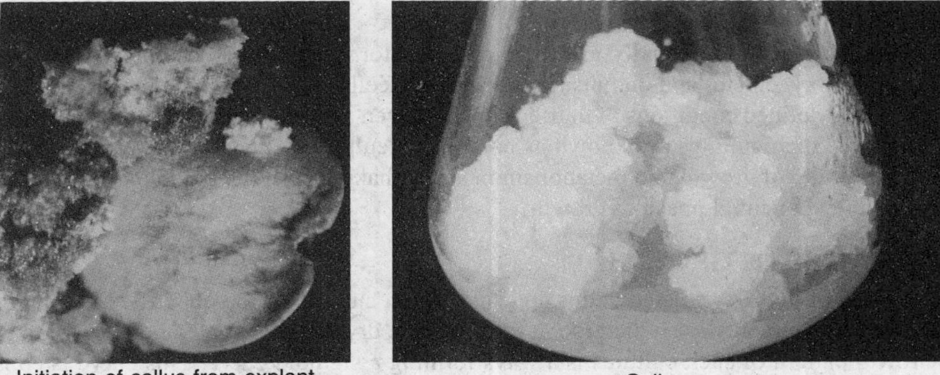

Initiation of callus from explant

Callus

Two tier rotary shaker for cell cultures

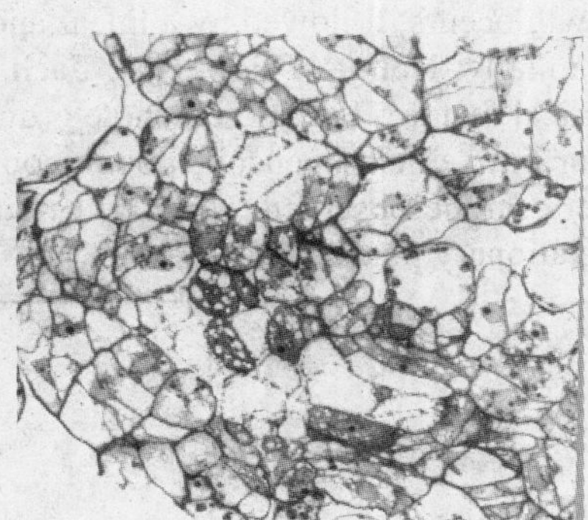

Section though compact callus showing cells

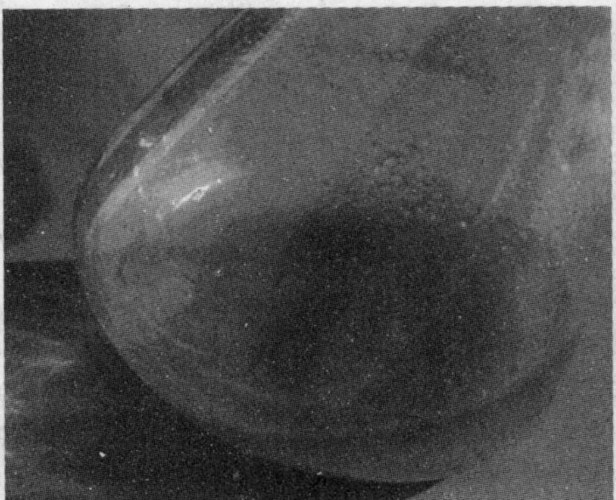

Cells culture in flask

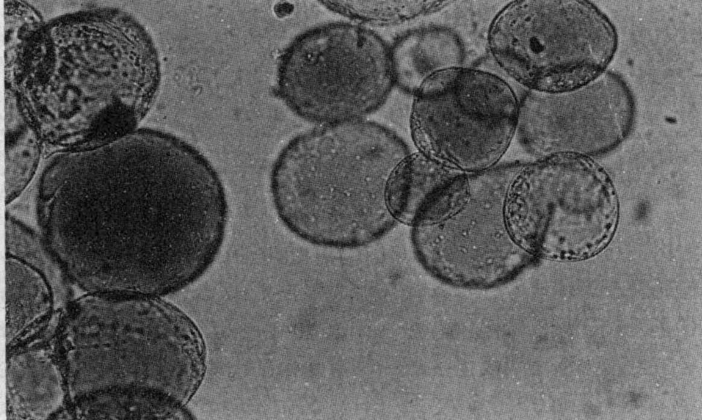

Cells from suspension culture under microscope

Fig. 26.2. Development and growth of callus and cell cultures and its structure.

3.1. Batch Culture

Whether cultures are grown in flasks (100 ml to 2 litres capacity containing 20 to 500 ml liquid medium) or in a bioreactor, system is used as a batch. Inoculated cultures are harvested after a definitive growth period (determined by growing cultures earlier and recording the growth phases during 1 to 6 weeks growth) at end of stationary phase and cultures are harvested for a specific objective. Such cultures are grown again and again in batches for the purpose of experiments and known as batch cultures.

3.2. Continuous Culture

Continuous cultures are usually grown in a bioreactor, (also known as chemostate) when inputs (nutrients, oxygen etc) and outputs (spent medium, cells) are precisely controlled and cells are grown under a defined conditions of nutrients, pH, oxygen and cell density continuously for the purpose of study. Such cultures are known as continuous cultures. This type of work is possible only after basic study of growth and metabolism in small shake flasks. In continuous culture, cells are always in an exponential phase of growth.

4. TIME COURSE OF GROWTH

The growth of undifferentiated callus and suspension culture can be quantified in a number of ways, but most commonly as an increase in fresh weight, dry weight or cell numbers over a time course. Growth pattern of such cultures is always forming a sigmoid curve. This curve is consist of a lag phase, in which there is a little or no cell division, an exponential phase in which maximum cell growth occurs, followed by a linear and stationary phase, in which the proportion of dividing cells gradually declines (Fig. 26.3). Each stage can be characterized by distinct structural and biochemical features but, infect whole growth cycle is a continuous physiological process. At any particular time a heterogeneous population of cells differing in morphological and physiological characters can be observed. Usually the cultures are subcultured at end of exponential phase or during stationary phase.

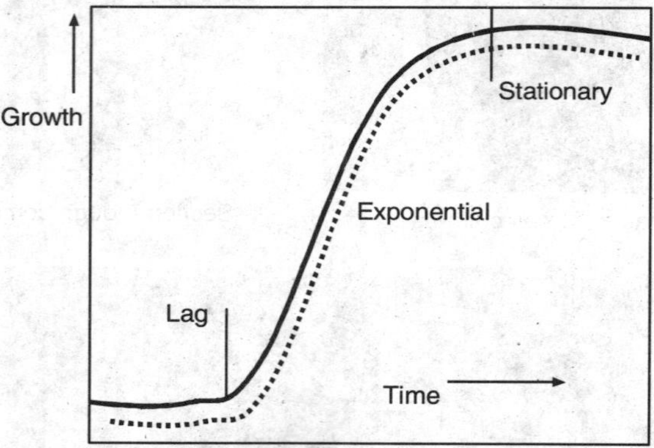

Fig. 26.3. Time course of growth showing different phases of growth in a growing system.

Fresh weight of callus is determined by carefully removing the adhering agar from callus and immediately weighing the callus in a sensitive balance. Cell cultures are filtered in a butchner funnel lined with filter paper, washed with distilled water (under mild vacuum) and weighed. Dry weights are determined by drying the cells to a constant weight at 60 °C in an oven and weighing in a balance. Packed cell volume (PCV) and cell culture is determined by centrifuging it at 1000 rpm for 5 min in a graduated centrifuge tube and measuring the volume occupied by the cells out of total volume of the cell suspension. Cells are counted by a haemocytometer under compound microscope during growth period.

5. SINGLE CELL CULTURES

5.1. Nurse Tissue Culture

Single cell cultures are obtained from established cell suspension cultures. Cell suspension cultures are comprised of mixture of single cells and small colonies of cells. These latter strains of cells are not necessarily derived from single cells. Thus, strains desired from such cultures are not true single cell clones. Single cell clones are provided the means for pure culture studies for higher plants comparable to those used for microorganisms. Several methods are developed for obtaining single cell cultures and establishing single cell clones of higher plants. The first successful higher plants single cell isolations were made by Muir and co-workers in 1954 at Wisconsin University, Madison, USA, using nurse tissue culture technique for marigold and tobacco callus cultures. Isolated single cells provide a unique system to study differentiation and divisions in a cell by monitoring changes under a microscope in a cell.

A. Nurse tissue culture technique: Single cells are observed and picked-up from cell suspension cultures under the microscope with the help of micropipette. Several days previous to the isolation of single cells, small sterile pieces (8 × 8 mm) of filter papers are places aseptically on the top of established callus of same or different species. By this time, filter paper place on nurse callus tissue is wetted by liquid and nutrients released by the cultures. The single cell, then placed on the filter paper raft, may divide and produce a small colony of cells within a few days or weeks. This small colony of cells or microcalli can be transferred to fresh medium, where it survives (Fig. 26.4). A number of previously recalcitrant (difficult to grow in culture) species, notably monocotyledons such as rice and maize have been grown by such technique.

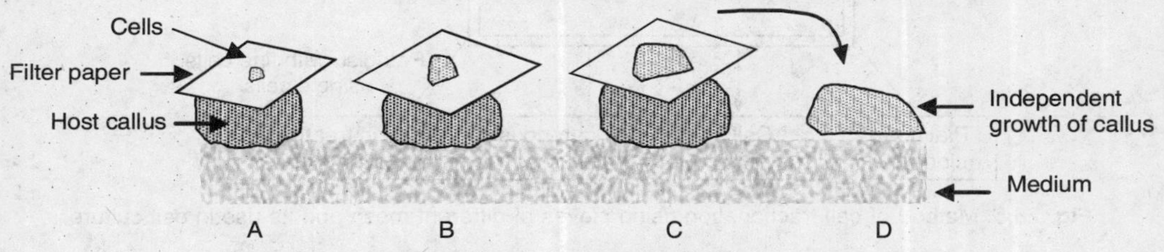

Fig. 26.4. Nurse tissue culture technique.

5.2. Cell Sieving

Cell suspensions are filtered through sieves of varying mesh (nylon or stainless steel wire gauge of different porosity) depending upon the size of the cells and aggregates in the suspension culture. A series of 1-3 mesh (80-150 mesh) may be used to obtain cells of desired size and number (Fig. 26.5). Different fractions so obtained contains single cells and clumps of cells depending upon the mesh size. These fractions can be used directly for initiating cell suspensions or may be used after concentrating the fraction by centrifugation. Fractions may be used to raise clones by plating the cells of these fractions. The appropriate fractions contain cells of similar shape, size and metabolic state. Such cultures are used to obtain synchronous cultures. Synchronous cultures are those cultures in which all the cells divide at the same time. Therefore, at one time all the cells divide and next time all the cells are in a stationary phase. This time interval between two cycles of cell division is called 'doubling time' (Fig. 26.6). Such cultures are useful in developing embryonic cultures as embryos of similar age can be obtained, which is also a necessity for the production of seeds at large scale of a species.

Cell plating and cell sieving methods are alternative methods to obtain strains and clones of cells. Microscopic observations and precautions are necessary to ensure that a colony of cells is originating from a single cell. By this method a true single cell clone can be obtained. Cell plating and cell sieving methods have been extensively used to obtain a large number of single cells and their colonies. Bergman (1959-1960) was the pioneer in establishing the cell plating technique.

5.3. Microculture Chamber

Another method of isolating and establishing single cell clones of higher plant cells is the microculture chamber method. This technique was developed in the laboratory of Prof. A.C. Hildebrandt at Wisconsin University, Madison, USA by Jones et al. (1960). For the study of single

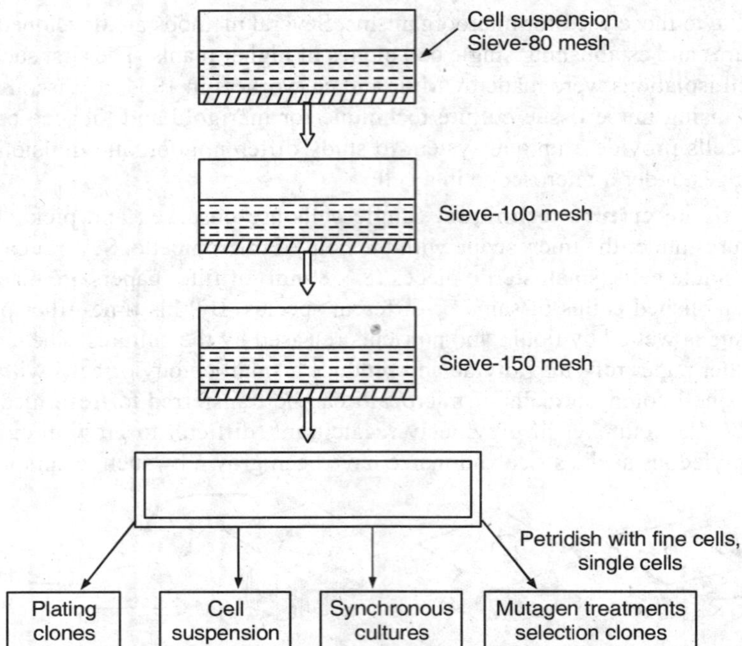

Fig. 26.5. Method of cell fractionation using sieves of different mesh and its use in cell culture.

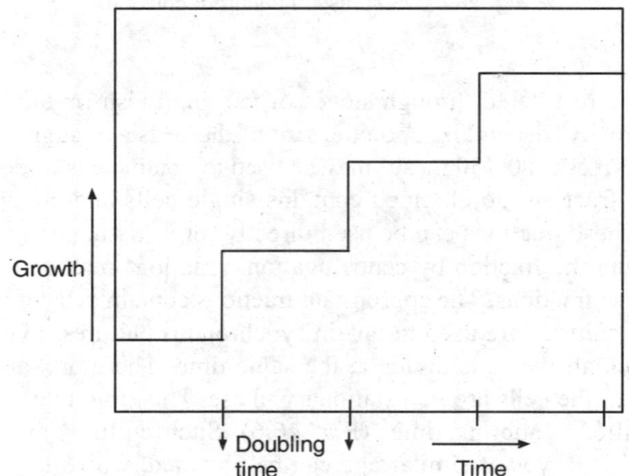

Fig. 26.6. Growth of synchronous cultures, all cells divide at a time.

isolated cells with this method, the single cell is cultured aseptically in a drop of liquid medium surrounded by sterile mineral oil on a microscope slide. A mini chamber is prepared using three cover glasses as shown in the Figure 26.7. The drop of medium containing cells is placed on a sterile microscope slide and ringed with sterile mineral oil. All the cover glasses are sealed with mineral (Paraffin) oil ensuring aseptic growth. This micro-chamber is placed in a sterile petridish and incubated. Cells grow in this system and can be observed periodically. Mineral oil prevents the water (medium) loss. Small colonies thus developed in chamber are aseptically transferred to fresh medium.

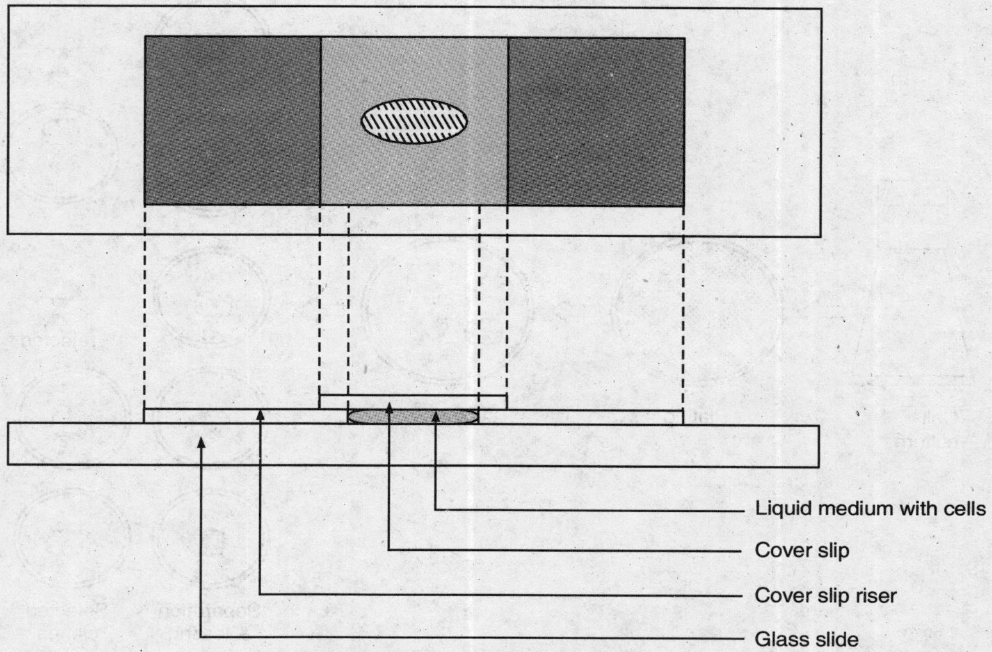

Fig. 26.7. Microculture chamber.

5.4 Applications of Cell Culture

1. Cell culture offers enormous opportunities in the study of single cells and group of cells.
2. In the isolation of protoplasts.
3. In cell cloning by the plating technique with or without specific treatment, e.g., mutagens, amino acid analogues etc.
4. Development of cell lines for various types of resistance e.g., salt and drought, and toxin resistant lines.
5. In scale-up technology using bioreactors of various types.
6. In providing an excellent system of micropropagation on mass-scale by somatic embryogenesis in cell culture and use of bioreactor. Such system can be used for artificial seed production.
7. In the study of nutrition.
8. In cellular differentiation, single cells constitute an excellent system to study cytodifferentiation in a cell leading to tracheiry element formation and plastids differentiation etc.
9. In secondary product formation, regulation and biosynthesis.
10. In the process of cell division and factors affecting cell divisions and related process.

6. CLONING AND SELECTION OF CELLS

6.1. Plating Technique

Clones of single cells or of cell aggregate origin are obtained by plating appropriate cultures. Single-cell clones can be obtained by isolating single cells or by isolating protoplasts. Therefore, in both cases a product of single cell is obtained which is supposed to be the best method for obtaining single-cell product.

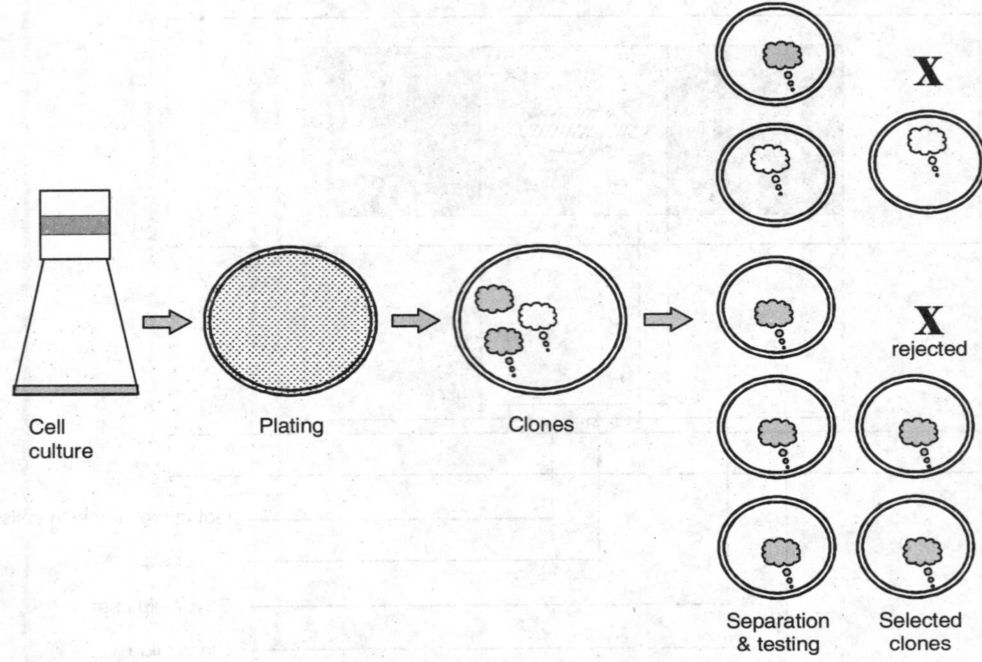

Fig. 26.8. Plating of cell suspension in the plates and selection of clones.

This way a definite selection of producer cells can be made. Plant cells grow in suspension in units ranging from single cells to clusters of more than 1000 cells. Such cultures can be made uniform in cell size by filtration through screens of appropriate size (Sigma chemical Co., USA). A screen of appropriate size is selected on the basis of size of the cells in the suspension. A uniform and defined, 1 ml suspension is spread over the surface of a 25 ml medium in 100 × 15 mm plastic (presterilized) petri dishes. The concentration of cells in the suspension determines the amount of inoculum. Generally, 100 mg fresh weight is reasonable to conduct plating. Single cells, smaller or larger aggregates can be used for plating and selection, depending on the growth of species in the cell cultures (Fig. 26.8). Obtaining a suspension of single cells for plating has been usually achieved by filtration through a nylon net (150-250 μm). In practice, it is impossible to get a pure single cell fraction and mostly inoculum consists of single cells and 3-4 cell aggregates. It is assumed that the aggregates are derived from single cells. By obtaining protoplasts, true single-cell clones can be obtained. Optimization of culture conditions is a prerequisite for inducing cell diversions in single-cell cultures.

Chemically defined cell culture media may not support cell division at low cell densities of less than 9000-15,000 cells per ml, a critical inoculum density. However, in order to achieve single cell clones, cells must be plated at low densities in order to prevent overlapping of the growing colonies. Use of conditioned medium in varying proportion with fresh medium has been suggested to achieve growth of single cells. A conditioned medium is one used for the growth of cells previously. Once colonies are developed, they are grown separately by regular sub cultures of whole colony on fresh medium as deemed necessary for that species. When sufficient callus is produced (usually after 4-6 subcultures), half the callus is used for subculture and half for analysis of secondary metabolites. Selected colonies (clones) are grown and their growth and primary/ secondary metabolite contents are determined. Low productive cells/clones are discarded. Quicker analytical methods are helpful in early selection of clones.

6.2. Cell Lines Selection for Disease Resistance

Plant tissue culture provides a powerful technique to assist the plant breeder in improving the propagation and performance of agricultural, horticultural and forest species. Plants are affected by environmental stresses biotic (fungi, bacteria, nematodes and viruses etc) and abiotic (salinity, draught, toxic metals etc). Resistance to these pathogens may be monogenic or polygenic. Generally, monogenic resistance is high. Plant cells are grown with a pathogen or pathogenic fungi (dual culture) or on a medium containing fungal toxins. Cells survive on such media are supposed to be resistant to fungi or its toxin. These cultures are regenerated and plants are again tested for resistance against the pathogens using standard pathological protocols. Pearl millet cells were selected for downy mildew resistance (*Sclerospora gramnicola*) using dual culture techniques. Similarly, resistance cell lines of rice, tobacco, tomato, wheat, barley, sugarcane were selected for toxins of *Fusarium*, *Helminthosporium*, *Alternaria* and *Phytophthora*. In vitro selection for disease resistance presents an excellent opportunity to assess the direct application of tissue culture to crop improvement.

6.3. Cell Line Selection for Draught and Salinity Resistance

The need for plants with increase tolerance to environmental stress is well known. At least 25% of currently cultivated land suffers from salinity, mainly from NaCl. An additional 25% from acidity, most often caused by aluminium ions, and 40-60% from draught. Establishment of totipotency in plants resulted in regeneration from selected material. Cells are grown on media containing NaCl or polyethylene glycol (PEG 4000 or 6000) for several passages, then grown on medium without selection pressure (PEG, NaCl) and again on the media containing selection pressure. Cells survive on such conditions are supposed to be resistant to the stresses. These cultures are regenerated and again tested for resistance against draught and salt. Such plants have been developed in barley, potato, rice and tobacco.

6.4. Cell Line Selection for Nutritional Quality

Selection of cells for nutritional quality includes improved levels of total amino acids, individual amino acid like leusine, methionine, tryptophan, threonine, and vitamins. Such cell lines were developed by mutations and use of amino acid analogue. These analogues block the pathway preventing metabolism of the amino acid. This results in increased amino acid accumulation. Various types of mutants for amino acids are known in Arabidopsis, tobacco and datura. This line did not produced many improved crop plants therefore other methods like gene transfer are employed, resulting in production of golden rice, improved for vitamin A.

6.5. Selection Parameters

Clones can be selected on the basis of visual (morphological) characters, growth characteristics and physiological parameters (protein banding, secondary metabolite production). If colonies are colored (pigmentation: anthocyanins, chlorophyll: presence-absence) or their fluorescence properties evident under ultraviolet light (secondary products), it is much easier to select producer colonies. This procedure has been effectively used in selection of high-ajmalicine producing cells of *Catharanthus roseus* and pigmented anthocyanin and shikonin producing cells in *Vitis vinifera* and *Lithospermum erythrorhizon* cell cultures, respectively.

Once the clone is grown in size, sufficient to divide and subculture, part of the clone is used for subculture while the other part is retained for qualitative/quantitative secondary compound determination. Once all clones are quantified for their useful metabolite contents, then high secondary metabolite-containing clones are selected and the poorly productive may be discarded or screened for new product/s. Any of the following criteria is used to select the clones depending on the nature of secondary metabolites.

Visible markers: Coloured pigments and compounds fluorescent under UV light may be used as visible markers for selection of clones.

Invisible compounds as markers: for such compounds a quick analytic system sensitive enough to detect minor quantities of secondary metabolites is required to evaluate the clones. These are:

- Chemical tests like Dragendorff's reagent for alkaloids or Libermann-Buchard test for steroids etc.
- Thin layer chromatography of crude samples and use of specific reagent as used for cultures of that particular species.

More sensitive methods such as GLC, HPLC, RIA may also be used which requires partial purification of samples before analysis is carried out.

QUESTIONS

1. How will you initiate cultures from an explant and measure the growth of the cultures?
2. Write short notes on:
   (a) Explant sterilization (b) Callus culture
   (c) Cell culture (d) Cell sieving
   (e) Microculture chamber (f) Nurse tissue culture
   (g) Growth curve (h) Disease resistant cell lines
   (i) Selection of cell lines for drought and salinity resistance
   (j) Cell line selection for nutritional quality (k) selection parameters
3. What are applications of nurse culture, microchamber and conditioned medium?
4. Compare cell cultures with callus cultures.
5. Compare batch cultures with continuous cultures.
6. Compare asynchronous cultures with synchronous cultures.
7. Describe a time course of growth in asynchronous cultures and compare it with synchronous cultures.
8. Describe plating technique for isolation of clones.

CHAPTER 27

Applications of Plant Tissue Culture in Agriculture, Medicine, Horticulture, Forestry, Germplasm Preservation and Industry

1. IMPACT ON AGRICULTURE

Plant organ, tissue and cell culture procedures have developed rapidly in the last half-century since the pioneering efforts of Gautheret, White and Nobecourt. The potential application of the methods of tissue culture are of special significance in crop improvement since conventional methods involve several difficulties, including heterozygosity and a long span between successive generations, hence many investigators are devising methods whereby tissue culture could be fully exploited to improve crop varieties.

The role of tissue culture in crop improvement could be identified in four areas:

(a) As an aid to conventional breeding programme;

(b) As a tool of unconventional breeding programme;

(c) In clonal propagation, and

(d) In obtaining disease-free plants.

Before launching a large-scale programme of crop improvement by tissue culture, it should be ensured that the method is economically viable. Monocotyledonous plant material has never been a favourable system for tissue culture though most crop plants are monocots. In recent years success has been achieved in growing graminaceous crops in cultures using root, *Triticum aestivum*; shoot primordia, *Sorghum bicolor*; mesocotyl segments, *Panicum miliaceum*; internodes of leaf segments, *Saccharum officinarum*; anther culture *Oryza sativa* and endosperm *Lolium perenne*. Obtaining embryogenic cultures and regeneration from protoplasts obtained from such cultures are the final steps to achieve crop improvement through plant biotechnological methods. Regeneration from protoplast has been achieved in *Pennisetum americanum*, *Oryza sativa*, *Triticum duram*, *Zea mays* and several others. Various approaches of plant biotechnological methods to improve the crop plants is presented in

1.1. Tissue Culture: An Aid to Conventional Breeding Programme

In a conventional breeding programme, a plant breeder comes across several obstacles. How plant breeder has advantageously used the techniques of tissue culture to overcome such obstacles is described in the following paragraphs.

i. Embryo Culture- Interspecific crosses may fail because of several reasons, but when the development of embryo is arrested owing to the degeneration of the endosperm, or when the embryo aborts at an early stage of development, embryo culture is the only technique to recover hybrid plants. Incidentally, embryo has been the first plant part successfully cultured in isolation. It was in 1904 that Hanning could rear seedlings from cruciferous embryo excised from immature pods. Laibach (1925) for the first time, employed embryo culture for obtaining hybrids from a cross between *Linum perenne* and *L. austriacum*. More recently, a number of hybrids have been successfully raised through embryo culture : *Hordeum vulgare* X *Secale cereale*, *H. vulgare* X *Agropyron repens*, *H. vulgare* X *Triticum* sp. inter-specific hybrids in *Abelmoschus* and *Secale*.

Embryo culture has helped in overcoming self-sterility of seeds, especially of crop plants propagated vegetatively, when the seeds do not germinate in nature. In a wild relative of commercial banana, *Musa balbisiana* for example, the seeds do not germinate under natural conditions. However, if the embryos are excised and grown on a simple culture medium of mineral salts, seedlings are readily obtained. Likewise, tuber crop *Colocasia esculenta* propagates only vegetatively and the seeds do not germinate in nature. This natural sterility barrier could be overcome by resorting to culture of embryos.

In vitro zygotic embryo culture continues as a means of rescuing hybrid embryos during interspecific gene transfer. Parker and Michaels (1986) employed this technology to recover hybrids of *Phaseolus vulgaris* and *P. acutifolius*. *P. acutifolius* has improved tolerance to common bean bacterial blight (*Xanthomonas compestris*) than *P. vulgaris* (common field bean) but hybrid and early backcross embryos from crosses between these species frequently abort. Embryos rescue assisted the transfer of improved bacterial blight tolerance to economically important field beans. In another case, embryo rescue technique has been utilized in the transfer of cytoplasmically-inherited herbicide tolerance from *Brassica napus* to *B. oleracea.*

The orchid industry owes a great deal to the technique of embryo culture. The culture of orchid embryos was initiated at the Singapore Botanic Gardens in 1928. To date, they have succeeded in producing 1,431 hybrids, and 141 selfed (plants raised through self pollination) species. Callus induction from embryos of orchids can be very useful in obtaining more hybrids.

ii. Ovary and Ovule Culture- Embryos at very early stages of development have not been amenable to culture, primarily because of the problems involved in excision and the complex nutrient requirements. Where such a problem exists, it is logical to culture either the entire ovary or the ovule. This facilitates the availability of maternal tissue and eliminates the cumbersome problem of dissecting very young embryos and selecting a very exacting medium. The culture of pollinated ovaries and ovule *in vitro* has been successful in a number of plants (Fig. 27.1).

The credit for developing and firmly establishing the technique of ovary culture goes to J.P. Nitsch (1951), who experimented with the ovaries of *Phaseolus vulgaris*, *Cucumis anguria*, *Fragaria chilensis*, *F. virginiana*, and *Lycopersicon esculentum*. Ovaries excised several days after pollination showed satisfactory growth on a simple medium. However, the size of the fruit development *in vitro* was smaller than that of field grown fruits. Later on ovaries of tomato, wheat, Pisum and Phlox were grown in culture.

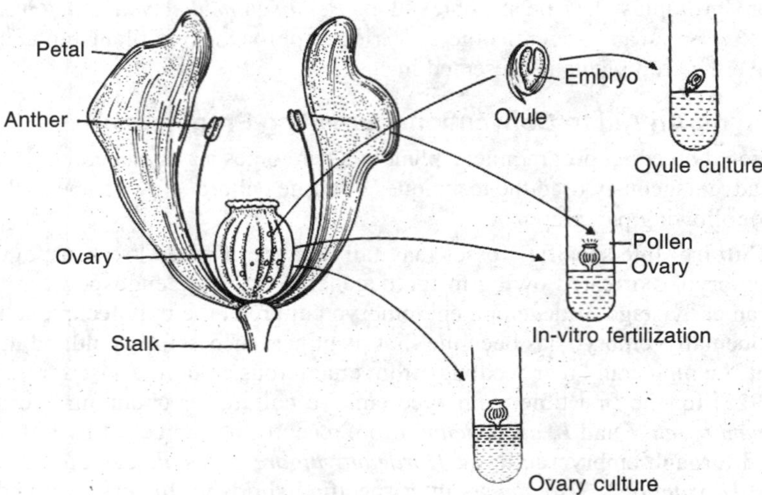

Fig. 27.1. Ovary and ovary culture.

Beasley (1971) observed good growth of fibres of cotton (*Gossypium hirsutum*) ovules, excised 2 days post-anthesis, and grown on liquid medium. The unfertilized ovules were found to shrivel and turn brown on the basal medium. However, the addition of GA_3 or IAA (0.5 to 50 μM) resulted in larger ovules and the production of fibres. Cotton breeders have long been interested in producing hybrids between *Gossypium arboreum* and *G. hirsutum* in order to combine several desirable characters. The crosses between these species are often unsuccessful because of the abortion of hybrid endosperm and embryo at early stages of development. The technique of culturing *in vitro* offers an alternative to overcome such handicaps in making the crosses successful. Ovules and embryos of the hybrids can be cultured on suitable media to obtain viable plants.

iii. Nucellus Culture- The culture of nucellar tissue has also yielded encouraging results and is now being employed in the improvement of *Citrus* crop. Rangaswamy (1958) evolved a novel technique of studying nucellar polyembryony. He excised the nucellus from pollinated ovules of *Citrus japonica* (*C. microcarpa*) and cultured on different media. The micropylar-half of the nucellus proliferated on white's medium containing casein hydrolysate and callus differentiated into pseudo bulbils, which eventually developed into plantlets.

During recent years nucellus culture has been attempted in many vegetable and fruit crops towards clonal propagation of selected materials, e.g., *Aegle marmelos*, *Cucumis melo*, *Luffa cylindrica*, *Trichosanthes anguina*, *Vitis vinifera*. Nucellar culture also helps in the elimination of viruses, e.g., in Citrus species.

iv. Control of Fertilization- In his attempts to evolve a desirable variety, the plant breeder is often confronted with several barriers. The common ones are: (a) disharmony in the flowering periods of the parents, (b) short life span of pollen, (c) failure of pollen germination, (d) slow growth of pollen tube, (e) inhibition or bursting of pollen tube in style, and (f) failure of male gametes to cause fertilization. Using the techniques of tissue culture, embryologists have successfully evolved methods for overcoming these barriers to crossability. Test-tube pollination and fertilization, first employed at the University of Delhi, have now been shown to have wider implications. With the success of intra-ovarian pollination by Kanta (1960), investigations were extended to demonstrate the feasibility of the technique in pollinating and fertilizing excised pistils and ovules *in vitro*. For successful pollination and fertilization *in vitro*, the following considerations are important: (a) culture of pollen grains and ovules at the right stage of development, (b) a suitable nutrient medium, and (c) appropriate cultural conditions. Success has been achieved in *Nicotiana tabacum*, where the entire pistil is cultured and stigma pollinated artificially. Consequent to such test-tube pollination, pollen germination, growth of pollen tubes, entry of tubes into ovules, fertilization, development of embryo and endosperm, viable seed-setting and seed germination take place *in vitro*. This technique has been used for cross pollination in *N. rustica* and *Petunia violacea*. In another method of ovule from unpollinated ovaries are aseptically excised and cultured on a suitable nutrient medium. Pollen from freshly dehisced anthers is then dusted the next day on cultured ovules. The pollen grains germinate, and growth of pollen tube and fertilization are brought about followed by normal development of endosperm, embryo and seed. This has already been successfully achieved in *Argemone mexicana*, *Eschscholzia californica* and *Papaver somniferum* (Fig. 27.1).

v. Endosperm Culture- Endosperm, usually a triploid tissue in angiosperms, can be suitable explant for raising triploids of cereals. As early as 1949, LaRue reported the proliferation of immature endosperm of maize. In *Oryza* induction of callus was observed at the cellular stage of endosperm containing starch grains. Endosperm culture to obtain triploid plants have been used in *Putranjiva roxburghii*, *Nigella damascena*, *Petroselinum crispum*, *Citrus*, apple and *Vitis vinifera*. The successful culture of endosperm tissue has opened up new avenues for plant breeders who can apply this technique to economically important crop plants, especially where conventional breeding methods prove futile.

1.2. Tissue Culture as a Novel Tool

i. Anther Culture- *In vitro* gametic embryogenesis has become an efficient means of producing haploids in a growing number of species since the success of Guha and Maheswari (1964) with cultured anthers of *Datura*. Success in anther culture and more recently in isolated microspore culture have stimulated interest in this technology for plant breeding and applications in both monocots and dicot species of plants. The utility of haploid production in the rapid production of homozygous breeding lines from highly heterozygous parents is well recognized including reduced time requirements per cycle of plant breeding due to immediate phenotypic expression of the recessive traits in haploid derived homozygous plants.

Anther culture undoubtedly offers a unique tool for obtaining haploid plants. Already, a number of economically important crop plants have yielded encouraging results: *Aegilops*, rice, *Brassica hybrid, B. napus, Capsicum annuum, Coffea arabica, Hordeum vulgare, Lycopersicon esculentum, Secale cereale, Solanum melongena, S. tuberosum, Triticum aestivum* and *Vitis vinifera*. In China the programme of large scale production of haploid plants through anther culture is reported to be progressing well and almost every crop is reported to be under test for obtaining haploids through anther culture. More details about haploid plants are given in Chapter 31.

ii. Somatic Hybridization and Genetic Modifications- Hybridization, involving both sexual fertilization and a variety of protoplast-protoplast and cytoplast-protoplast fusion have been successfully demonstrated *in vitro* in a few crop species. Plant protoplasts cultures offer a potential system to achieve somatic hybridization and to create genotypically and cytoplasmically novel hybrids (hybrids and cybrids; see Chapter 30). Plant protoplasts also offer a system for genetic modification to attain the objective of crop improvement. This could be possible due to the remarkable property of plant protoplasts for the uptake of macromolecules such as proteins, DNA, ferritin, latex, polystrene, virus, bacteria and even isolated chloroplasts and whole nuclei.

Several investigators have reported interspecific and intergeneric protoplasts fusion in *Petunia, Atropa belladonna, Vicia, Glycine, Bromus*, carrot, lily, oat x maize, carrot x tobacco, *Torenia fournieri x T. flava* and *Brassica chinensis X B. oleracea*. Melchers et al. (1978) reported regeneration of somatic hybrid plants of potato and tomato from fused protoplasts. Details of genetic modifications through cell and protoplasts cultures and given in the Chapters 30, 16, 17.

iii. Propagation through tissue culture- Clonal propagation by vegetative methods is a practice followed since man started cultivating plants. The main objective of clonal propagation has been to reproduce plants of selected, desirable qualities uniformly and in bulk. The traditional propagation methods, requires long duration, whereas tissue culture helps in rapid plant multiplication. A selected examples of clonal multiplication of trees and horticultural plants are as follows: *Aegle marmelos, Tecomella undulata, Musa* sp., citrus, peach, prunus, *Malus, Prunica granatum* etc. (for details on this topic see the Chapter 31). Improved cultivars developed by biotechnological methods are then clonally multiplied to replace inferior cultivated varieties, e.g. As in *Mentha arvensis, m. Piperata, m. Spicata* and several other aromatic plants.

iv. Production of disease free plants- A majority of the commercial crop plants propagated vegetatively through tubers, bulbs, cuttings and grafting which may contain systemic bacteria, fungi and viruses. Such infections affect the yield and quality of the crop, unless it is readily detected and the plants made free from infection. The traditional method of eliminating viruses by heat treatment is applicable only to a few varieties. Generally, the apical meristem of a virus diseased plant contains very little virus, or is altogether free. Hence, plants obtained from apical meristem are usually free of viruses. There are a number of instances where meristem and/or floral bud culture has resulted in elimination of the virus and restoration of healthy plants, e.g., *Solanum tuberosum*, *Saccharum officinarum*, *Petunia* spp *Allium sativum*, *Ananas comosus*, *Brassica oleracea*, *Musa* sp and *Pisum sativum*.

2. IMPACT ON FORESTRY

Forest resources are disappearing at an alarming rate. An important component of forest ecosystem is trees, which provide food, fuel, construction and industrial products. In addition, trees are reorganized as the critical elements in maintaining stability in the world's atmosphere. Erosion of genetic variability is also a matter of concern now a days. Therefore, both *in situ* and *ex situ* management is necessary for biodiversity, conservation and consequently for tree crop improvement.

Tree improvement involoves managing genetic resources, and includes conservation, selection and breeding and propagation. Biotechnology has the potential to accelerate the genetic improvement of forest yield.

The traditional methods for propagation of trees are by seeds and rooted cuttings or grafting. Micropropagation offers several advantages over conventional vegetative propagation. viz., small area requirement, large number of plant formation in short duration, cloning of selected material, easy to transport etc. Rapid progress has been made in tree tissue culture after production of first true plantlet via organogenesis from *Populus tremuloides* by Winton (1970). Conifers are soft wood plants used in pulp and paper industry. Sommer and co-workers (1975) produced complete plantlets from *Pinus palustris* for the first time.

The ability to regenerate complete plants from individual protoplasts, cells, or tissue is a necessary requirement that must be established prior to the application of most biotechnological methods for crop improvement. The methods employed in micropropagation and improvement of tree species are-

A. Clonal propagation
B. Organogenesis and embryogenesis
C. Protoplasts culture and genetic engineering

2.1. Clonal Propagation

There are two main objectives for developing clonal techniques for forest trees. Firstly, viable large scale clonal systems can be an asset for selected trees and secondly, reliable regeneration protocols are necessary for further genetic manipulations. Some of the requirements for a large scale system are simplicity, reproducibility, cost effectiveness and adaptability to nursery practices.

Table 27.1. Vegetative propagation of some tree species by cloning, somatic embryogenesis and through protoplasts culture.

A. By clonal multiplication	B. By somatic embryogenesis
Populus tremula	*Abies balsamea*
Sterculia urenes	*Biota orientalis*
Tectona grandis	*Cupressus orizonica*
Prosopis cineraria	*Dalbergia sisso*
Tecomella undulata	*Dalbergia latifolia*
C. Protoplasts culture and genetic engineering	
Eucalyptus grandis	*Picea glauca*

Asexual multiplication of both hard woods and softwoods using tissue culture methods can be achieved by axillary bud breaking, production of adventitious buds and somatic embryogenesis.

The first two approaches lead to the development of a bipolar embryo. A large number of forest trees have been multiplied by these methods (Table 27.1). Axillary bud break is common in angiosperm while adventitious shoots formation is used in gymnosperm for multiplication. Clonal multiplication is the most frequently used method is tree propagation.

2.2. Somatic Embryogenesis

To date, shoot cultures are the most successful method of propagation for hardwood trees. However, somatic embryogenesis is alternative method providing large number of plantlets at low cost, the potential for generating artificial seeds and possibility of automation. Suspension cultures can provide embryogenis protoplasts which potentially can be used for genetic engineering. Somatic-embryogenic tissue can be generated directly of treated explants or obtained indirectly by manipulating non-embryogenic callus *in vitro*. In confers, this is the principal method of propagation. The examples of forest tree species propagated by somatic embryogenesis are -*Larix decidua*, *Picea abies* and others (Table 27.1)

2.3. Protoplast Culture and Genetic Engineering

Protoplast culture provides an alternative method of improving plants, and is one of the few systems which can utilize somaclonal variation, somatic hybridization, cybridization and genetic engineering. Protoplasts have been isolated from both tropical and temperate species of hard woods, e.g., sandal wood, *Eucalyptus grandis*, *E. urophylla*, *E. saligna*. Transformation of forest species has been successful with only a few genera like *Populus* and *Juglans* mainly using the Agrobacterium vector system.

One of the major problems preventing a wider use of micropropagation technology is the difficulty in using explants from mature trees. Success has been achieved in culturing juvenile explants like excised zygotic embryos and seedling parts, and by the time trees are old enough for evaluation, they are often recalcitrant (not responding) in culture. Therefore, tree propagation for the most plant is carried out with unproven (juvenile) material. Contamination and secretion of phenolics are the other problems associated with the tree species culture. Lastly, the cost of regenerating plantlets is very high.

3. IMPACT ON HORTICULTURE

In vitro culture techniques have had numerous applications to fruit crops beginning nearly 60 years ago with embryo rescue techniques for stone fruits. Later on, the method has been applied successfully to produce commercially acceptable early-ripening peach and nectarine cultivars. The methods have been adapted to other crops as well, e.g. in breeding programmes to produce both early-ripening and seedless grapes.

Historically, the second application of in vitro culture methods to fruit crops was to eliminate disease causing viruses from strawberries. Meristem-tip culture has since become an integral part of virus- indexing programmes for a number of fruit crops, usually in conjunction with thermotherapy. In some cases, it has been necessary to micrograft the meristem tip into an in vitro-grown seedling as is done for citrus.

Within the past 25 years, micropropagation of fruit crops has become an important application of in vitro technology (Table 27.2). Strawberry was the first fruit crop for which the method was developed. Now many fruit crops are being micropropagated commercially. More recent uses of in vitro culture emphasize application for genetic improvement of fruit crops. These applications include production of hybrid plants from fused protoplasts, somaclonal variation and mutagen-driven changes in regenerated plants, haploids plants from anther culture and transfer of specific genes via *Agrobacterium*-mediated transformation.

Table 27.2. Micropropagation of fruit crops.

A. Temperate fruit crops	B. Tropical fruit crops
Apple (*Malus*)	Kiwifruit (*Actidinia chinensis*)
Pear (*Pyrus communis*)	Pineapple (*Ananas cosmosus*)
Peach (*Prunus persica*)	Cashew (*Anacardium occidentale*)
Cherry, sweet (*Prunus avinum*)	Papaya (*Carcica papaya*)
Cherry, sour (*Prunus cerasus*)	Lime (*Citrus aurantifolia*)
Strawberry (*Fragaria*)	Lemon (*Citrus limon*)
Black berry (*Rubus sp.*)	Mulberry (*Morus alba*)
Rasp berry (*Rubus sp.*)	Banana (*Musa sp.*)
Grape (*Vitis vinifera*)	Date palm (*Phoenix dactylifera*)
	Guava (*Psidium guajava*)
	Pomegranate (*Prunica granatum*)

Following methods are used to improve and multiply the horticultural crop plants. Details about these methods are given in respected chapters, therefore not presented here.

1. Micropropagation
2. Virus eliminaton
3. Genetic improvement
4. Germplasm conservation
5. Haploid production

The above mentioned methods have been employed for fruit crops (Table 2) plantation crops and vegetable crops. Examples of plantation crops are—Pineapple, palms, banana, cacao (*Theobroma cacao*), coffee (*Coffea* sp), tea (*Camellia* sp.) rubber (*Hevea brasiliensis*), Vanilla (*Vanilla planifolia*), clove (*Syzygium* sp.) ginger (*Zingiber officinale*), turmeric (*Curcuma longa*), cardamom (*Elleteria cardamomum*). Some of these plants are produced in millions by plant tissue culture industry. In vegetable crops, major emphasis has been given on - cabbage (*Brassica oleracea*), mustard (*B. campestris*), artichoke (*Helianthus tuberosus*), garlic (*Allium sativum*), onion (*A. cepa*), tomato, carrot, melon (*Cucumis melo*), cucumber (*Cumis sativus*) pepper (*Capsicum annum*), and watermelon (*Citrullus vulgaris*).

Micropropagation of ornamental plants (foliage and flower plants) has become a major plant tissue culture based industry on developed and developing countries. Detail of this and examples are given in other chapters.

4. PLANTATION CROPS

Crops may be called plantation crops because of their association with a specific type of farming economy. Most of these involve a large landowner, raising crops with economic value rather than for subsistence, with a number of employees carrying out the work.

High value food crops- Plantings of a number of trees or shrubs grown for food or beverage, including tea, coffee, and cacao are generally called plantations. Some spice (black pepper, green cardamom, large cardamom) and high value crops (rubber, banana, sugarcane) grown from permanent perennial stock, such as black pepper may also be so called. When the holding belongs to a single individual, that person may be called a *planter*.

4.1. Industrial Plantations

Industrial plantations are established to produce a high volume of wood in a short period of time. Plantations are grown by state forestry authorities (for example, the Forestry Commission in

Britain) and/or the paper and wood industries and other private landowners (such as Weyerhaeuser and International Paper in the United States, Asia Pulp & Paper (APP) in Indonesia). Christmas trees are often grown on plantations as well. In southern and southeastern Asia, rubber, oil palm, and more recently teak plantations have replaced the natural forest.

Plant biotechnology is useful in clonal propagation of these materials and millions of plantation crop plants are produced annually to replace old stocks. There is decline in productivity in old plants therefore; such crop plants are periodically replaced by new materials. The details of some plants produced by different companies are given in next section (plant tissue culture industry).

The Central Plantation Crops Research Institute (CPCRI, Kasargod, Kerala) is established in 1916 and later brought under Indian Council of Agricultural Research (ICAR) during 1970. The initial mandate of the institute was on crop husbandry of coconut, arecanut, cocoa, oilpalm, cashew and spices. The restructuring process during VII and VIII Plan resulted in the establishment of separate Research Institute/Centres for Spices, Cashew and Oilpalm, but the institute continue to maintain strong linkage with these institutes. At present the Institute has a countrywide research network of four regional stations, four research centres and 17 Centres under AICRP on palms. Besides, the institute also hosts the headquarter of Indian Society of Plantation Crops.

5. APPLICATIONS IN MEDICINE

Medicinal plants or their extracts are used by humans since time immemorial for different ailments and provided valuable drugs such as analgesic (morphine), antitussive (codeine), antihypertensive (reserpine), cardiotonic (digoxin), antineoplastic (vinblastine and taxol) and antimalarial (quinine and artemisinin). Some of the plants which continues to be used from Mesopotamian civilization till today are *Cedrus* species, *Cupressus sempervirens*, *Glycirrhiza glabra*, *Commiphora wightii*, and *Papaver somniferum*. About two dozen new drugs derived from natural sources have been approved by FDA and put in market during 2000-2005, which include drugs for cancer, neurological, cardiovascular, metabolic and immunological diseases, and genetic disorders. Seven plant derived drugs currently used clinically for various types of cancers are taxol from *Taxus* species, vinblastine and vincristine from *Catharanthus roseus*, topotecan and irinotecan from *Camptotheca accuminata*, and etoposide and teniposide from *Podophyllum peltatum*. Most of the natural products are compounds derived from primary metabolites such as amino acids, carbohydrates and fatty acids, and are generally categorized as secondary metabolites. Secondary metabolites are considered as end product of primary metabolism and are generally not involved in metabolic activity viz., alkaloids, phenolics, essential oils and terpenes, sterols, flavonoids, lignins, tannins etc. These secondary metabolites are the major source of pharmaceuticals (Table 27.3), food additives, fragrances, and pesticides. Plant tissue culture provides excellent system to study and produce medicinally important compounds. Details of cell cultures grown in bioreactor and various medicines obtained from plants are given in Chapters 28, 29.

Table 27.3. Plant derived medicines used in Western medicine.

Ajmalicine	Digoxin	Papain	Reserpine
Atropine	Ephedrine	Papaverine	Scopolanine
Caffeine	Hyoscyamine	Pseudoephedrine	Sennosides
Codein	L-Dopa	Quinidine	Vincaleukoblastine
Colchicine	Morphine	Quinine	

6. TISSUE CULTURE INDUSTRY IN INDIA

India is an agricultural country and its economy is largely dependent on agriculture. Therefore, technological developments in the agriculture sector have a direct impact on country's economy. The abundance of genetic base (biodiversity), diverse agroclimatic zones and highly qualified man-power offer a potential scope for technological advances in agricultural biotechnology.

India has been traditionally a leading exporter of primary agricultural produce such as tea, coffee, rubber, cardamon, cotton and several spices, flavouring, aromatic and medicinal plants. This status is declining due to low productivity, non-competitive prices and sometimes, poor quality of the material. Therefore, our urgent need to improve plant cultivars for quality and quantity is required using biotechnological methods.

Most of the agricultural produce in India is produced as rain fed crops. The shortage of healthy seeds/planting material and lack of disease resistant clones have often been affecting the agricultural economy. Large scale production of elite clones through micropropagation or somatic embryogenesis, production of disease free clones and improvement of varieties through cellular, molecular and conventional breeding methods hold the promise of improving some of the problems of Indian agriculture, horticulture and forestry.

6.1. Scope for Development

i. Diverse Agroclimatic Regions- India has rich resources of land spread over diverse agroclimatic zones that have the potential to produce a wide variety of agricultural and commercially high value crops. It has large biodiversity, germplasm of rare ornamentals and medicinal plants, most of them native to this continent. Therefore, India offers many advantages over other countries in the world.

ii. Forest Wealth- The native forests all over the country are well known for their richness and variety in foliage and ornamental plants. These plants can be propagated by tissue culture method and introduced in the market.

iii. Different Climatic Zones- In India all the climatic zones, from temperate to tropical are available. Therefore, any plant growing in any climate can be multiplied and grown in the country. On the same time, cool climatic zones are less expensive for micropropagation work as extensive air conditioning for plant growth is not required.

iv. Scientific Talent- In India, the scientific talent and the operational skills of the people is at par with the best in the world. The salaries are low in India. This makes it an ideal place for production facility and that is why, some companies have been shifted to India from USA and Europe. It is worth mentioning here that anther and pollen culture for haploids and in-vitro fertilization like plant biotechnologies were discovered in India. Embryo culture and rescue of hybrid embryo is of quite significance in this aspect.

Propagation of plants through tissue culture has become an important and popular technique to multiply crops that are otherwise difficult to propagate conventionally by seeds and/or vegetative methods. Differentiated cells or organs are manipulated to give rise to multiple copies of the parent plant under optimum aseptic environmental conditions and appropriate stimuli. It offers several advantages over conventional propagation methods such as rapid multiplication of valuable genotypes, rapid improvement of crop plants, production of disease and virus free plants, production throughout the year possible, germplasm conservation and easy transport and transfer (international) of material (in vitro material is contamination free). There are several methods developed in last 3 decades for vegetative multiplication of plant species involving axillary budding, adventitious budding, somatic embryogenesis and organogenesis from callus cultures. In fact, the genetic stability of the plant copies (population) produced is likely to vary according to the procedure chosen. However, these methods have certain limitations.

6.2. Problems Associated with Agricultural Biotechnology

Development and use of any technology is dependable on infra-structural facility and simultaneous development of other related technologies. Agricultural biotechnology requires infrastructure like electricity, air-conditioned rapid transport system, skilled labour and related biotechnology like innovative and novel packaging material, plastic ware and glass appliances etc. Followings are major hurdles-

(a) Industry is a capital intensive (requires elaborate establishment money).

(b) Seasonal activity with peak demand; delivery of plant material at one time requires large facility which may be under utilized at other times.

(c) High production cost applicable to selected crop plants with established protocols for multiplication.

(d) Management of personnel and organization of work schedule in chain oriented manner.

(e) Contamination problem adds a big cost factor.

(f) Somaclonal variation at large multiplication rate.

In context of India, on export scene, lack of variety of packing material at affordable price, high air freights, unsuitable air connections, lengthy phytosanitory, custom and export documentation formalities make this non-traditional, highly perishable export commodity less attractive as compared to potential it has. However, recently the government of India has set up suitable cargo-handling cold storage facility at New Delhi and Bangalore for export of biotech products.

6.3. Commercialization

Commercialization of agricultural biotechnology began in India in real sense in 1987 when various private sector companies adopted this technology as a means of increasing the productivity of their primary produces or as a part of their diversification programme. This includes A.V. Thomas group of companies, Indo-American Hybrid seeds (IAHS), Hindustan Lever, Tata Tea, Unicorn Biotech, Nath seeds, RP Goenka Enterprises, India Tobacco company, Maharashtra Hybrid Seeds Company, Hindustan Agrigenetics Ltd. and some others.

The first professionally managed commercial plant tissue culture unit in the country was set up in 1987-88 by A.V. Thomas, at Cochin in export processing zone, and remained one of the largest in Asia and only one of its kind in India (producing 190 million plants in 1996), until IAHS began production in its Bangalore based unit. IAHS has 48 Laminar flow work stations, hundred growth rooms (each can accommodate 400,000 culture at one time) and a green house of 30,000 sq. ft. (2790 sq m) with most advanced equipment such as automatic shade system, fogging, blackout and irrigation systems, and movable benches for hardening of tissue culture plants.

Subsequently, advancement in commercialization of plant tissue culture and acceptance of plantlets raised through tissue culture have led to continued exponential growth within the industry in terms of number of new units as well as number of plants produced by these units (Table 27.4). Most of these units are set up with foreign know how. Most of our commercial units prefer to seek the entire technology package on a turn-key basis. The major foreign collaborators (mostly with buy back arrangements) are M/S Cultiss, Holland; M/S Green Teek, Holland; M/S Microplants Ltd., UK; Centre de Research Agronomique, Belgium; Planex, Australia; and Agrobio, France etc.

Table 27.4. Plant Micropropagation Industry in India.

State	No. of Industries	Annual production capacity in million
Maharashtra	25	More than 31
Karnataka	9	30.1
West Bengal	9	5
Andhra Pradesh	6	16.25
Tamil Nadu	4	26.35
Kerala	4	25.35
Haryana	3	13.25
Gujarat	3	11.25
U.P.	3	25.0
Punjab	1	1.7
H.P.	1	5
Orissa	1	Not known
Rajasthan	1	Not known

Indo-American Hybrid Seeds (IAHS), A.V. Thomas & Co., Harrisons Malayam Ltd., Godrej Plant Biotech Ltd. and SPIC have established ultramodern micropropagation facilities. IAHS and A.V. Thomas have exported millions of tissue cultured plants to Europe and released several lakhs banana and cardamon plantlets to farmers in domestic market. A.V. Thomas have developed a cardamon variety with yield of 250 kg/hectare as compared to normal 70 kg/hectare and can be cropped in two years instead of usual three years. A.V. Thomas have also developed high glucovanillin producing and early flowering vanilla plants through tissue culture. Cost effective micropropagation system has also been developed in tea and coffee.

IAHS has won several contracts from the Netherlands, UK, and Denmark to develop new plant varieties by somatic hybridization, selection of somaclonal variants, endosperm culture, *in vitro* pollination and embryo rescue method in combination with conventional plant breeding. IAHS is providing service for screening plants for 40 viruses, 2 viroids and several bacterial, fungal and mycoplasmal diseases for 19 ornamental and 26 horticultural and agricultural crops. This company had exported cut flowers worth Rs. 14 million to Holland, Denmark and UK in 1991.

Beena Nursery (P) Ltd., Trivandrum offers high quality and disease free plantlets of orchids suitable for the cut flower industry. Hindustan Lever Ltd, Mumbai is involved in clonal multiplication of high yielding cultivars of plantation crops such as cardamon through meristem culture, virus-free microtubers of potato. It has also designed a bioreactor to scale-up the microtuber production. Godrej plant Biotech Ltd. (earlier known as Unicorn Biotech), multiply trees and plants by apical and axillary meristems and somatic embryogenesis. Micropropagation protocols for 2 cultivars of strawberry suitable for cultivation in tropical plains and conventional hilly regions have been developed. The company sold over 2 lakhs plants to farmers in Maharashtra, Gujarat and Andhra Pradesh during 1992-94.

Micropropagated banana, *Chrysanthemum*, rose, carnation and foliage ornamental plants are not only being produced at large scale but also being exported. Dhampur Sugar Mills, Dhampur and EID Parry, Chennai are involved in production of virus-free sugarcane through tissue culture.

SPIC science foundation, Chennai have introduced sex-typed female papaya plants in commercial tissue culture market for the first time in India. SPIC Agrobiotech centre is working on coffee, banana, potato, sunflower, rose, gerbera, orchids and tree species (Callistemon, Simarouba, jack fruit, mango and date palm).

National Agricultural and Scientific Research foundation, Calcutta is engaged in large scale production of *Musa*, *Spathiphyllum*, *Anthurium*, *Elettaria* and orchids. It has already developed protocols for tea, jojoba, lilies, gladiolus, gerbera, *Ficus* and *Philodendron*. In all there are 75 units of small, medium and large capacity, capable of producing 190 million plants per annum. The major 4 units are producing 15-25 million plants per annum. The others are capable of producing plants from 1-10 million per annum.

6.4. Plants Produced

Indian units are currently producing fruit crops, forest trees, ornamental (foliage and flowering plants), vegetative crops and plantation crops (Table 27.5). The major emphasis is on the production of ornamentals.

Table 27.5. Plant species currently micropropagated at commercial level in India.

Fruit crops	Banana, Citrus, Grapes, Papaya, Pomegranate, Strawberry, Pineapple, Cherry apple, Peach, Datepalm, Mango, Apple.
Forest Trees	Callistemon, Eucalyptus, Teak, Kadamb, Jack fruit, Simarouba, Salvadora, Bamboo.
Ornamentals- Foliage	Aglaonema, Calathea, Cordyline, Diffenbachia, Ficus, Philodendron, Spathyphyllum, Syngonium.
Ornamental - Flowering plants	Alstroemeria, Anthurium, Callalily, Carnation, Chrysanthemum, Gerbera, Impatiens, Rose, Dracaena, Gladiolus, Iris, Orchids, Lily, Dendrobium.
Vegetative crops	Potato, Asparagus, *Brassica napus*, *B. juncea*, tomato, cauliflower, cabbage.
Plantation crops	Cardamom, Sugarcane, Vanilla, Coffee, Ginger, Turmeric, Tea, Black pepper

6.5. Governmental Efforts

The Indian Government has identified micro propagation of plants through tissue culture as an industrial activity under Industrial (Development and Research) act, of 1951 which was made effective in 1991. The units engaged in floriculture can avail duty free imports under export oriented units (EOU)/ export processing zone (EPZ) scheme even if they sell 50% in domestic market and export the remaining 50%. The import duty on live trees, plants, bulbs, roots, cut flowers as well as ornamental and foliage plants has been reduced from 55% to 10%. Import duty on seed development machinery for soil preparation, greenhouses, pre-cooling unit and refrigerated transport units has been reduced to 25%. A subsidy on air-freight to a maximum of 25% has been allowed for export of cut flowers and tissue culture plants.

Government of India has formed National Board of Biotechnology in 1982 to promote research and development work in the area of Biotechnology in the country. Since then several well defined programmes of National priorities began in the country through National Institutes and Universities. Later on, Department of Biotechnology (DBT) was set up in February 1986 for better functioning under the Ministry of Science and Technology. This department is planning, promoting and coordinating biotechnological programmes in the country.

DBT has set up task forces in various biotechnological areas. The research projects funded by DBT are in various research themes like micropropagation for fuel, fodder and biomass, medicinal and endangered plant propagation, protoplasts culture and genetic engineering, and crop improvement. Plant species have been identified for Research and Development programmes. Two pilot plant for micropropagation have been set up at National Chemical Laboratory, Poona and Tata Energy Research Institute, Delhi.

Government of India has liberalized rules for import of basic materials, provides tax exemptions and subsidies on air freight of plants raised through tissue culture, cold storage facilities have been created at Delhi, Mumbai and Bangalore for storage of consignment awaiting shipment. Tissue culture consignments are given quarantine clearance within 6 hours. Due to these reasons, Indian export of floriculture products increased from Rs. 738 lakhs in 1990-91 to Rs. 1139 lakhs in 1991-92.

7. GERMPLASM PRESERVATION

In the recent years, with the tremendous increase in the population, pressure on the forest and land resources have increased which has resulted in the decline in the population of medicinal and economically important plant species. Even some of the plant species are at the verge of vanishing because of severe threat to their natural habitat due to human interference. Such species are termed as 'threatened' species.

Thus, attempt has been made in the recent years at the national and international levels to protect and preserve plant species, threatened plants as well to those plant species which are not in use today but may serve as important resource for future breeding programme.

In India, Botanical Survey of India has released three volumes (Red Data Book) enlisting the threatened plants of Indian subcontinent. There are several other organisations, which are engaged in the protection and preservation of such plants. Some of these organisations are Indian Council of Agricultural Research, New Delhi; International Bureau of Plant Genetic Resources, Nottingham, U.K.; National Bureau of Plant Genetic Resources, Pusa Campus, New Delhi; Council of Scientific and Industrial Research, New Delhi and Botany Departments of various Universities. Attempts have been made to conserve the germplasm by preserving the genetic material, which can be done by two ways:

A. *In situ* preservation

B. *Ex situ* preservation

A. ***In situ* preservation:** It aims at the preservation of the germplasm in their natural environment by establishing biosphere reserves, national parks, gene sanctuaries etc.

Limitations:The limitation of this type of preservation is the risk of declination of the preserved species due to environmental hazards.

B. ***Ex Situ* Preservation:** This is the chief mode of preservation of germplasm. Providing the suitable condition in the gene bank preserves the genetic materials in the form of seed or in vitro cultures. But for being successful in establishment of gene bank considerable knowledge of genetic structure as well as elements influencing them are necessary.

Usually seeds form the most common material to conserve plant germplasm in the seed propagated plants. But this method has certain limitations such as -

Loss of seed viability with passage of time.

Seed destruction due to seed born pathogens, pest etc.

This method is confined only to seed propagating plants.

In contrast to seed propagating plants, the vegetatively propagated plants are preserved in vitro as shoots, meristems, embryos etc. The advantages of in-vitro preservation over in situ preservation are -

Large amount of material can be preserved in small area.

It overcomes the destruction due to environmental hazards.

It provides large amounts of plant material for culturing.

Using either of the following techniques, preservation of germplasm can be achieved.

7.1. Gene Bank

Germplasm or genetic resources are stored in the form of 'seeds' in a specialized place known as 'gene bank'. Thus, a gene bank contains seeds (and not DNA/genes) of a plant species for future use. Collection of genetic resources at one place for use and distribution to others has a long history. By the end of nineteenth century, the US department of Agriculture had setup a plant exploration section. This started collection and import of new germplasm to the US. This leads to cultivation of soybean in the US, which has become a multi-million dollar soybean oil and protein industry.

As concern about the conservation of genetic resources increased, the collection and storage of material for immediate use was changed to storage for future use. In addition, the aim was not only to collect popular cultivated varieties, but also the land races and wild relatives of crop plants. This idea was initiated by N.I. Vavilov, who set up a large gene bank in Leningrad. By 1980s there were dozens of gene bank in seventy countries around the world. Some of these collections were started at international research institutes like IRRI, Phillipines, CIMMYT, Mexico, ICRISAT, India, IBPGR, Rome and IARI, India. All are under the general co-ordination of the international board for plant genetic resources (IBPGR) at Food and Agriculture Organization, Rome. All materials sent to gene bank are given an identification number called an accession number. The bank grown all seeds and compare their characters. This is a big task to manage seeds, data and information and computers are used for these purposes. New germplasm of existing cultivated plants was collected by IBPGR, from Africa and west of Asia, e.g., perennial chick pea from Morocco, perennial oat from Algeria, and wild sorghum from Namibia.

For most important crops, the plant part that is stored in a gene bank is the seed. In vegetatively propagated crops, the seeds are either not produced or are not suitable. Therefore in such plants cuttings (many fruit trees) or tissue cultures cells/tissues are stored. Clonal propagation of stored tissue culture cells is the way, forest genetic resources are stored. Seeds with the high oil content or large size are usually stored in bags at –18 °C. These seeds must be tested for germination every 10 years. Small seeds with less oil (e.g., cereal grains) are stored in liquid nitrogen at –196°C. Storage at such a low temperature reduces the cost of germination work; as such seeds may be stored for decades without germination testing.

7.2. Cryopreservation

Cryopreservation [preservation in the frozen state (cryo means extreme cold, derived from Latin word kruos = 'frost')] is based on the reduction and subsequent arrest of metabolic functions of biological material by imposition of ultra-low temperature. At the temperature of liquid nitrogen (–196°C) almost all the metabolic activities of cells are ceased and the sample can then be preserved in such state for extended periods. However, only few biological materials, in their natural state, can be frozen to sub-freezing temperatures without adversely affecting the cell viability. Knowledge about the chemicals having cryopreservative properties such as glycerol and dimethyl sulfoxide facilitated the development of effective cryopreservation technique.

7.3. Need for Cryopreservation

The main application and objective of developing cryopreservation technology is preservation of valuable genetic resources, especially of vegetatively propagated and also of those species which have short lived seeds. Therefore, it is imperative that use of cryopreservation technology requires efficient regeneration protocols through tissue culture of the species. Cryopreservation of endangered species is one of the most important objectives of NBPGR, New Delhi. Endangered species for which micropropagation techniques have been developed in India are listed in Table 27.6. The regeneration methods using apical meristems is advantageous because these are simple and there are no chances of genetic variation. It is essential that tissue culture methods do not create genetic variability.

Table 27.6. Threatened species for which micropropagation methods have been developed.

Boswellia serrata	Commiphora wightii	Nepenthes khasiana
Calligonum polygonoides	Coptis teeta	Podophyllum hexandrum

In contrast to above, the cases where genetic variability has been induced in the cultures, this variability need preservation for use in the future. Therefore, cryopreservation technique is equally good for preserving the genetic resources of existing genotypes and also of new variants.

7.4. Procedure of Cryoprotection and Pretreatment

Some form of cryoprotection is necessary for cryopreservation of plant material unless they are naturally dehydrated, as in the case of dormant vegetative buds in the winter, or artificially cold acclimated. Several chemicals such as dimethyl sulphoxide (DMSO), glycerol, various sugars and sugar alcohols protect living cells against damage during freezing and thawing (means to become unfrozen or warm after preservation at ultra low temperatures). These compounds lower the temperature at which freezing first occurs and can alter the crystal habit of ice when it separates. The colligative properties of the cryoprotectants minimize the harmful action of electrolyte concentration resulting from conversion of water into ice. High solubility in aqueous phase and low toxicity to the cells are the two essential characteristics for cryoprotectants, e.g., DMSO, methanol, glycerol and sugars, sugar alcohols, high molecular weight polymers as dextran, polyvinyl pyrrolidone, hydroxy ethyl starch.

The cells require different pretreatment periods with different compounds for proper cryoprotection. DMSO enters more rapidly than glycerol and therefore requires shorter period for treatment. Most of the cryoprotectants exhibit varying degree of cytotoxicity at higher concentrations. Generally, DMSO at 5 to 10% and glycerol at 10 to 20% are used as cryoprotectants. Sometimes, a mixture of cryoprotectants can also be used to improve the efficacy. Addition of osmotically active compounds in the culture medium such as mannitol, sorbitol, sucrose and proline increase the freezing resistance of the cells. These compounds mainly act by their dehydration effect Sorbitol has been successfully used as an osmotic agent as well as cryoprotectant for *Glycine max, Datura innoxia, Brassica napus,* and *Daucus carota*.

7.5. Freezing Methods

Methods used by different workers for cryopreservation of various plant materials can be categorized as - (i) slow freezing, (ii) rapid freezing and (iii) droplet-freezing.

i. Slow freezing - Methods of slowly freezing biological specimens are based on the physicochemical events occurring during the freezing. When plant cells are cooled progressively, ice crystal formation is usually initiated extra-cellularly. It is presumed that plasma membrane acts as a barrier that prevents the ice crystal formation in the cytoplasm. In absence of ice crystals, cytoplasm remains super cooled. On further lowering of temperature, the concentration of extra-cellular liquid increases as more water is converted to ice. Since the vapour pressure of the frozen solution is lower than the same concentration of the super cooled liquid, the vapour pressure of slowly cooled cells reach equilibrium with external ice by efflux of water. Thus, slow freezing prevents the intracellular ice formation and consequently freezing injury is prevented. It is believed that slow freezing increase the concentration of cytoplasm and increased dehydration increases the survival of cells. There are different methods of obtaining protective dehydration, including slow cooling at constant or varying cooling rates, or keeping the samples at one or more intermediate subzero temperatures.

The development of efficient slow freezing method depends upon several factors like cooling rates, pretreatment and cryoprotection, type and physiological state of the material, and the temperature prior to immersion in liquid nitrogen. The most commonly used methods for the cryopreservation of plant cells generally involves regulated slow cooling at a constant rate of 0.5 to 2 °C/min. to terminal temperature between –30 °C to –40 °C followed by storage in liquid nitrogen (Table 27.7).

ii. Rapid freezing - Rapid freezing is unsuitable for the cryopreservation of cell cultures, it is employed to cryopreserve shoot tips of carnation, potato, strawberry and several others. Rapid freezing is accomplished by direct immersion of the cryoprotectant-treated specimens in liquid nitrogen. The cooling rate in this method is very high, usually several hundred degrees per minute. At such high cooling rates, the intracellular fluids do not have sufficient time to equilibrate with the external ice with the possibility of intracellular ice formation, which is considered to be lethal for cells and somatic embryos of oil palms.

Table 27.7. Plant species in which callus, cells, protoplasts and meristem/shoot tips have been used for cryopreservation.

Callus cultures	*Acer pseudoplatanus, Atropa belladonna, Brassica napus, Capsicum annuum, Catharanthus roseus, Datura innoxia, Dioscorea deltoidea, Digitalis lanata, Daucus carota, Glycine max, Hordeum vulgare, Hyosyamus muticus, Malus domestica, Medicago sativa, Nicotiana tabacum, Oryza sativa, Panax ginseng, Populus sp., Rosa, Saccharum sp., Zea mays.*
Cell cultures	*Lavandula vera, Medicago sativa, Populus sp., Saccharum sp., Triticum aestivum, Ulmus Americana.*
Protoplasts	*Datura innoxia, Daucus carota, Glycine max, Hordeum vulgare, Triticum aestivum, Zea mays, Meristem/shoot tips, Arachis hypogaea, Brassica napus, Cicer aerietinum, Dianthus caryophyllus, Lycopersicon esculentum, Malus domestica, Manihot esculenta, Pisum sativum, Solanum tuberosum.*

iii. Droplet freezing - In this method the cryoprotactant treated meristems are dispensed in droplets of 2-3 μl on an alluminium foil in a petriplate. The specimens are frozen by slow cooling (0.5°C/min.) to a subzero temperature between –20 to –40 °C prior to immersion in liquid nitrogen.

iv. Storage, thawing and regrowth - Material can be kept stored in liquid nitrogen (–196 °C) or in its vapour (–150 °C). Rapid thawing is recommended for most cryopreservation methods. The basis of applying rapid thawing is to avoid the damaging ice recrystallization which may occur during slow warming. In general thawing is carried out by removing the sample by liquid nitrogen storage and transferring in a water bath (34-40 °C) for about 1-2 min or until material is warmed up.

Regrowth of cryopreserved specimens is the most reliable and accurate estimate of viability. The other technique used for vitality test of the cells such as fluorescein diacetate stain, triphenyl tetrazolium chloride test etc. may provide a quicker method of testing cell viability. The further growth and regeneration of the cryopreserved cells after thawing will be like normal cells. However, care should be taken in handling, plating and subculture of such cells.

v. Freezing Apparatus - Various types of cryostate and freezing units are available by which different rates of cooling can be easily regulated (Fig. 27.2).

The culture subjected to ultra cooling can be stored at –196 °C in the liquid nitrogen container for various lengths of time, and can be taken out and thawed when required.

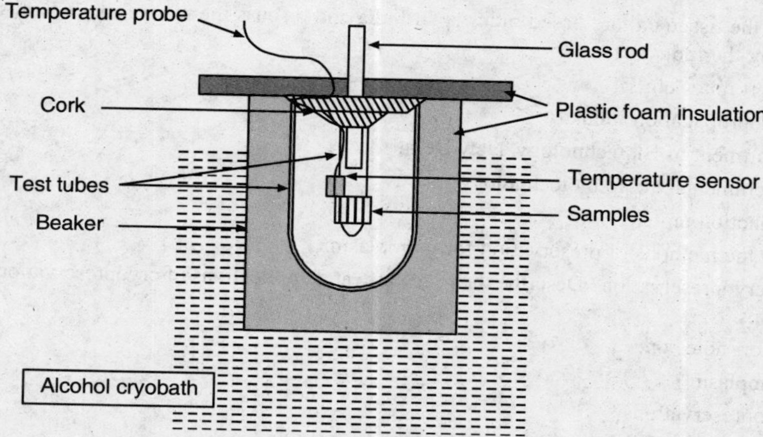

Fig. 27.2. Apparatus used for freezing the plant samples in cryobath for further immersion in the liquid nitrogen.

vi. Applications of cryopreservation

1. Conservation of genetic uniformity.
2. Preservation of rare genomes.
3. Freeze storage of cell cultures and cell lines.
4. Maintenance of disease free material.
5. Cold acclimation and frost resistance.
6. Retention of morphogenetic potential in long-term cultures.
7. Slow metabolism and aging.

In spite of the information and technology available for *in vitro* storage of several crop species, *in vitro* active gene banks exist so far only for potato at International Potato Cenre, Lima, Peru and casava at Centro International de Agricultura Tropical (CIAT), Cali, Colombia. Recently, efforts have been made under the auspices of International Board for Plant Genetic Resources (IBPGR) to exploit the full potential of in vitro storage methods employing meristem and shoot tip cultures. The initiation of a pilot project by collaborative efforts of IBPGR and CIAT, using cassava as a model system, for assessing the potential and feasibility of establishing and running an *in vitro* active gene bank is a first significant step. Under the project more than 4000 clones of cassava are being maintained in vitro under minimal growth conditions. In India, storage facility has been created at National Bureau of Plant Genetic Resources (NBPGR), New Delhi for germplasm preservation using *in vitro* methods.

QUESTIONS

1. Discuss the role of plant biotechnology in agriculture.
2. Describe the impact of plant biotechnology on crop improvement.
3. Write short notes on:
   (a) Anther culture.
   (b) Ovary and ovule culture.
   (c) Embryo rescue.
   (d) In vitro fertilization.
   (e) Genetic modifications.

4. Describe the tissue culture based industry in India and discuss the role of the federal government in supporting the Industry.
5. Write short notes on:
   (a) Micropropagation Industry
   (b) Department of Biotechnology, New Delhi
   (c) Governmental aid to biotechnology
   (d) Production units.
6. Describe the methods of preservation for germplasm.
7. What is cryopreservation? Describe the procedure and applications of cryopreservation for storage of germplams.
8. Write short notes on:
   (a) Germplasm
   (b) Cryopreservation
   (c) In situ preservation
   (d) Ex situ preservation
   (e) Thawing
   (f) Gene bank

CHAPTER 28

Prospects of Drug Production in Cell Cultures and Bioreactor

1. INTRODUCTION

Plant cell and callus cultures have been extensively used in the past three decades to explore the possibility of producing useful secondary metabolites through biotechnological methods. These empirical approaches generated enormous data about the plant species producing secondary metabolites in culture, physico-chemical factors like light, temperature, plant growth regulators and nutrients affecting the production and lastly, selection of cells with high yield of secondary metabolites. However, all the cultures investigated were not commercially viable systems. Therefore, two way approaches was made: (1) to generate the basic know-how for the production of secondary metabolites, and (2) to develop technology for specific product using plants like *Catharanthus roseus*, *Berberis*, *Coptis* and *Panax* etc., as a model system. Investigations were focused to develop technology for these plants and also to better understand the process as a whole.

Plant cell in the agitated liquid medium produces secondary metabolites characteristics of the parent plant, a process governed by several genes (i.e., genetically controlled) and hence it is a multi-steps reaction. To produce secondary metabolites at commercial level there are certain pre-requisites like- high demand, high product cost, and availability of uniform raw material without interruption. Technology should be cost effective otherwise the production will not be commercially viable. Only a few selected compounds can fulfil these criteria because of high production cost. It is imperative that the production cost can be reduced by increasing the yield of secondary metabolites by improving the bioreactor technology to reduce the capital cost of inputs. Serious attempts are being made to increase the yield by manipulating the biosynthetic pathways, removing the limiting factors barriers, and incorporating the techniques of genetic engineering. Here, we are presenting the details of bioreactor system and various cultures used to produce secondary metabolites in these bioreactors, towards developing technology for the industrial-level production.

Application of plant cell tissue or organ culture for the production of secondary metabolites was originated in 1947, when James first reported the occurrence of alkaloids in meristem culture of solanaceous plants. Scale-up (process at large-scale as compared to small-scale process in laboratory) problems are common in all kinds of engineering sciences. The classical processes of the fermentation industry have been developed along the time with trial and error method. Bioreactors are advantageous then shake flasks as they provide better control on the system (i.e., pH, dissolved gas concentration, cell growth, etc) there are examples of cultivation of plant cells in large fermenter demonstrating feasibility of the technology for the scale-up production of secondary metabolites. It is imperative that a bioreactor is used when basic studies related to optimization of product yield have been completed. In most cases, a growth medium and a production medium are used to obtain maximum yield of the secondary metabolite.

Tulecke and Nickel (1959) successfully developed a 10L system in a simple carboy for the cultivation of plant cells. Noguchi and co-workers (1977) cultivated *Nicotiana tabacum* cell suspensions in a 20,000 L stirred tank reactor while Schiel and Berlin (1987) studied *Catharanthus roseus* cells in a 5,000 L stirred tank reactor. The first commercial process used *Lithospermum erythrorhizon* cell suspensions grown in stirred tank reactors of 200 and 750L. Various published results show that plant cell cultures can be grown in the microbial type bioreactors, thus suggesting the use of existing fermentation facilities and conventional equipment for industrial production of plant products using the cell culture technology. In the contrary to above, Archanbault *et. al.* (1998) suggested that specifically designed and operated bioreactors might be more appropriate for plant cell cultures. It is worth mentioning that different types of bioreactor designs and processes are currently in use for the production of secondary metabolites (Table 28.1).

Table 28.1. Some examples of large-scale production of plant cells.

Plant species	Capacity in litres	Compound produced
Catharanthus roseus	85	Serpentine
Coleus blumei	450	Rosmarinic acid
Lithospermum erythrorhozon	750	Shikonin
Nicotiana tabacum	20,000	Biomass
Panax ginseng	200-20,000	Saponins

2. BIOREACTOR PROCESS

Bioreactor is a large culture vessel made up of glass for use at laboratory-scale (up to 10L) but large-scale bioreactors are made up of stainless steel. Laboratory-scale bioreactors can be designed and fabricated in the laboratory or can be purchased as a functional unit (Fig. 28.1). The commercially available systems are sophisticated equipment fitted with microprocessor control unit (or a computer) to control the pH, dissolved oxygen, gas flow rate, agitation speed, nutrients, temperature inside the vessel and cell density for the optimal growth and/or the production. All the units have a culture vessel and a control unit. The details of a typical stirred type bioreactor are given in the figure. 28.2. There are different types of stirrers available to suit the requirement for tissue shear pressure and effective agitation. The modern bioreactors are fitted with various sensors like pH, temperature, foaming and nutrient concentrations. The control unit records the changes in the composition of the medium and then transfer the desired chemical/nutrient to adjust the medium according to set parameters. Samples are taken, at time intervals to study the time course of the growth, with the help of sample tubes fitted with sterilized membrane filter and syringe to create a suction. For sterilization, small reactors are autoclaved while commercial scale reactors are sterilised *in situ* by passing steam at appropriate pressure. Due care should be taken to sterilize the system according to its volume. Various kinds of bioreactors have been designed depending upon the requirements and the systems, which are discussed later in this chapter. However, all the bioreactors require a cooling unit (cryobath with water circulating pump) to circulate cold water (being a tropical country, in temperate countries, tape water is quite cold and can serve the purpose), a membrane type air pressure pump to supply oil-free air at a desired pressure, an autoclave of desired dimensions to sterilise the unit, disposable or autoclavable air filters, and autoclavable tubings to connect the system besides general facility for tissue and cell culture. A sufficient amount of stock of cell culture is required to inoculate the bioreactor.

This system is used to develop technology for the industrial production of useful secondary metabolites or for the production of large number of plantlets using somatic embryogenesis. Depending upon these factors for the accumulation/release of secondary products, different process modes are used viz., batch culture, fed batch culture, two-stage batch culture, and continuous chemostat type cultures. In batch cultures vessels are inoculated and harvested after growth period. In fed-batch cultures, the cultures are fed once or twice with additional nutrient supply during the growth period. Additional supply of nutrients boost the production of metabolites. In continuous cultures, a balance is maintained between continuous supply of nutrients and harvest of cells and cultures and in steady state of growth.

3. FACTORS FOR GROWTH IN BIOREACTOR

3.1. Gas-Liquid Mass Transfer

Maintaining a constant oxygen mass transfer coefficient (K_La) is a basis of scale-up of many bio-processes. The oxygen transfer requirements of cultured plant cells are low as compared to that of bacterial cultures because unlike microorganisms, plant cells have lower respiration rates. For instance, if cells which respires at a rate of 0.2 mM g^{-1} h^{-1} are to be grown to 10 gl^{-1} without allowing the dissolved oxygen concentration to fall below 20% of saturation. Plant cell culture in bioreactors typically requires KLa values between 10-30 h-1. Operation of bioreactor at higher KLa values results in poor cell growth or production of secondary metabolites. This may be due to either increased shear associated with high KLa conditions or due to enhanced CO_2 stripping from the medium. To maintain a constant dissolved oxygen level, bioreactors should be equipped with a dissolved oxygen probe to adjust aeration and agitation rate.

Fig. 28.1. A stirred tank bioreactor and harvest of cultures from vessel.

3.2. Shear

Plant cells are shear sensitive. Shear refers to forces exerted on the surface of a body in a direction parallel to surface. This is contrasted to normal forces, which are exerted on a surface but, perpendicular to the surface. Mathematically, shear can be described by the equation [$\tau = \eta\lambda$], where τ is the shear stress (a force per unit area), γ is the shear rate (a change in velocity across a distance), and η is a viscosity (a coefficient which describes the resistance to flow). Earlier, efforts were made to describe the shear damage to eukaryotic cells growing in a mechanically agitated bioreactor system by calculating impeller tip speed by the equation [Tip speed = πND_i], where N is the impeller speed (*rpm*) and D_i is the impeller diameter. Sinskey *et al.* (1981) proposed that it may be better to use an integrated shear factor (ISF) to correlate shear damage to cells by the equation [ISF = $\pi ND_i/(D_t - D_i)$] where, $(D_t - D_i)$ is the measure of the distance between the impeller and the tank wall. D_t is the tank diameter. ISF thus represents a pseudo-shear rate, which exists between the impeller tip and the vessel wall. Even in the absence of mechanical agitation, gas sparging (profuse bubbling of air/gas) can also exert shear.

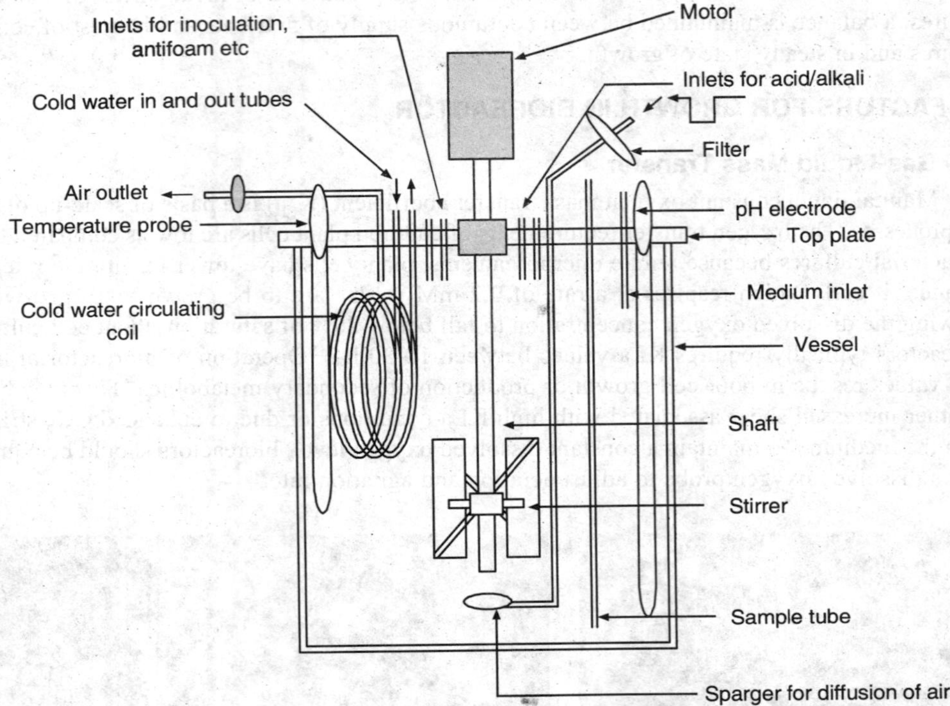

Fig. 28.2. A schematic presentation of various parts of a stirred type bioreactor.

3.3. Mixing

Mixing of the dissolved nutrients of the culture medium is generally not been a problem in suspension cultures. But, large size of plant cells and especially the cell aggregates settle into dead zones or unmixed regions at the bottom of the bioreactor. Moreover, cells get adhere to the surface of the tank above the level of the medium and become deprived of nutrients. These exhibit a serious problem. Recently, there has been a trend towards using very high cell concentrations [packed cell volume (PCV of 90%)]. Under these conditions incomplete mixing can be a serious problem.

4. TYPES OF BIOREACTORS

Depending upon the mode of agitation, bioreactors can be basically classified into following two types:

4.1. Mechanically Agitated Bioreactors

The mechanically agitated bioreactors are most commonly used for large-scale culture of plant, animal and microbial cells. In this type of bioreactor, the medium is agitated with the help of a mechanically driven impeller. Various types of impellers are in use, depending upon the requirement. Two traditional types of impeller used in this system are shown in figure. 28.3.

Flat blade turbine impeller (Fig. 28.3A) with high speed is generally used in bacterial culture. High agitation breaks the incoming air into small bubbles. Since plant cells cannot tolerate high shear conditions and mixing of air may be a more serious problem with plant cell cultures, an alternate impeller, capable of inducing low shear have been used. Marine propeller impeller is better suited for low shear mixing (Fig. 28.3 B). It provides axial mixing of the medium. On the

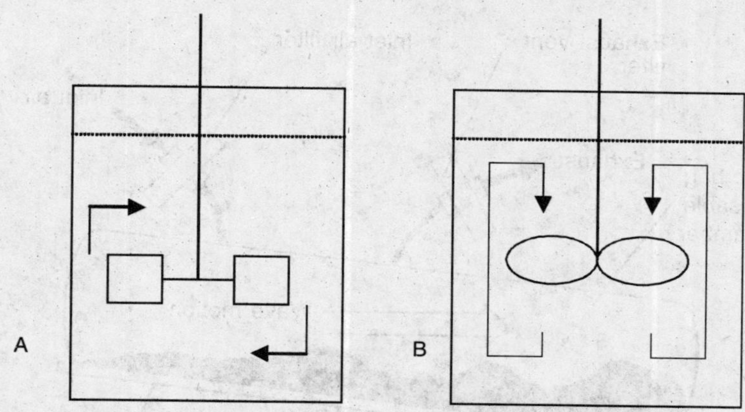

Fig. 28.3. Types of impellers: (A) Flat-blade turbine impeller and (B) Marine propeller.

other hand, flat-blade turbine promotes radial mixing. For large-scale systems, neither of these two types of impeller is used. Alternatively, low-shear impellers (e.g., paddle and helical types) have been shown to more useful for plant cell cultivation. Since low agitation is insufficient to break incoming gas into small bubbles, incoming gas stream is dispersed as fine bubbles using an appropriate gas distributor. Enrichment of incoming gas with oxygen is beneficial.

4.2. Pneumatically Agitated Bioreactors

Pneumatically agitated bioreactors are of two types viz., the bubble column and air-lift (Fig. 28.4). These bioreactors are tall and thin as compare to mechanical agitation type reactors. Typically, the height-to-diameter ratio in pneumatically agitated bioreactors is high. In bubble columns, air is bubbled at the base of the column and medium is agitated with this. In air-lift bioreactors, gas is sparged in the riser section and after the gas is disengages at the top of the column; the medium then flows downward in the down-corner section. These two sections may be separated using a baffle, a concentric cylinder, or an external loop. Circulation in the air-lift bioreactor promotes better mixing and therefore, has advantages in suspending cells and clumps more uniformly. But, in large-scale air-lift bioreactors, oxygen transfer rate is low in the down-corner section. It should be emphasized that the performance of an air-lift bioreactor is strongly dependent upon the geometry of the system. Ratio of the cross-section area of the riser section and down-corner sections is of special importance because this ratio affects mixing and oxygen transfer.

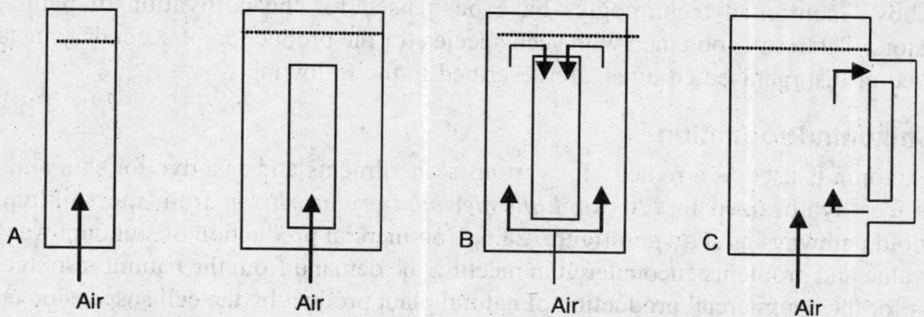

Fig. 28.4. Pneumatically agitated bioreactors; A) Bubble column, B) Draft tube air-lift and C) External loop air-lift.

4.3. Wave bioreactor

Wave bioreactor consists of a pre-sterilized flexible, plastic chamber that is partially filled with medium and inoculated with cells. The remaining part of the chamber is inflated with air. The

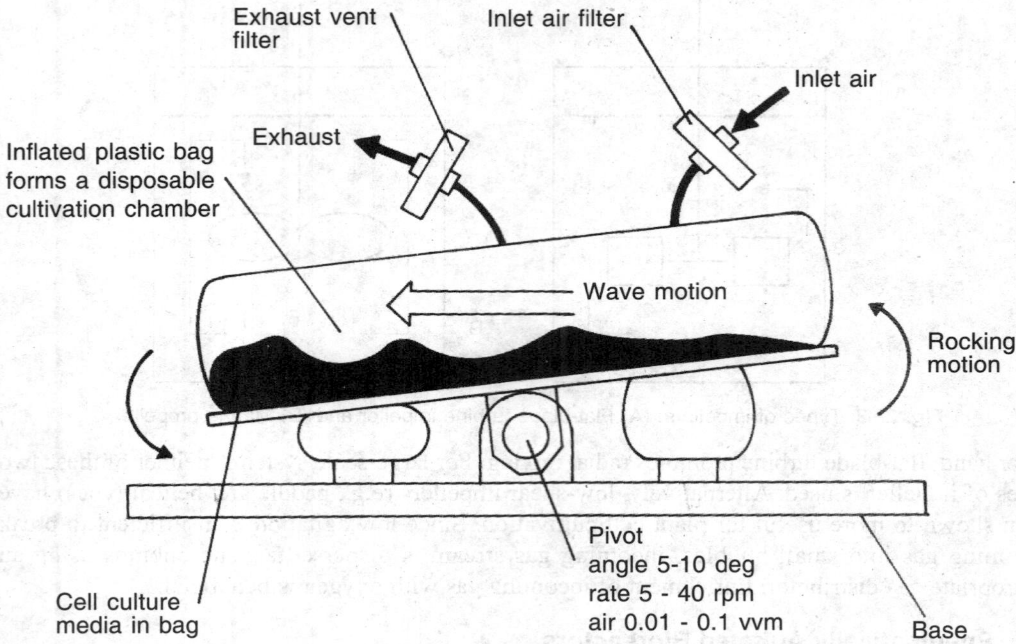

Fig. 28.5. Wave bioreactor.

air is continuously passaged through the head space during the cultivation mixing and mass transfers are achieved by rocking the chamber back and forth. This rocking motion generates waves at the liquid air interface, greatly enhancing oxygen transfer. The concept of using rocking for agitation is used extensively for the agitation of liquids in laboratory assay plates and gels. In this disposable bioreactor, the cultivation chamber is discarded after harvest, eliminating any need for cleaning or sterilization. The chambers are made of FDA approved biocompatible polyethylene. Special ports were developed to allow sterile addition or to withdraw samples without the need to place the bioreactor inside the laminar air flow cabinet. Theses bioreactors are available up to a working volume of 500L and due to wave motion known as Wave Bioreactor (Fig. 28.5).

5. PRODUCTION OF SECONDARY METABOLITES

Above mentioned technologies have been used for the cultivation of plant species in bioreactors. The results obtained with such species for the production of secondary metabolites, in organised or disorganised cultures, are described in the following.

5.1. Shikonin Production

Shikonin is used as a remedy for various skin ailments and as a dye for skin and cosmetics. Shikonin is synthesized in cells of *Lithospermum erythrorhizon* from the shikimic acid and isoprenoid pathways as shown in figure. 28.6. Commercial production of shikonin was stimulated by its value and problems encountered in meeting the demand from the natural resources. The first process for the commercial production of natural plant product by the cell suspension cultures was developed in Japan for the production of the naphthoquinone shikonin. Through visual inspection it was possible to identify the high-producing cultures.

Since the growth of cells was low in the medium (designated as M-9) which supports high levels of shikonin production, a two-stage culture process was developed. The cells are first grown in a growth promoting medium (designated as MG-5) for 9 days, filtered and pumped to a second vessel in which the M-9 production medium is added (Fig. 28.7). The cells are cultured for

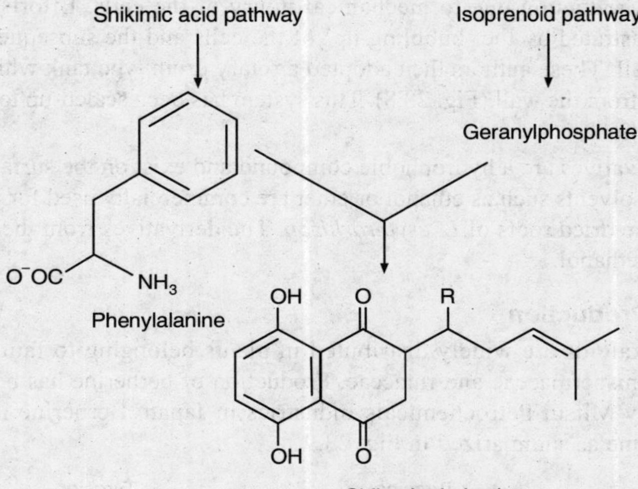

Fig. 28.6. Pathway of shikonin and its derivatives.

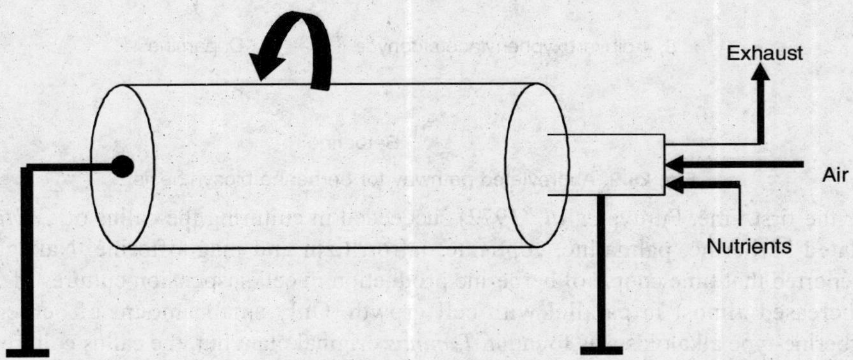

Fig. 28.7. Schematic diagram of process of commercial production of shikonin.

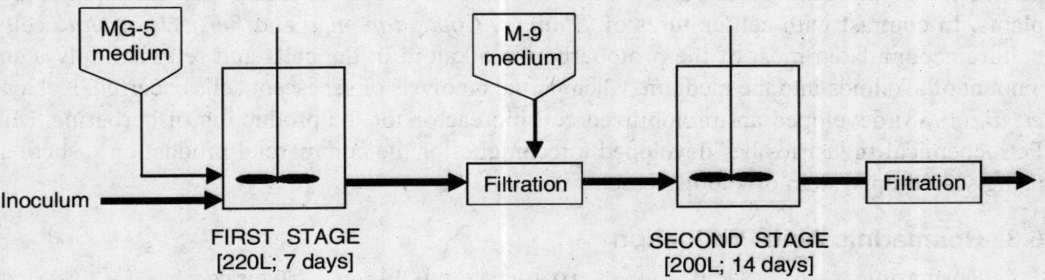

Fig. 28.8. Rotary-drum bioreactor for the production of shikonin (Tanaka et al. 1983).

14 additional days, after which they are recovered by filtration, and the shikonin is extracted from the cells. Skikonin levels at that time were reported to be 4g l^{-1}. They reported the effect of oxygen on shikonin production. Growth and production were enhanced by increased K_La for values upto 18 h^{-1}. In a fermenter with a paddle-type impeller, shikonin production was enhanced by increased K_La upto 10 h^{-1}. Paddle-type impeller favours cell growth however, shikonin production was reduced as compared to that in shake-flask cultured cells. High speed of the impeller further

reduced shikonin production due to mechanical injury of the cells. Efforts to use air-lift type bioreactor were frustrated by the 'bubbling-up' of the cells and the subsequent adherence of the cells to the tank wall. These authors then adopted a rotary drum type tank which revolved slowly, thus washing cells from the wall (Fig. 28.8). This system has been scaled-up to 1000L without loss in shikonin production.

Shikonin derivatives are a hydrophobic compound and exist on the surface of the cells as oil particles. Organic solvents such as ethanol or ether are commercially used for the extraction of the derivatives from the dried roots of *L. erythrorhizon*. The derivatives from the dried cultured cells are extracted with ethanol.

5.2. Berberine Production

Berberine alkaloids are widely distributed in plants belonging to families berberidaceae, ranunculaceae, menispermaceae and rutaceae. Production of berberine has been scaled-up to the 400L bioreactor by Mitsui Petrochemicals Industries in Japan. Berberine is derived from two molecules of tyrosine as summarized in Fig. 28.9.

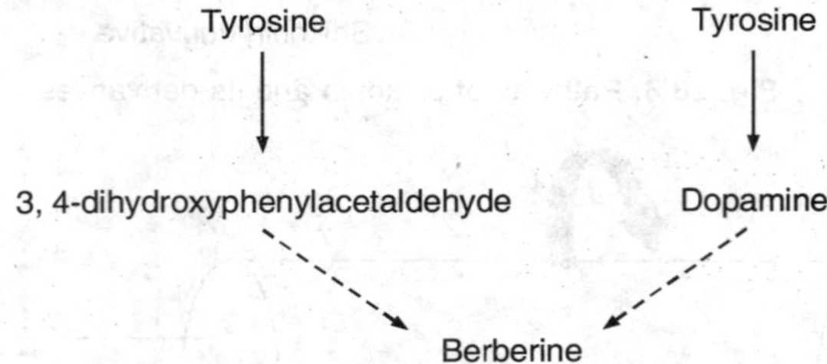

Fig. 28.9. Abbreviated pathway for berberine biosynthesis.

For the first time, Furuya *et. al*. (1972) succeeded in culturing the callus of *Coptis japonica* and isolated berberine, palmatine, coptisine, jatrorrhizin and magnoflorine. Nakagawa *et. al*. (1984) reported that time course of berberine production in cell suspension cultures of *Thalictrum minus* increased almost in parallel with cell growth. Only small amount of berberine as the protoberberine-type alkaloids was found in *T. minus* original plant but, the callus cultures produced several protoberberine-type alkaloids, moreover, berberine was the main product. Berberine production in the cultured cells of *Thalictrum* was about 350 times higher than that in field grown plants. In contrast with cell cultures of *T. minus*, *Coptis japonica* and *Berberis aristata* cells in culture accumulated most of the protoberberine alkaloid in the cells and released only a small amount of alkaloids into the medium, when partial autolysis of senescent cells occurred. Kobayashi *et. al*. (1988) developed an immobilized cell bioreactor for the production of berberine. Mitsui Petrochemical Industries has developed a technique for the commercial production of berberine using scaled-up system of 4000L vessel.

5.3. Rosmarinic Acid Production

Rosmarinic acid (α-o-caffeoyl-3,4-dihydroxypharyllacetic acid) has been most commonly found within the families lamiaceae and borigenaceae, and seems largely restricted to the tubiflorae. Rosmarinic acid is produced from the two distant precursors, phenylalanine and tyrosine (Fig. 28.10). Although most of the rosmarinic acid was reported to be produced during the period of rapid growth, production only occurred in *Coleus blumei* cell cultures when detectable levels of (PAL) was present.

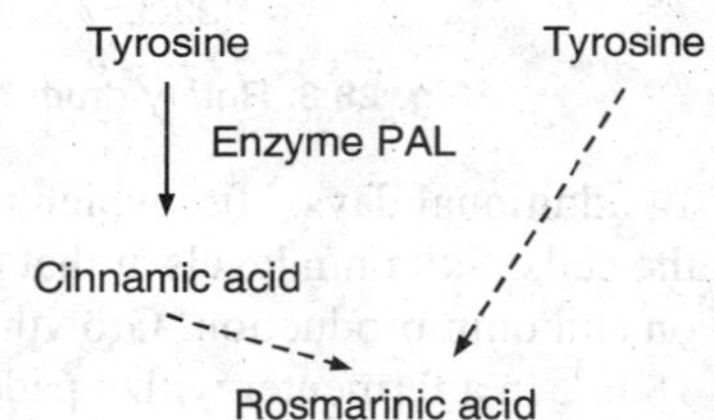

Fig. 28.10. Pathway of biosynthesis of rosmarinic acid.

Oxidized rosmarinic acid displayed anti-thyrotropic activities in tests with human thyroid membrane preperations and the pure compound has been shown to effectively suppress the complement-dependent components of endotoxin shock in rabbits.

De-Eknamkul and Ellis (1985) examined the dynamics of growth and rosmarinic acid production by the cultures of *Anchusa officinalis*. They reported that the conditions which enhanced growth and also enhanced the accumulation of rosmarinic acid. Zenk *et. al*. (1977) cultured cells of *C. blumei* in a single-stage air-lift bioreactor but the production of rosmarinic acid was much low as compared to that in shake-flask cultures. Latter, they rported serious mixing problems in a 32L air-lift bioreactor system. Therefore, they opted to use a mechanically agitated-type bioreactor system with helical impeller. They reported production of 5.5g of rosmarinic acid per liter in a two-stage batch culture system. Firstly, cells were grown for 8 days in a glucose fed batch system and after suitable growth, cells were filtered out and pumped into a second culture vessel in which they were diluted with a sucrose solution to initiate production of rosmarinic acid. Incorporation of the precursors of rosmarinic acid biosynthesis inhibits both, growth and production.

5.4. Indole Alkaloids Production

A number of indole alkaloids are produced by *in vitro* cultures of *Catharanthus roseus* including anti-tumour agents. Since *C. roseus* cells enlarge as they age, they may become more shear sensitive, and this increased shear sensitivity may explains low production of secondary metabolites. Due to this shear sensitivity of cells, several workers have adopted pneumatically agitated type bioreactors for culturing *C. roseus* cells. Both, external loop and draft tube airlifts as well as bubble column systems have been used. In contrast, others used a mechanically agitated bioreactor with flat blade turbine impellers, and the alkaloid production in this type of bioreactor was less than that observed in shake flasks even when same inoculum was used in both the systems.

Since the medium which favours growth of cells do not favours production of the alkaloids in *C. roseus* cells, a two-stage culture system was used similar to that for the production of shikonin. By coupling filtration to aeration, it was possible to filter the cells from the medium without the need for operations outside the bioreactor. This internal filtration operation reduces contamination risk and permits the use of a single bioreactor for both, growth and production phases. In a similar system, the spin filter-coupled agitation with filtration to achieve cell separation within a mechanically agitated bioreactor was used. It has been recorded that when NAA is used in place of 2,4-D, a single medium can be used for growth and production and medium exchange not required.

The relationship between the ajmalicine production rate by *C. roseus* and various constant glucose concentrations have been investigated. Highest ajmalicine production rate was greatly enhanced when glucose concentration was reduced from a constant 57 gl^{-1} to a constant 32 gl^{-1}. A relationship between the age of cells in the inoculum and the ajmalicine production by *C. roseus* has been established. The cells in late stationary phase produced 5 times more ajmalicine than cells in the early stationary phase. Young cells were not in appropriate state to withstand the osmotic shock of the high glucose concentration (80 gl^{-1}). Similarly, positive effect of 4% CO_2 in the inlet gas stream on the production of indole alkaloids has been recorded Schltmann *et. al*. 1993. The importance of gaseous metabolites on ajmalicine production by feeding the exhaust air to the culture in an air circulation bioreactor has been established. With the use of such an experimental setup, the timing, amount and type of gaseous metabolite is under natural control of the cell suspension, in contrast to the addition of CO_2 or ethylene to the aeration gas. By removing CO_2, ethylene, or both from the air circulation stream, it was demonstrated that both gases play a negligible role in ajmalicine production. These investigations on the negative effect of high aeration rate on secondary metabolites were performed with relatively low biomass concentration (< 15 gl^{-1}).

5.5. Organ Culture

Micropropagation of a number of plant species by somatic embryogenesis, shoot culture and microbulbil formation is well established. Conventional micropropagation methods require hundreds or even thousands of cultures to produce plantlets at commercial scale. The process is also labour intensive, time consuming and expensive. Micropropagation in suspension cultures are potential cost-effective as less material and labour may be required and shoot and root can be grown in the same vessel (see previous chapters for somatic embryogenesis and cell culture). But, the growth of organogenetic cultures using bioreactor system is not well established so far. There are only a few reports on the production of somatic embryos, bulbils, or microtubers in bioreactor as compared to large number of reports on micropropagation through organogenesis in callus culture. Timmis *et. al.* (1998) constructed a bioreactor comprising four independently controlled 1.5L vessels as a research tool for the improvement of somatic embryogenesis using several embryogenic lines of Douglas-fir (*Pseudotsuga menziesii*). Bapat and colleagues at Bhabha Atomic Research Centre, Bombay have successfully produced large number of somatic embryos of *Santalum album* (sandal wood) using a stirred-type bioreactor system of one litre. Similarly, Bamboo shoots are produced in an air lift bioreactor at National Chemical Laboratory, Poona. Production of *Lillium longifolium* bulblets, *Gentiana* shoot culture and *Artemisia* shoot culture are other examples of organ cultures raised in bioreactor.

The culture of differentiated plant tissues is strongly affected by the four main factors in all bioreactors: moisture, temperature, soluble nutrients and gases. Light is also a critical factor required for some cultures. Assuring efficient gas exchange is a particular challenge, especially in densely packed reactors containing more than 50% biomass volume. Nutrient mist technology, or aeroponics, is one approach to alleviate this problem. Inventions in the design of nutrient mist reactors, includes the use of acoustic windows, have made the technology simple and cost effective. Other then angiosperms, a few fern species has also been cultured in bioreactors like e.g., gametophytes of *Pteridium aquilinum* and *Anemia phyllitidis* in 8L air-lift type bioreactor.

5.6. Hairy Root Culture

Hairy root cultures, as an alternative approach to produce secondary metabolites *in vitro*, have received attention during the last decade. It has been claimed that hairy root cultures have a stable and high biosynthetic capacity for the secondary metabolite production as compared to unorganised cultures. Furthermore, hairy roots can be easily retained within the bioreactor, facilitating continuous operation. An additional feature of hairy roots is the low inoculum density required for growth, as compared to cell suspension cultures and immobilized cell systems where high density is required. During incubation, the structured nature of hairy root cultures leads to formation of interconnected, non-homogeneous material unevenly distributed throughout the bioreactor, which made it necessary to develop novel process strategies and modified bioreactors. Furthermore, the tendency of hairy roots to grow in a large cluster results in mass transfer limitations inside the cluster and in inefficient exploitation of the reactor volume. Scale-up of processes using hairy roots will be hampered by these technological drawbacks.

A large number of investigations have been carried out on the growth of roots and production of secondary metabolite using bioreactor systems. *Catharanthus roseus* roots were grown in been grown in 20L stirred-tank-reactor and the production of indole alkaloids in such cultures was reported. The other works on root culture includes effect of reactor types on hairy root growth of *C. roseus*, cultivation of hairy roots of *Datura stramonium* in 14 L stirred tank bioreactor, the production of thiophene from hairy root cultures of *Tagetes*, the production of anthraquinone pigments by hairy roots of *Rubia* and Carrot hairy roots in 1.5L stirred tank bioreactor. In 1990, a 500L bioreactor was developed, which could be operated as nutrient mist reactor as well as with submerged hairy root cultures. Unfortunately, the scale-up of hairy root culture is still troublesome.

Despite of technological disadvantages, hairy roots can be grown at a relatively high biomass densities in small bioreactors. Biomass yield of *D. stramonium* cultures grown in 1.5L bubble column reactor was achieved up to 40 gl^{-1}.

5.7. Commercialization

Commercial-scale plant cell culture for the production of secondary metabolites has acquired much attention in Japan. Shikonin, a natural dye, is produced by Mitsui Petrochemcal Industries. Berberine is sold exclusively in Asia and can be produced commercially in a continuous-flow bioreactor in which cells are retained by a membrane and harvested periodically. This system has been operated at a scale-up to 4000 L. Production of ginseng roots as organized tissues (tuberous-roots) is being exploited commercially. In Japan, Ushiyamà (1989) successfully cultured ginseng roots in 20,000 L bioreactor and efforts are being made to exploit this system for the production of ginsenosides. The production of a natural vanilla from cell culture is being developed by Escagenetics in USA. Elicitation is apparently a key component of the production strategy. A 300 L air-lift type bioreactor system has been developed by Vipont Research Laboratories, U.S.A. for the production of sanguinarine from *Papaver somniferum*. Colgate has signed a formal arrangement with Vipont, which suggests strong interest in completing the development of commercialization of this project. A new start-up company, Phyton Catalytic, U.S.A. has been founded to make commercial scale levels of taxol and other plant products. In Europe, several processes have seriously considered for the commercialization of natural plant products but, ultimately put on hold or abandoned. However, a 75,000 L reactor facility has been established near Hamburg in Germany for use and exploitation of secondary metabolites production.

In addition to these efforts to produce secondary metabolites, whole plants are being produced. Levin *et. al.* (1988) reported that Plant Biotechnology Industries in Israel uses a bioreactor as an important component of a micropropagation work that has already been successfully applied to several species. A collaborative venture between Kirin Brewery and the US Firm Plant Genetics has produced a large-scale lettuce and cereal crops from immobilization of somatic embryos.

Technology for the production of useful secondary metabolites has come to a threshold level. A final push is required to make this technology widely applicable for the production of pharmaceuticals, dyes and fragrances. demand for natural products for fragrances, dyes and pharmaceuticals (anti-AIDS, anti-aging, anti cancer etc.), technology should be ready to meet the demand in coming decades.

QUESTIONS

1. What is a bioreactor? Describe various types and the factors governing the cell growth in a bioreactor.
2. What is the effect of air flow and mixing on cell growth in a bioreactor?
3. Write short notes on the following
   (a) Bioreactor
   (b) Immobilized cell bioreactor
   (c) Stirred Tank bioreactor
   (d) Air-lift bioreactor
   (e) Wave bioreactor
   (f) Organ culture in bioreactor
   (g) Secondary metabolites production in bioreactor.
   (h) Shikonin, berberine production
   (i) Compare wave bioreactor with stirred tank reactor.

CHAPTER 29

Secondary Metabolites: Alkaloids and Dioscorea

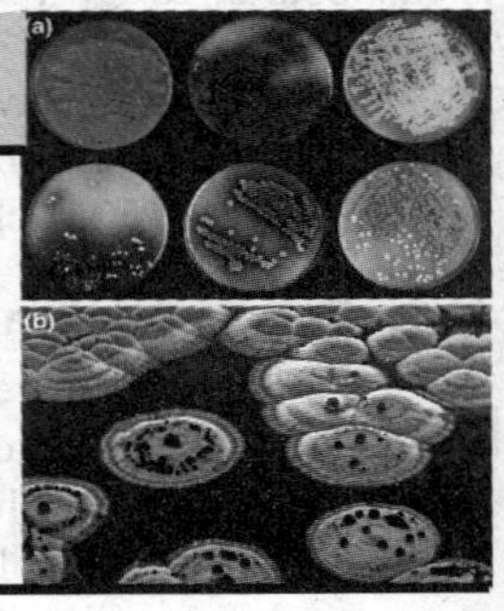

1. INTRODUCTION

A plant cell produces two types of metabolites: primary metabolites involved directly in growth and metabolism, viz., carbohydrates, lipids and proteins, and secondary metabolites considered as end products of primary metabolism and are in general not involved in metabolic activity, viz., alkaloids, phenolics, essential oils, steroids, lignins, tannins etc. Primary metabolites are produced as a result of photosynthesis and these products are further involved in the cell component synthesis.

In general, primary metabolites obtained from higher plants for commercial use are high-volume, low-value bulk chemicals. They are primarily used as industrial raw materials, foods or food additives, examples: vegetable oils, fatty acids (used for making soaps and detergents), and carbohydrates (sucrose, starch, pectin and cellulose). These materials cost Indian Rs. 15-150 per kg (or US $ 0.5 to 4 per kg) and are readily available in large quantities. However, some primary metabolites such as myoinositol and B-carotene are expensive because their extraction, isolation and purification are difficult.

The medicinal plants rich in secondary plant products are termed as 'medicinal' or 'officinal' plants. These secondary metabolites or products exert in general a profound physiological effect on the mammalian system and, thus are known as active principles of plant. The physiological effect of these active principles is used for curing ailments and therefore, these are drugs of plant origin or natural drugs. The use of crude drugs of plant origin (unpurified preparations of active principles, plant extract or sometimes powdered plant material) is used in the Indian system of medicine or 'Ayurveda'. A large number of drugs of plant origin are still used in Western medicine (Table 29.1). Though some of the drugs are obtained by synthesis for commercial uses, others are still obtained from natural sources.

Table 29.1. Plant derived medicines used in Western medicine.

Ajmalicine	Digoxin	Papain	Reserpine
Atropine	Ephedrine	Papaverine	Scopolanine
Caffeine	Hyoscyamine	Pseudoephedrine	Sennosides
Codèin	L-Dopa	Quinidine	Vincaleukoblastine
Colchicine	Morphine	Quinine	

Synthesis of various classes of secondary metabolites from primary metabolites is presented in a schematic form in Figure 29.1. All the necessary carbon skeletons are derived from carbohydrates synthesized from photosynthesis. The other major primary products are amino acids. Acetyl co-enzyme and mevalonic acid play a key role in the synthesis of various terpenoids, while the shikimic acid pathway is involved in the synthesis of lignins and indole alkaloids.

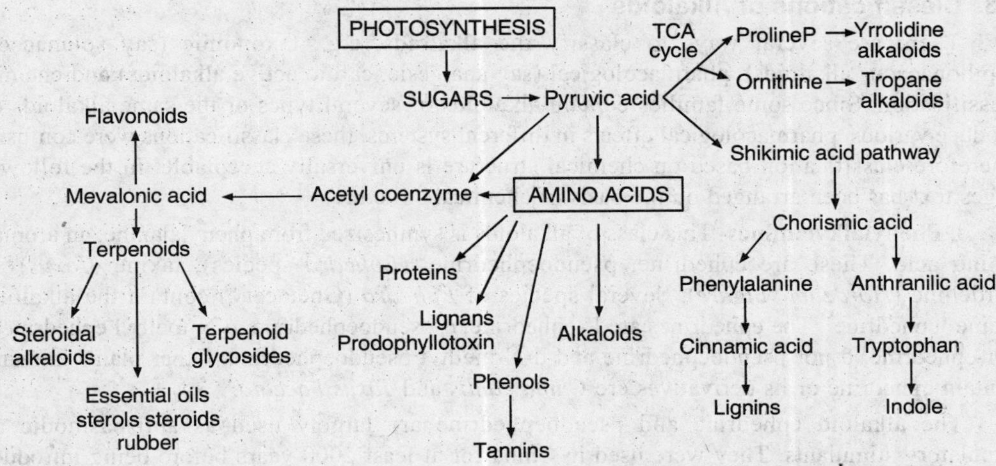

Fig. 29.1. Synthesis of major classes of secondary metabolites from primary metabolites.

2. ALKALOIDS

Alkaloids have been known to man for several centuries. Morphine, an alkaloid of latex of the opium poppy was isolated by F.W. Serturner in 1806. Latter other alkali-like active principles were isolated and identified, e.g., narcotine in 1817 by Robiquet, emetine in 1817 by Pelletier and Magendie and so on. The term 'alkaloid' was coined by W. Meibner, a German pharmacist, meaning 'alkali like'. Latter it was demonstrated that the alkalinity was due to the presence of a basic nitrogen atom. The first alkaloid to be synthesized was coniine in 1886 by Ladenburg, which had already been isolated in 1827.

2.1. Definition

The alkaloids are defined as 'basic nitrogenous plant products, mostly optically active and possessing nitrogen heterocycles as their structural unit, with a pronounced physiological action'. Pelletier (1982) suggested following definition for alkaloids: 'an alkaloid is a cyclic organic compound containing nitrogen in a negative oxidation state which is of limited distribution among living organisms'. Examples of true alkaloids according to definition are: morphine (*Papaver somniferum*), the first alkaloid isolated, quinine (*Cinchona species*), coniine (poison hemlock, the first alkaloid synthesized, from *Conium maculatum*; this plant alkaloid was given to Socrates in 400 B.C.) and reserpine (*Rauwolfia serpentina*).

2.2. Occurrence of Alkaloids

Around 5000 alkaloids of all types have been known to occur in 15% of all land plants and in more than 150 families. The important families are: Apocynaceae, Papaveraceae, Papilionaceae, Ranunculaceae, Rubiaceae, Rutaceae and Solanaceae and less common lower plants and fungi (ergot alkaloids).

In plants, alkaloids generally exist as salts of organic acids such as acetic acid, oxalic, citric, malic, lactic, tartaric, tannic acid etc. Some weak basic alkaloids (nicotine etc.) occur free in nature. A few alkaloids also occur as glycosides of sugar (e.g., glucose, rhamnose and galactose) e.g., alkaloids of the solanum group (solanine) as amides (piperine) and as esters (atropine, cocaine) of organic acids.

2.3. Classifications of Alkaloids

There are several ways to classify the alkaloids, e.g., taxonomic (say solanaceous, papilionaceous alkaloids), pharmacological (say analgesic, cardio active alkaloids) and chemical classification. Since some families contain alkaloids of several types or the same alkaloids can produce various pharmacological effects in different systems, these classifications were confusing. Therefore classification based on chemical structure is universally acceptable. In the following pages text has been arranged on the basis of chemical structure.

i. Phenylalkylamines- This class of alkaloids is synthesized from phenylalanine, an aromatic amino acid. These are ephedrine, pseudoephedrine, (*Ephedra* species), taxine (*Taxus*) and hordenine (*Hordeum vulgare*). Several species of *Ephedra* (Gnetaceae) contain the alkaloidal amine 'ephedrine'. The ephedrines are L-ephedrine, d-pseudoephedrine, *p*-N-methyl ephedrine, p-nor-ephedrine, d-nor pseudoephedrine and d-N-methyl pseudoephedrine. Other plants known to contain ephedrine or its derivatives are *Catha edulis* and *Taxus baccata*.

The alkaloid ephedrine and pseudoephedrine are largely used as antispasmodic and circulatory stimulants. They were used in China for at least 2000 years before being introduced into Western medicine in 1924. Ephedrine is largely used as substitute for epinephrine against bronchial asthma of allergic and reflexive types. It is also used orally and locally in patients suffering from hay fever, urticaria and other allergic reactions. *E. foliata* grows in the Thar desert contains traces of pseudoephedrine, *E. geraradiana*, a cold desert species growing in western Himalayas contains 2% alkaloids.

ii. Pyrrolidines, Piperidines and Pyridines- Piperidine alkaloids such as coniceine, coniine and N-methyl coniine are present in *Conium maculatum*, but not detected in the cultures of this plant. Lobeline and other piperidine alkaloids are present in *Lobelia inflata* tissue cultures and intact plants. The alkaloids belonging to these groups are known to have divergent physiological activities.

Apart from **tobacco alkaloids**, nicotinic acid and its derivatives are the major pyridine alkaloid present in plants. The most commonly occurring compound is trigonelline (N-methyl-nicotinic acid) present in *Trigonella foenum-graecum*. Plants and cultures of *Nicotiana tabacum* contain nicotine, anatabine, anabasine, myosmine and nicotelline. Nicotine has been a major target of study among them. Nicotine and nicotinic acid are one of the most extensively studied secondary metabolites in plant tissue culture. Though nicotine production through plant tissue culture in not a viable programme, tobacco cultures were used as model system to develop technology for the production of secondary metabolites. The practice of tobacco smoking was made known to Europeans about the year 1492 when they visited West Indies, after the discovery of New World (America). Since then tobacco was introduced in several countries of the world including India where it is extensively cultivated for its leaves used as tobacco in cigarette, bidis and other tobacco preparations. When plant is cultivated, plant tissue culture cannot compete for biomass production with a crop plant. Plant issue culture is used to know biosynthetic pathway of nicotine and production of low nicotine containing plants. Such plants will have aroma of tobacco but low tar and nicotine, which is injurious to health. In plants, nicotine is synthesized in roots and accumulated in leaves. This means nicotine is transported from roots to leaves and is energy dependent process. About a third of CO_2 fixed in photosynthesis is used in nicotine synthesis. Tobacco cultures have been used for all types of studies ranging from cell cultures, cloning, biosynthesis of nicotine, hairy roots cultures, and growth in bioreactors up to 2000 litres.

iii. Tropane alkaloids- The alkaloids hyoscyamine, atropine and hyoscine (scopolamine) are found principally in plants of the family Solanaceae and categorized as anticholinergics. More than 30 alkaloids are known to be present in *Datura*. *Datura stramonium* and *D. innoxia* are the main source of hyoscyamine and scopolamine, respectively. Other species known to contain tropane alkaloids are *Dubosia* hybrids, Hyoscyamus niger (henbane), *H. muticus*, *D. myoporoides*, *D. leichhardtii*, and *Atropa belladona* (deadly nightshade). Cocaine in coca (*Erythroxylon coca*) was the first local anaesthetic to be discovered. Leaves of a few species of *Erythroxylon* indigenous to Peru and Bolivia, contain 0.6-1.8% cocaine. The leaves of the plant have been used for centuries by the natives to increase endurance and to promote a sense of well-being. Cocaine was isolated in 1859 by A. Niemann for its Central Nervous System (CNS) stimulatory activity, which can lead to dependence liability, hence cocaine has been used as drug of. These alkaloids are synthesized from tropic acid.

iv. Quinolizidine and pyrrolizidine- Quinolizidine alkaloids are common natural products of many Fabaceae and commonly called as lupin alkaloids because of their presence in all species of the genus *Lupinus*. This group of alkaloids is synthesized from lysine via cadaverine, e.g., lupanine, sparteine etc. Isoquinoline-type alkaloids show strong pharmacological activities like those of morphinane-, protoberberine-, and benzophenanthridine-type alkaloids; they are widely distributed in the plant kingdom, mainly in Papaveraceae, Berbidaceae, Ranunculaceae and Menispermaceae.

Papaver alkaloids require a special mention in isoquinoline alkaloids. The opium poppy, *Papaver somniferum*, is one of the man's oldest cultivated plants. The therapeutic use of latex obtained from unripe capsules of poppy was recorded by Theophrastus in the third century B.C. Discorides (A.D.77) described the curative properties of the opium poppy and presented various

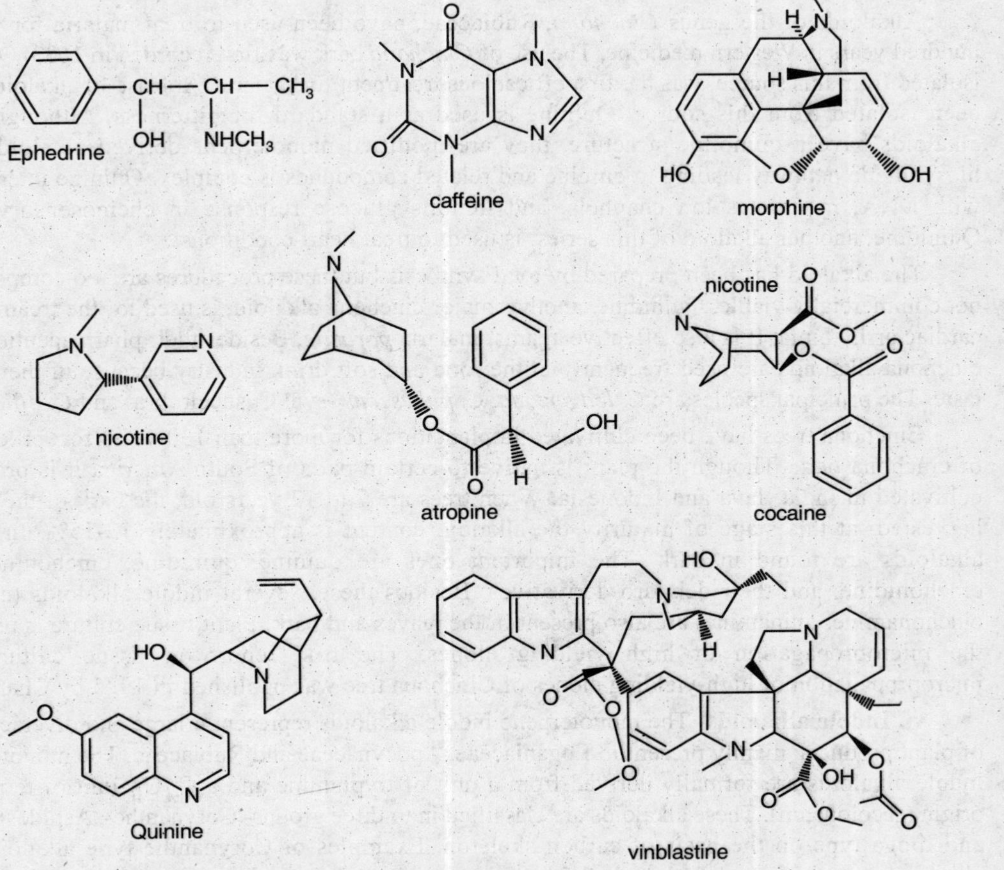

Fig. 29.2. Chemical structures of principal compounds.

uses for both latex and extracts of whole plants. Purified alkaloids are used in modern medicine for treatment of pain, cough and diarrhoea. Opium is the dried cytoplasm of a specialized internal secretoary system, the laticifer. Morphine content increases in the morning with decrease in codeine and thebaine content in the latex. So capsules are incised in the morning to obtain morphine-rich latex. When the green unripe capsule (fruit) is cut, milky latex oozes out, which turns dark brown on drying and is collected as raw opium? This is a labour intensive manual process and requires strict vigilance and control to check illegal drug transport. Alkaloids are purified in laboratories (factories controlled by the Government).

More than 40 alkaloids have been identified from *Papaver somniferum* including 25 from latex. However, from the medicinal point of view, benzyl isoquinolines (papaverine and narcotine) and phenanthrines (morphanians: morphine and codeine) are important. The demand for legal opium is estimated to be more than 1000 tons per annum. In India, cultivation and alkaloid isolation is a business completely controlled by the Government. World trade of opium is governed by the United Nations Opium Conference Protocol (1953).

Morphine was the first alkaloid isolated in pure form. Synthesis of morphine takes place from tyrosine via DOPA (Dihydroxy phenylalanine), nor-landanosoline to papaverine or reticuline. Reticuline is converted into thebaine, thebaine in to codeine and then morphine as the end product. Thus codeine is an intermediate product, not present in large quantities in the latex.

v. Quinoline alkaloids of *Cinchona*- Extracts obtained from the bark of *Cinchona* species have been of great therapeutic value.

Alkaloids of the genus *Cinchona*, Rubiaceae, have been used to treat malaria for several hundred years in Western medicine. The use of *Cinchona* bark was first recorded in 1633. Quinine, isolated from this source, was the first efficacious treatment for malaria. At least 40 alkaloids have been isolated from this source. Quinine is used as a standard for bitterness. Although these alkaloids have a quinoline structure, they are modified monoterpene-derived alkaloids. The biosynthetic pathway leading to quinine and related compounds is complex. Quinine intercalates with DNA, modulates ion channels, and inhibits glucose response in chemosensory cells. Quinidine, another alkaloid of this series, is used to treat heart conditions.

The alkaloid has been prepared by total synthesis but these procedures are too complex and not commercially viable. Quinidine, another major cinchona alkaloid, is used for the treatment of cardiac arrhythmia. It is also effective against malaria *parasite*. Besides their pharmaceutical use, cinchona alkaloids are used frequently in the food and soft drink industry because of their bitter taste. The principal species are *C. ledgeriana*, *C. pubescens* (syn. C. succirubra) and *C. officinalis*.

Cinchona trees have been cultivated in plantations for more than 130 years for production of cinchona bark. Though the plant is native to certain parts of South America it is presently cultivated in India, Java and Indonesia. When trees are 7 to 12 years old, the bark of the tree is harvested; at this stage of maturity the alkaloid content is approximately 12-15%. In all 35 alkaloids are found in bark. The important ones are quinine, quinidine, cinchonine, and cinchonidine, and their dehydro derivatives. Besides these, several indole alkaloids (aricine, cinchonamine, quinamine) are also present in the leaves and bark. Plant tissue culture is used for the micropropagation of high yielding clones. The first report on tissue culture and micropropagation of high-yielding clones of Cinchona tree was published in 1975 by Chatterjee.

vi. Indole alkaloids- The monoterpene indole alkaloids represent a large and diverse group of plant products, mainly present in Loganiaceae, Apocynaceae and Rubiaceae. The monoterpene indole alkaloids are formally derived from a unit of tryptamine and a C_9/C_{10} unit of terpenoid origin (secologanin). These alkaloids are classified in to three groups- Corynanthe-, Aspidosperma- and Iboga type on the basis of carbon skeleton. Examples of Corynanthe type alkaloids are ellipticine, reserpilline and other alkaloids of this type in *Ochrosia elliptica*, *reserpine* and ajmaline in *Rauwolfia serpentina* and *Catharanthus roseus*, and cinchonamine, quinamine, aricine in *Cinchona*. Conoflorine, tubotiwine are Iboga type alkaloids present in *Tabernanthe iboga*. Tabersonine, ichnericine and minovincine are Aspidosperma alkaloids present in *Voacanga africana*.

Large number of indole alkaloids produced by *Catharanthus roseus* (synonym *Vinca roseus*) have been identified. The plant is native to Madagascar but now cultivated in India. Several of these have been found to be valuable agents in the treatment of hypertension and others of cancerous growths. In particular, vincristine and vinblastine, two dimeric indole alkaloids (0.0005% on dry weight basis obtained from roots) are used for the treatment of leukaemia and Hodgkin's disease. The antineoplastic properties of the alkaloidal constituents of Catharanthus were independently described by Canadian and American scientists. It contains more than 75 alkaloids, the other indole alkaloids present in large quantities are ajmalicine, serpentine, lochneridine, tabersonine and vindoline.

Medicinal properties of *Catharanthus roseus* have been described in traditional and folk medicine of several countries. Beneficial effect of its extract in diabetes mellitus was known, but later on active principles suppressing neoplasm were also isolated. The extracts yielded four active dimeric monoterpenoid indole alkaloids- vinblastine, vincristine, vinleurosine and vinrosidine. The catharanthus alkaloids are cell-cycle specific agents, similar to colchicine and podophyllotoxin, block mitosis and cause metaphase arrest. Though vincristine and vinblastine have anti-proliferative properties, but both have different patterns of cytotoxic effect and are used in combination for the last 40 years (received FDA approval in 1963 and 1965, respectively).

The plant is extensively investigated for the production of indole alkaloids but, ajmalicine and serpentine are major alkaloids present in cell cultures. Shoot cultures of Catharanthus are reported to contain vincristine and vinblastine. However, the cell culture have been associated with various technological developments for the production of secondary metabolites as whole, such as cloning, two stage culture system, gene cloning and transfer to other indole alkaloid containing plants, enzymes of biosynthetic pathway and elicitors use.

The Catharanthus alkaloids, navelbine (trade name Vinorelbine), vinblastine (trade name Velban), and vincristine (trade name Oncovin) are currently used clinically. Vinblastine introduces a wedge at the interface of two tubulin molecules, thus interfering with tubulin assembly.

vii. Purine alkaloids- Purine alkaloids are widely distributed within the plant kingdom (Fig. 2.12) and have been detected in at least 90 species belonging to 30 genera. Their occurrence, however, is limited to dicots. Caffeine and threobromine, methylated derivatives of xanthine, are generally the main purine alkaloids and are regularly accompanied in low concentrations by the two methylxanthines- theophylline and paraxanthine as well as by methylated uric acids such as theacrine, methylliberine and liberine.

Since purine alkaloids are present in tea and coffee are widely consumed in human diet across the continents. Plant species from different families are made in to a pleasant stimulant, e.g., coffee (*Coffea arabica*, *C. robusta*), tea (*Camellia sinenesis*), Cocoa (*Theobroma cacao*), mate (*Ilex paraguayiensis*) and cola (*Cola nitida*).

viii. Tropolone alkaloids- The neutral alkaloid colchicine present in the bulbs of *Colchicum autumnale* and tubers of Superb Glory (*Gloriosa superba*), Liliaceae is an example of a tropolone-type alkaloid. Colchicine is synthesized from phenylalanine via cinnamic acid and sinapic acid.

Other important secondary metabolites categorized as alkaloids are acridone alkaloids present in plants of the family Rutaceae and steroidal alkaloids present in plants of the family Solanaceae.

2.4. Biological Functions of Alkaloids

The biological functions of alkaloids within the plants are not clearly understood but it is clear that they are not produced in plants for a single function, but for many functions. The following functions have been observed in different plant species. Alkaloids are considered as:

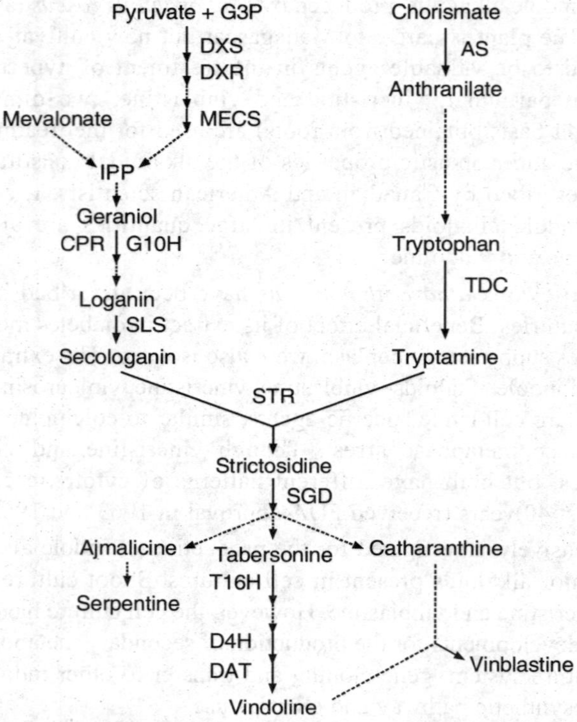

Fig. 29.3. Biosynthetic pathway of indole alkaloid. Broken lines represents several steps.

1. Reserve substances to supply nitrogen, but very little evidence is available about this function.
2. End products of the detoxification mechanism, otherwise their accumulation in plants might cause damage to the plants, e.g., in tobacco, 10% of carbon metabolism is directed to synthesize nicotine biosynthesis. Thus, it is an energy expensive process.
3. Poisonous substances to protect the plant itself from insects and animals. Nicotine has insecticidal properties. Sheep avoid grazing lupin plants with high alkaloid content. Some cacti repel fruit fly but *Drosophila pachea* is resistant and breeds on cactus.
4. Plant stimulants or regulators, e.g., alkaloids inhibit rye and oat seedling growth. Colchicine inhibits cell division.
5. Reservoirs for protein synthesis.
6. Excretory products of plant.
7. Inhibition of enzymatic activity by alkaloids is also known.

3. PROSPECTS OF DRUG PRODUCTION IN CULTURE

Plant cell cultures have great potential for the production of secondary metabolites. In recent years, considerable success has been achieved in increasing the secondary metabolites using cell suspension cultures in several plant species. Plant cells grown in culture have potential to produce and accumulate chemicals similar to the parent plant from which they were derived. There are numerous reports describing the production of diverse secondary metabolites, viz., anthocyanins, alkaloids, carotenoids, flavones, coumarins, naphthaquinones, saponins, sesquiterpenes, steroidal alkaloids, sterols, tannins, terpenoids and several others.

3.1. Why Tissue Culture

Several pharmacologically active novel compounds are extracted from plants (Table 29.2). In plant systems, they accumulate in leaves (nicotine in *Nicotiana*), roots (ajmalicine in *Catharanthus roseus*), bark (quinine in *Cinchona*) or in the whole plant (ephedrine in *Ephedra*). Sometimes these products are produced in specialized differentiated tissues such as resin in resin ducts and latex in laticifers. Except the herbaceous cultivated plants (e.g., *Papaver somniferum*), most of the secondary metabolites are accumulated after a certain age or maturity of the plant. In the case of tree or shrub species, e.g., *Cinchona, Rauwolfia, Camptotheca, Ochrosia*, etc., plants attain maturity in a few years before they accumulate the active principle in high amounts. It is difficult to increase the area under plantation for a particular species and growth of plants takes it own time. To meet the ever-increasing demand (e.g., vincristine) the natural resources are not sufficient. The world political scenario may also affect the supply of a particular raw material. To overcome all these hurdles, the industry requires alternative methods of assured supply of uniform material throughout the year. Harvesting of plants (except cultivated species) from natural forest resources is not only difficult, but also makes them endangered species; e.g., *Ephedra gerardiana* and several other Himalayan plants.

Table 29.2. Important alkaloids of plant origin and their pharmacological activity.

Alkaloid	Source	Pharmacological property
Ajmalicine	*Catharanthus roseus*	hypotensive
Atropine	*Atropa belladonna*	anticholinergic
Berberine	*Berberis* spp., *Coptis japonica*	antispasmodic, antiprotozoal
Codeine	*Papaver somniferum*	sedative, analgesic
Colchicine	*Colchicum autumnale*	antimitotic
Caffeine	*Coffea arabica, Camellia sinensis*	stimulant
Camptothecine	*Camptotheca acuminata*	Antitumour
Emetine	*Cephaelis ipecacuanha*	Emetic
Ellipticine	*Ochrosia elliptica*	antitumour
Ephedrine	*Ephedra gerardiana*	spasmolytic
Morphine	*Papaver somniferum*	analgesic, sedative
Papaverine	*Papaver somniferum*	spasmolytic
Quinine, quinidine	*Cinchona ledgeriana*	antimalaria
Reserpine	*Rauwolfia serpentina*	hypotensive
Vinblastine, vincristine	*Catharanthus roseus*	anticancer

When plant material is not available throughout the year in a quantity sufficient for industrial production and chemical synthesis is not possible, particularly for large complex molecules, biotechnological methods offer an excellent alternative. But before implementing this approach, cost of the product and its demand should justify production by biotechnological means.

Production of various alkaloids in cultures, their yield and approximate price are given in Table 29.3. Demand and cost of colchicine is not very high, but both criteria are very high for vincristine. Alkaloids such as ephedrine are obtained by chemical synthesis, a cheap method, while nicotine is obtained from field-grown plants (again a cheap source); therefore, their production through plant tissue culture is not commercially viable.

Table 29.3. Alkaloids produced in culture and their market value.

Alkaloid	Culture type	Yield in culture	Price US $/g
Ajmalicine	Cell suspension	0.2 g/l	37.0
Ajmaline	Cell suspension	0.04 g/l	10.0
Camptothecine	Cell suspension	0.00025% D.W.	480.0
Colchicine	Callus culture	0.0006% D.W.	50.0
Ellipticine	Cell suspension	0.005% D.W.	2630.0
Emetine	Root culture	0.3-0.5% D.W.	26.30
Morphine	Cell suspension	0.25 g/l	340.0
Quinine	Cell suspension	Traces	0.50
Reserpine	Cell suspension	0.002 g/l	8.30
Scopolamine	Hairy root	0.08 g/l (*Duboisia*)	
		0.4% D.W. (*Hyoscyamus*)	17.30
Vincamine	Cell suspension	3.3 g/l	19.70
Vinblastine	Shoot culture	Traces	10530.00
Vincristine	Shoot culture	Traces	25200.00

D.W., Dry weight basis.

3.2. Optimization

Optimization of alkaloid production is done using physical factors (light, temperature, vessel type), nutrients (carbon source, nitrogen source - nitrate versus ammonium nitrogen, organic (reduced) nitrogen, precursor molecules, phosphate, etc.), plant growth regulators (auxin type and concentration, cytokinin) and perhaps complex organic supplements (casein hydrolysate, coconut milk, yeast extract, etc.). It has been established from a large number of reports that fast-growing cultures accumulate alkaloids in low amounts during exponential phase of growth and in high amounts during the stationary phase. During this phase nutrients are exhausted and primary

Table 29.4. Alkaloid production in cell and callus cultures.

Plant species	Alkaloid
Atropa belladonna	Atropine
Camellia sinensis	Caffeine, theobromine
Catharanthus roseus	Ajmalicine, serpentine, vincristine, vinblastine
Cinchona succirubra, C. ledgeriana, C. officinalis	Quinine, quinidine
Cocculus pendulus, C. hirsutus	Cocculine, coccutrine
Coffea arabica	Caffeine, theobromine
Coptis japonica	berberine
Datura stramonium, D.innoxia	Scopolamine, hyoscyamine
Ephedra sinica, E. gerardiana	Ephedrine, pseudoephedrine
Hyoscyamus muticus, H. niger	Scopolamine, hyoscyamine
Nicotiana tabacum	Nicotine,
Ochrosia elliptica	Ellipticine, 9-methoxy ellipticine
Papaver somniferum	Papaverine, morphine, codeine
Rauwolfia serpentina	Serpentine, ajmalicine
Ruta graveolens	Platydesminium, ribalinium, rutalinium
Thalictrum minus	berberine

metabolism comes to a halt and the stored pool of primary metabolites is diverted to the synthesis of secondary products. Alkaloids production in various plant species is presented in Table 29.4.

Zenk *et. al.* (1977) used a two step culture system: medium for optimal growth (growth or maintenance medium) and medium for production of alkaloids (production medium). In the latter medium, growth is practically arrested by manipulating the nutrients, viz., high sucrose (4-10%) and low phosphate, 2,4-D and nitrogen. On transfer of growing cells to the production medium, product yield is increased several-fold.

Various factors affecting the production of secondary metabolites in plant tissues grown in culture are presented in Figure 29.4. Initially, growth and production of secondary metabolites are optimized by manipulating the physico-chemical factors followed by selection of high-productive cells. Afterwards, use of the production medium and specific techniques such as elicitation, hairy roots and immobilization are applied. When cultures are considered suitable for commercial production, cultures are grown in a bioreactors of increasing capacity to develop technology.

There are two approaches to enhance product synthesis and accumulation. The most widely applied empirical approach is optimization wherein medium and environmental factors of the cultures are manipulated. The second approach is to select suitable cells/tissues under normal or selective conditions. The most recent method is to combine both the approaches, i.e., selection of cells/tissues on the optimal medium condition for high product yield. For large scale production of a compound, stable high-producing cell lines of plants of interest have to be obtained.

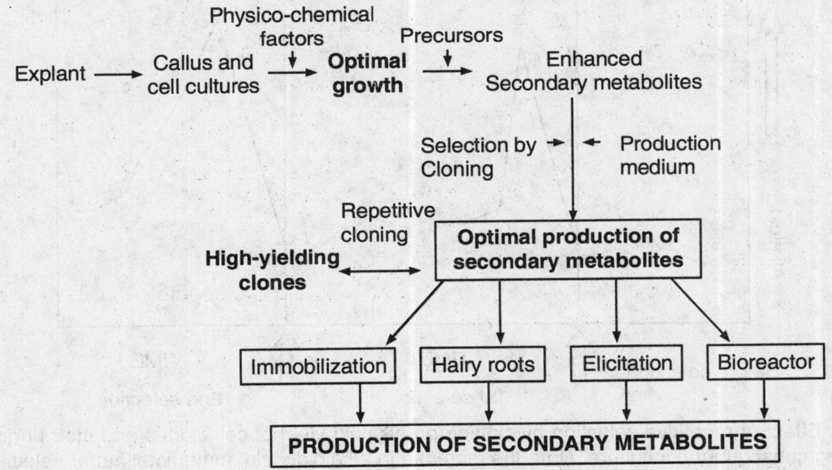

Fig. 29.4. Various approaches used for the optimal production of secondary metabolites.

The conditions for high growth and conditions for high productions are different. High growth is associated with low yield of secondary metabolites and slow growth is associated with high yield of secondary metabolites. Therefore, all those nutrients and environmental conditions which support high growth are not suitable for the production of secondary metabolites such as low sucrose, high auxin (particularly 2,4-dichlorophenoxyacetic acid) and high phosphate. Therefore, growth medium and a medium for production re used for the production of secondary metabolites. Cells are grown for a period in growth medium then transferred in to production medium where high production is obtained. Increased synthesis of secondary products occurs during the stationary phase of cultures when primary metabolism and cell proliferation come to a halt. The production medium contains low or no auxin, moderate or high cytokinin, low phosphate and high to very high concentration of sucrose. In this medium, cells grow slowly and accumulate more secondary metabolites as nutrients in high concentrations are always available.

3.3. Selection of Cells

An *in vitro* growing cell system has two components: the cell and its environment. In the above paragraphs, the details of manipulating physical and chemical environment of the cell are presented. Here, we discuss how selection procedures are helpful in increasing the yield of cultures. Before producing secondary metabolites at the industrial/commercial level, it is a prerequisite to achieve optimal yield of secondary metabolites through optimization and select the cells for maximal yield. Though the production of a secondary metabolite is a genetically controlled phenomenon, cells can be selected for high yield of secondary metabolites from a heterogeneous population to improve the overall quality of the cultures, in relation to the production of active principle. Before starting the selection procedures, it is necessary that heterogeneous nature of the cultures be established and a sensitive method of analysis is available to analyze a large number of clones. The 'stability' of such selected clones is of paramount importance for developing further method to achieve industrial production. Cell suspension are plated on Petri dishes, growing colonies are subcultures and clones are selected as described in plating technique (Chapter 26). Variability in secondary metabolite content is determined and high productive clones are selected. These clones may not be stable in production of secondary metabolites so, plating and selection is performed repetitively to obtain stable clones. This is presented in simplified form in the Figure 29.5. Repetitive selection yields high secondary metabolite productive clones.

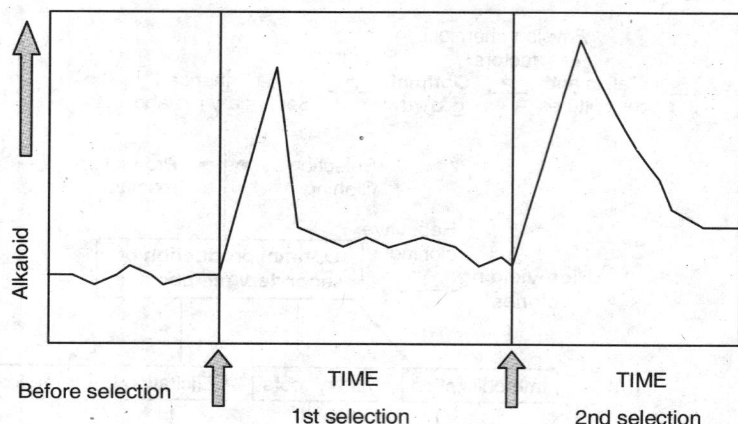

Fig. 29.5. Effect of repetitive selection by cloning on alkaloid yield of cell clones and their performance during subsequent subculture. Note the increase in alkaloid yield immediately after selection.

4. SPECIAL TECHNIQUES

Use of special techniques such as elicitation and hairy roots cultures not only provide a method to enhance the production of secondary metabolites but also a method to clearly understand the mechanism under lying the production of secondary metabolites. Plant cells produce secondary metabolites when treated with elicitors. This phenomenon is used to understand the gene expression and gene involved in the production of secondary metabolites.

4.1. Elicitors

Strictly speaking elicitors are compounds of biological origin involved in plant microbe interaction. In the context of product accumulation by plant cell cultures, elicitors are mediator compounds of microbial stress (biotic elicitors) or are stress agents like UV light, alkalinity, osmotic pressure, or heavy metal ions (abiotic elicitors). Molecules, which stimulate secondary metabolism leading to the induction of stress metabolites, are called elicitors and those derived from fungi may be referred to as fungal elicitors. The discovery of 'elicitation response' in 1972 by Keen and co-workers demonstrated that enhanced production of phytoalexins may also be achieved by treating plant cells with extracts prepared.

It has been demonstrated repeatedly that fungal-induced stress of normal, intact plant tissues leads to the induction and accumulation of phytoalexins. Although alkaloids have not generally fallen into the phytoalexin category, in the mid 1980s a number of investigators demonstrated that the production of certain alkaloids in plant cell cultures could be induced by treating cells with elicitors derived from a variety of pathogenic fungi. The ability of cultures plant cells to produce certain metabolites in response to molecules isolated from fungi appears to be a general phenomenon, and must be correlated with physiological stress imposed on the cells.

Chemically defined elicitors (actinomycin-D, arachidonic acid sodium salt, chitosan, nigeran, poly-L-lysine and ribonuclease-A) or fungal preparations (whole extract or cell wall fraction) from common pathogenic fungi (*Pythium*, *Alternaria*, *Helminthosporium*, *Fusarium*, *Colletotrichum*, *Sclerotinia* etc) are used. Fungus is grown on fresh culture medium from stored stock cultures, homogenized and autoclaved at 121°C for 20 minutes and suitably diluted fungal preparations or chemicals are added in to medium to evaluate the elicitation effect. Certain examples of elicitation are given in the Table 29.5.

4.2. Hairy Root Culture

In the production of hairy root cultures, the explant material is inoculated with a suspension of *Agrobacterium rhizogenes*. This culture is generated by growing bacteria in yeast maltose broth (YMB) medium for two days at 25 °C with rotary shaking, pelleting by centrifugation (5×10^3 rpm, 20 min.) and resuspending the bacteria in YMB medium to form thick suspension (approximately 10^{10} viable bacteria per ml). Transformation may be induced on aseptic plants grown from seed, or on detached leaves, leaf discs, petioles or stem segments from greenhouse plants followed by sterilization of the excised tissues. In some species a profusion of root may appear directly at the site of inoculation, but in others a callus will form initially and roots emerge subsequently from it. In either case, hairy roots normally appear within one to four weeks. The susceptibility of species to infection is very variable. Addition of acetylsyrigone, the compound produced during the wounding response of plants, activates the Vir genes of *Agrobacterium* adding plasmid T-DNA transfer. An alternative approach to transformation is to use cocultivation of plant protoplasts with *A. rhizogenes*. Cultures may be cleared of bacteria by several passages in media containing 200 mg/l cephalosporin and 500 mg/l ampicillin. The infection of plants with *A. rhizogenes* causes one or both of two pieces of T-DNA (Tt and Tg) to be inserted into the plant genome. Integration alters the auxin metabolism of transformed tissues in such a way that the hairy root phenotype is expressed and amino acid metabolism is modified in such a way that specific metabolites such as opines are produced. Both TL-DNA and TR-DNA have rhizogenic functions but in most species the TL-DNA appears to be more important in determining the hairy root phenotype. Although synthesis in opines (manopine or agropine) is the fine indication that roots are transformed, the expression of the opine genes in hairy root tissue may be unstable over time. Thus, detection of the T-DNA by Southern hybridization is necessary to confirm that the cultured root tissue has transformed. Hairy roots are characterized by a high degree of lateral branching, a profusion of root hairs and an absence of geotropism. Hairy root cultures do not require conditioning of the medium. A feature of hairy root systems of paramount importance for their commercial exploitation is their stable, high-level production of secondary metabolites. Although the productivity of untransformed root cultures may be similarly exploited, establishment and maintenance of such cultures is difficult and the auxin supplement needed for optimal growth can depress productivity. The transformed hairy root cultures utilized in the production of secondary metabolites are listed in Table 29.6. It is important to note that several products are synthesized in roots (e.g., nicotine) and in such cases hairy root cultures offer a unique fast-growing organized culture system.

Table 29.5. A variety of physiological responses produced under chemically defined and fungal elicitor stress.

Plant species	Fungus	Preparation	Secondary metabolite	Effect
Apium graveolens	*Fusarium* sp.	—	Furanocoumarins	Increase
Catharanthus roseus	*Pythium aphanidermatum*	Homogenate	Indole alkaloid and related enzymes	Increase
Catharanthus roseus	*P. aphanidermatum*	Homogenate	Indole alkaloid and phenolics (TDC,PAL)	—
Daucus carota	*P. aphanidermatum*	Elicitor	Anthocyanins and enzymes	Increase
Glycine max	*Phytophthora megasperma*	Elicitor Chitosan	Phytoalexin Isoflavonoids	Production production
Papaver somniferum	*Botrytis* sp.	Mycelium Actinomycin	Codeine Sanguinarine	26 times increase Production
Petroselinum hortense	*Alternaria* sp.	Cell wall	Bergapten	Lowering cytoplasmic phosphate
Petroselinum hortense	*Alternaria* sp.	Cell wall	PAL , mRNA	Inhibition of uptake of phosphate and amino acid
Phaseolus vulgaris	*Alternaria* sp.	Poly-L-lysine RNAse	Phaseollin Phaseollin	Production Production
Sanguinaria canadensis	*Penicillium expansum*	Glucan	Benzophenanthridine alkaloids	Role of calcium in signal transduction
Solanum tuberosum	*Phytophthora infestans*	Elicitor	Phenolic and PAL activity	Increase in enzyme activity

Table 29.6. Development of hairy root cultures in an attempt to regulate the production of secondary metabolite.

Family	Species	Secondary products
Solanaceae	***Atropa belladonna***	Atropine
	Datura stramonium	Hyoscyamine
	Hyoscyamus muticus	Hyoscyamine
	Nicotiana rustica	Nicotine, anatabine
	N. tabacum	Nicotine, anatabine
	Scopolia japonica	Hyoscyamine
	Solanum laciniatum	Steroidal alkaloids
Apocynaceae	*Catharanthus roseus*	Ajmalicine, serpentine, catharanthine, vindolinine
Chenopodiaceae	*Beta vulgaris*	Betacyanin, betaxanthine
Boraginaceae	*Lithospermum erythrorhizon*	Shikonin
Asteraceae	*Tagetes patula*	Thiophenes
Rubiaceae	*Cinchona ledgeriana*	Quinoline alkaloids
Lamiaceae	*Mentha piperita*	Monoterpenes
	M. vulgaris	Monoterpenes

4.3. Biotransformations

A bioconversion can be defined as the conversion of one chemical into another, i.e., of a precursor (or substrate) into a final product, using a cell suspension acting as biocatalyst (Fig. 29.6). The biocatalyst can be microorganisms, plant or animal cells, either growing or in a quiescent state, or an extract from such cells or a purified enzyme. The biocatalyst may be free, in solution, immobilized or on solid support or entrapped in a matrix (To understand these processes, see chapter on cell immmobilization). In bioconversion by whole cells or extract a single enzyme or several enzymes may be involved. By means of recombinant DNA technology genes encoding relevant enzymes can be introduced in the host cells.

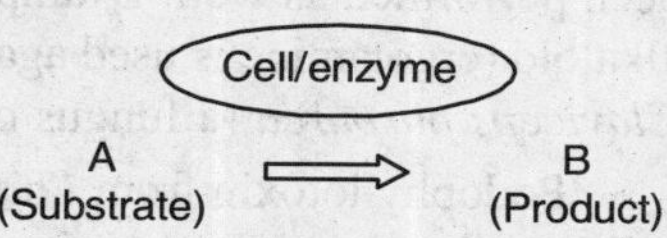

Fig. 29.6. Bioconversion of substrate into product in presence of biocatalyst.

Each individual cell contains many enzymes, which can display different catalytic properties depending on the conditions to which they are exposed. A plant cell contains 800-1000 different enzymes belonging to primary and secondary metabolism. Enzymes has two properties similar to catalysts; the ability to speed up a reaction, and to remain unchanged at the end of the reaction, they 'recycle' in the reaction. The principle for this process is that enzymes efficiently bind to the transition state of a precursor. Consequently, the free energy of activation for the reaction is strongly reduced and the bioconversion facilitated. The reaction is speeded up without changing the equilibrium. Once the product is formed it is released and the active site becomes available for further biocatalysis.

Since each enzyme has its own specific site, enzymes are selective towards their precursors (substrate specificity). However, a number of enzymes with broad specificity are known, e.g., phenoloxidases, lipases, and esterases. Most biocatalyst are active at room temperature ensuring stability of the precursor and the product.

Freely suspended cells form the most simple biocatalytic system, as precursors can be supplied directly to the cultures. Often batch-grown cultures consisting of undifferentiated plant cells are used. The only barriers a precursor has to pass are the cell wall and cell membrane. A wide range of potentially valuable compounds has been produced by adding precursors to various culture species. To produce one particular compound the best way is to use plant cells that can perform a one-step bioconversion. Unfortunately, often the precursor undergoes more than one bioconversion resulting in complex mixtures of (unknown) products.

Nevertheless, a number of one-step bioconversions by freely suspended cells have been described. The bioconversion rates have been calculated in terms of mg product formed per liter culture per day (mg l^{-1} d^{-1}). This way of calculation enables comparison of bioconversion rates of freely suspended with those obtained by immobilized cells. The bioconversion of papaverine (an opium alkaloid with spasmolytic properties) has been studied by Dorisse *et. al.* (1988) using cell suspensions of *Glycyrrhiza glabra*. After 21 days, 31.5% of this precursor was hydroxylated, mainly into papaverinol. The bioconversion rate of 3.7 mg l^{-1} d^{-1} was rather low. Addition of gitoxigenin (belonging to the cardenolides with action on the insufficient heart) to a culture of Daucus carota yielded a single bioconversion product, 5β-hydroxygitoxigenin (Table 29.7). The precursor was added at 0 h or at 72 h after inoculation of the cells. In the first case 50% of the precursor was completely hydroxylated into 5-hydroxygitoxigenin after 51h, in the latter case this level was already reached after 24h, corresponding to a high bioconversion rate of 50 mg l^{-1} d^{-1}.

Besides hydroxylation, glycosylation is one of the most interesting bioconversions. Glycosylation, particularly glucosylation (glucose coupling), occurs readily in plant cells, but only on a limited scale in micro-organisms and the organic chemist has problems with sugar coupling as well. Consequently many glucosyltransferases with different substrate (precursor) specificities for several hydroxyl bearing compounds have been investigated. Examples are glucosylations by cells of *Ruta graveolens*, *Perilla frutescens*, *Catharanthus roseus*, *Lithospermum erythrorhizon*, *Mallotus japonicus* and *Eucalyptus perriniana*. Glucosylations by cells of *Glycyrrhiza echinata*, *Aconitum japonicum*, *Coffea arabica*, *Dioscoreophyllum cumminsii* and *Nicotiana tabacum* have been performed as well. Examples of other glycosylation reactions are the fructosylation of ergot alkaloid (ergotamine is used against migraine, ergometrine provokes uterus contraction) by cells of *Claviceps purpurea* (a fungus of class ascomycetes). Other examples are listed in Table 29.7.

Podophyllotoxin from *Podophyllum* resin (podophyllin is highly active against several cancer cell lines but its general toxicity is too high to allow clinical application. Chemical modification of the naturally occurring lignan has led to the clinically applied anticancer drugs teniposide and etoposide. The organic synthesis of podophyllotoxin-like lignans is difficult, particularly with regard to their chiral centres and, therefore, alternative routes to lignans are needed. More than 30 lignans are known to exhibit powerful cytotoxic action. Lignans are clearly candidates for plant cell biotechnological production. For the formation of podophyllotoxin and 5-methoxypodophyllotoxin by cultures of *Podophyllum hexandrum* and *Linum flavum* respectively, a series of bioconversions in the cells are required.

Table 29.7. Selected one-step bioconversions by freely suspended cells.

Cell culture species	Reaction type	Precursor	Product	Remarks
Glycyrrhiza glabra	Hydroxylation	Papaverine	Papaverinol	Long period of bioconversion (21 d)
Daucus carota	Hydroxylation	Gitoxigenin	5-hydroxygitoxigenin	Precursor added after 72 h of cultivation
Mentha species	Reduction	(–) menthone	(+)-neomenthol	Stereospecific bioconversion
Rauwolfia seperpentina	Glucosylation	Hydroquinone	Arbutin	Precursor added continuously
Coffea arabica	Glycosylation	Vanillin	vanillin-D-glucoside	High bioconversion percentage *85%)
Choisya ternata	Hydroxilation	ellipticine	5-formyl ellipticine	In static medium

5. ANTITUMOUR AGENTS

Plants have been used in the treatment of cancer for over 3500 years. But it is only since 1959 that a concentrated systematic effort has been made to screen crude plant extract for their inhibitory activity against tumour systems. Under the programme sponsored by National Cancer Institute, USA, around 2,00,000 plant extracts from 2,500 genera of plants have been screened.

The number of medicines used in chemotherapy is around 20. These comprise the synthetic drugs and drugs obtained from plants. A large number of plant products have been shown to possess cytotoxic or antineoplastic properties (Table 29.8).

Lomax and Narayanan (1988) of the National Cancer Institute (NCI), USA have presented a list of potent antitumour agents of interest to the Institute, which are at a developmental stage and the clinical trial level. Information about the new compounds of cancer therapy can be obtained from NCI, USA. The institute also helps in evaluating the chemicals against experimental tumours. Plant tissue culture provides an excellent system to produce such compounds in controlled conditions and to study their production, regulation, biosynthesis and bio-transformation. Unavailability of pure compounds in sufficiently large quantities is a limiting factor in their evaluation on animal systems. The drugs are cleared for human consumption only after extensive clinical trials to evaluate their efficacy and side effects.

Table 29.8. Atitumour agents produced in culture.

Plant species	Compound
Baccharis megapotamica	Baccharin
Brucea antidysenterica	Bruceantine
Camptotheca acuminata	Camptothecine
Catharanthus roseus	Vincristine, vinblastine
Cephalotaxus harringtonia	Harringtonine, homoharringtonine
Fagara zanthoxyloides	Fagaronine
Heliotropium indicum	Indicine = N-oxide
Maytenus bucchananii, Putterlickia verrucosa	Maytansine
Ochrosia elliptica	Ellipticine, 9-methoxy-E
Podophyllum peltatum, P. hexandrum	Podophyllotoxin
Taxus brevifolia, T. baccata	Taxol
Thalictrum dasycarpum	Thalicarpine
Tripterygium wilfordii	Tripdiolide, tryptolide
Withania somnifera	Withanolide-A, Withaferine

Antitumour compounds produced using various techniques of plant tissue culture are presented in Table 29.8. As stated earlier, production of secondary metabolites in the cultures is very low compared to the parent plant of the same species. Therefore, it is desirable to apply various factors to optimize the product yield and use special techniques for obtaining high-yields. These techniques were helpful in increasing the production of ellipticine, podophyllotoxin and tripdiolide in the cultures of *Ochrosia elliptica*, *Podophyllum peltatum* and *Tripterygium wilfordii*, respectively.

Among the anticancer agents of natural origin, vincristine, vinblastine, podophyllotoxins and taxol are currently used clinically. All these products are in great demand, production is low, difficult to isolate and purify and hence very high cost forced the world scientific community to search for alternative means of these drugs production. If newer methods are available it will not only reduce the cost of drug but also relieved the pressure on plants which have become endangered because of over-exploitation, e.g., *Taxus* and *Podophyllum*. Therefore, efforts are being to develop efficient protocols for large scale multiplication of endangered medicinal plants like *Podophyllum* and *Taxus* and toprotect the habitat to conserve the plants *in situ*. The conservation programmes in different countries are part of global efforts to save medicinal and other plant species from being extinct.

6. *DIOSCOREA*

Dioscorea deltoidea (family Dioscoreaceae) related to *D. batata* commonly known as Ratalu. This has attained importance due to population explosion in the Asian countries. The plant contains diosgenin used in the biosynthesis of hormones for contraceptive pills for females. Micropropagation technology has been developed by National botanical research institute, Lucknow in the leadership of Dr H.C. Chaturvedi.

Diosgenin, a steroid sapogenin, is the product of hydrolysis by acids, strong bases, or enzymes of saponins, extracted from the tubers of Dioscorea wild yam. The sugar-free (aglycone), diosgenin is used for the commercial synthesis of cortisone, pregnenolone, progesterone, and other steroid products. Diosgenin is converted into natural progesterone scientifically. There are no enzymes in the human body that will convert diosgenin, which is the active component of wild yams into progesterone. Diosgenin is still very useful in the body and has been used by phytotherapists for centuries as an adaptagen. Whatever the effects of wild yam, it does not have the same benefits as Natural progesterone.

Diosgenin, which is a sapogenin chemical very similar to cholesterol, progesterone and DHEA - the precursor to testosterone. Diosgenin provides about 50% of the raw material for the manufacture of cortisone, progesterone, and many other steroid hormones and is a multi-billion dollar industry.

The major breakthrough in the development of progesterone for birth control was the synthesis (or semi-synthesis) of progesterone from diosgenin, the major spirostan found in yams. In 1934, Schering Laboratories, a major player in the pharmaceutical applications of steroids then and now, was able to isolate 20 mg of progesterone from 625 kg of ovaries obtained from 50,000 sows. This was hardly an economically feasible process. However, in 1940 Russell E. Marker developed the process, initially for Parke Davis and Co., for degrading sapogenins to C21 steroids. This process was applied by Marker to convert diosgenin obtained from a Japanese *Dioscorea* species into progesterone. Marker's investigation of the Mexican yam, *Dioscorea macrostachy* known locally as *cabeza de negro*, showed that it contained large amounts of diosgenin, and that this diosgenin could be easily converted into progesterone by his method, known as the Marker degradation. Marker was unable to convince Parke Davis to apply for foreign patents for this process and establish a hormone industry in Mexico.

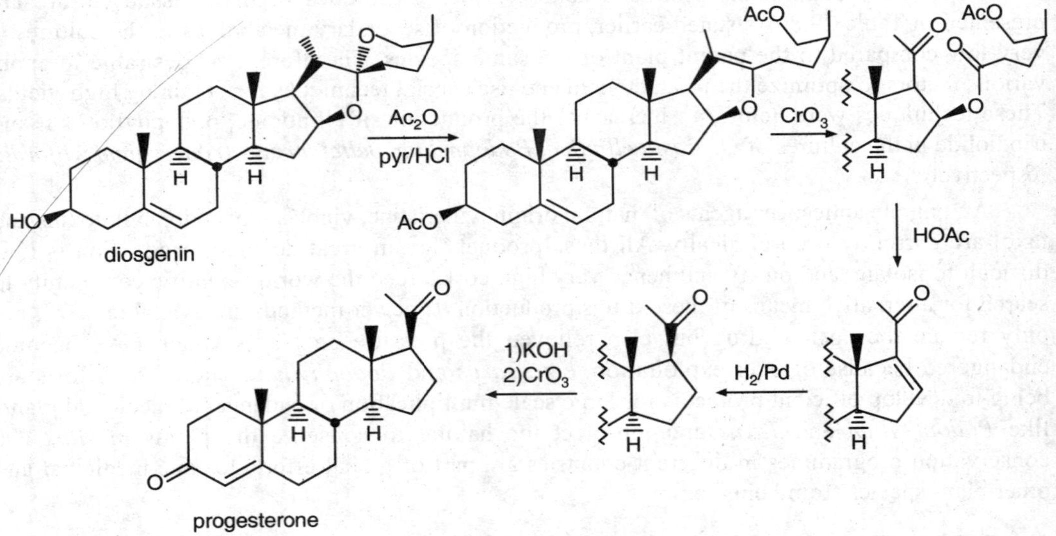

The Marker Process

Fig. 29.7. Synthesis of progesterone from diosgenin by Marker process.

QUESTIONS

1. What is biotransformation? Describe different systems used to obtain products of interest.
2. Write short notes on:
   (a) Biotransformation
   (b) Hairy root cultures
   (c) Ri plasmid d. Elicitors
   (d) Optimization
   (e) Selection,

(f) Cloning
(g) Production medium
(h) Dioscorea
(i) alkaloids,
(j) Antitumour agents
(k) Tobacco alkaloids
(l) Catharanthus alkaloids

3. Describe the various factors affecting the production of secondary metabolites.

4. How will you improve upon production of secondary metabolites in a given cell culture system?

5. Why recurrent selection of clones is required for stable production of secondary metabolites?

6. What are secondary plant products? Describes various alkaloids.

7. What are alkaloids? Describe the methods of production in culture with suitable examples.

8. Describe the secondary metabolites of natural origin with special reference to alkaloids.

9. Write short notes on the followings. A. true alkaloids B. Primary v/s secondary metabolites C. Classification of alkaloids

(a) Biological functions of alkaloids
(b) Biosynthesis of indole alkaloids
(c) Cinchona alkaloids
(d) Shikimic acid pathway.

10. Describe in brief classification, biosynthesis and major classes of alkaloids.

CHAPTER 30

Plant Protoplasts: Isolation, Culture and Somatic Hybridization

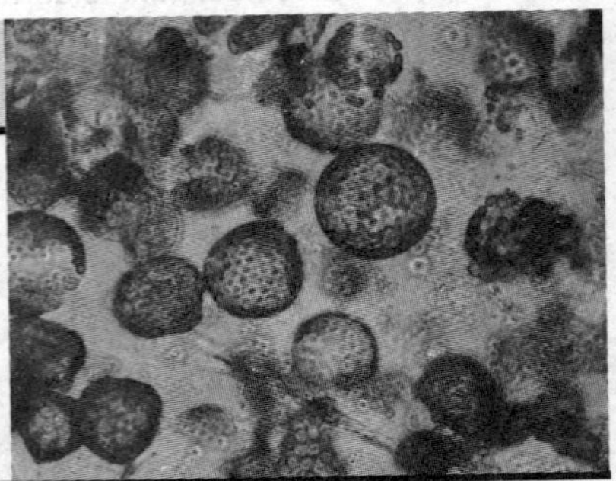

1. INTRODUCTION

Plasma membrane of plant cells is surrounded by cellulose wall and adjacent cells are joined together by a thick pectin rich matrix. Separation of plant cells and removal of the cell wall experimentally by an enzymatic process, results in the production of isolated, nacked plant cells termed 'protoplasts'. Protoplasts thus can be called as a functional individual cell with plasma membrane as outermost layer and the only barrier between environment and the cell contents.

The real start of plant protoplasts research started in 1960's with the enzymatic removal of cell wall by E.C. Cocking at the University of Nottingham, England. It became a widely used technique with the availability of commercially produced enzymes. Takabe and coworkers using commercially available enzymes isolated protoplast in 1968 and subsequently achieved regeneration of whole tobacco plant from mesophyll protoplasts in 1971. Carlson's and coworkers were successful in obtaining the first interspecific somatic hybrids by protoplast fusion between *Nicotiana glauca* and *N. longsdorffii*. During these experiments polyethylene glycol (PEG) was established as a potent fusogenic agent. Subsequently, the improved fusion methods have resulted in the production of several somatic hybrids including the first intergeneric hybrid (called pomato) between potato and tomato by Melchers and coworkers in 1978. The first interspecific hybrid between protoplasts of normal *Nicotiana sylvistris* and X-ray irradiated protoplast of male sterile *N. tabacum* was obtained by Zeleer and coworkers in 1978. Next year in 1979, stimulation of protoplasts fusion by electrical stimulus and gene transfer through protoplast fusion were demonstrated. Zimmerman and Scheurich (1981) improved the electric fusion method for protoplasts isolation. The 1980s witnessed many protoplasts based works particularly those reporting novel protoplast to plant systems for genetic manipulation. During the 1990s protoplast based technologies for gene transfer were overshadowed by *Agrobacterium* and Biolistics® mediated gene delivery to plants. However, public antagonism (especially in Europe) to rDNA technology renewed interest in exploiting protoplast in somatic hybridization, cybridization, protoclonal variation study, proteomix and metabolomics.

2. APPLICATIONS OF PROTOPLASTS IN PLANT RESEARCH

Plant cells are normally connected to each other via many plasmodesmata in multicellular tissues. Plant protoplasts cultures provide an excellent system to obtain single cells of higher organisms and study its functioning in a controlled environment. Protoplasts isolation provides millions of cells, which is comparable to microbial system and can be used many ways. Protoplasts stage is for a small time before cell wall is regenerated and protoplast becomes a cell. Therefore, these studies can be done during a short duration of time when cell is without cell wall or in protoplasts stage.

(a) Isolation of mutants, developed spontaneously or through mutagens, is easier in single cell derived colony.

(b) Single cell cloning can easily be performed with protoplasts.

(c) Somatic cell fusion can only be performed with protoplasts.
(d) Genetic transformations through DNA or organelles uptake can be achieved using protoplasts.
(e) Complete plants of single cell origin can be obtained, as plant cells are totipotent.
(f) It is a unique experimental system to study *de novo* cell wall formation.
(g) Protoplasts are very useful objects of membrane studies e.g., membrane transport or uptake processes and membrane structures.
(h) Protoplasts are also excellent experimental material for ultrastructural studies e.g., in the study of microtubules formation during cell division.
(i) Protoplasts are very suitable for isolation cell organelles and chromosomes because of absence of thick cell wall barrier.

3. ISOLATION OF PLANT PROTOPLASTS BY ENZYMATIC METHOD

Protoplasts are isolated by two methods, mechanical and enzymatic. Mechanical isolation method of protoplasts is no more practically used. It is only a historical method. This involves cutting of plasmolysed cells (in fact cell walls) with a sharp blade to release protoplasts.

This involves four steps -
(a) Sterilization of leaves
(b) Peeling off the lower epidermis
(c) Incubation in enzyme solution and
(d) Isolation and cleaning of the protoplasts.

Healthy leaves are obtained from green house grown plants and leaves are sterilized as described earlier (sterilization of explant). After sterilization lower epidermis is removed with the help of fine forceps and the stripped leaves are cut into small pieces. Flaccid leaves facilitate peeling. *In vitro* grown cultures are aseptic material and therefore, used directly for the isolation of protoplasts. Actively growing cultures in the exponential phase are good material for the isolation of protoplasts.

Usually one gm material is used to obtain protoplast and conditions are optimized empirically by changing enzyme concentration and composition, concentration of osmoticum, duration of incubation and types of tissues.

Enzymes - Different enzyme preparations are available in the market but the idea is to combine one middle lamella dissolving and one cell wall digesting enzymes in proper composition to achieve maximum protoplasts release from one gm material. Following enzymes are used -

1. Macerozymes R-10
2. Cellulase – Onozuka R-10
3. Hemicellulase
4. Pectinase
5. Drieselase

A combination of these enzymes in a concentration of 0.5-2% is used. In many cases only macerozyme and cellulase are sufficient to obtain protoplasts in significant number. The enzyme solution (pH 5.5) is prepared in 10-15% sorbitol or mannitol containing small amount of $CaCl_2$ (7 mM) for membrane stability. This solution is sterilized through a membrane filter (cold sterilization) and leaf or callus tissues are placed in it. Petriplates containing tissue and enzyme mixture are sealed with parafilm and incubated for 4-12 hours (sometimes 0.5 to 20 hrs) on a rocking shaker at 24-26 °C. After incubation, solution is filtered through a wire or nylon mesh

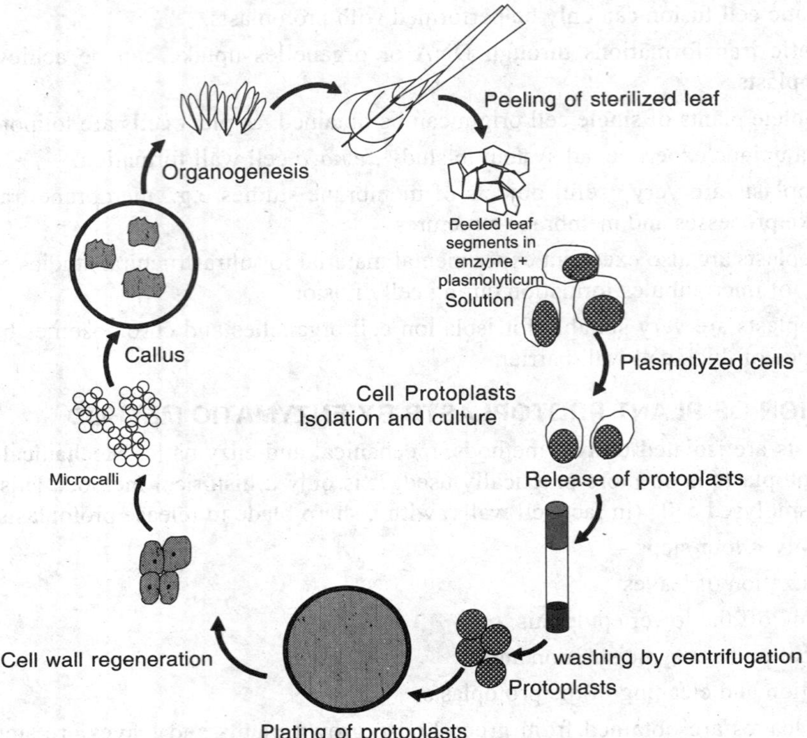

Fig. 30.1. Protoplasts isolation procedure.

(50-100 μm) to remove debris (undigested cells, tissues, broken cells etc.), transferred into screw capped small centrifuge tubes (sterilized), and centrifuged at 100 g. The protoplasts formed a pellet while the debris in the supernatant is carefully removed. Fresh sterilized sorbitol solution (no enzyme) is added to tube and centrifuged. By repeating the process 2 to 3 times, protoplasts are cleaned (debris is removed). If 20% sucrose solution is used, protoplasts will float and debris will settle during centrifugation at 200g for 1 min. Floating protoplasts are carefully removed with the help of sterilized pipette and bulked together for further use. Protoplasts are counted by haemocytometer and then diluted to proper strength (number per ml) in the culture medium containing osmoticum (Fig. 30.1).

4. PROTOPLASTS CULTURE

Isolated protoplasts require somatic protection in the culture medium until they generate a strong wall (Fig. 30.2). Osmolarity in the medium is adjusted to the same level as in the enzyme and washing solution. Prolonged culture at high osmotic pressure can result in browning of the cultures and inhibition of callus growth. The osmolarity of the medium is gradually decreased with cell wall formation and cell divisions. Abrupt decrease in osmolarity may cause bursting of protoplasts and cells and also affects the growth of cells.

Generally the basic constituents in protoplast culture medium are similar to those used for cell cultures viz. MS or B_5 medium. Some modifications can have beneficial effects on protoplasts survival and colony formation. Usually carbon, nitrogen, vitamins and plant growth regulators are suitably modified. Successful culture of plant protoplasts often depends on the appropriate medium volume per cell ratio (density). The optimal protoplasts density usually has to be experimentally determined. The usual density range from 0.5 to 2×10^5 protoplasts per ml of culture medium.

5. PROTOPLASTS FUSION AND SOMATIC HYBRIDIZATION

In nature, gametic fusion forms zygot with a new combination of genes. Diploid nature is restored by haploid gametic fusion. Sexual crossing is one of the major sources of genetic variability. Incompatibility at various levels, before fertilization and after fertilization, restricts recombination between distantly related species and species specific characters are maintained. A limited success has been achieved in obtaining interspecific and intergeneric hybrids by conventional breeding programmes. These hybrids are useful in further breeding programmes and in the understanding of cytological and genetical behaviour of chromosomes in these hybrids. However, it is not possible to obtain hybrids by conventional method in all the desired plant materials. Protoplasts fusion provides a novel approach overcoming sexual incompatibility and obtained somatic hybrids. Fusion of somatic cells (2n) and production of hybrids is known as somatic cell fusion or somatic hybridization.

Plant membranes are flexible, asymmetric structure of membrane proteins integrated into phospholipid molecules. The fluid mosaic model with lipid bilayer is the established structure of intact membranes. The process of protoplasts fusion requires the direct contact between membrane surfaces and integration of membranes of two different cells. The adhesion of protoplasts depends on surface charge. According to electrophoretic studies outer membrane surface is negatively charged. This change varies from –10 mV to –50 mV and can significantly be reduced by Ca^{2+}. Due to these fusion properties of the membranes, protoplast fusion has been achieved between dicot and dicot, dicot and monocots and plant and insect cells (*Drosophila*), amphibian cells and mammalian cells. Protoplast fusion between haploid (gametic) and diploid (somatic) cells has been achieved which clearly demonstrate that fusion is independent of cell type.

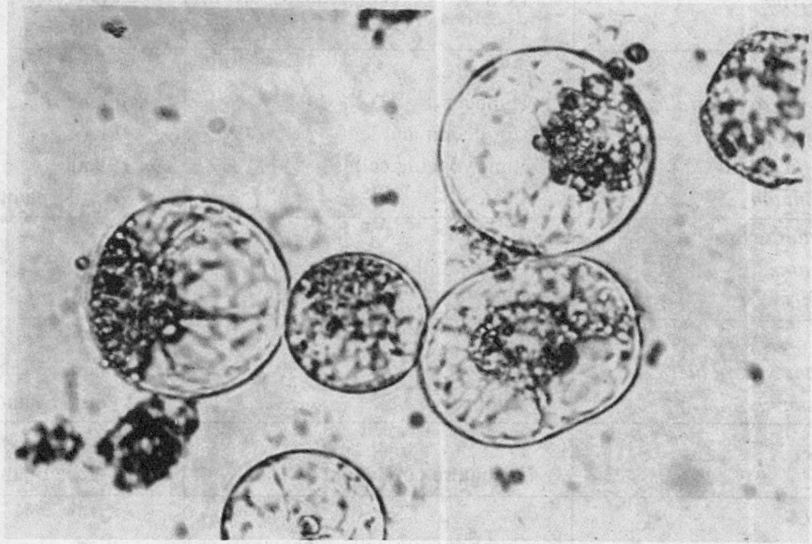

Fig. 30.2. Freshly isolated protoplasts of *Ruta graveolens*.

Protoplasts of both types are mixed in an equal proportion attaining a density of 5×10^4 to 2×10^5 protoplasts per ml. Protoplasts density is usually determined by haemocytometer.

There are several chemical and physical agents who enhance fusion between protoplasts, known as fusogen or fusion agent, e.g., sodium nitrate, polyethylene glycol of different molecular weight (PEG 1500, PEG 4000, PEG 6000), dextran sulphate and gelatin etc. High calcium ion concentration and pH of the incubation mixture have profound effect on fusion percentage. Fusion is stimulated under electric charge as membranes are negatively charged.

Isolated protoplasts are suspended in an aggregation mixture (5.5% sodium nitrate /20-40% PEG in 10% sucrose solution), kept at 35 °C for 5 minutes and then centrifuged at 200 g for 5 min to obtain dense pellet. The tube with pellet of protoplasts is once again transferred to a water bath maintained at 30 °C for 30 minutes during which most of the protoplasts undergo fusion. The protoplasts are gently washed and transferred to liquid medium or plated. The fusion and fusion products are observed under inverted microscope.

6. REGENERATION FROM PROTOPLASTS

Protoplasts form cells by regenerating cell wall and subsequently form microcalli by cell divisions. Thus, regeneration from protoplasts is through callus formation. Both, organogenesis or embryogenesis can be obtained as from callus cultures of the given species. In some cases, protoplasts can directly form somatic embryos, e.g., in *Medicago*, *Brassica juncea* and *Citrus*. Regeneration from protoplasts and mode of regeneration are dependent upon the species, genotype, and all the factors governing regeneration in callus or cell cultures. The successful regeneration from protoplasts has been reported in several species. A list of selected species is presented in Table 30.1.

Table 30.1. Plant species regenerated from protoplasts.

Species	Plant source	Regeneration
Cereals		
Oryza sativa var. *japonica*	Embryogenic cell suspension	Plant
Zea mays	ECS	Sterile plant
Triticum aestivum	ECS	Plantlet
Hordeum vulgare	ECS	Albinoplant
Vegetables		
Brassica oleracea var. *stalica*	Hypocotyl	Plant
Cucumis sativus	Cotyledon and leaf	Plant
Asparagus officinalis	Embryogenic callus	Plant
Capsicum annuum	Leaf	Fertile plant
Tuber and root crops		
Ipomoea batatas	Petiole and stem	Plant
Beta vulgaris	Leaf	Plant
Oil crops		
Brassica napus var. *oleifera*	Leaf	Plant
Helianthus annuus	Hypocotyl	Fertile plant
Legumes		
Glycine max	Immature cotyledons	Fertile plant
Woody trees		
Larix eurolepsis	Embryogenic cell suspension	Plant
Picea glauca	Embryogenic cell suspension	Plantlet
Populus sp.	Leaf	Plant
Prunus avium	Cell suspension (root)	Plant
Coffea cansphora	Somatic embryo	Plantlet
Ornamentals		
Rosa sp.	Embryogenic cell suspension	Plantlet
Pelargonium sp.	Cell suspension (leaf, petiole)	Plantlet
Chrysanthemum sp.	Cell suspension (leaf)	Plant

7. FUSION TECHNIQUES

While several methods were proposed to induce protoplasts fusion, earliest success was achieved using sodium nitrate (Carlson et al, 1972). Plant protoplasts have uniformly negatively charged membranes. Divalent cations, especially Ca^{2+}, modify the electrophoretic mobitility of protoplasts. Aggregation mediated by neutralization of plasmalemma negative charges by 50 mM Ca^{2+} mixed with 0.4 M mannitol in a solution of high pH (10.5) has led to the genuine cytoplasmic fusions of tobacco protoplasts and *Petunia* protoplasts. The other fusogens are as follows.

7.1. PEG

Various fusogenic agents have been used for getting protoplasts fusion. The discovery of the powerful aggregation effect of polyethylene glycol (Kao and Michayluck, 1974) provided the awaited tool. A few minutes after being immersed in a 20-40% (w/v) solution of PEG (1500 to 6000 mw) virtually all protoplasts exhibit adhesion. However, actual fusion occurs upon dilution of PEG. A combination of PEG induced adhesion followed by elution with PEG, high Ca^{2+}, high pH was even more reliable. Actual role of PEG in the process of membrane fusion remains yet to be clearly determined. Addition of dimethyl sulphoxide (DMSO) with PEG has stimulatory effect on protoplast fusion. It has been suggested that bridge formation mediated by Ca^{2+} and PEG through its slightly negative polarity might be the cause of PEG induced fusion of protoplasts.

7.2. Dextran

Similarly fusogenic properties of dextran and of polyvinyl alcohol suggest that weak non-ionic surfactant properties be involved. It seems perturbation of negative charges of the protoplast plasmalemma seems essential. This is illustrated by fusion induced by a positively charged phospholipid and above all electric field induced fusion.

7.3. Electrofusion

In electrofusion method of Zimmerman (1982) protoplasts are brought in close contact by a non-uniform alternating field between two electrodes, then fusion is initiated by field pulse of high voltage (500 V - 3 KV, cm^{-1}) for a very short time (10-100 micro seconds). This is the most widely used apparatus (Fig. 30.3) and method in the studies on membrane fusion and for the production of somatic hybrid colonies or plants. Electrofusion is a biophysical process that involves two steps. In the first step the protoplasts are exposed to alternating electric field (0.5-1.5 MHz) that generates dipoles through dielectrophoresis. In this non-uniform field, the protoplasts acts as dipoles, move in the direction of the increasing field as a function of the field strength, cell size and the mobility of charge within the protoplast and in the surrounding solution. As a result of mutual dielectrophoresis in non-uniform as well as uniform field a protoplasts clump, and bind together to form the chain. The second step for electrofusion is the application of one or more short (10-100 μs) direct current (DC, 10^3 KV/cm) pulses that cause reversible membrane breakdown. This results in pore formation in the membranes touching to each other. This way contacting

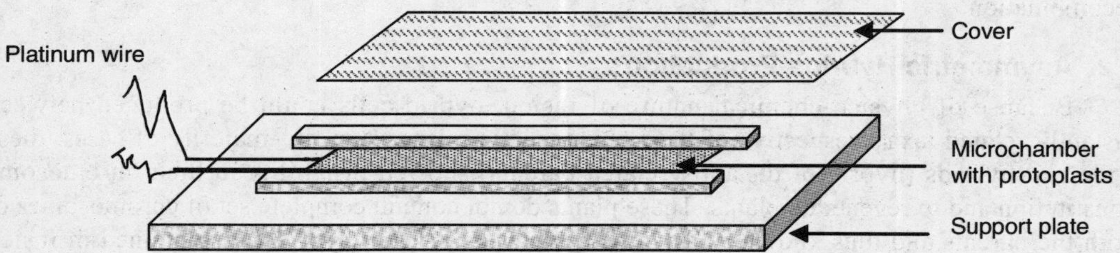

Fig. 30.3. Zimmerman's apparatus for protoplast fusion by electric current.

membranes can be fused and this fusion opens the way for hybrid cell formation. Generally several cycles of AC and DC current are given to increase the fusion frequency.

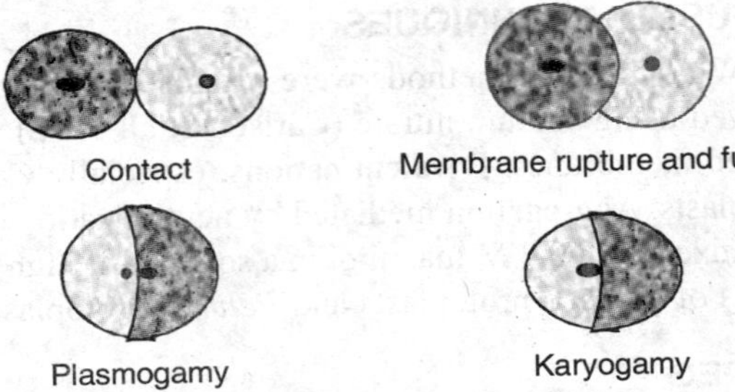

Fig. 30.4. Protoplast fusion and production of somatic cell hybrids.

The fusion chamber is a small apparatus having two parallel running platinum wires for passing the current (Fig. 30.3). Protoplasts suspension is placed in this microchamber having platinum electrodes (wires). This can be placed on inverted microscope and fusion process can be observed simultaneously (Fig. 30.4). PEG fusion efficiency is given as the best 5-30% heterospecific fusion products under optimal conditions. The electric fusion method appears very promising, as its efficiency is very high.

8. APPLICATIONS OF PROTOPLASTS HYBRIDIZATION (Somatic Cell Hybridization)

A question comes in the mind of students that when we regenerate cell wall and protoplasts become a cell then why we isolate and culture protoplasts? Following applications can be done only in protoplasts and not in intact cell. Therefore, these techniques are employed when protoplast stage is available for a short duration during isolation and culture of protoplasts. There are three major applications

a. Production of somatic hybrids and cybrids

b. Asymmetric hybrids

c. Direct DNA uptake

8.1. Hybrids and Cybrids Production

Protoplasts fusion technique during 1970s established that parasexual hybrids can be produced by this method. Intra and inter-specific hybrids have been produced by this method in several Solanaceous plants like *Datura*, *Petunia* and *Nicotiana* species. By this method complete hybrid plants of intermediate characters were produced. Mostly amphidiploid plants were obtained.

The partial transfer of cytoplasm (cytoplasmic characters), organelles uptake or loss of one set of chromosomes of one of the parent protoplasts leads to the formation of cybrid plants. This is not possible by conventional breeding programme. The donor-recipient protoplasts fusion system has been used to transfer chloroplasts and mitochondria in several crop plants. It is established that mitochondria and chloroplasts have their own DNA and therefore, characters governed by this DNA are transferred and expressed. Under selective selection pressure, recombination between chloroplasts DNAs from different origin can occur with low frequency. In comparison to chloroplasts DNA, mitochondrial DNA undergo more frequent and extensive recombination.

8.2. Asymmetric Hybrids Production

Because of physico-chemical nature of fusion, hybrid cells could be produced between distantly related taxa irrespective of their phylogenetic relationship. In majority of cases, these wide cell hybrids (hybrid of distantly related parents) showed instability in their chromosome constitution and to regenerate plants. These plants do not contain complete set of chromosomes of both the parents and thus known as asymmetric hybrids, e.g., in hybrid cells obtained in fusion between soybean x *Nicotiana glauca*, soybean x *N. tabacum*, carrot x tobacco, carrot x rice, *Duboisia* x tobacco and soybean x rice etc.

In some intergeneric fusion combinations, hybrid plants with additive chromosome number were regenerated e.g., after fusion between potato and tomato, *Arabidopsis* and *Brassica*, *Datura innoxia* and *Atropa balladonna*, *Lycopersicon esculentum* and *Solanum lycopersicoides*, *Eruca sativa* and *Brassica juncea* and *Lycopersicon esculentum* and *Solanum nigrum*. In some of the cases, two parental diploid genome coexisted in the cybrid cells (amphidiploid). However, morphological abnormalities and sterility occurred in several of these intergeneric hybrids.

8.3. Direct DNA and Macromolecules Uptake

Cell membranes are semi-solid structures as per the fluid mosaic model and fluidity of membrane is physiological characteristics. Transient aqueous pores in the bilayer portion can make the membrane permeable for macromolecules. The formation of pores and repair of membrane pores (sealing) opens the new approach for introduction of nucleic acid molecules into cytoplasm and the nucleus. The reversible, non-destructive permealization of membranes can be achieved by both the chemical treatments and action of short duration high electric pulses. Uptake of DNA molecules into a variety of plant protoplasts was successfully promoted by application of PEG treatment or electroporation.

Plant protoplasts have the remarkable property to take up particles and macromolecules, like polystrene, latex particles, ferritin, proteins, DNA, viruses, bacteria or even whole chloroplasts and nuclei can be transplanted. Mayo and Cocking (1969) reported that the uptake is by pinocytosis, and demonstrated this by differential staining effects of phosphotungustic acid on the plasmalemma.

i. DNA uptake - The discovery that *Diplococcus pneumoniae* transforms itself by taking up exogenous DNA into its genomes has made genetic engineering a routine task in bacteria. New genes can be synthesized and DNA-RNA hybrids can be obtained.

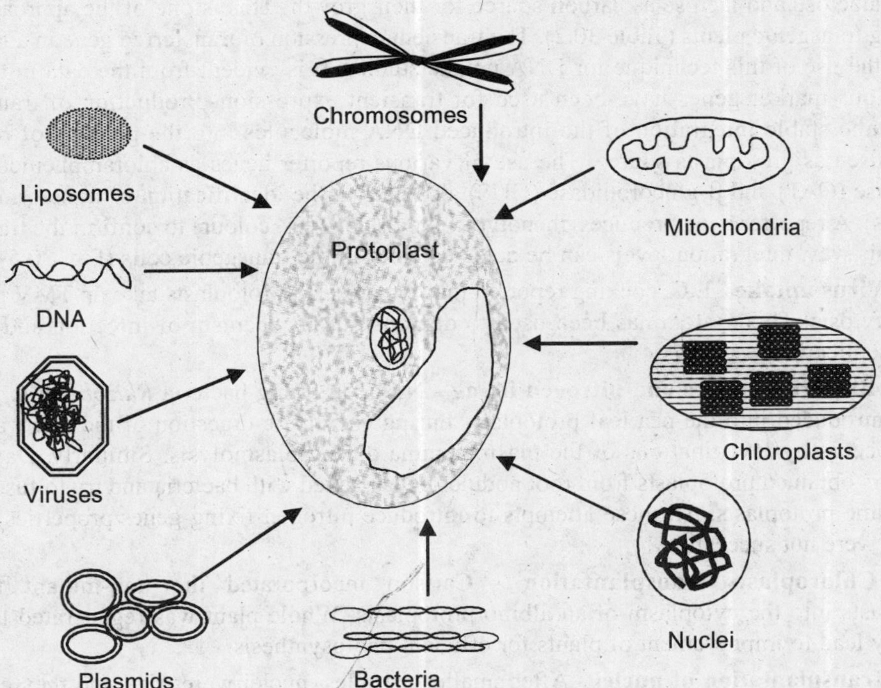

Fig. 30.5. Various applications of protoplasts technology.

Table 30.2. Transgenic plants produced through DNA uptake into protoplasts.

Species	Promoter/gene	Method	Regeneration
Oryza sativa	Promoter and the first intron of the alcohol dehydrogenase gene of maize/β-glycuronidase gene.	Polyethylene glycol	Plant
	Cauliflower mosaic virus 35S gene promoter/ neomycin phosphotransferase gene.	Electroporation	Plant
	-do-	-do-	-
	Cauliflower mosaic virus 35S gene promoter (35S) hygromycin phosphotransferase gene (H+P)	Electroporation	Fertile plant
Zea mays	Promoter and the first intron of the alcohol dehydrogenase gene of maize/β-glycuronidase gene.	Electroporation	Sterile plant
Citrus sp.	Mannopine synthase gene promoter of *Agrobacterium tumefaciens* (chloramphenicol acetyl transferase gene	PEG	Plant
Solanum tuberosum	Nopaline synthase gene of A. tumefaciens/ neomycin phosphotransferase gene (neo.)	PEG	Plant
Lactuca sativa	35S/neo	Electroporation	Fertile plant
Arabidopsis thaliana	35S/hph	PEG	Fertile plant

DNA can now be incorporated into the higher plant cells protoplasts. Doy *et. al.* (1972-73) transformed haploid tomato cell cultures by phage lambda (bacterial virus). Transformed cells can utilize galactose and lactose as carbon source for their growth. This is one of the approaches for obtaining transgenic plants (Table 30.2). The transient expression of transferred gene in a few days permits the use of this technique for DNA uptake studies. It is evident from the data in the table that various marker genes have been used for transient expression. Production of transgenics requires the stable integration of the introduced DNA molecules into the genome of recipient species used as protoplasts source. The use of various reporter genes as chloramphenicol acetyl transferase (CAT) and β-glucoronidase (GUS) also help in the identification of transformed cells (colonies). A reporter gene produces phenotypic expression (e.g. colour) to confirm the transfer of gene. This way, integration levels can be detected early in the transgenic cells (Fig. 30.5).

ii. Virus uptake - E.C. cocking reported that tomato fruit protoplasts take up TMV particles by pinocytosis. This system has been used to demonstrate mechanism of infection and nucleic protein replication.

iii. Bacterial uptake and nitrogen fixing - Nitrogen fixing bacteria *Rhizobium japonicum* can be introduced into the pea leaf protoplasts during enzymatic digestion of the cell wall. This uptake occurs by invaginations of the plasmalemma during plasmolysis. Similarly, Davey and coworkers obtained protoplasts from root nodules cells packed with bacteria and try to fuse it with non-legume protoplasts. All these attempts to introduce nitrogen fixing genes/properties to non-legumes were not succesful.

iv. Chloroplasts transplantation - Carlson incorporated the non-mutant tobacco chloroplasts into the cytoplasm of an albino protoplasts. Whole plant was regenerated later on. This may lead to improvement of plants for efficient photosynthesis.

v. Transplantation of nuclei - After smaller particles, nuclei were tried and they could be incorporated through plasmalemma without bursting it. Potrykus and colleagues transplanted nuclei from *Petunia hybrida* into protoplasts of *Zea mays*, *Petunia hybrida* and *Nicotiana glauca*.

Further development of this technology led to the development of cloned sheep and monkey by nuclear transplantation.

9. IDENTIFICATION AND SELECTION OF HYBRID CELLS

Following fusion treatment, the protoplasts in liquid culture medium regenerate cell walls and undergo mitosis resulting in a mixed population of parental cells, homokaryotic fusion products and heterokaryotic fusion products or hybrids. Hybrid cells must be distinguished from the other cells present. Heterokaryotic recognition techniques as well as colonies selection methods are thus highly desirable. For this purpose, several methods are available. These are dependent on -

a. Physical properties of fused cells.

b. Biological properties of fused cells.

c. Biological properties of colonies derived from fused cells. Following methods have been adopted for various types of protoplasts.

(I) Fusion partners such as non-green protoplasts (albino protoplasts from tissue culture) and leaf derived chloroplastic protoplast allow microscopic visual detection of heterokaryotes. A few hours or a few days after fusion products are pipetted and cultivated individually. Because visual techniques applied are tedious, other non-visual methods have been sought.

(*ii*) By centrifugation in density gradient, enriched bands of fusion products can be recovered at an intermediate density.

(*ii*) Vital staining, specially using fluorescent dyes is another possibility to detect heterokaryotes providing a means of automatic separation with cell sorter.

(*iii*) Controlled fusion by electric field should provide a possible way to cultivate only true heterokaryotes.

(*iv*) A method adapted from animal cells to plant protoplasts allows the recovery of fusion products by complementation between inactivated protoplasts by different metabolic inhibitors. Cybrids as well as hybrids selection *in vitro* should be even more effective by use of complementary markers selectable *in vitro*.

(*v*) One of the principal features to characterize somatic hybrid is the chromosome complement. Chromosome number and morphology may be compared to the patterns displayed by the fusion parents to determine whether or not presumptive hybrids are true hybrids.

(*vi*) Supportive evidences for hybridity can be obtained from an electron microscopic morphometric determination of chromatic texture in hybrid nuclei. This technique was used for the verification of *Arabidopsis-Brassica* somatic hybrids and it may prove useful for identifying other interspecific and intergeneric hybrids.

(*vii*) Morphological characters of the hybrid plants and parents, are used to ascertain the hybrid nature of the regenerated hybrids.

(*viii*) Isoenzymes-proteins ultimately manifest the diversity laid down in the genetic programme of each individual. Isoenzymes (multiple molecular forms of an enzyme exhibiting identical and similar catalytic properties) are frequently used for this purpose. Isoenzymes used for hybrid identification include esterase, peroxidase, amylases, phosphatase, alcohol dehydrogenase, lactate dehydrogenase, aspartate and aminotransferase etc.

(*ix*) More recent techniques are - RFLP (Restriction Fragment Length Polymorphism), use of reporter gene and RAPD (Random Amplified Polymorphic DNA) for looking into genetic variability and transformations in the plants regenerated after hybridization or DNA uptake studies.

10. STABILITY OF HYBRIDS

Unfortunately somatic hybrids of plants display chromosomic instabilities. Even for sexually compatible species, regenerated somatic hybrids exhibit quite variable chromosome numbers relative to normal parental additions. Nevertheless, it has been shown in several cases that hybrids are capable of morphogenesis, although resulting plants are abnormal and sterile.

Table 30.3. Transfer of useful agronomic traits by protoplasts technology.

Species	agronomic trait transferred
Brassica oleracea	C3-C4 photosynthesis intermediate trait
Raphanus sativus	C3-C4 photosynthesis intermediate trait
Citrus reticulata	Virus resistant trait
Citrus sinensis	Phytophthora and tristeza virus tolerance
Solanum melongena	Bacterial and fungal wilt resistance
Solanum tuberosum	resistance to Phytophthora

Protoplasts cultures have their own advantages and disadvantages. The problems faced by scientists in recovering hybrids, regeneration of hybrids, sexually viable hybrids etc. are the main difficulties and disadvantages with the technique. Therefore, other methods of transferring genetic material have been developed. Therefore, protoplast culture is being used selectively for specific purposes for gene transfer and several agronomic traits (characters) have been transferred by this technique (Table 30.3).

QUESTIONS

1. Describe the technique of protoplasts isolation, culture and regeneration with the help of suitable diagram.
2. Describe the protoplast technology for crop improvement.
3. What is somatic cell hybridization? Describe the procedure, application and stability of somatic cell hybrids
4. Write short notes on the following:
   (a) Protoplast culture.
   (b) Fusion agents.
   (c) Somatic hybridization.
   (d) Selection of hybrid cells.
   (e) Application of somatic hybridization.
5. What are interspecific and intergeneric hybrids? Give suitable examples.
6. How somatic hybrids are identified and selected?
7. What are different applications of protoplasts in plant sciences?
8. What is the role of osmoticum in protoplasts culture?

CHAPTER 31

Regeneration
(Somatic embryogenesis, Androgenesis, Somaclonal variation)

1. REGENERATION

Potentiality of a plant cell to regenerate the entire organism (plant) is termed as 'totipotency'. This potentiality has been exploited through the culture of protoplasts, cells, tissues and organs in vitro. In cultured materials it has been possible to study such processes as differentiation of a parenchyma cell into tracheid (cell differentiation or cytodifferentiation), organ formation (leaf, shoot bud, root; organogenesis) and somatic embryo formation (somatic embryogenesis). The term 'morphogenesis' is defined as 'the origin of form' and can be examined by through manipulation of physico-chemical (physical and chemical) environment of the cell. Morphogenesis is measured by description, physiological changes, biochemical changes or even at molecular level. Besides the study of fundamental process of differentiation, the capacity of cell to form organs and embryos can be exploited to regenerate plantlets for clonal propagation (clone is vegetatively obtained group of plants from an individual). The basic concept mentioned above led to the establishment of diverse multiplication techniques. The genetic stability of vegetatively produced plantlets is likely to vary according to the procedure chosen.

A

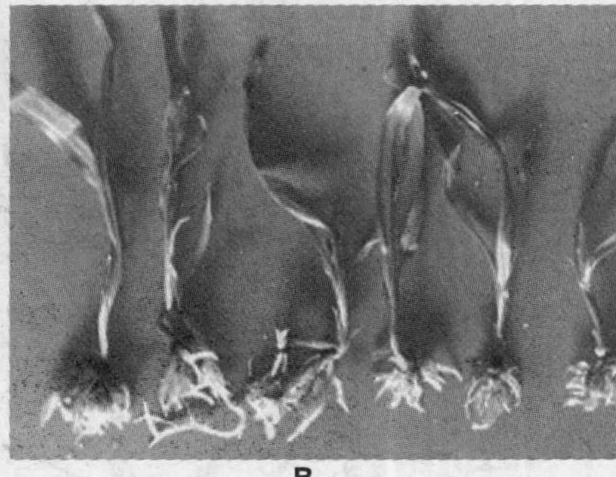

B

Fig. 31.1. Mode of regeneration, A. Organogenesis from callus, B. Adventitious shoot formation from leaf explants.

1.1. Organogenesis and Clonal Propagation

This is the process by which cells and tissues are forced to undergo changes which lead to the production of a unipolar structure, namely a shoot or root primordium, whose vascular system is often connected to the parent tissue (Fig. 31.1). This system is commonly produced in callus cultures, but can be produced directly from the explant. Organogenesis from callus and explants became possible because of discovery of auxin by F. Went and K.V. Thiman (1939) and subsequently kinetin from DNA hydrolysate by F. Skoog in 1955.

The earliest reports on controlled organogenesis *in vitro* were by White 1939 who obtained shoots on callus of a tobacco hybrid and by Nobecourt 1939, who observed root formation in carrot callus. The finding of White was confirmed and extended by Skoog in 1944, who showed that auxin could stimulate rooting and inhibit shoot formation. Further studies of Skoog and co-workers conclusively established that a balanced combination of auxin and cytokinin controls the root and shoot formation (Skoog and Miller, 1957). Studies with tobacco callus showed that a high ratio of auxin to cytokinin in the medium favoured root formation, a reverse ratio having high cytokinin favours shoot formation and that intermediate ratios promoted callus formation. As an outcome of this approach, several hundred-plant species have been reported to form shoot and/or root *in vitro*.

The earliest report on controlled somatic embryogenesis *in vitro* was with carrot reported simultaneously in 1958 by J. Reinert and F.C. Steward. The process occurs naturally in a wide range of species from both reproductive and somatic tissues. Somatic embryos can be formed on callus, in cell suspensions and protoplasts cultures, or directly from cells of organized structures, such as stem segments or zygotic embryo.

There are several advantages of plantlet regeneration through plant biotechnological methods using organogenesis or embryogenesis, as compared to conventional methods of propagation. The advantages include the efficiency of process (the formation of plantlet in fewer steps, simultaneously with reduction in labor, time and cost), the potential for the production of much higher number of plantlets and the morphological and cytological uniformity of the plantlets. Today about 150 species from angiosperms and gymnosperms have been reported to produce somatic embryos in culture.

The first plantlet formation *in vitro* was reported as early as mid 1940s; Ernest Ball (1946) reported in *Tropaeolum* and *Lupinus*. It was only during 1960s that Morel reported plantlets formation in orchids, which became commercially viable programme. Today this use of tissue

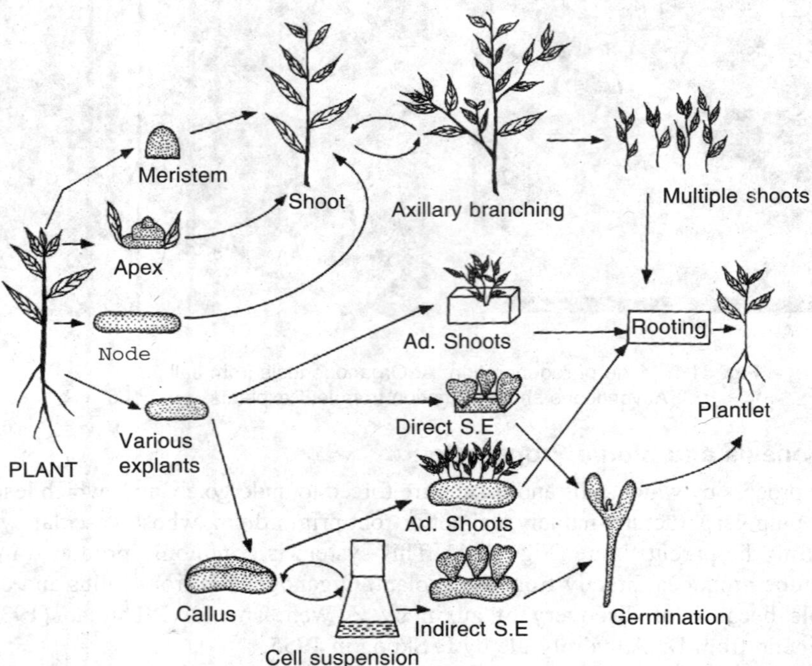

Fig. 31.2. Different methods of regeneration from explant and callus/cell cultures.

culture technology is its widest application, with over 300 commercial operations. The plant multiplied through tissue culture includes ornamental food crops, vegetables, spices, fruits and nut crops, medicinal plants and forest trees. There are four ways by which micropropagation can be achieved (Fig. 31.2).

Somatic embryogenesis leads to the production of bipolar structure containing a root/shoot axis, with a closed independent vascular system. Somatic embryogenesis can be produced directly from the explant (termed as direct embryogenesis) or can be produced via callus formation from explant (indirect embryogenesis).

Axillary budding ensures genetic stability, the production of plants from axillary buds has proved to be the most reliable method of *in vitro* propagation. Yielding about 90% of the current production it is now by far the most widely used multiplication technique.

Shoots and subsequently plantlets formation from axillary bud (pre-existing meristem) ensures genetic fidelity and termed as '**clonal propagation**' (clone constitutes vegetatively derived individuals from a plant, e.g., by cuttings, explants). The plantlets produced by this method constitute the clone of the parent plant from which they were derived.

Adventitious budding in response to external stimulus, pre-existing meristem as well as meristematic cells of the explant produce shoot buds. Shoot buds originating from other then existing meristems are known as adventitious buds such as those produced by stem cortex, petiole. In highly proliferating explants, adventitious shoots are produced giving a gregarious appearance to the explant.

Table 31.1. Selected examples of regeneration from different explants and cultures.

Root culture • *Aegle marmelos* • *Albizzia lebbeck* • *Dalbergia sissoo* **Tree species from callus cultures** • *Betula pendula* • *Citrus grandis* • *Citrus sinensis* • *Populus deltoides* • *Tecomella undulata* **Stem culture** • *Eucalyptus camaldulensis* • *Psidum guajava* • *Syzygium cuminii* • *Rose hybrida* • *Camellia sinensis* • *Tecomella undulata* • *Ziziphus mauritiana* **Endosperm culture** • *Dendrophthoe falcata*	**Bud culture** • *Atropa belladonna* • *Chrysanthemum monifolium* • *Dalbergia latifolia* • *Morus indica* • *Tectona grandis* • *Terminalia bellerica* **Leaf culture** • *Curculigo orchioides* • *Dioscorea floribunda* • *Rauwolfia serpentina* • *Petunia hybrida* • *Solanum melongena* • *Artemisia annua* • *Saccharum officinarum* **Flower culture** • *Rannunculus scleratus* • *Tagetes erecta* • *Utricularia inflexa* **Embryo culture** • *Hordeum vulgare*	• *Coccus nucifera* • *Arachis hypogaea* • *Costus speciosus* • *Podophyllum hexandrum* **Seed and seedling callus** • *Capsicum* spp. • *Cajanus* cajan • *Vigna mungo* • *Coptis teeta* • *Brassica juncea* • *Azadirachta indica* • *Carthamus tinctorium* **Inflorescence culture** • *Pennisetum americanum* • *Sorghum almum* • *Triticum aestivum* • *Triticale* • *Zea mays* • *Musa* sp. • *Brassica oleracea var. botrytis*

1.2. Regulation of Regeneration

Using a suitable explant and a proper medium callus or explants can be induced to form organ or embryos. Both embryogenesis and organogenesis may occur in the same cultures e.g. in pumpkin. But usually one or the other pattern of organization is produced. Similarly either root or shoot is formed depending upon the growth regulators. The manifestation of organogenesis or embryogenesis in the culture tends to be species specific. Plant growth regulators play a most important role in these processes.

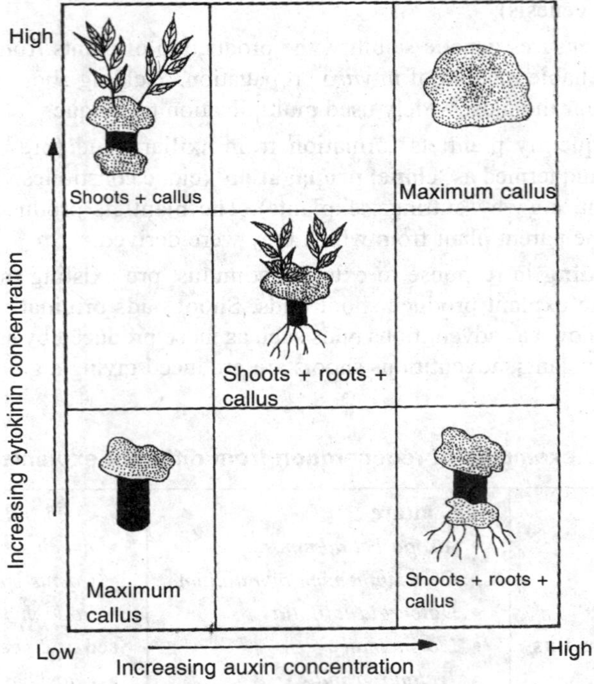

Fig. 31.3. Effect of a combination of auxin and cytokinin ratio.

Table 31.2. Plants micropropagated through tissue culture.

Ornamentals	Wood species	Vegetables, crop species
Anthurium	*Araucaria*	*Actidinia*
Bromeliads	*Betula*	*Allium*
Cordyline	*Coffea*	*Apium*
Cymbidium	*Eucalyptus*	*Asparagus*
Dianthus	*Hevea*	*Beta*
Draceana	*Malus*	*Brassica*
Syngonium	*Prunus*	Cardamom
Chrysanthemum	Cryptomeria	*Musa*
Gladiolus	*Populus*	*Trifolium*
Gerbera	*Pyrus*	
Pelargonium	Rose	
Tulipa	Salix	
Orchids (many)	*Santalum*	
Rhododendron	*Sequeoia*	
Saintpaulia	*Tectona*	
Spathiphyllum	*Vitis*	

Many growth active substances, plant growth regulators as well as other nutrients, have been included in the culture medium to regulate organogenesis *in vitro*. As stated earlier, most of the plant species respond to a suitable auxins and cytokinin balance by forming shoots and roots (Fig. 31.3). Evans and co-workers (1981) found that for 75% of the species forming shoots, either kinetin or BA was used in a concentration range of 0.05-46 m. Auxins such as IAA and NAA were used in concentrations of 0.06-27 μm. Graminaceous species tend to have a lower requirement for cytokinin for shoot formation than other species. IBA and NAA are the most commonly used auxins for rooting. In some case a mixture of two cytokinins or two auxins has proven superior to a single cytokinin or auxin. Among the auxins, 2,4-D promotes cell proliferation and suppresses cellular and organ differentiation in dicots. A large number of plants micropropagated through tissue culture are presented in Table 31.2.

2. SOMATIC EMBRYOGENESIS

Plant cells are totipotent and can produce whole new plants under favourable conditions of nutrients and plant growth regulators. Steward and co-workers in U.S.A. and Reneirt in Germany almost simultaneously reported for the first time somatic embryo formation in carrot cell suspension cultures in 1958-59. These somatic embryos (called somatic embryos because originate from somatic cells instead of gametic fusion which forms zygotic embryo) were similar to zygotic embryos in development and structure (Fig. 31.4). Although the origin of somatic embryos produced in carrot cell suspension cultures was uncertain, these somatic embryos morphologically

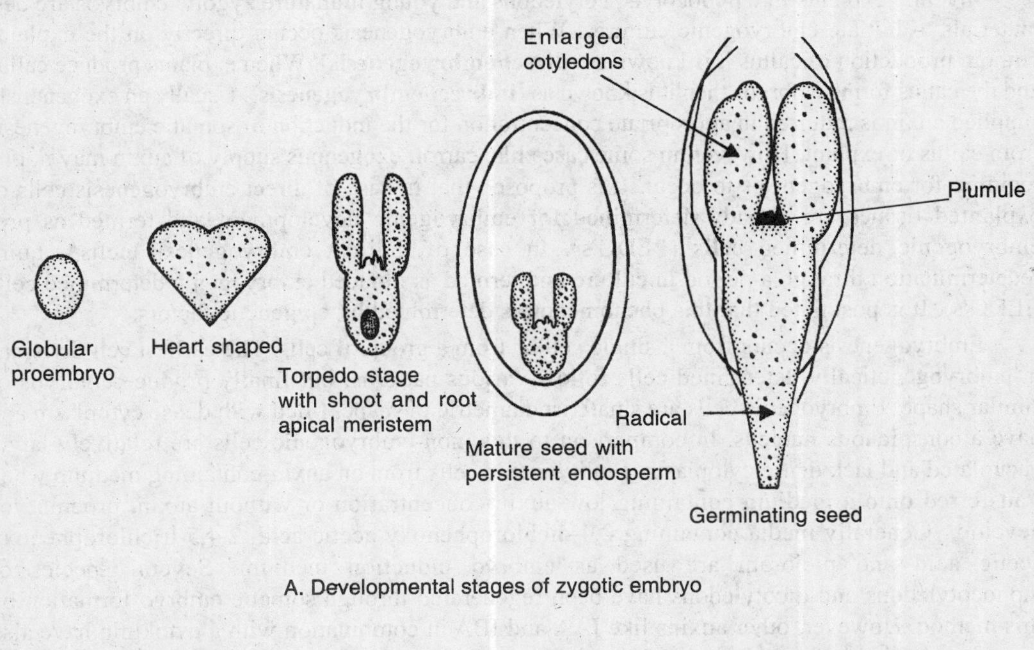

Fig. 31.4. Development of zygotic and somatic embryos.

developed through the globular, heart and torpedo stages. The emergence of cotyledons then could be used to distinguish between mature cotyledonary stage and heart/torpedo shaped embryo. At torpedo stage cell differentiation occurs establishing root and shoots meristem. Initially it was observed that these embryo originated from cell aggregates, however, McWilliam *et al.* (1974) reported that the "young embryoids have always a clearly defined outline indicating discontinuity with the surrounding cells". The aggregates are found because cells fail to separate following cell divisions.

After the initial observation of somatic embryos, there were considerable efforts devoted to define the medium conditions, which permitted embryogenesis, and characterising the morphological and developmental events leading to embryo formation (Fig. 31.5, 6). Embryogenesis is a two step process. The first stage is the induction of embryogenesis and the second step is the development of embryo, ultimately leading to germination. The requirements for embryo induction and embryo development are different, and thus separate media are used for each step. Ammirato (1987) described four stages viz., induction, early growth, embryo maturation and germination of conversion. These stages not only differ in their morphological structure but also in their physico-chemical requirements. Carrot somatic embryogenesis is the most extensively studied system for various aspects of somatic embryogenesis.

Induction - From place of origin point of view embryogenesis is of two types; **direct embryogenesis** and **indirect embryogenesis**.

Juvenile explants like hypocotyls, cotyledons and young immature zygotic embryos are best materials to initiate embryogenic cultures. When embryogenesis occurs directly on the explants without production of callus it is known as '**direct embryogenesis**'. When explants produce callus and the callus forms embryos then it is known as '**indirect embryogenesis**'. Usually an exogenously supplied auxin is required in appropriate concentration for the induction of somatic embryogenesis from callus or explant. However, in some cases like carrot, exogenous supply of auxin may not be required for embryogenesis to occur. It is proposed that in case of direct embryogenesis cells of explanted tissues are already determined for embryogenic development and termed as pre-embryogenic determined cells (PEDC's). In case of indirect embryogenesis cells require redetermination through a period in culture and termed as induced embryogenic determined cells (IEDC's). It is postulated that this phenomenon is determined by epigenetic factors.

Embryogenesis occurs from a single cell or from a group of cells. The earliest cell divisions in embryogenetically determined cells follow various patterns, but finally produce embryos of similar shape. Embryogenic cells are small, isodiametric in shape, filled with dense cytoplasm and have a conspicuous nucleus. In comparison to this, non-embryogenic cells are relatively large, vacuolated and lack dense cytoplasm. Embryogenic cells from an auxin-containing medium when transferred onto a medium containing low auxin concentration or without auxin, proembryos develop. Generally media containing 2,4-dichlorophenoxy acetic acid, 2,4,5-trichlorophenoxy acetic acid and picloram are used as embryo induction medium. Several species of monocotyledons and dicotyledons have been regenerated through somatic embryo formation by this method. However, other auxins like IAA and IBA in combination with a cytokinin have also been found suitable for embryo induction in many dicotyledonous species. This method is useful for continuously obtaining embryos in large number e.g., in *Atropa belladona, Carrot, Ranunculus, Pennisetum purpurium* and *Pannicum maximum.*

Development- During somatic embryogenesis in cell suspension cultures embryos of different sizes are produced. The development and maturation of somatic embryos is similar to zygotic embryos. For any experimental or micropropagation method, embryos of uniform size are required. This can be achieved by sieving or fractionation of suspension with appropriate sieve size. Such cultures may be fully synchronised for their growth.

Embryogenesis as a model of developmental regulation- With the establishment of somatic embryogenesis in synchronous cell culture, the system is used as a model to study organ differentiation and regulation.

1. Callus specific and embryo specific proteins have been identified. It was observed that embryo specific genes were expressed long before the morphologically visible differentiation.
2. A glycoprotein was released into the medium when embryogenic carrot cell aggregates were transferred into a 2,4-D medium and embryo development was initiated. A second glycoprotein was released into the medium from non-embryogenic tissues.
3. Rapid increases in the rates of protein and RNA synthesis after transfer embryogenically competent carrot cells to an auxin free medium were recorded.
4. Polyamine synthesis and its role in embryogenesis suggested embryogenic cultures had higher level of polyamines, inhibitors of polyamines biosynthesis could suppress embryogenesis whereas the addition of the polyamine spermidine could reverse this inhibition and levels of enzymes for polyamine synthesis were higher under embryogenic conditions.

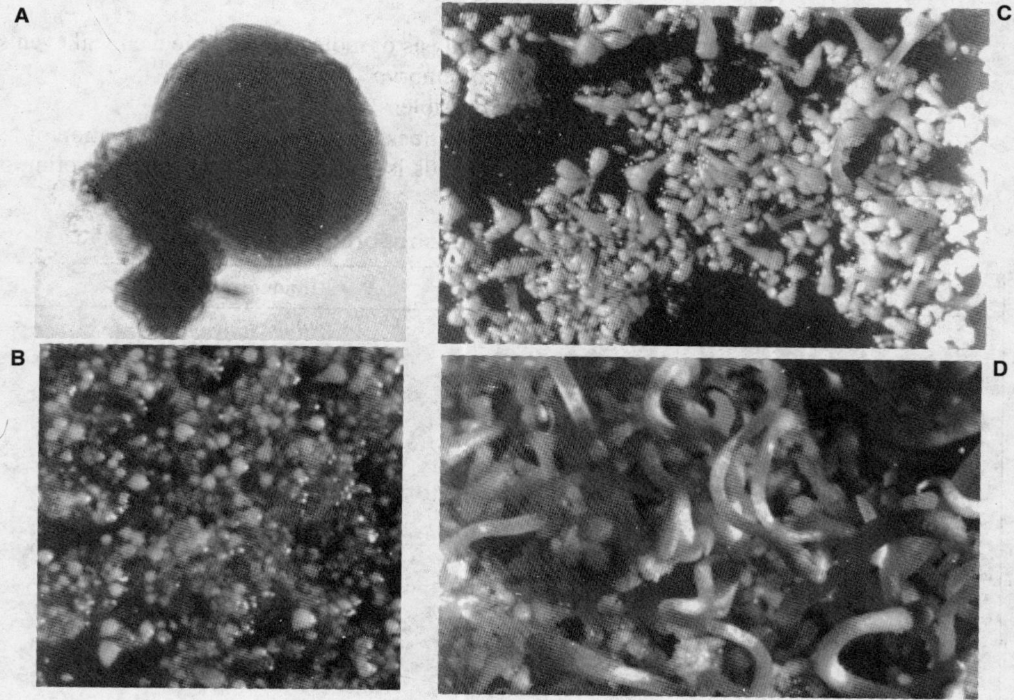

Fig. 31.5. Somatic embryogenesis in carrot: under microscope globular stage (A), macroscopic globular (B), torpedo and heart shaped (C) and germinated somatic embryos (D).

Applications- Following applications are suggested:

1. Somatic embryogenesis provides potential plantlets in the form of somatic seeds. These are available in enormous number once the method is established. These somatic embryos can be used for the production of synthetic seeds for direct sowing in the field. Synthetic seeds are produced by encapsulation in agarose or calcium alginate as described latter in this book (Chapter 19).

2. Somatic embryo provides organized culture system comparable to intact plant in sterile liquid cultures. Such cultures produce organ specific or differentiation related compounds in higher amounts compared to cell culture of that species, e.g., as studied in *Digitalis* and *Theobroma cacao*.

Advantages- Somatic embryo is a versatile technique for micropropagation of plant species. A large number of herbaceous dicots and monocots have been regenerated through somatic embryogenesis (Table 31.3). However, woody trees are still difficult materials to regenerate via somatic embryogenesis. This method of micropropagation offers several advantages over organogenesis and clonal propagation through explants.

1. Rapid multiplication through production of somatic embryogenesis in cell cultures and use of bioreactors for scale-up technology.
2. Presence of bipolar structures in the same unit (presence of both root and shoot) avoids the rooting step required in organogenesis.
3. Somatic embryos grow individually making the system easy to manipulate (to sub-culture) and develop scaling-up methods.
4. It is possible to induce dormancy and store the culture for long duration.
5. Possibilities of encapsulation and other methods of packing and direct delivery system can be employed.
6. Provides an important resource for the analysis of molecular and biochemical events that occur during induction and maturation of embryo.
7. Isolation of specific storage protein is possible.
8. It shortens the breeding cycle of deciduous trees and increases the germination of hybrid embryos where delayed germination of seeds is a significant handicap in rooting of the plants of horticultural interest.

Table 31.3. Somatic embryogenesis reported in tissue culture.

Dicotyledons	Monocotyledons
Albizzia lebbeck	*Allium sativum*
Atropa belladonna	*Asparagus officinalis*
Brassica oleracea	*Chlorophytum*
Carica papaya	*Avena sativa*
Citrus spp.	*Bambusa arundinacea*
Coffea arabica	*Hordeum vulgare*
Dancus carota	*Oryza sativa*
Solanum melangena	*Panicum maxicum*
Theobroma cacao	*Pennisetum americanum*
Vitis vinifera	*Saccharum officinarum*
	Secale cereale
	Sorghum bicolor
	Zea mays

3. HAPLOIDS

Haploid plants have the gametophytic number of chromosomes, a single set (n) of chromosomes in the sporophyte. These are of great significance for plant improvement, useful in mutations and for the production of homozygous plants. Conventional breeding programmes are tedious and are not very efficient for the production of homozygous lines or haploids. These methods are not very successful in cross-pollinated crops for obtaining homozygous plants. Use of anther and pollen cultures for the induction of haploids by Guha and Maheshwari in 1964 at Delhi University opened new avenues for obtaining homozygous plants in short duration and in large number for breeding programmes.

In 1953 Tulecke, for the first time observed that mature pollen grains of a gymnosperm *Ginkgo biloba* can be induced to proliferate in culture to form haploid callus. This work demonstrated that haploid callus can be produced from highly differentiated cells like pollen instead of normal process of pollen germination. It was not until 1964 that Sipra Guha and Satish C. Maheshwari reported the direct development of embryoids from mirospores of *Datura innoxia* by the culture of excised anther. Later Bourgin and Nitsch in France (1967) obtained complete haploid plants of *Nicotiana tabacum*. Since then the production of haploids has become a major area of research interest and haploid cultures have been produced in a large number of species of diverse taxa.

Plant tissue culture technique for obtaining haploid cultures is simple and efficient. Closed flower buds, which have anthers containing uninucleate microspores, are most suitable for the induction of androgenosis. Excised flower buds are sterilized like other explants and cut open longitudinally with the help of scalpel or can be opened with a pair of fine forceps and anthers are transferred on the medium. Before transfer filaments should be removed from the anthers. Anthers should not be damaged during excision and transfer. Damaged anthers should be discarded, as they often tend to produce callus from parts other than pollen. Anthers can be grown on solid or in liquid medium. The cultures are incubated at 24-27 °C and under illumination of about 2000 lux for 12-16 hours/day.

Depending upon the plant species it takes about 1-8 weeks for pollen plantlets to emerge from the anthers. Plantlets of appropriate size can be transferred directly into pots after necessary washing to remove agar. All the care should be taken as desirable for young micropropagated plants.

Haploid plants can be diplodized to produce homozygous plants by two methods:

1. By treating the young plantlets for 24-48 hours with 0.5% aqueous colchicine solution while still attached to the anther. After washing they are transplanted (Colchicine is a neutral alkaloid obtained from *Colchicum antumnale*). Mature plants are treated with colchicine lanolin paste (0.4%) to obtain diploid shoots. This leads to production of homozygous seeds by selfing or diploid callus can be obtained from such branches.
2. Haploid stem explants are cultured. Such cultures tend to become diploid by endomitosis. Diploid homozygous plants are obtained from such cultures.

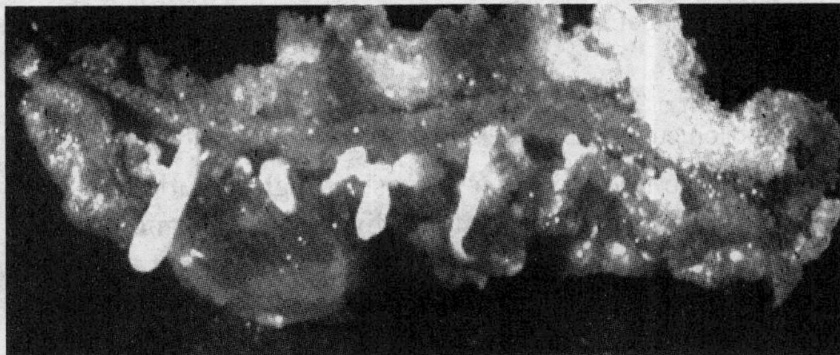

Fig. 31.6. Anther culture and production of haploids; (A) Haploid embryos formed from pollens in an anther; (B) Haploid plantlets formation from germinating embryos.

Androgenesis - There are several pathways of androgenesis which leads to the formation of haploids either directly by embryogenesis or indirectly through callus formation (Fig. 31.6,7). Formation of somatic embryos directly from pollen termed as direct androgenesis while formation embryos from pollen callus is termed as indirect androgenesis. From breeding point of view direct embryogenesis is preferred over indirect embryogenesis involving callus formation which may produce variation. There are several hundred plant species in which haploid callus or complete plantles have been produced. A few selected species are listed in the Table 31.4.

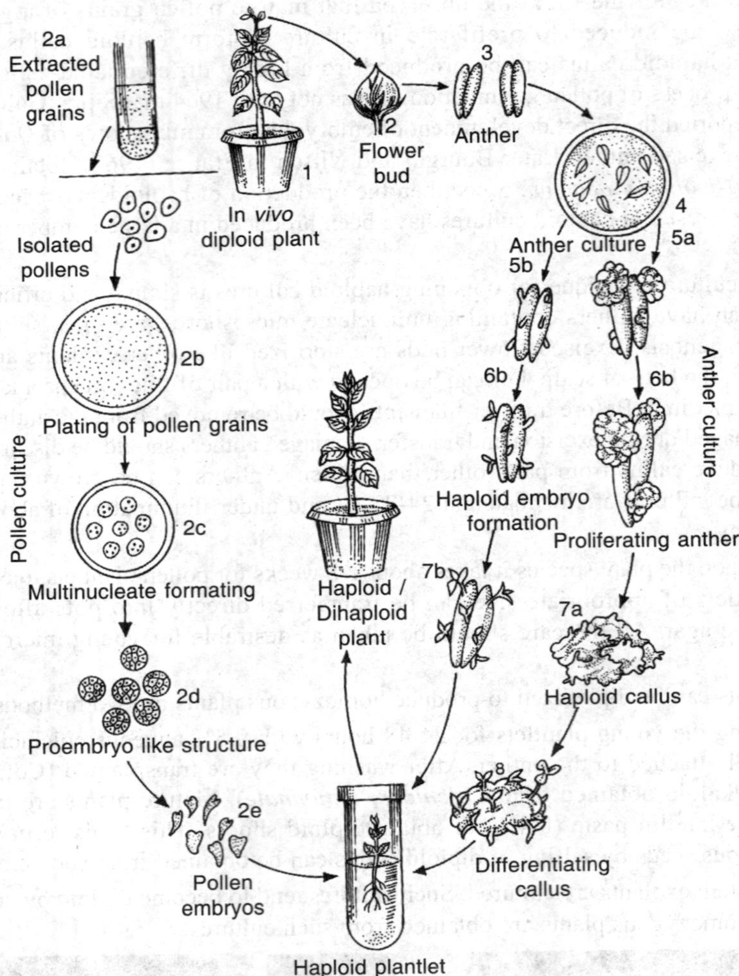

Fig. 31.7. Schematic representation for production of haploid and dihaploid plants through anther and pollen culture.

In nature, pollen mother cells (P.M.C.) give rise to pollen tetrads through meiotic division. These microspores are released in the atmosphere. The first mitotic division in the microspores produces a large vegetative cell and small generative cell. The former remains quiescent while the later divides to form two male nuclei or sperms. In culture although androgenesis can be induced in anthers in the tetrad stage at the binuclear pollen stage, but microspores just before or at the time of first mitosis, are most suitable at the induction of androgenesis. Some of the modes of androgenesis are presented in Figure 31.8 on the basis of results obtained with various species, notably Datura, tobacco and barley. Androgenesis may be produced by pollen grains through mitosis producing equal or unequal nuclei. In all cases, they may give rise to somatic embryos directly or may produce callus. Direct androgenesis, behaviour of microspores as zygote leading to embryo formation, has been observed in *Atropa belladonna*, *Datura innoxia* and *Nicotiana tabacum*. Indirect androgenesis is the most common method (96%) of producing haploids. This type of development (indirect androgenesis) may be due to disturbed polarity and/or complex nutrient medium.

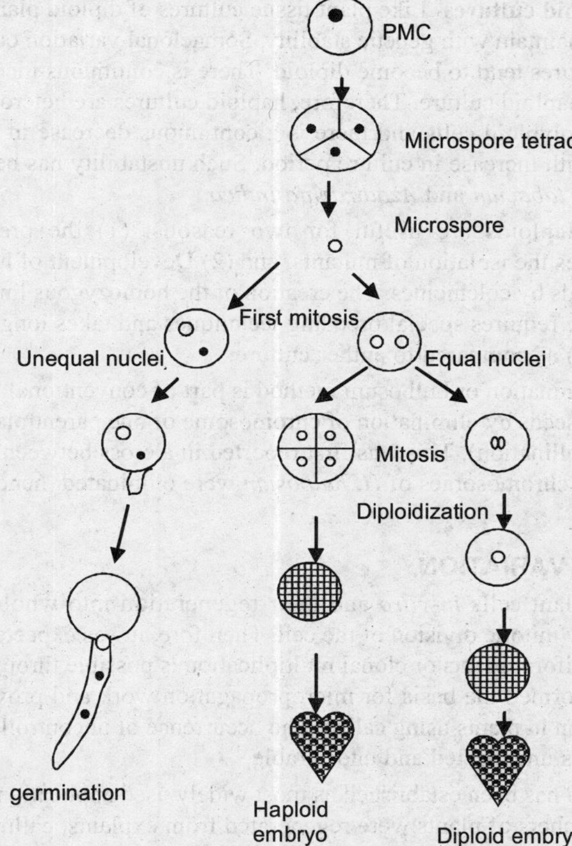

Fig. 31.8. Microsporogenesis and development of haploid or diploid embryo or normal germination of pollen grain.

Table 31.4. Some economically important species in which haploid plants and calli have been obtained by excised anther/pollen.

Plant species	Family	Inoculum	Response
Asparagus officinalis	Liliaceae	Anther	Callus, plant
Brassica oleracea × *B. alboglabra*	Cruciferae	Pollen	Callus
B. napus	Cruciferae	Anther	Callus, embryos
Capsicum annuum	Solanaceae	Anther	Callus, embryos
Coffea arabica	Rubiaceae	Anther	Callus, embryos
Hordeum vulgare	Graminae	Anther	Callus, plants.
Lycopersicon esculentum	Solanaceae	Anther	Callus, plants
Nicotiana tabacum	Solanaceae	Anther, pollen	Embryos, plants
Oryza sativa	Graminae	Anther	Callus, embryos, plants
Petunia hybrida	Solanaceae	Anther	Callus, embryos, plants.
Secale cereale	Graminae	Anther	Callus, embryos, plants
Solanum melongena	Solanaceae	Anther	Callus, plants
S. tuberosum	Solanaceae	Anther	Embryos, plants.
Triticum aestivum	Graminae	Anther	Callus, embryos, plants.
Zea mays	Graminae	Pollen	Embryos

Stability of haploid cultures- Like plant tissue cultures of diploid plants, cultures of haploid origin are difficult to maintain with genetic stability. Somaclonal variation occurs during long-term culture as well as, cultures tend to become diploid. There is continuous increase in the proportion of diploid cells in the haploid culture. Therefore, haploid cultures are heterogeneous population of haploid, diploid and polyploid cells and there is a continuous decrease in percentage of haploid cells in this mixture with increase in culture period. Such unstability has been reported in *Atropa belladonna, Nicotiana tabacum* and *Azadirachta indica.*

Applications- Haploids are useful for two reasons: (1) the presence of one set of chromosomes facilitates the isolation of mutants, and (2) Development of homozygous lines after diplodization of haploids by colchicines. The creation of the homozygous lines from which hybrids are produced, normally requires special breeding techniques and takes long time (several years in conventional breeding) as compared to anther culture.

Chromosome elimination or bulbosum method is part of conventional breeding. This method also produces haploid seeds by elimination of chromosome of one parent plant in hybrid seed after hybridization (cross pollination). This was first reported in a cross between *Hordeum vulgare* and *H. bulbosum* in which chromosomes of *H. bulbosum* were eliminated, hence known as bulbosum method.

4. SOMACLONAL VARIATION

The growth of plant cells *in-vitro* and their regeneration into whole plants is an asexual process, involving only mitotic division of the cell. Therefore, it was expected that the process will produce genetically uniform plants or clonal multiplication is possible through callus regeneration. This expectation has formed the basis for micropropagation work and provided a technical basis for genetic manipulation in plants using callus. The occurrence of uncontrolled variation during the callus regeneration was unexpected and undesirable.

Micropropagation has been established as most widely used application of plant tissue culture almost and a large numbers of plants were regenerated from explants, callus and cell cultures and lastly from protoplasts cultures. In these regenerated plants, chromosomal instability and phenotypic variations were observed. As changes in agronomic trails, were observed in crop plants, interest in exploiting this variation through plant tissue culture for the improvement in crop plants increased. In 1981, Larkin and Scowcroft, named the phenomenon of variation found in plants regenerated from cell cultures as "somaclonal variation". The term is now universally adopted for variation found or created through use of somatic cells and similar terms like 'protoclonal' and 'gametoclonal' are used for describing variation from protoplasts and anther (pollen) culture, respectively. Pollen grains are produced through meiotic division and pollens in an anther are theoretically of different types. When these pollens are cultures they produce a clones of different types termed as gametoclonal variation.

During the processof regeneration, a cell divides and redivides several hundred times to produce callus and subsequently organs. During this process of division, several internal and external factors influence the cell. This leads to creation of variations in cell culture, called somaclonal variation. This is the reason that, direct shoot formation from explant through existing meristems and meristem culture are used for clonal propagation and considered that no genetic variation occur in such system. In comparison to this, callus and cell cultures, organogenesis from cultures and indirect somatic embryogenesis are sources of somaclonal variations. Somaclonal variation is common in long term cultures.

In culture, plant cells divide under the influence of plant growth regulators. A balanced combination of plant growth regulators is required to sustain growth and division. Abnormalities in cell division can occur such as multi-polar spindle formation, bridges and fragments, laggards, micronuclei and chromosome fragmentation. These errors result in numerical and structural chromosome changes in regenerated plants. Growth regulators, temperature, light, osmolarity and agitation rate of the culture medium are all known to affect the cell cycle *in vitro* in plants. Therefore, it may be concluded that inadequate control of the cell cycle *in vitro* is one of the causes of somaclonal variation. It has been proposed that the various mutational events are directly or indirectly related to the modification of DNA, specially DNA hypo/hyper methylation. It is clear that base modifications and chromatic structural changes are responsible physiological and phenotypic changes associated with somaclonal variation.

Improved plants through somaclones - Improved plants have been produced in many crop varieties by exploitation of somaclonal variation, similar to that produced by conventional breeding programme. It takes about 5 years to grow and test the improved material in different localities. There are a few varieties already released in the market developed through somaclonal variation, e.g., ultra-crisp celery and sweeter carrots produced by DNA plant technology in the USA; tomato with higher solid content, altered pigment; potato with disease resistance, improved cooking quality, and variants of strawberry, lettuce and maize. 'Somaclonal snowstorm' as its name suggests, is a new variety of Paulownia tomentosa derived by somaclonal variation by Marcotrigiano and Jagnnathan in 1988.

QUESTIONS

1. Write short notes on:
   (a) Organogenesis
   (b) Somatic embryogenesis
   (c) Regulation of embryogenesis
2. Describe the various methods of achieving regeneration in vitro and the factors affecting regeneration.
3. Write short notes:
   (a) Stages of embryo development.
   (b) Bud breaking.
   (c) Advantages and disadvantages of micropropagation.
4. What do you understand by somaclonal variation? Discuss its importance in crop improvement.
5. What is somatic embryogenesis? Describe various stages of somatic embryogenesis and factors influencing somatic embryos formation with the help of suitable diagrams.
7. Write short notes on:
   (a) Haploids
   (b) Anther culture
   (c) Homozygous lines
   (d) Androgenesis
8. Describe the technique and applications of anther and pollen culture for the production of homozygous lines.

CHAPTER 32

BIOSENSORS

BIOSENSORS

Biosensors make use of biological components in order to sense a compound of interest. On the other side, chemical sensors not using a biological component but placed in a biological matrix *are not biosensors* by definition. Enzymes are still indispensable tools in biosensing. An algal biosensor using alkaline phosphatase for determination of heavy metals was reported. It exploits the inhibitory action of heavy metals on the alkaline phosphatase present in the membrane of *Chlorella vulgaris* microalgae. The microalgal cells were immobilized on removable membranes placed in front of the tip of an optical fiber bundle. This section comprises sensors for biotechnological, industrial, environmental, food, pharmaceutical, medical, and related applications. Optical sensors for oxygen have been used for noninvasive analysis of dissolved oxygen in shake flasks for cell cultures. The oxygen-sensitive element is a thin, luminescent patch affixed to the inside bottom of the flask. Both intensity and decay time may be measured; the latter said to be more reliable, particularly with respect to repeated sterilization. Mass-transfer coefficients are reported as well. The sensors are fast, do not consume oxygen, and are affordable.

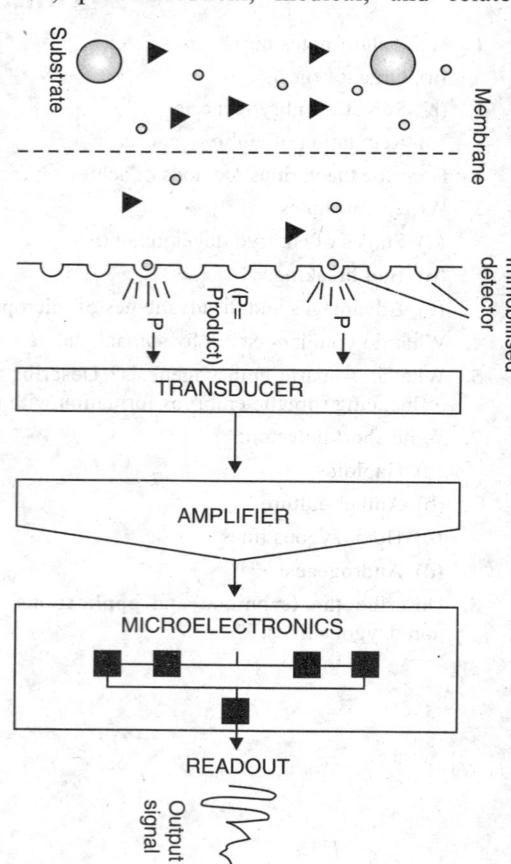

Fig 32.1 Principle of biosensor

Principle

Biosensors are a combination of the biological components with electronic gadgets or transducer to make direct transduction and measurement of signals. Any change in the response of biological components, which in turn is dependent on the environment, is recorded as a change in the signal. The sample is allowed to pass through a membrane so that the selection may be exercised and the interfering molecules are retained outside the membrane. The sample then interacts with the biological sensor and forms a product which may be an electric current/charge, heat, gas or a suitable chemical. The product then passes through another membrane and reaches the transducer, which converts the biological signal into an electric signal that can be further amplified and can be either read or on digital panel or can be recorded over recorders. This allows the measurement of the concentration of the substrate without processing and thus avoiding consumption of the sample. The selectivity, sensitivity and stability of the biosensors are determined by the biological components. The latter are generally enzymes, antibodies, organelle, cells and microorganisms. The physicochemical transductors are electrodes, thermisters, transisters, and optical instruments (spectrometer, fluorimeter, etc.).

The biosensors can be used for environmental analysis: in two ways (i) the application in continuous measurement systems for monitoring and control of processes, and *(ii)* the on-field analysis by portable equipment and biosensors. Table 1 summarises some biosensors and relevant biocomponents for the detection and monitoring of environmental toxicants.

Table 1 Biological and physical components of some biosensors and their applications

Biological sensor	Transducer device (physical device)	Substance or parameter measured
I. Enzymes		
Glucose oxidase	Oxygen electrode	Glucose
NADH and dehydrogenase	Redox electrode	Ethanol
Creatinine iminohydrolase	NH_3 field effect transistor (FET)	Creatinine
Cholinestase	Chem FET	Nerve gas
Labelled hCG Catalase	Oxygen electrode	Human chorionic gonadotropin (hCG)
II. Receptors		
ACH receptor	Conductimeter	Nerve gas
III. Antibodies		
Antibody to parathion	Piezoelectric crystal	Parathion
Antibody to ¡—globulin	Polarized light	¡—globulin
IV. Microbial cells		
Pseudomonas fluorescens	Oxygen electrode	Glucose
Azobacteria vinelandi	NH_3 electrode	Nitrate
Methylomoans flagella	Oxygen electrode	Methane
Bacillus subtilis	Oxygen electrode	Mutagen screening
Citrolacter fluendli	pH electrode	Cephalosporin

ENZYMATIC BIOSENSORS

Because of the high specificity of enzymes, enzyme biosensors are very useful for the detection of individual substances. In most of the cases, the effect is measured in terms of enzyme catalysis i.e. degradation rate of the specific substrate or production rate of the reaction product. The urease, peroxidase and tyrosinase sensors have been proved useful to detect and, in some cases, quantify the specific substrates. The tyrosine sensor (tyrosine-electrode) effectively quantifies phenol through sensing the product. In the presence of oxygen, phenol is oxidised to polymer products. The sensitivity of such sensors can be substantially increased by the measurement of a soluble intermediate reaction product, instead of quantifying the end-product. Wollenberger *et al.* (1992) developed a multi-enzyme system for quantification of inorganic phosphates. Such systems are effective, especially for quantification and monitoring of electrochemically inactive reaction products. A biologically active membrane, used by Wallenberger and his co-workers, consisted of nucleoside -phosphorylase and xanthinoxidase enzyme sequence. The former catalyses inosine-phosphorylation reaction, that utilises inorganic phosphate, and subsequently the phosphorylated intermediate is converted to hypoxanthin by xanthoxidase. The rate of phosphorylation vis-a-vis hypoxanthin production is determined by the concentration of inorganic phosphate. Such a sensor has been found useful to detect inorganic phosphate within the range of 0.5-500 mmol / l. In order to increase the sensitivity of the measurement, a xanthin sensor containing co-immobilised alkaline phosphatase and nucleoside-phosphorylase was integrated. It could increase the sensitivity of the sensor, for about 20 times that of the initial sensor.

PREGNANCY TEST SENSOR

A **pregnancy test** attempts to determine whether a woman is pregnant. Markers that indicate pregnancy are found in urine and blood, and pregnancy tests require sampling one of these substances. The first of these markers to be discovered, human chorionic gonadotropin (hCG), was discovered in 1930 to be produced by the trophoblast cells of the fertilised ova (eggs). While hCG is a reliable marker of pregnancy, it cannot be detected until after implantation. this results in false negatives if the test is performed during the very early stages of pregnancy. the first home test kit for hCG was invented in 1968 by Margaret Crane in New York. She was granted two U.S. patents. The kits went on the market in the United States and Europe in the mid-1970s.Pregnancy test is a qualitative test giving positive or negative indication. Urine containing hormone react to a mediator and then to antibody giving colour. If there is no hormone, no colour is produced. Improper usage may cause both false negatives and false positive.

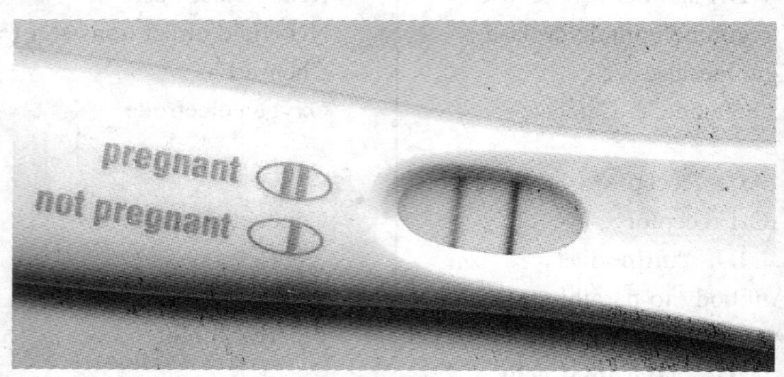

Fig 32.2 Pregnancy test sensor

Glucose Biosensor-glucometer

Diabetes mellitus is the most common endocrine disorder of carbohydrate metabolism. Worldwide, it is a leading cause of morbidity and mortality and a major health problem for most developed societies. The prevalence of diabetes continues to increase. Self-monitoring of blood glucose (SMBG) has been established as a valuable tool for the management of diabetes. The goal of SMBG is to help the patient achieve and maintain normal blood glucose concentrations in order to delay or even prevent the progression of complications. Due to such recommendations for maintaining normal blood glucose levels, a series of suitable glucose-measuring devices have been developed. Biosensor technology has developed rapidly and can play a key role providing a powerful analytical tool.

Today's biosensor market is dominated by glucose biosensors (Fig 32.3). In 2004, glucose biosensors accounted for approximately 85% of the world market for biosensors, which had been estimated to be around $5 billion USD. Generally, glucose measurements are based on interactions with one of three enzymes: hexokinase, glucose oxidase (GOx) or glucose-1-dehydrogenase (GDH). Glucose biosensors for SMBG are usually based on the two enzyme families, GOx and

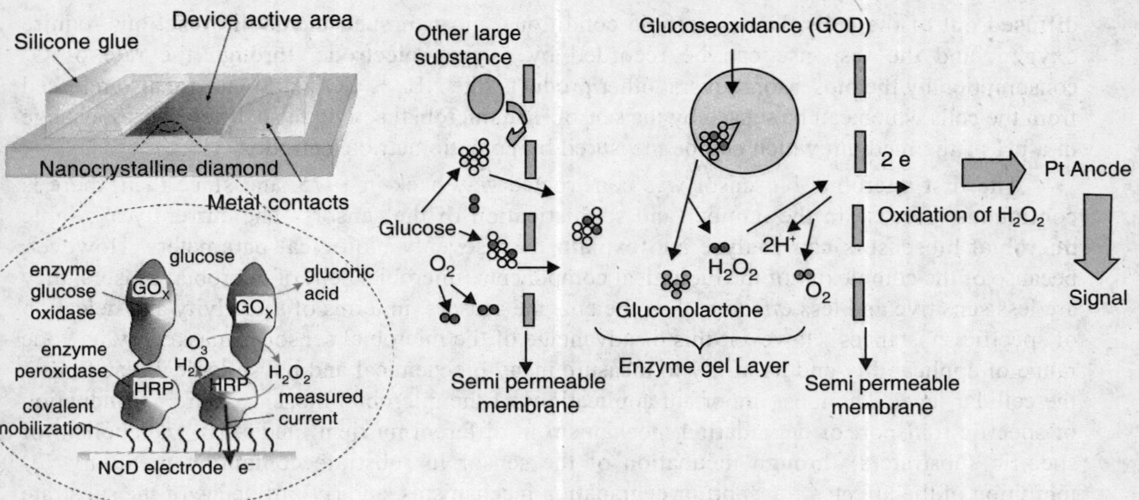

Fig 32.3 a. Mechanism of glucose biosensor and b. over all working of a glucose biosensor

GDH. These enzymes differ in redox potentials, cofactors, turnover rate and selectivity for glucose. GOx is the standard enzyme for biosensors; it has a relatively higher selectivity for glucose. GOx is easy to obtain, cheap, and can withstand greater extremes of pH, ionic strength, and temperature than many other enzymes. The basic concept of the glucose biosensor is based on the fact that the immobilized GOx catalyzes the oxidation of â-D-glucose by molecular oxygen producing gluconic acid and hydrogen peroxide. In order to work as a catalyst, GOx requires a redox cofactor Ã ('Ã flavin adenine dinucleotide (FAD). FAD works as the initial electron acceptor and is reduced to $FADH_2$. The cofactor is regenerated by reacting with oxygen, leading to the formation of hydrogen peroxides. Hydrogen peroxide is oxidized at a catalytic, classically platinum (Pt) anode. The electrode easily recognizes the number of electron transfers, and this electron flow is proportional to the number of glucose molecules present in blood.

For analysis of the carbamates and organophosphates, the acetyl- or butyl-cholinesterase biosensors are highly effective. These enzyme sensors, not only measure the effect, but are also effective in determining the toxicants concentration in the medium. The simplest and easily made sensor consists of a membrane containing the cholinesterase enzyme, immobilised on the membrane and connected to an electrode, which can electrochemically oxidise the thiocholine produced by the enzyme reaction and measure the activity in terms of the O_Z consumed. Another way of measuring the activity of the enzyme, in the presence of the toxicants, is through use of the artificial substrate fluoriscein diacetate (FDA) and quantification of the activity, fluorimetrically, by the measurement of the fluorescence signal from the fluoriscein.

An alternative way of using the cholinesterase/cholineoxidase system is by the use of an extremely low concentration of acetylcholinesterase enzyme in the sensor, for the purpose of examining the presence of the inhibitors. Such sensors are activated after 1 min of incubation of the sensor in the toxicant containing medium and are allowed to run only for three minutes.

Microbial Biosensors

The microbial sensors cover a wide range of the substrates of the biological system and are connected with highly sensitive and reliable physical components. The microorganisms in the sensor utilise the substrate or the substrate mixture in the medium and absorb them to cause intracellular metabolism. Some of the products of such metabolic reactions are excreted or diffused out of the cells. Under aerobic conditions, most of such metabolic reactions require oxygen, and the response can be recorded by oxygen electrode, through the rate of O_2 consumption by the biosensor. Several other products like NH_4 +, CO_2, H_2S, etc. are also released from the cells which can be sensed by the sensor. The microbial activity also, often causes, change in a pH of the medium which can be measured by potentiometric electrode.

The first microbial biosensor was constructed way back in 1975, and since then, there is continuous increase in the number and sophistication of the sensors. The currently available microbial biosensors can analyse and explain over seventy biological parameters. However, because of the complexity of the biological components (microbial cell) of microbial sensors, they are less sensitive and less effective than the enzyme sensors, in terms of selectivity and detection of specific substances. However, this disadvantage of the microbial sensor is masked by its wide range of applicability and its utility to measure many biochemical and physiological reactions at the cellular level. The most important applications of the microbial biosensors are: (i) induction of specific transport or degradation mechanism in different nutrient media, (ii) metabolism of specific substrate(s) through incubation of the sensor in substrate containing medium, *(iii)* inhibition of the affected transport or degradation mechanisms, and (iv) influence of the substrate transport rate on specific degradation of the substrate by the sensor.

Companella and colleagues (1987) developed a microbial biosensor using *Saccharomyces cerevisiae* to measure the toxic effects of heavy metals on the cellular uptake of organic carbon. The effect was electrochemically analysed through the rate of O_Z consumption.

Because of the adaptation of microorganisms to different environmental conditions, microbes having the ability to metabolise specific environmental contaminants can be isolated and used as biological components in microbial biosensors. These sensors are especially meant for determining the biodegradability of the environmental contaminants. An example of such a sensor is the *Pseudomonas* sensor for detection and metabolism of chlorinated phenols, and the *P. diminutum*-sensor, for detection and metabolism of organ phosphorus insecticides. Similarly the cyanobacterial sensors are used for detection of herbicides, as well as for monitoring the effect of pesticides photosynthesis. Hansen (1992) prepared a cyanobacterial electrode by immobilising *Synechococcus* cells on a graphite electrode. The activity of the cyanobacterium was measured through ferricyanide conversion reaction under dark-light cycle. Fluorescence bacteria have also been used as biocomponents of the microbial biosensors and have been applied to measure the effect of chromium on luminescence.

Immunosensors

While enzyme sensors are useful to determine the effect of contaminants on enzymatic reactions and substrate metabolism, immunosensors detect the antigen-antibody reactions. Any antigen-antibody reaction can be potentially affected by environmental contaminants and, therefore, in principle, such reactions can be useful to detect the presence, as well as the concentration of the toxicant. In the last decade, numerous immunoassays have been developed and some of them are commercially available as ELISA-kit. However, the immunosensors have not yet reached the laboratories for analysing most of the environmental chemicals. The transductors include different optical instruments of different degrees of sensitivity, which can measure the decrease in antibody production and analyse the effect by comparing it with a reference value. Such sensors are usually equipped with fluorescence detection equipment and the antibodies used here are tagged with a fluorescent marker.

The construction of an electrochemical biosensor can be made by immobilising the antibodies (or antigens) in a column and connecting the biocomponent with the electronic part, which may be optical, fluorimetric or amperometric in nature. The intensity of the signal is determined by the rate of metabolism of the substrate by the enzyme. The antibodies can be marked by more than one probe that can be catalysed by more than one enzyme, and this reaction gives a more intense signal due to the production of more products. The next step in the development of electrochemical immunosensors is the spatial integration of antibody-antigen reactions through enzymatic catalysis and electrode.

The other examples of useful sensors are:

1. Oxygen gradients have been determined in engineered tissue using a fluorescent sensor spot. Optical sensor patches containing a luminescent O2-sensitive indicator were placed on the bottom of a Petri dish, and oxygen supply in a tissue of chondrocytes was monitored over a 3-week culture period. Two-dimensional pO2 images across the tissue section were acquired over the duration of the experiment.
2. Proteins in the eluate of a preparative continuous annular chromatograph were detected with a quartz fiber-optic system that measures the intrinsic absorption of two aromatic amino acids (Tyr, Trp) in proteins. Two types of optical multichannel detectors were developed. The first is a multichannel detector, and the second is a circular optic device. UV absorption is recorded at 280 nm. Calibration plots were established for a series of stock solutions of known concentrations of proteins.
3. Novel infrared optical probes have been introduced for process monitoring and analysis based on silver halide fibers. Compared to near-IR spectroscopy, for which quartz fiber probes can be applied, the application of previously used mid-IR fiber materials was restricted due to deficiencies with regard to their optical transmission and mechanical properties. Several flexible probes of different geometries were constructed that are suitable for process monitoring. Oil aerosols from natural pipelines can be detected with an optical sensor assembly.
4. Sensors for moisture and pH value in reinforced concrete were reported. The dye pyridinium-*N*-phenolate shows a moisture-dependent absorption. It was embedded in a polymer matrix. Increasing moisture causes a shortwave shift of the absorption. pH in concrete was measured with a fiber-optic sensor consisting of a pH indicator dye immobilized in a hydrophilic polymer matrix. The sensor system is reported to display long-term stability even in media of pH 12-13.
5. Rugged, low-cost diode laser sensors for water and temperature were described. These provide fast, accurate, and nonintrusive methods for monitoring species and temperature in combustors.
6. A commercial gas analyzer based on tunable diode laser spectroscopy has successfully been applied to control and monitoring applications in the waste incineration industry. The analyzer was optimized for process control in the industrial environment and provides real-time signals for water vapor, oxygen, and temperature.
7. An optical algal biosensor for heavy metals exploits the inhibitory effect of heavy metals on alkaline phosphatase. The enzyme is present in the external membrane of *C. vulgaris* microalgae. Oxygen in underground air can be analyzed with a plastic fiber sensor array. Oxygen, consumed in a lignite mine tailing affected by acid mine drainage formation, displays a strong concentration gradient to a depth of 6 m.
8. Fiber optics can be used to monitor the dissolution of drugs such as rifampicin and the concentration of adriamycin in rabbit blood.
9. The excitement about the possibility of monitoring glucose noininvasively via near-IR spectroscopy appears to have ceased since the number of respective articles has fallen dramatically. A novel glucose fiber probe was reported that consists of one central optical fiber around which several others were arranged in circle. Light was shone onto the skin

surface through the circle fibers, and scattered light reaching the central detecting fiber was collected and transmitted to the detection system. Glucose intake experiments showed that there is good correlation between the optically determined glucose levels and the values determined with blood samples, particularly if measured at around 1600 nm.

10. Breath ammonia, resulting from *Helicobacter pylori* infection, can be analyzed with an ammonia sensor. Before urea ingestion, pylori-positive subjects had significantly lower breath ammonia levels than negative subjects, but had a significantly larger increase in breath ammonia after urea ingestion.

QUESTIONS

1. What are biosensors? Give illustrated account of glucose biosensor.

2. How pregnancy is determined at an early stage?

3. What is basic principle of a biosensor? Describe different types of sensors available in the market.

4. Write short notes: a. biosensor, b. immunosensor, c. enzymatic sensor, d. microbial sensor, e. glucometer, f. pregnancy test.

CHAPTER 33

NANOTECHNOLOGY

1. INTRODUCTION

The emerging field of nano-scale engineering and science has generated considerable excitement recently. This field is focused on materials, devices, structures, and processes that occur on the scale of atoms or groups of atoms. The size regime is on the order of nanometers. Another common term used to describe activity and research in this field is Micro-Electro-Mechanical Systems (MEMS).

Microarray technology allows for the parallel synthesis of large numbers of nano-structured materials. These materials can then be tested using the same chip-based technology. These activities are focused on materials that could be useful for applications in the fields of electronics, energy conversion, energy storage, catalysis, and others. New approach enables the synthesis of hundreds to thousands of materials in parallel followed by testing of all of those materials at once, thus allowing for greater efficiency.

Although a metre is defined by the International Standards Organization as 'the length of the path travelled by light in vacuum during a time interval of 1/299 792 458 of a second' and a nanometre is by definition 10^{-9} of a metre, this does not help scientists to communicate the nanoscale to non-scientists. It is in human nature to relate sizes by reference to everyday objects, and the commonest definition of nanotechnology is in relation to the width of a human hair.

There is a commonly held belief that nanotechnology is a futuristic science. In the last 15 years over a dozen Nobel prizes have been awarded in nanotechnology, from the development of the scanning probe microscope (SPM), to the discovery of fullerenes. According to CMP Científica, over 600 companies are currently active in nanotechnology, from small venture capital backed start-ups to some of the world's largest corporations such as IBM and Samsung. Governments and corporations worldwide have spent over $4 billion into nanotechnology in the last year alone.

Many of the initial applications of nanotechnology are materials related, such as additives for plastics, nanocarbon particles for improved steels, coatings and improved catalysts for the petrochemical industry. All of these are technology based industries, but industries with multi-billion dollar markets.

2. THE NANOTECHNOLOGY INDUSTRY

Many of the companies working with nanotechnology are simply applying our knowledge of the nanoscale to existing industries, whether it is improved drug delivery mechanisms for the pharmaceutical industry, or producing nanoclay particles for the plastics industry. In fact nanotechnology is an enabling technology rather than an industry in its own right. While it is possible to buy a packet of nanotechnology, a gram of nanotubes for example, it would have no value. The real value of the nanotubes would be in their application within existing industry.

3. BUILDING ATOM BY ATOM

One of the defining moments in nanotechnology came in 1989 when Don Eigler used a SPM to spell out the letters IBM in xenon atoms. For the first time we could put atoms exactly where we wanted them, even if keeping them there at much above absolute zero proved to be a problem. While useful in aiding our understanding of the nanoworld, arranging atoms together one by one is unlikely to be of much use in industrial processes.

Zooming down to the microscale we still have far more complexity than we would like to attempt to replicate, with cells, cytoplasm, mitochondria, chromosomes, ribosomes and many other highly complex items of natural engineering. Moving closer to the nanoscale, we still have to deal with nucleic acids, nucleotides, peptides and proteins, none of which we fully understand, or expect to even have the computing power to understand in the near future.

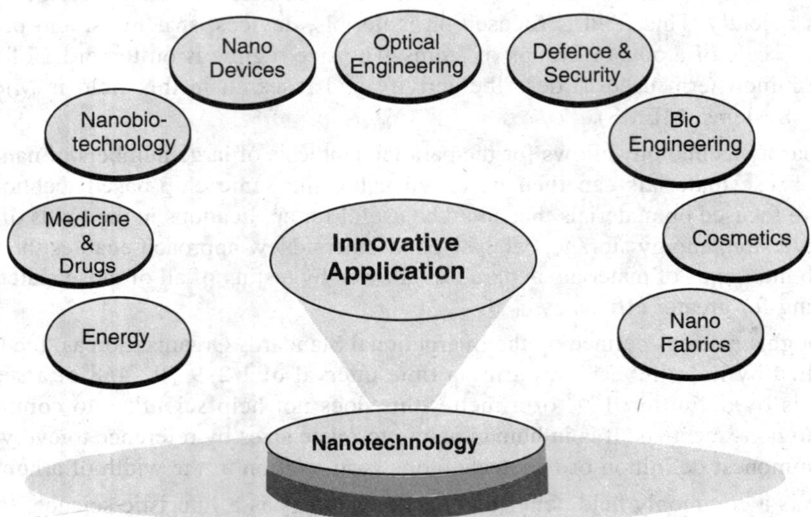

4. APPLICATIONS OF NANOTECHNOLOGY

Medicine- The biological and medical research communities have exploited the unique properties of nanomaterials for various applications (e.g., contrast agents for cell imaging and therapeutics for treating cancer). Terms such as *biomedical nanotechnology*, *bionanotechnology*, and *nanomedicine* are used to describe this hybrid field. Functionalities can be added to nanomaterials by interfacing them with biological molecules or structures. The size of nanomaterials is similar to that of most biological molecules and structures; therefore, nanomaterials can be useful for both in vivo and in vitro biomedical research and applications. Thus far, the integration of nanomaterials with biology has led to the development of diagnostic devices, contrast agents, analytical tools, physical therapy applications, and drug delivery vehicles.

Diagnostics- Nanotechnology-on-a-chip is one more dimension of lab-on-a-chip technology. Biological tests measuring the presence or activity of selected substances become quicker, more sensitive and more flexible when certain nanoscale particles are put to work as tags or labels. Magnetic nanoparticles, bound to a suitable antibody, are used to label specific molecules, structures or microorganisms. Gold nanoparticles tagged with short segments of DNA can be used for detection of genetic sequence in a sample. Multicolor optical coding for biological assays has been achieved by embedding different-sized quantum dots into polymeric microbeads. Nanopore technology for analysis of nucleic acids converts strings of nucleotides directly into electronic signatures.

Drug delivery- The overall drug consumption and side-effects can be lowered significantly by depositing the active agent in the morbid region only and in no higher dose than needed. This highly selective approach reduces costs and human suffering. An example can be found in dendrimers and nanoporous materials. They could hold small drug molecules transporting them to the desired location.

Tissue engineering- Nanotechnology can help to reproduce or to repair damaged tissue. This so called "tissue engineering" makes use of artificially stimulated cell proliferation by using suitable nanomaterial-based scaffolds and growth factors. Tissue engineering might replace today's conventional treatments like organ transplants or artificial implants. On the other hand, tissue engineering is closely related to the ethical debate on human stem cells and its ethical implications.

Chemistry and environment- Chemical catalysis and filtration techniques are two prominent examples where nanotechnology already plays a role. The synthesis provides novel materials with tailored features and chemical properties: for example, nanoparticles with a distinct chemical surrounding (ligands), or specific optical properties. In this sense, chemistry is indeed a basic nanoscience. In a short-term perspective, chemistry will provide novel "nanomaterials" and in the long run, superior processes such as "self-assembly" will enable energy and time preserving strategies. In a sense, all chemical synthesis can be understood in terms of nanotechnology, because of its ability to manufacture certain molecules. Thus, chemistry forms a base for nanotechnology providing tailor-made molecules, polymers, etcetera, as well as clusters and nanoparticles.

Catalysis- Chemical catalysis benefits especially from nanoparticles, due to the extremely large surface to volume ratio. The application potential of nanoparticles in catalysis ranges from fuel cell to catalytic converters and photocatalytic devices. Catalysis is also important for the production of chemicals.

Filtration- A strong influence of nanochemistry on waste-water treatment, air purification and energy storage devices is to be expected. Mechanical or chemical methods can be used for effective filtration techniques. One class of filtration techniques is based on the use of membranes with suitable hole sizes, whereby the liquid is pressed through the membrane. Nanoporous membranes are suitable for a mechanical filtration with extremely small pores smaller than 10 nm ("nanofiltration") and may be composed of nanotubes. Nanofiltration is mainly used for the removal of ions or the separation of different fluids. On a larger scale, the membrane filtration technique is named ultrafiltration, which works down to between 10 and 100 nm. One important field of application for ultrafiltration is medical purposes as can be found in renal dialysis. Magnetic nanoparticles offer an effective and reliable method to remove heavy metal contaminants from waste water by making use of magnetic separation techniques. Using nanoscale particles increases the efficiency to absorb the contaminants and is comparatively inexpensive compared to traditional precipitation and filtration methods.

Reduction of energy consumption- A reduction of energy consumption can be reached by better insulation systems, by the use of more efficient lighting or combustion systems, and by use of lighter and stronger materials in the transportation sector. Currently used light bulbs only convert approximately 5% of the electrical energy into light. Nanotechnological approaches like light-emitting diodes (LEDs) or quantum caged atoms (QCAs) could lead to a strong reduction of energy consumption for illumination.

Increasing the efficiency of energy production- Today's best solar cells have layers of several different semiconductors stacked together to absorb light at different energies but they still only manage to use 40 percent of the Sun's energy. Commercially available solar cells have much lower efficiencies (15-20%). Nanotechnology could help increase the efficiency of light conversion by using nanostructures with a continuum of bandgaps. The degree of efficiency of the internal combustion engine is about 30-40% at the moment. Nanotechnology could improve combustion by designing specific catalysts with maximized surface area. Scientists have recently developed tetrad-shaped nanoparticles that, when applied to a surface, instantly transform it into a solar collector.

Displays-The production of displays with low energy consumption could be accomplished using carbon nanotubes (CNT). Carbon nanotubes can be electrically conductive and due to their small diameter of several nanometers, they can be used as field emitters with extremely high efficiency for field emission displays (FED). The principle of operation resembles that of the cathode ray tube, but on a much smaller length scale.

Consumer goods- Nanotechnology is already impacting the field of consumer goods, providing products with novel functions ranging from easy-to-clean to scratch-resistant. Modern textiles are wrinkle-resistant and stain-repellent; in the mid-term clothes will become "smart", through embedded "wearable electronics". Already in use are different nanoparticles based improved products, especially in the field of cosmetics, such novel products have an excellent potential.

Household- The most prominent application of nanotechnology in the household is self-cleaning or "easy-to-clean" surfaces on ceramics or glasses. Nanoceramic particles have improved the smoothness and heat resistance of common household equipment such as the flat iron.

Optics-The first sunglasses using protective and antireflective ultrathin polymer coatings are on the market. For optics, nanotechnology also offers scratch resistant surface coatings based on nanocomposites. Nano-optics could allow for an increase in presicion of pupil repair and other types of laser eye surgery.

Textiles- The use of engineered nanofibers already makes clothes water- and stain-repellent or wrinkle-free Textiles with a nanotechnological finish can be washed less frequently and at lower temperatures. Nanotechnology has been used to integrate tiny carbon particles membrane and guarantee full-surface protection from electrostatic charges for the wearer. A military application could be in camouflage where nanocameras mixed with nanodisplays could create an "invisibility coat", acting like the skin of a chameleon. Many other applications have been developed by research institutions such as the Textiles Nanotechnology Laboratory at Cornell University

Cosmetics- One field of application is in sunscreens. The traditional chemical UV protection approach suffers from its poor long-term stability. A sunscreen based on mineral nanoparticles such as titanium dioxide offer several advantages. Titanium oxide nanoparticles have a comparable UV protection property as the bulk material, but lose the cosmetically undesirable whitening as the particle size is decreased.

Nanotechnology, like any other branch of science, is primarily concerned with understanding how nature works.

For business, nanotechnology is no different from any other technology: it will be judged on its ability to make money. This may be in the lowering of production costs by, for example, the use of more efficient or more selective catalysts in the chemicals industry, by developing new products such as novel drug delivery mechanisms or stain resistant clothing, or the creation of entirely new markets, as the understanding of polymers did for the multi-billion euro plastics industry.

'Mercury sponge' for removal of mercury

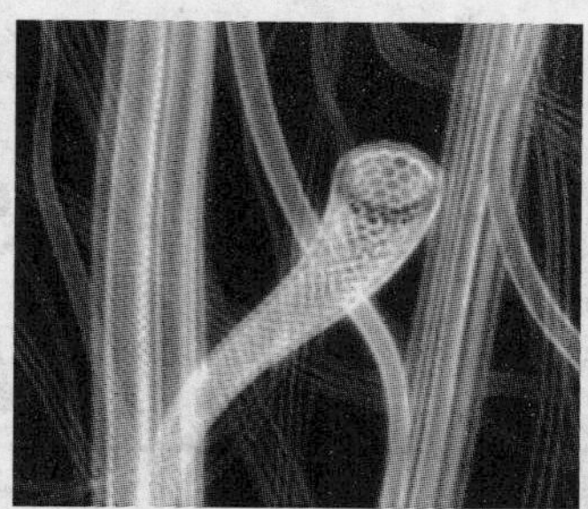
Nanotubes

microarryas

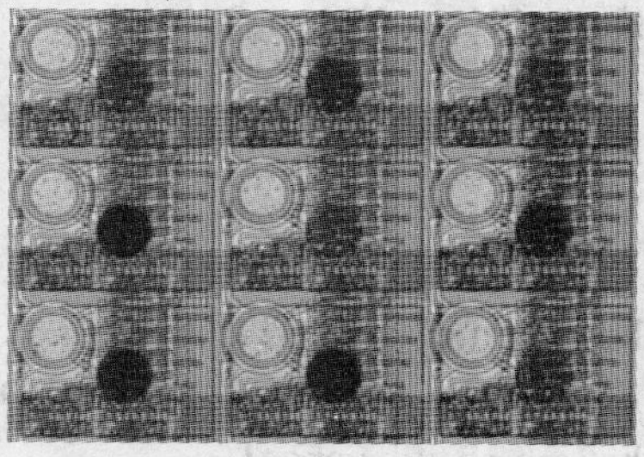
Details of microarray

In Indian context, There are only 10 companies based on nanotechnology. The government of India has earmarked 200 crores for research development and production of nanotechnology based materials. The principal companies involved in this direction are-Raymonds for use in fabric manufacture, Samsung in electrical hardwares, TATA, Mahindra, Nicolas Piramal and Intel in various fields like materials, drugs, and softwares. The major problem is deficiency of trained man power in this field.

QUESTIONS

1. Define nanotechnology. What are its applications?
2. What do you understand by Nanotechnology? Discuss its applications in industry and medicine.
3. Write short notes on: a. Nanotechnology, b. microarray, c. mercury sponge, d. nanotubes.

SHORT ANSWER QUESTION

(Also for Viva Voce)

1. What is a nucleosome?
2. What is Chargaff's rule?
3. Name proteins forming octomer.
4. What is a coding region?
5. What are introns?
6. Give three bases of initiation codon.
7. Who gave the concept of one gene one enzyme?
8. Who gave the concept of 'Central Dogma'?
9. What are pseudogenes?
10. How do you define a gene?
11. What is a gene on a DNA stretch?
12. Who gave the double helix model?
13. Does wheat flour/chapati contain DNA?
14. What are 'overlapping genes'?
15. What are 'split genes'?
16. What are 'cryptic genes'?
17. What is a perithecium?
18. How mutants are recognised in Neurospora?
19. How heavy DNA can be obtained?
20. Which enzyme is responsible for the production of cDNA ?
21. Which enzyme seals nick between two short DNA fragments?
22. What is the name of DNA fragments on lagging strand?
23. What is role of SSB proteins in DNA replication?
24. How much DNA is contents are different in a GMO/crop?
25. What is a frame shift mutation?
26. How natural recombination takes place in bacteria?
27. What is capsid?
28. What is a prophage?
29. What is horizontal gene transfer?
30. What is a genetic map?
31. What are differences between a DNA and RNA molecule?

32. What are biopolymers?
33. Give two examples of biopolymers.
34. Does the changed DNA content in a GMO make difference from nutrition point of view?
35. Who proposed the name mRNA?
36. What is ribozyme?
37. What is Wobble hypothesis?
38. What is an antisense DNA strand?
39. What is a sigma factor?
40. What is holoenzyme?
41. What is a promoter region in DNA?
42. What are consensus sequences?
43. What is TATA box?
44. What is a spliceosome?
45. Where autocatalytic splicing is involved?
46. What is inteins?
47. What is a polyribosome?
48. What is protein targeting?
49. What is an 'operon' on DNA strand?
50. What are structural genes?
51. What are down-stream and up-stream sequences?
52. How preferential utilization of a substrate is explained?
53. What is an attenuator site?
54. What is DNA alkylation?
55. Name two alkylating agents.
56. What are 'reactive oxygen species'?
57. What do you understand by 'Photoreactivation' in DNA repair?
58. Name a few free living nitrogen fixing bacteria.
59. Give components of *Nif* gene.
60. Name the enzyme which is expressed in anaerobic conditions.
61. What are 'insertion sequences'?
62. What are transposons?
63. What are 'jumping genes'?
64. Why Barbara McClintock is famous?
65. What is important function of transposons?
66. What is 'Klenow fragment' of DNA polymerase I?
67. What is thermostable DNA polymerase?
68. What is TAQ polymerase?
69. Why Thermus aquaticus is famous for?
70. What is ribonuclease?
71. What is function of ribonuclease?
72. What are 'exonucleases'?

73. What is function of exonuclease?
74. What is function of polynucleotide kinase?
75. What are restriction enzymes?
76. What are palindrome sequences?
77. How sticky ends are generated in a DNA?
78. What are plasmids?
79. What are conjugative plasmids?
80. What is plasmid incompatibility?
81. What is 'fosmid'?
82. What are cosmids?
83. What are phagemids?
84. What are 'cos' sites?
85. What is the size of DNA which can be packed in head of bacteriophage?
86. How plaque is formed in Petri plates?
87. What is T-DNA?
88. What is 'nos' site?
89. What is acetosyringonc?
90. What is the role of acetosyringone?
91. What do you mean by rDNA molecule?
92. What is a genomic library?
93. What is colony hybridization?
94. What is electroporation?
95. What are liposomes?
96. What is a particle gun?
97. Name two DNA transfer methods for plant cells.
98. What is role of vir genes?
99. What are binary vectors?
100. What are cointegrative vectors?
101. What are selectable markers?
102. What is a reporter gene?
103. What are transgenic plants?
104. Expand Bt.
105. Why Bacillus thuringiensis is so famous?
106. What is transgene?
107. What is Bt toxin?
108. What is difference between first and third generation transgenic plants?
109. What is mid gut pH of insects?
110. What are 'cry' toxins?
111. Name three transgenic Bt crops.
112. What id glyphosate?
113. Which compound inhibits activity of EPSP synthase?
114. What is cross protection in plants?

115. What is antisense RNA?
116. How coat protein gene imparts resistance in plants?
117. What is 'Golden rice'?
118. Why golden rice is famous?
119. Who created the golden rice?
120. What is terminator gene?
121. What are edible vaccines?
122. What is traitor seed technology?
123. Which problem is caused by Norwalk virus in humans?
124. What are transgenic animals?
125. What is transfection?
126. Why Dolly sheep is famous?
127. What is physical transfection?
128. What is chemical transfection?
129. What is DNA finger printing?
130. What are applications of DNA finger printing?
131. What is gene therapy?
132. How human insulin is now a day produced?
133. Name a few recombinant organisms used for the production of insulin.
134. What are attenuated organisms?
135. How BCG vaccine is produced?
136. What is nature of Hepatitis B vaccine?
137. What is copyright?
138. What is TRIPS?
139. What is GATT?
140. What is genetic linkage?
141. What are linked genes?
142. What is recombination frequency?
143. What is genetic map of a bacterium?
144. What is RFLP?
145. What is RAPD?
146. What is T-RFLP?
147. What is AFLP?
148. What is ISSR?
149. Which is better technique RFLP or RAPD?
150. Who discovered bacterial recombination?
151. What is Red Biotechnology?
152. What is Green Biotechnology?
153. What is Grey or white Biotechnology?
154. How components of cells can be separated?
155. Which is heavier, nucleus or mitochondria?

156. What is FISH technique?
157. What is GISH technique?
158. CTAB is used for the isolation of which biomolecules?
159. What for ethidium bromide gradient centrifugation is used?
160. What for sucrose density gradient centrifugation is used?
161. What is Southern blotting technique?
162. What for Southern blotting technique is used?
163. What is Western blotting technique?
164. Why Prof Sanger is famous for?
165. What for Maxum and Gibert method is used?
166. What for Pal Nyren's method is used?
167. What is an automated DNA sequencer?
168. What is capillary gel electrophoresis?
169. What is PCR?
170. What is thermal cycler?
171. Who proposed the cell theory?
172. What is in vitro fertilization in plants?
173. Who gave the concept of totipotency?
174. Who conceived the idea of micropropagation technology and its commercialization?
175. Who cultured for the first time plant's meristem?
176. Who is known for tomato root culture for the first time?
177. Who reported for the first time somatic embryogenesis in carrot?
178. Who discovered the cytokinin?
179. Who isolated protoplasts for the first time?
180. Who reported haploids embryos in Datura for the first time?
181. What is pH?
182. At what temperature autoclave sterilizes materials?
183. At what pressure autoclave sterilizes materials?
184. How much a vessel should be filled with medium?
185. What is HEPA filter?
186. What is resolving power of a microscope?
187. What is TEM?
188. What is SEM?
189. What is speed of light?
190. What is numerical aperture?
191. What is resolving power of lens/microscope?
192. What is a colorimeter?
193. What is photomultiplier?
194. What is lmax?
195. Give Lambert's law.
196. What is Beer's law?

197. What is Optical density?
198. What is transmittance?
199. What are applications of a colorimeter?
200. What is RCF?
201. What is sedimentation coefficient?
202. What is adsorption chromatography?
203. What is paper chromatography?
204. What is TLC?
205. What is hygrometer?
206. What is use of hygrometer?
207. What is a Lux meter?
208. How light intensity is measured?
209. What are different units of light intensity?
210. What do mean by asepsis?
211. Name a few compounds used for surface sterilization.
212. What is mode of action of chlorine to kill microorganisms?
213. What are membrane filters?
214. What are microbial glass filter?
215. What is MS medium?
216. What is source of carbon in MS medium?
217. What compounds are used as source of nitrogen in MS medium, name three?
218. Name three micronutrients used in MS medium.
219. Name three vitamins used in MS medium.
220. What is full name of 2,4-D?
221. What is positive pressure in relation to asepsis?
222. What is principle behind nurse tissue culture?
223. Which is fast growing system, callus or cell culture?
224. What do you understand by exponential phase of growth?
225. What a sigmoid curve of growth represent.
226. What are synchronous cultures?
227. What is cells separation by using sieves?
228. What is a microculture chamber?
229. What is use of a microculture chamber?
230. What is cell plating?
231. What is a clone in botany?
232. What is cloning in plant tissue culture?
233. What is use of cell clones?
234. What is cell selection?
235. What is embryo culture?
236. What is ovary and ovule culture?
237. What do you understand by industrial plantations?

238. Name a few fruit plants produced through micropropagation.
239. Name a few flower plants produced through micropropagation.
240. Name a few foliage plants produced through micropropagation.
241. Name a few plantation crop plants produced through micropropagation.
242. Name a few vegetatively propagated plants produced through micropropagation.
243. What is germplasm?
244. What is in situ preservation?
245. What is ex situ preservation?
246. What is gene bank?
247. What is cryopreservation?
248. What are threatened species?
249. What are endangered species?
250. What is red data book?
251. At what temperature plant material is stored for cryopreservation?
252. What are secondary metabolites?
253. What is an active principle of plant?
254. What are bioreactors?
255. What is a sparger?
256. What are air lift bioreactors?
257. What is a wave bioreactor?
258. What are hairy root cultures?
259. Name three plant derived natural products used as drug.
260. What is Shikimic acid pathway?
261. Define an alkaloid.
262. Name various classes of alkaloids.
263. From where ephedrine is obtained?
264. From which plant vinblastine is obtained?
265. From which plant morphine is obtained.
266. Name the alkaloid present in tea.
267. Name the alkaloid obtained from Cinchona.
268. Why tissue culture is established for production of secondary metabolites?
269. Name two anticancer compounds of plant origin.
270. Name two sedative/analgesic compounds of plant origin.
271. What is biotransformation of compounds?
272. Give two examples of biotransformation using cell cultures.
273. Name two compounds of plant origin produced through hairy roots.
274. Name compound/s of Dioscorea.
275. What is the use of diosgenin?
276. What are protoplasts?
277. What enzymes are used to isolate protoplasts?
278. Why osmotic solution is used for isolation of protoplasts?

279. What is electrofusion?
280. What are cybrids?
281. What is anther culture?
282. What is use of anther and pollen cultures.
283. Can anther wall produce haploids?
284. What is use of PEG in protoplasts cultures?
285. What are somatic cell hybrids?
286. What is organogenesis?
287. What is direct somatic embryogenesis?
288. What do you understand by scale up technology?
289. What are advantages of somatic embryos over axillary bud break?
290. Name the growth regulator responsible for shoot formation.
291. Name the rooting hormones.
292. Which compound arrest the cell division?
293. Which alkaloids treatment cans double the chromosomes?
294. Name three species where haploids have been produced.
295. What is somaclonal variation?
296. How variation in callus and cell cultures is caused?
297. What are possible causes of somaclonal variation?
298. Provide three advantages of plant tissue culture over in vivo culture.
299. What is most widely used application of plant tissue culture?
300. What is government support for plant tissue culture industry?

Glossary

Acclimatization- Adaptation of an organism to a new environment.

Adventitious, adventive- Developing from unusual points of origin, such as shoots or roots arising from a leaf or stem tissues other than the axils or apex; often dependent on close physical or temporal association with organized or semi-organized tissues or cells. Compare to "*de-novo*".

Aerobic- Needing oxygen for growth.

Agrobacterium tumefaciens- A common soil bacterium used as a vector to create transgenic plants.

Allele- Any of several alternative forms of a gene.

Alkaptonuria- A human genetic defect caused by a single recessive mutation that caused the accumulation of homogentisic acid, a metabolic by-product of amino acid tyrosine.

Amino acids- Long chains of amino acids make up proteins. About 20 are known. Some amino acids are made in the body; those that are not called essential amino acids and must be supplied in food.

Amplification- The process of increasing the number of copies of a particular gene or chromosomal sequence.

Anaerobic- Growing in the absence of oxygen.

Androgenesis- Development of haploid plants from the male gametophyte following a developmental pattern resembling embryogenesis, resulting from the culture of anthers or microspores.

Anther culture- The aseptic culture of anthers for the production of haploid callus or plants from the microspores, see "androgenesis".

Antibiotic- Chemical substance formed as a metabolic byproduct in bacteria or fungi and used to treat bacterial infections. Antibiotics can be produced naturally, using microorganisms, or synthetically.

Antibody- Protein produced in body tissues in response to the presence of a specific antigen. See antigen.

Anticodon- Triplet of nucleotide bases (codon) in transfer RNA that pairs with (is complementary to) a triplet in messenger RNA. For example, if the codon is UCG, the anticodon is AGC. See also Base; Base pair; Complementarity.

Antigen- A substance that, when introduced into the body, induces an immune response by a specific antibody.

Antiparallel- The relative polarity of two strands in a DNA double helix, one strand goes in the 5′ to 3′ direction and other strand goes in the 3′ to 5′ direction. The numbers refer to the carbon atoms in the deoxyribose.

Antisense- A piece of DNA producing a mirror image ("antisense") messenger RNA that is opposite in sequence to one directing protein synthesis. Antisense technology is used to selectively turn off production of certain proteins.

Anti-terminator- A protein that antagonizes the termination of transcription at the termination site.

Anti-tumour agents- Compounds active against or inhibit neoplasm or cancerous growth.

Apoptosis- Programmed cell death, a process whereby cells kill themselves for the good of the organism.

Aseptic- The absence of contaminating fungi, bacteria, viruses, mycoplasma and other competing microorganisms in cultures of eukaryotic cells or tissues.

Assay- Technique for measuring a biological response; a test.

Attenuated- Weakened; with reference to vaccines, made from pathogenic organisms that have been treated so as to render them avirulent.

Attenuator- A region or sequence of nucleotide in the leader region of several operons that act in the presence of the product of the operon (e.g., tryptophan) to reduce the rate of transcription from the structural genes.

Autoimmune disease- A disease in which the body produces antibodies against its own tissues.

Autosome- Any chromosome other than a sex chromosome.

Avirulent- Unable to cause disease.

Axenic- A pure culture of single organism that is not contaminated by symbionts or parasites.

Bacillus- Singular for a rod-shaped bacterium (plural, bacilli). Also used as the name of a genus of bacteria, including the species *Bacillus thuringiensis*.

Bacillus subtilis- A bacterium commonly used as a host in recombinant DNA experiments. Important because of its ability to secrete proteins.

Bacillus thuringiensis (Bt)- Bacterium that produces a protein called Bt toxin, a biological insecticide. Bt toxin is used to control insect pests by dusting the crop with Bt bacteria. When ingested, Bt toxin kills certain insect larvae, but is regarded as harmless to humans, pets and most beneficial insects such as bees. Inserting a copy of the Bt gene into plants enables them to produce Bt toxin protein. Such plants can resist some insect pests. See biological control, microbial insecticide.

Bacteriophage- Virus that lives in and kills bacteria. Also called phage.

BAC- A bacterial artificial chromosome, a plasmid that is capable of replicating in a bacterial host (usually *E.coli*) with a very large insert of foreign DNA, perhaps as large as 300 kb of known plasmid DNA.

Bacterium- Class of single-cell organisms (plural, bacteria). One member, *E. coli*, is commonly used in recombinant DNA technology for producing proteins and other chemicals.

Base- On the DNA molecule, one of the four chemical units that are linked in a series to make a strand of DNA. The four DNA bases are: adenine (A), cytosine (C), guanine (G), and thymine (T). In RNA, uracil (U) substitutes for thymine. See DNA finger printing, nucleotide.

Base anlogues- Compounds that are structurally similar to the four natural bases and can be incorporated in to DNA. Base analogues can cause mutations by in precise base pairing during replication.

Base pair- Two nucleotide bases on different strands of the nucleic acid molecule that bond together. The bases can pair in only one way: adenine with thymine (DNA), or uracil (RNA) and guanine with cytosine.

Batch culture- A suspension culture in which cells grow in a finite volume of nutrient medium and follow a sigmoidal pattern of growth between subcultures.

B-chromosone- A small chromosome found maize that has no known function and carries no known gene.

B-DNA- The naturally occurring form of DNA that is the same as the model proposed by Watson and Crick.

Bioassay- Determination of the effectiveness of a compound by measuring its effect on animals, tissues or organisms in comparison with a standard preparation.

Biocatalyst- In bioprocessing, an enzyme that activates or speeds up a biochemical reaction.

Biochip- An electronic device that uses organic molecules to form a semiconductor.

Bioconversion- Chemical restructuring of raw materials by using a biocatalyst.

Biodegradable- Capable of being reduced to water and carbon dioxide by the action of microorganisms.

Bioinformatics- The science of informatics as applied to biological research. Informatics is the management and analysis of data using advanced computing techniques. Bioinformatics is particularly important as an adjunct to genomics research, because of the large amount of complex data this research generates.

Biolistic device- A device that shoots microscopic DNA-coated particles into target cells.

Biomass- The totality of biological matter in a given area. As commonly used in biotechnology, refers to the use of cellulose, a renewable resource, for the production of chemicals that can be used to generate energy or as alternative feed stocks for the chemical industry to reduce dependence on nonrenewable fossil fuels, callus or cells biomass in culture.

Biomaterials- Biological molecules, such as proteins and complex sugars, used to make medical devices, including structural elements used in reconstructive surgery.

Bioprocess- A process in which living cells, or components there of, are used to produce a desired product.

Bioreactor- A vessel, such as a fermenter, used for fermentation or other biological processes.

Biosynthesis- Production of a chemical by a living organism.

Biotechnology- Development of products by a biological process requiring engineering technologies such as fermentation or controlled environments, or utilizing biochemical or genetic technologies such as recombinant DNA techniques for the modification and improvement of biological systems.

Biotransformation- (Biochemical transformation or bioconversion) - The use of cultured cells to convert substrates into desired organic compounds by virtue of an endogenous enzyme which catalyzes the reaction.

Callus (plural, calli)- (1) An undifferentiated growth of plant cells *in vitro* on a culture medium, (2) A wound response *in vivo* resulting in a tumorous growth.

Cancerous growth- Uncontrolled and abnormal cell division of cells leading to tumour formation.

Carbohydrate- A type of biological molecule composed of simple sugars such as glucose. Common examples include starch and cellulose.

Carcinogen- Cancer-causing agent.

CAAT box- A DNA sequence located above 75 bp upstream of the translation start site in eukaryotic gene. It affects the level of gene expression.

Catabolite activator protein- A protein, which, together with camp, activates operon in bacteria. These operons are subject to catabolite repression.

Catalyst- An agent (such as an enzyme or a metallic complex) that facilitates a reaction but is not itself changed during the reaction.

Cell- Smallest unit of living matter able to grow and reproduce independently. Cells contain DNA for storing information, ribosomes for making proteins, and mechanisms for converting energy.

Cell culture- The growing of cells *in vitro*, including the culture of single cells or small aggregates of cells in a liquid medium.

Cell differentiation- The process by which descendants of a common parental cell achieve specialized structure and function.

Cell fusion- *See* Fusion.

Cell line- Derived from a primary culture and selected at the time of the first subculture, or derived from a cell population with particular attributes.

Cell- The smallest structural unit of a living organism that can grow and reproduce independently.

Chargaff's rule- The concentration of adenine always equals the concentration of thymine, and the concentration of cytosine always equals the concentration of guanine.

Chemical genomics- Using structural and functional genomic information about biological molecules, especially proteins, to identify useful small molecules and alter their structure to improve their efficacy.

Chemically defined medium- A nutritive solution for culturing cells in which each component is of known chemical structure.

Chimera- The individual (animal or lower organism) produced by grafting an embryonic part of one individual onto an embryo of either the same or a different species.

Chromosomes- Thread like components in the cell that contain DNA and proteins. Genes are carried on the chromosomes.

Chromosome scaffold- A central core structure of a condensed chromosome. It is composed of protein.

Cis acting site- A genetic site represented by a binding site on the DNA, that influences genes immediately adjacent to the site such as promoters, enhancer, and operators.

Cistron- A genetic unit defined by the cis-trans test. It is used interchangeably with the term gene.

Clinical studies- Human studies that are designed to measure the efficacy of a new drug or biologic. Clinical studies routinely involve the use of a control group of patients that is given an inactive substance (placebo) that looks like the test product.

Clonal propagation- A sexual reproduction or vegetative propagation of plants considered to be physiologically and/or genetically uniform and which originated from a single individual or explant.

Clone- *as noun* (*i*) Population of plants derived from a single individual by vegetative propagation or by regeneration of plants from *in vitro* cultures. (*ii*) Population of cells mitotically derived from a single cell. (*iii*) Population or recombinant DNA molecules carrying the same inserted sequence. *as verb-* (*i*) To propagate a clone. (*ii*) To insert a particular DNA sequence into a vector. Researchers have achieved laboratory cloning using genetic material from adult animals of several species, including mice, pigs and sheep.

Cloning- Technique of creating a group of genetically identical cells or DNA molecules from a single ancestor. In horticulture, cloned plants are reproduced asexually from a single parent.

Co-cultivation- In transformation experiments, the incubation of host cells or tissues with Agrobacterium or another organism acting as a vehicle for a genetic vector.

Codon- A sequence of three nucleotide bases that specifies an amino acid or represents a signal to stop or start a function.

Co-enzyme- An organic compound that is necessary for the functioning of an enzyme. Co-enzymes are smaller than the enzymes themselves and sometimes separable from them.

Co-factor- A nonprotein substance required for certain enzymes to function. Co-factors can be co-enzymes or metallic ions.

Combinatorial chemistry- A product discovery technique that uses robotics and parallel synthesis to generate and screen quickly as many as several million molecules with similar structure in order to find chemical molecules with desired properties.

Complementary base pairing- The relationship of the nucleotide bases on two different strands of DNA or RNA. When the bases are paired properly (adenine with thymine [DNA] or uracil [RNA]; guanine with cytosine), the strands are complementary.

Complementary DNA (cDNA)- DNA synthesized from a messenger RNA rather than from a DNA template. This type of DNA is used for cloning or as a DNA probe for locating specific genes in DNA hybridization studies.

Computational biology- A subdiscipline within bioinformatics concerned with computation-based research devoted to understanding basic biological processes.

Conjugation- Sexual reproduction of bacterial cells in which there is a one-way exchange of genetic material between the cells in contact.

Consensus sequence- The average of several similar sequences.

Continuous culture- A suspension culture held at constant volume and continuously supplied with nutrients by the inflow of fresh medium.

Cosmid- A plasmid vector which contains the cos site of phage and one or more selectable markers such as an antibiotic resistance gene.

Culture- Cultivate cells or living organisms in a prepared medium under laboratory conditions. "Culture" is both the process and the growing cells.

Culture medium- Any nutrient system for the artificial cultivation of bacteria or other cells; usually a complex mixture of organic and inorganic materials.

Cybrid- A cytoplasmically hybrid cell with organelles from both parental sources, e.g., achieved by fusion of a cytoplast with a whole cell. During subsequent cell divisions random or nonrandom segregation of organelles may occur, resulting in variable parental combinations in cybrid cell lines.

Cystic fibrosis- Disease of mucous glands throughout the body that usually develops during childhood, and makes breathing increasingly difficult. If a child receives two copies of the defective gene called the CF gene - one copy from each parent - then the child will develop the disease.

Cytotoxic- Able to cause cell death.

D-Loop- The structure resulting from the displacement of a region of one strand of duplex DNA by a single stranded invading strand in a rec-A mediated recombination process.

Dalton- The mass of hydrogen atom.

De-novo- Literally to "arise a new", with reference to plant regeneration or developmental processes, arising from unorganized cells or tissues or from undetermined cells. Compare to "adventitious".

Deoxyribonucleic acid (DNA)- The molecule that carries the genetic information for most living systems. The DNA molecule consists of four bases (adenine, cytosine, guanine and thymine) and a sugar-phosphate backbone, arranged in two connected strands to form a double helix. See also Complementary DNA; Double helix; Recombinant DNA.

Differentiation- The process of biochemical and structural changes by which cells become specialized in form and function.

Diploid- A cell with two complete sets of chromosomes. Compare Haploid.

DNA analysis- See polymerase chain reaction (PCR) and RFLP mapping. Both PCR and RFLP analysis can be used in DNA fingerprinting for genealogical studies and forensics. See next entry.

DNA chip- A small piece of glass or silicon that has small pieces of DNA arrayed on its surface.

DNA fingerprinting- Detecting patterns in DNA that indicate the presence of a gene for a trait. The pattern resembles a bar code printed on a commercial product so computers can scan the price. Forensics experts can use this distinct pattern to link or clear an Individual suspected of being involved in a crime, like they compare fingerprints. Breeders can use these patterns to find and select breeding stock with traits such as disease resistance.

DNA hybridization- The formation of a double-stranded nucleic acid molecule from two separate strands. The term also applies to a molecular technique that uses one nucleic acid strand to locate another.

DNA library- A collection of cloned DNA fragments that collectively represent the genome of an organism.

DNA polymerase- An enzyme that replicates DNA. DNA polymerase is the basis of PCR-the polymerase chain reaction.

DNA probe- A small piece of nucleic acid that has been labeled with a radioactive isotope, dye or enzyme and is used to locate a particular nucleotide sequence or gene on a DNA molecule.

DNA repair enzymes- Proteins that recognize and repair certain abnormalities in DNA.

DNA sequence- The order of nucleotide bases in the DNA molecule.

DNA vaccines- Pieces of foreign DNA that are injected into an organism to trigger an immune response.

Double haploid- A plant with two sets of haploid chromosomes.

Double helix- A term often used to describe the configuration of the DNA molecule. The helix consists of two spiraling strands of nucleotides (a sugar, phosphate and base) joined crosswise by specific pairing of the bases. See also Deoxyribonucleic acid; Base; Base pair.

Electrofusion- The application of electroporation techniques for protoplast fusion.

Electrophoresis- Technique for analyzing and separating molecules based on the movement of charged particles in an electric field.

Electroporation- The creation of reversible small holes in a cell wall or membrane through which foreign DNA can pass. This DNA can then integrate into the cell's genome.

EMBO J. 2000;19:3159-3167.

Embryo- Early stages of an animal's development that results when a sperm fertilizes an egg. Embryo cloning produces many calves from one embryo, for example.

Embryo rescue- Application of embryo culture techniques to obtain interspecific hybrid plants prior to abortion of the embryos due to post-fertilization hybridization barriers.

Embryogenesis (embryogeny)- Initiation and development of embryos from zygotes, or of bipolar structures from somatic cells that parallel the developmental path of zygotic embryos. In the later case, such embryos may be either adventitious or *de novo* in origin.

Enzyme- A protein catalyst that facilitates specific chemical or metabolic reactions necessary for cell growth and reproduction.

Escherichia coli (E. coli)- Common bacterium found in human and mammalian digestive tracts. Some strains of *E. coli* are used in recombinant DNA work because they have been genetically well-characterized and are easily grown in laboratory fermenters.

Eukaryote- A cell or organism containing a true nucleus, with a well-defined membrane surrounding the nucleus. All organisms except bacteria, viruses and cyanobacteria are eukaryotic. Compare Prokaryote.

Exon- In eukaryotic cells, that part of the gene that is transcribed into messenger RNA and encodes a protein. *See* also Intron; Splicing.

Explant- The tissue taken from a plant or seed and transferred to a culture medium to establish a tissue culture system or regenerate a plant.

Expression- In genetics, manifestation of a characteristic specified by a gene. In industrial biotechnology, production of a specific protein by inserting a gene into a new host organism.

Expression sequence tag- Short cDNA sequences used to link physical maps of genome.

Expression vector- A plasmid that possesses a strong promoter, ribosome binding site adjacent to a start codon, but only a small fragment of the original gene. The coding sequence for a gene can be inserted in the correct reading frame to allow the production of the gene product in large quantity.

Fermentation- Chemical reaction induced by a living agent - yeast, bacterium or mold - that splits complex organic compounds to simple ones. For example, yeast converts sugar to alcohol and carbon dioxide. In biotechnology, the process of growing microbes to produce chemical or pharmaceutical compounds. Also referred to as classical biotechnology, traceable back 6,000 years.

Functional foods- Foods containing compounds with beneficial health effects beyond those provided by the basic nutrients, minerals and vitamins. Also called nutraceuticals.

Functional genomics- A field of research that aims to understand what each gene does, how it is regulated and how it interacts with other genes.

Gametoclonal variation- Phenotypic variation apart from normal segregation occurring during gamete formation, both epigenetic and genetic, arising from the culture of gametophytic cells or tissues, such as in anther culture or unfertilized ovule culture. Related to "somaclonal variation".

Gel electrophoresis- A process for separating molecules by forcing them to migrate through a gel under the influence of an electric field.

Gene amplification- The increase, within a cell, of the number of copies of a given gene.

Gene flow- Movement of genes by migration from one population to another.

Gene gun- A particle acceleration device for physically delivering recombinant DNA, typically precipitated onto a microprojectile, into a cell using an explosive propellant.

Gene knockout- The replacement of a normal gene with a mutated form of the gene by using homologous recombination. Used to study gene function.

Gene machine- A computerized device for synthesizing genes by combing nucleotides (bases) in the proper order.

Gene mapping- Determination of the relative locations of genes on a chromosome.

Gene sequencing- Determination of the sequence of nucleotide bases in a strand of DNA. *See* Sequencing.

Genetic marker- Genes, usually mutant, used to experimentally track an individual during a genetic cross or selection procedure.

Gene therapy- The replacement of a defective gene in an organism suffering from a genetic disease. Recombinant DNA techniques are used to isolate the functioning gene and insert it into cells. More than 300 single-gene genetic disorders have been identified in humans. A significant percentage of these may be amenable to gene therapy.

Generic name for an enzyme that phosphorylates tyrosine molecules in proteins.

Genetic analysis- Studying how traits and genes for traits are passed from generation to generation, and how genes and the environment interact to result in traits.

Genetic code- Information coded within nucleotide sequences of RNA and DNA that specifies the amino acid sequence in protein synthesis and on which heredity is based.

Genetic counseling- Providing current or prospective parents with information on the probabilities of inherited diseases occurring in their children, and on diagnosis and treatment of such diseases.

Genetic engineering- Using recombinant DNA techniques and related methods to move one or several genes from one organism to another, to rearrange one or several genes within a cell, or to alter gene-controlled processes. Transferring a DNA segment from one organism and inserting it into the DNA of another organism to modify, amplify, transform and express genetic information. The two organisms can be totally unrelated. *See* Recombinant DNA.

Genetic modification- A number of techniques, such as selective breeding, mutagenesis, transposon insertions and recombinant DNA technology, that are used to alter the genetic material of cells in order to make them capable of producing new substances, performing new functions or blocking the production of substances.

Genome- The total hereditary material of a cell, comprising the entire chromosomal set found in each nucleus of a given species.

Genomics- The study of genes and their function. Recent advances in genomics are bringing about a revolution in our understanding of the molecular mechanisms of disease, including the complex interplay of genetic and environmental factors. Genomics is also stimulating the discovery of breakthrough healthcare products by revealing thousands of new biological targets for the development of drugs and by giving scientists innovative ways to design new drugs, vaccines and DNA diagnostics. Genomic-based therapeutics may include "traditional" small chemical drugs, as well as protein drugs and gene therapy.

Genomic library- A collection of DNA fragment in a cloning vector that is representative of total genome of a organism.

Genotype- Genetic makeup of an individual or group. Compare Phenotype. **Germ cell** Reproductive cell (sperm or egg). Also called gamete or sex cell.

Germplasm- The total genetic variability, represented by germ cells or seeds, available to a particular population of organisms.

Glycoprotein- A protein conjugated with a carbohydrate group.

Gne splicing- Inserting new genetic information into a chromosome using recombinant DNA techniques.

Haploid- A cell with half the usual number of chromosomes, or only one chromosome set. Sex cells are haploid. Compare Diploid.

Heredity- Transfer of genetic information from parent cells to progeny.

Homologous- Corresponding or alike in structure, position or origin.

Homozygous- Having the same allele of a gene in homologous chromosomes.

Host- A cell or organism used for growth of a virus, plasmid or other form of foreign DNA, or for the production of cloned substances.

Host-vector system- Combination of DNA-receiving cells (host) and DNA-transporting substance (vector) used for introducing foreign DNA into a cell.

House-keeping genes- Genes that code for proteins that are used by nearly all cells.

Human Genome Project- An international research effort aimed at discovering the full sequence of bases in the human genome. Led in the United States by the National Institutes of Health and the Department of Energy.

Human immunodeficiency virus (HIV)- The virus that causes acquired immune deficiency syndrome (AIDS).

Hybridization- Production of offspring, or hybrids, from genetically dissimilar parents. The process can be used to produce hybrid plants (by crossbreeding two different varieties) or hybridomas (hybrid cells formed by fusing two unlike cells, used in producing monoclonal antibodies). *See* DNA hybridization.

Immobilization- Entrapment /attachment of cells or protoplasts into/onto a matrix.

Inbreeding- The mating of related individuals.

In situ- In its original or natural place or position.

***In situ* hybridization**- A hybridization of a DNA probe to homologous sequences in a chromosome while still in the cell. The location of hybridization in the chromosome indicates the location of DNA sequences complementary to the probe.

In vitro- Any process carried out in sterile cultures, or the measurement of biological processes outside the intact organism such as enzyme reactions. Literally, "in glass".

In vivo- In a living organism.

Inducer- A molecule or substance that increases the rate of enzyme synthesis, usually by blocking the action of the corresponding repressor.

Insulin- Protein hormone that regulates blood sugar, made in cells of the pancreas. In the laboratory, microbes given a copy of the gene for human insulin can make insulin to treat diabetes mellitus, a shortage of insulin.

Interferon- A class of lymphokine proteins important in the immune response. There are three major types of interferon: alpha (leukocyte), beta (fibroblast) and gamma (immune). Interferons inhibit viral infections and may have anticancer properties.

Intron- In eukaryotic cells, a sequence of DNA that is contained in the gene but does not encode for protein. The presence of introns "splits" the coding region of the gene into segments called exons. *See* also Exon; Splicing.

Isoenzyme- One of the several forms that a given enzyme can take. The forms may differ in certain physical properties, but function similarly as biocatalysts.

Isogenic- Of the same genotype.

Lactose- Milk sugar; also white crystal sugar made from whey used in baby food, baked goods, candies and pharmaceuticals.

Leader region- A sequence of bases on the transcript between the promoter and start codon.

Library- A set of cloned DNA fragments that taken collectively contain the entire genome of an organism. Also called a DNA library.

Ligase- An enzyme used to join DNA or RNA segments together.

Linkage- The tendency for certain genes to be inherited together due to their physical proximity on the chromosome.

Linker- A fragment of DNA with a restriction site that can be used to join DNA strands.

Lipoproteins- A class of serum proteins that transport lipids and cholesterol in the bloodstream. Abnormalities in lipoprotein metabolism have been implicated in certain heart diseases.

Lysogenic bacteria- A bacterium harbouring a temperate phage.

Lysogenic cycle- A phage life cycle in which the phage is incorporated in to the bacterial genome and is propagated with the bacterial genome.

Lytic cycle- The phage life cycle in which the phage infects the bacterial cell and produces more phage particles that are released by lysis of the bacterial cell.

Marker gene- Gene that is easy to find or observe. Attaching a marker gene to another gene that is hard to find, such as an antisense gene, is like putting a reflective collar on a dog so you can see the collar at night - even though you cannot see the dog, you know where it is. *See* antisense.

Meiosis- Process of cell reproduction whereby the daughter cells have half the chromosome number of the parent cells. Sex cells are formed by meiosis. Compare Mitosis.

Messenger RNA (mRNA)- Nucleic acid that carries instructions to a ribosome for the synthesis of a particular protein.

Metabolism- All biochemical activities carried out by an organism to maintain life.

Microbiology- Study of living organisms that can be seen only under a microscope.

Microinjection- The injection of DNA using a very fine needle into a cell.

Microorganism- Any organism that can be seen only with the aid of a microscope. Also called microbe.

Mitosis- Process of cell reproduction whereby the daughter cells are identical in chromosome number to the parent cells. Compare Meiosis.

Molecular genetics- Study of how genes function to control cellular activities.

Morphogenesis- (1) The ontological development of differentiated structures. (2) The development of a structure from an unorganized state to a differentiated and organized state.

Multigenic- Of hereditary characteristics, one that is specified by several genes.

Mutagen- Agent that causes biological mutation. Examples include chemicals, radioactive elements and ultraviolet light. *See* pentachlorophenol.

Mutant- A cell that manifests new characteristics due to a change in its DNA.

Mutation- A heritable alteration in a gene; the process by which such changes occur.

Noncoding DNA- DNA that does not encode any product (RNA or protein). The majority of the DNA in plants and animals is noncoding.

Northern blotting- The transfer of separated RNA from an agarose gel to a membrane for eventual hybridization using a labeled probe.

Nuclease- An enzyme that, by cleaving chemical bonds, breaks down nucleic acids into their constituent nucleotides.

Nucleic acids- Large molecules, generally found in the cell's nucleus and/or cytoplasm, that are made up of nucleotides. The two most common nucleic acids are DNA and RNA.

Nucleotide- Organic compound composed of a sugar, a phosphate and a base. Two nucleotides joined crosswise by specific pairings of the bases make one rung in the DNA molecule's double helix "ladder."

Oligonucleotide- A polymer consisting of a small number (about two to 10) of nucleotides.

Okazaki fragments- short segments of DNA synthesized on the lagging strand of DNA. They are usually ~1000-2000bp in length.

Open reading frame- A putative gene that is unirrupted by any stop codons.

Operator site- A sequence of nucleotide in an operon where the repressor protein binds.

Operator gene- A region of the chromosome, adjacent to the operon, where a repressor protein binds to prevent transcription of the operon.

Operon- Sequence of genes responsible for synthesizing the enzymes needed for biosynthesis of a molecule. An operon is controlled by an operator gene and a repressor gene.

Optimization- Manipulation of chemical and physical environment of cells to obtain optimal growth of the cells or their contents.

Organ culture- The maintenance or growth of organ primordia or whole parts of an organ *in vitro*, in a way that allows differentiation and/or preservation of its structure and/or function.

Organogenesis- The process of initiation and development of a structure which shows natural organ form and/or function, the initiation of which is temporally separated from the initiation of other organs. Such organs may be adventive or *de novo* in origin.

Origin of replication- A site consisting of a base sequence where DNA synthesis always starts. E.coli has one origin of replication, where as eukaryotes have many.

Packed cell volume- A measurement that estimates the total cell mass in a given volume of cell culture by centrifugation of the cells to a constant volume.

Palindrome- A sequence of DNA that is symmetrical and reads same in both the directions.

Parafilm- A stretchable film based on paraffin wax which is used to seal tubes and petri dishes.

Pathogen- Organism able to cause disease in a certain host.

Pedigree- A diagram showing the phenotype and relationship between member of a family.

Peptide- Two or more amino acids joined by a linkage called a peptide bond.

Pertaining to the genome, the entirety of the genetic information encoded by the nucleotide sequence of an organism, cell, organelle or virus.

Phenotype- Observable characteristics resulting from interaction between an organism's genetic makeup and the environment. Compare Genotype.

Plant regeneration- The process of producing plantlets from *in vitro* cultures through organogenic or embryogenic modes of development, either by adventitious or *de novo* origination. This term should not be used with reference to micropropagation by enhanced axillary branching.

Plaque- A clear area on a lawn of bacteria that represents lysed bacterial cells from a phage infection. A single plaque will represent a clone of a specific phage.

Plasmid- A small circular form of DNA that carries certain genes and is capable of replicating independently in a host cell.

Polycistronic mRNA- A mRNA synthesized in prokaryotes that is composed of agenetic information of more than one gene or cistron.

Polymerase chain reaction (PCR)- Multiplying a particular DNA segment in repeated cycles. The "copies" made in a previous cycle are used as "originals" or templates in the next cycle. For example, PCR enables forensics experts to do DNA testing on very small blood samples.

Polypeptide- Long chain of amino acids joined by peptide bonds.

Pribnow box- An *E. coli* promoter sequence located about –10bp upstream of the start of transcription and site of RNA polymerase binding.

Primer- A sequence of nucleotides that possess a free 3′OH group for extension of new DNA polymer.

Primary products- Compound produced by primary metabolism and directly involved in cell metabolism/ cell building processes.

Prion- An infectious agent that causes degenerative disorders of central nervous system (scrapie in sheep, Creutzfeldt-Jacob disease); thought to be a protein.

Probe- A fragment of DNA or RNA labeled with radioactivity, a dye, or an antigen for detecting a presence of complementary sequences of DNA or RNA.

Prokaryote- An organism (e.g., bacterium, virus, cyanobacterium) whose DNA is not enclosed within a nuclear membrane. Compare Eukaryote.

Promoter- A DNA sequence that is located in front of a gene and controls gene expression. Promoters are required for binding of RNA polymerase to initiate transcription.

Proofreading- The ability of DNA polymerases to remove mismatch nucleotide with 3′ to 5′ exonuclease activity during DNA synthesis.

Propagule- A plant part serving for propagation.

Prophage- Phage nucleic acid that is incorporated into the host's chromosome but does not cause cell lysis.

Protein- Composed of amino acids, proteins are a key component in many cell structures. In addition, many hormones and enzymes that regulate cells are proteins.

Protein A- A protein produced by the bacterium *Staphylococcus aureus* that specifically binds antibodies. It is useful in the purification of monoclonal antibodies.

Proteomics- Each cell produces thousands of proteins, each with a specific function. This collection of proteins in a cell is known as the proteome, and, unlike the genome, which is constant irrespective of cell type, the proteome varies from one cell type to the next. The science of proteomics attempts to identify the protein profile of each cell type, assess protein differences between healthy and diseased cells, and uncover not only each protein's specific function but also how it interacts with other proteins.

Protoplast- The cellular material that remains after the cell wall has been removed from plant and fungal cells.

Pseudogene- An inactive DNA sequence derived from an active gene by the accumulation of inactivating mutation.

Pulse field gen electrophoresis- A method for separating very large fragments of DNA or whole chromosomes up to 10000 kb in length. The procedure uses an alternating electric field during separation of DNA and it has been found that the rate of migration varies with the size of DNA.

Pure culture- *In vitro* growth of only one type of microorganism.

Radioimmunoassay (RIA)- A test combining radioisotopes and immunology to detect trace substances. Such tests are useful for studying antibody interactions with cell receptors, and can be developed into clinical diagnostics.

Recombinant DNA (rDNA)- Technique of isolating DNA molecules and inserting them into the DNA of a cell. This technique includes taking copies of genes from one organism and inserting them in another organism. The two organisms can be totally unrelated. Recombinant DNA has a variety of uses, such as studying how genes work, and producing medicines such as human insulin and other commercial products. *See* Genetic engineering.

Recombinant DNA (rDNA)- The DNA formed by combining segments of DNA from two different sources.

Rec A- Agene that encodes the recA protein which participates in homologous recombination.

Regeneration- Laboratory technique for forming a new plant from a clump of plant cells.

Regulatory gene- A gene that acts to control the protein-synthesizing activity of other genes.

Replication- Reproduction or duplication, as of an exact copy of a strand of DNA.

Replicon- A segment of DNA (e.g., chromosome or plasmid) that can replicate independently.

Repressor- A protein that binds to an operator adjacent to a structural gene, inhibiting transcription of that gene.

Restriction enzyme- An enzyme that breaks DNA in highly specific locations, creating gaps into which new genes can be inserted.

Restriction fragment length polymorphism (RFLP)- The variation in the length of DNA fragments produced by a restriction endonuclease that cuts at a polymorphic locus. This is a key tool in DNA fingerprinting and is based on the presence of different alleles in an individual. RFLP mapping is also used in plant breeding to see if a key trait such as disease resistance is inherited.

Restriction map- A map of the location of restriction endonuclease cut sites in a piece of DNA.

Retrovirus- A virus that contains the enzyme reverse transcriptase. This enzyme converts the viral RNA into DNA, which can combine with the DNA of the host cell and produce more viral particles.

RFLP (restriction fragment length polymorphism)- RFLP mapping detects patterns in DNA that can indicate the presence of a gene for a trait. Both RFLP and polymerase chain reaction (PCR) analysis can be used in DNA fingerprinting for genealogical studies and forensics.

Rhizobium- A class of microorganisms that converts atmospheric nitrogen into a form that plants can utilize for growth. Species of this microorganism grow symbiotically on the roots of certain legumes, such as peas, beans and alfalfa.

Ribonucleic açid (RNA)- A molecule similar to DNA that delivers DNA's genetic message to the cytoplasm of a cell where proteins are made.

Ribosome- Site of protein synthesis in the cytoplasm. selection and breeding - Manipulating microbes, plants or animals, and choosing desirable individuals or populations as breeding stock for new generations.

RNA (ribonucleic acid)- Molecule similar to DNA that functions primarily to decode the instructions that genes carry for protein synthesis.

RNA interference- A natural process used by organisms to block protein production.

Scale-up- Transition from small-scale production to production of large industrial quantities.

Secondary products- Metabolites produced as end products of primary metabolism and not involved directly in metabolic activity.

Selfing- A self cross in plants, fertilization of egg with pollen from the same plant.

Sepsis- The presence in the blood or other tissues of pathogenic microorganisms or their toxins; the condition associated with such presence.

Sequencing- Decoding a strand of DNA or gene into the specific order of its nucleotides: adenine, cytosine, guanine and thymine. This analysis can be done manually or with automated equipment. Sequencing a gene requires analyzing an average of 40,000 nucleotides.

Single-cell protein- Cells or protein extracts from microorganisms, grown in large quantities for use as protein supplements.

Somaclonal variation- Variation which occurs in cultures of cells and tissues that may be either genetic or epigenetic.

Somatic cell gene therapy- Somatic cell gene therapy involves the insertion of genes into cells for therapeutic purposes; for example, to induce the treated cells to produce a protein that the body is missing. It does not affect genetic makeup of a patient's offspring and generally does not change all, or even most, cells in the recipient. Somatic cell gene therapy is only one way of applying the science of genomics to improve healthcare.

Somatic cells- Cells other than sex or germ cells. High somatic cell counts found in milk indicate mastitis in dairy cows. Low somatic cell counts indicate better quality milk and net a premium price from dairy processors.

Species- Group of organisms with common or similar characteristics and capable of inter breeding.

Stop codon- One of three codons in messenger RNA that signal the end of the amino acid chain in protein synthesis.

Structural gene- A gene that codes for a protein, such as an enzyme.

Substrate- Material acted on by an enzyme.

Suicide gene- A gene that codes for an antibiotic that can kill the host bacterial cell. It is genetically modified into the bacterium along with a molecular switch that is controlled by a nutrient in the environment. When the nutrient disappears, the suicide gene is switched on and the bacterium dies.

Suppressor gene- A gene that can reverse the effect of a mutation in other genes.

Synchronous culture- A culture in which a specified proportion of the cells are at the same indicated phase of the cell cycle, enter the same indicated phase of the cell cycle simultaneously, and/or exhibit the same approximate duration of cell cycle.

Synthesis- The process whereby separate elements are combined to form a new complex product, synthetic chemical compound or material.

TATA element- Element present in eukaryotes at about –30 bp from the transcription start site, likely prompoter site. The consensus sequence is TATAAAAA.

Taxol(TM) paclitaxel- Natural product approved by the U.S. Food and Drug Administration in 1992 to treat ovarian cancer after failure of first-line chemotherapy; also used experimentally to treat breast, lung and other cancers. Manufactured by Bristol-Myers Squibb U.S. Pharmaceutical Group from bark of old-growth Pacific yew (cedar) trees. Because these forests are also home to endangered spotted owls, researchers are working to synthesize paclitaxel from twigs and needles, leaving the tree standing.

Technology transfer- The process of transferring discoveries made by basic research institutions, such as universities and government laboratories, to the commercial sector for development into useful products and services.

Telomerase- The enzyme responsible for synthesis of telomere. It is a ribonucleoprotein of about 220 kd and is characteristics of each species in specifying the sequence of the telomere repeat unit.

Telomere- The terminal region of a chromosome that prevents random fusions to other chromosomes. The telomere is composed of multuple repeats of a basic consensus sequence of C18(T/A)14.

Template- A molecule that serves as the pattern for synthesizing another molecule.

Terminator- Sequence of DNA bases that tells the RNA polymerase to stop synthesizing RNA.

Tertiary structure- The total three-dimensional shape of a protein that is essential to protein function.

Tissue culture- Growing plant or animal tissues or cells in test tubes or other laboratory glassware, without other contaminating organisms, for propagation, chemical production and medical research.

Topoisomerases- A class of enzymes that can confer twists in to closed circular DNA to give or relive supercoil. Topoisomerase I makes one incision in double stranded DNA and allows rotation of the other strands and reseals the end.

Totipotency- The property of normal cells that they have the genetic potential to give rise to a complete individual. Exceptions include terminally specialized cells.

Toxin- A poisonous substance produced by certain microorganisms or plants.

Transcription- Synthesis of messenger (or any other) RNA on a DNA template.

Transduction- Transfer of genetic material from one cell to another by means of a virus or phage vector.

Transfection- Infection of a cell with nucleic acid from a virus, resulting in replication of the complete virus.

Transfer RNA (tRNA)- RNA molecules that carry amino acids to sites on ribosomes where proteins are synthesized.

Transformation- The transfer and incorporation of DNA, especially recombinant DNA, into a cell. Cells, plants, or progeny resulting from this process are said to be transformed upon demonstration of the expression in the recipient organism of unique marker genes carried by the transferred DNA.

Transgenic- Refers to animals or plants in which novel DNA has been incorporated into the germ line.

Transgenic organism- An organism formed by the insertion of foreign genetic material into the germ line cells of organisms. Recombinant DNA techniques are commonly used to produce transgenic organisms.

Translation- Process by which the information on a messenger RNA molecule is used to direct the synthesis of a protein.

Transposon- A segment of DNA that can move around and be inserted at several sites in bacterial DNA or in a phage, thus alerting the host's DNA.

Vaccine- A preparation that contains an antigen, consisting of whole disease-causing organisms (killed or weakened) or parts of such organisms, that is used to confer immunity against the disease that the organisms cause. Vaccine preparations can be natural, synthetic or derived by recombinant DNA technology.

Vector- The agent (e.g., plasmid or virus) used to carry new DNA into a cell.

Vegetative propagation- Reproduction of plants using a nonsexual process, involving the culture of plant parts such as stem and leaf cuttings. Synonymous with "asexual propagation".

Virion- An elementary viral particle consisting of genetic material and a protein covering.

Virology- Study of viruses.

Virulence- Ability to infect or cause disease.

Virus- A submicroscopic organism that contains genetic information but cannot reproduce itself. To replicate, it must invade another cell and use parts of that cell's reproductive machinery.

Wild type- The form of an organism that occurs most frequently in nature.

Wobble hypothesis- The ability of a base in the third position of an anticodon to form hydrogen-bond to different bases, so as to read more than one codon.

X-ray crystallography- An essential technique for determining the three-dimensional structure of biological molecules. This information aids in the discovery of products that will interact with the biological molecule.

Yeast- A general term for single-celled fungi that reproduce by budding. Some yeasts can ferment carbohydrates (starches and sugars) and thus are important in brewing and baking.

Zinc finger- Configuration of a DNA-binding protein that resembles a finger with a base, usually cysteines and histidines, binding a zinc ion. Discovered in a transcription factor in Xenopus but present in a large number of different proteins.

Index

NOTES

NOTES

NOTES

NOTES